Nuclear Heat Transport

M. M. EL-WAKIL

Professor of Mechanical and Nuclear Engineering
University of Wisconsin

INTERNATIONAL TEXTBOOK COMPANY
An Intext *Publisher*
Scranton *Toronto* *London*

U-I-RC

ISBN 0-7002-2309-6

Library of Congress Catalog Card Number 77-117433

Preface

This text covers the processes of energy generation and transport from the core of a nuclear fission reactor to the thermal design of such a core.

The material is the outgrowth of classroom lectures given by the author over several years to a class of mixed senior-graduate engineering students. In a previous book [1], the author combined the salient and relevant features of physics, thermodynamics, heat transfer, and fluid flow to cover the subject of nuclear power engineering. However, there is a need for more thorough treatments of the processes of energy (heat) generation in nuclear processes, the transport of that energy by the reactor coolant to the power cycle, and the limitations imposed by the transport mechanism on the designer of nuclear reactor cores. This book was written to fill this need.

Because courses in nuclear and reactor physics are now more available to the engineer than before, the author chose to condense the physics portion of the text to the first three chapters. This portion is sufficient for the understanding of the material that follows, or it may serve as a refresher, or else be bypassed.

The first three chapters have the ultimate aim of developing the manner in which neutrons interact with the material in a reactor core and distribute themselves throughout a core volume. With this information as a necessary prerequisite, the remainder of the text tackles, in a continuous and orderly fashion, the topics leading to the thermal design of the core.

The heat generation caused by the interaction of neutrons in a core as a whole is covered in Chapter 4. Chapters 5 to 8 deal with the heat transfer and temperature distributions in individual reactor core fuel elements and other reactor bodies of various shapes. Steady- and unsteady-state, one- and two-dimensional, and other special cases, as well as various methods of solution (analytical, numerical, and analogical) are covered.

The mechanisms of heat removal by fluids (reactor coolants) are then covered. Nonmetallic and metallic coolants in single-phase flow are treated in Chapters 9 and 10. The important mechanism of change in phase and the limitations it imposes on the heat flux from the fuel of both nonboiling and boiling reactors is covered in Chapter 11. In Chapter 12 the phenomenon of two-phase flow and its effects on reactor hydrodynamics as well as the role it plays in loss-of-coolant accidents are treated.

Chapter 13 deals with the thermal design of the core, beginning with nominal temperature distributions. The maximum operating temperatures under expected deviations from design specifications are then statistically evaluated. A procedure is then outlined for core thermal design. Chapter 14 covers the hydrodynamics and design procedure for a boiling core.

A companion book [2] takes over at this juncture and treats the topic of nuclear energy conversion. This includes the engineering treatment of the various types of nuclear power plants such as fission (BWR, PWR, GCR, FBR, etc.), isotopic (with direct energy conversion), and fusion, in which the nuclear reactions produce the necessary energy, and lend unique and interesting characteristics to the plants.

The material is aimed at the senior or graduate student in engineering, physics, or chemistry, and the practicing graduate engineer, and may be covered in 3 to 4 semester hours. Because of the rapid advances in the field of nuclear power engineering, the student is encouraged to supplement his reading of this material with the latest information in the periodical literature, especially where design correlations are sought.

The author gratefully acknowledges the encouragement and help of the College of Engineering, University of Wisconsin, and of his many colleagues and students for their suggestions, assistance, and support during the preparation and writing of this book. He is indebted, in particular, to Professors C. W. Maynard, W. F. Vogelsang, W. A. Moy, and J. W. Mitchell who read and criticized portions of the manuscript, to Professor E. F. Obert who advised on various aspects of the preparation of the book, and to Mr. M. E. Sawan who checked galleys. I am also appreciative of the patience and moral support of Tania, Fred, and Leila who put up with me during the long arduous times of going through the manuscript and book.

M. M. EL-WAKIL

Madison, Wisconsin

Contents

CHAPTER 9. HEAT TRANSFER AND FLUID FLOW, NONMETALLIC COOLANTS . . . 231 (Single Phase)

CHAPTER 10. LIQUID METAL COOLANTS . . . 259

CHAPTER 11. HEAT TRANSFER WITH CHANGE IN PHASE . . . 291

CHAPTER 12. TWO-PHASE FLOW . . . 325

CHAPTER 13. CORE THERMAL DESIGN . . . 368

CHAPTER 14. THE BOILING CORE . . . 405

APPENDIX A. ALPHABETICAL LIST OF THE ELEMENTS . . . 427

chapter 1

Atomic and Nuclear Structure and Reactions

1-1. INTRODUCTION

Before studying energy-generation processes in nuclear reactors it is necessary to review the structure of the atom and nucleus, and the reactions that give rise to such energy generation. These include fission, fusion, and different types of neutron-nucleus interactions and radioactivity, [1-5].

In 1803 John Dalton, in an attempt to explain the laws of chemical combination, proposed his simple but incomplete *atomic hypothesis*. In it he postulated that all elements consisted of indivisible minute particles of matter, the *atoms**. These atoms were different for different elements and preserved their indentity in chemical reactions. In 1811 Amadeo Avogadro introduced the molecular theory based on the *molecule*, a particle of matter composed of a finite number of atoms.

It is now known that the atoms are themselves composed of subparticles, common among the atoms of all elements. An atom consists of a relatively heavy, positively charged *nucleus*, with a number of much lighter, negatively charged *electrons* that exist in various orbits around it. The nucleus in turn consists of subparticles called *nucleons*. There are primarily two kinds of nucleons: the *neutrons* which are electrically neutral, and the *protons* which are positively charged. The electrical charge on the proton is equal in magnitude, though opposite in sign, to that on the electron. The atom as a whole is electrically neutral, so that the number of protons in it equals the number of electrons in orbit.

An atom of one element may be transformed into an atom of another element by losing or acquiring some of the above subparticles. Such reactions result in a change in mass Δm, and therefore release (or absorb) large quantities of energy ΔE, according to Einstein's law,

* Much earlier, in the fifth century B.C., the Greek Democritus declared that the simplest thing out of which everything is made is an *atom*. (The Greek word *atomos* means *uncut*.)

$$\Delta E = \frac{1}{g_c} \Delta mc^2 \tag{1-1}$$

where c is the speed of light in vacuum and g_c is a conversion factor, Sec. 1-6. Equation 1-1 applies to *all* reactions in which energy is released or absorbed. Energy is, however, classified as *nuclear* if it is associated with changes in the atomic nucleus.

1-2. ATOMIC STRUCTURE

Figure 1-1 shows schematically three atoms. The first, hydrogen, has a nucleus composed of a single proton and no neutrons, and one

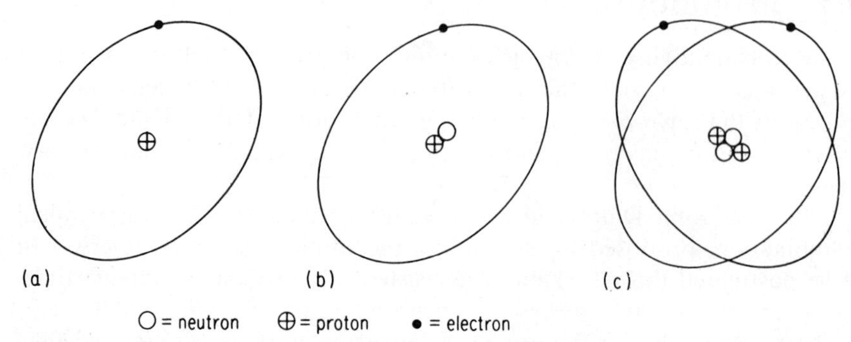

FIG. 1-1. Diagrammatic sketches of some light atoms. (a) Hydrogen; (b) deuterium, or heavy hydrogen; (c) helium.

orbital electron (hydrogen is the only atom that contains no neutrons). The second, deuterium, has one proton and one neutron in its nucleus and one orbital electron. The third, helium, contains two protons, two neutrons, and two electrons. The electrons, in the form of a unit of negative charge, exist in orbits and may be quantitized as a lumped charge as shown.

Most of the mass of the atom is concentrated in the nucleus. The masses of the three primary atomic subparticles mentioned above are

$$\begin{aligned} \text{Neutron mass } m_n &= 1.008665 \text{ amu} \\ \text{Proton mass } m_p &= 1.007277 \text{ amu} \\ \text{Electron mass } m_e &= 0.0005486 \text{ amu} \end{aligned}$$

The abbreviation amu means *atomic mass unit*, a unit of mass approximately equal to 1.66×10^{-24} g_m, or 3.66×10^{-27} lb_m. (See Appendix H for conversion factors.)

These particles are the three primary building blocks of all atoms.

When atoms differ in their mass, it is because they contain varying numbers of them. Atoms and nuclei that have the same number of protons have similar chemical and physical characteristics and differ mainly in their masses. They are called *isotopes.* For example, deuterium is an isotope of hydrogen. It is frequently called *heavy hydrogen.* Ordinary, or naturally occurring hydrogen contains 1 part in 6400 of deuterium. When combined with oxygen, ordinary hydrogen and deuterium form *ordinary water* (or simply *water*), and *heavy water,* respectively.

The number of protons in the nucleus is called the *atomic number, Z.* The total number of nucleons in the nucleus is called the *mass number, A.* Since the mass of a neutron or a proton is approximately equal to 1 amu, A is the integer nearest the mass of the nucleus, which in turn is approximately equal to the atomic mass of the atom. Isotopes of the same element thus have the same atomic number but differ in mass number. Nuclear symbols are written conventionally as

$$_Z\mathrm{X}^A$$

where X is the usual chemical symbol of the element. Thus the hydrogen nucleus is $_1\mathrm{H}^1$ while deuterium is $_1\mathrm{H}^2$ (but sometimes D), and ordinary helium is $_2\mathrm{He}^4$. For particles containing no protons, the subscript indicates the magnitude (in terms of proton charges), and sign of the electrical charge. Thus an electron is $_{-1}e^0$ and a neutron is $_0n^1$. Symbols are also often written in the form He-4, helium-4, etc. In a different system of notation the symbol is written as $^A_Z\mathrm{X}$, but this system will not be used in this book.

Many elements (as hydrogen, above) appear in nature as mixtures of isotopes of varying abundances. For example, naturally ocurring uranium, called *natural uranium,* is composed of 99.282 mass percent U^{238}, 0.712 mass percent U^{235}, and 0.006 mass percent U^{234}, where the atomic number has been deleted, being 92 in all cases. Many isotopes that do not appear in nature are produced in the laboratory or in nuclear reactors. For example, uranium is known to have a total of 14 isotopes having mass numbers ranging from 227 to 240.

The known elements, their chemical symbols, and their atomic numbers are listed alphabetically in Appendix A. Appendix B lists the same elements in the order of their atomic numbers, together with some of their properties. The third column of Appendix B gives the atomic masses of the elements as they appear in nature. With few exceptions, these fall in the same order as the atomic numbers. The following three columns list the mass numbers of the known isotopes, the isotope masses in atomic mass units, and the abundances in atomic percent of the isotopes as they appear in nature. The last three columns list properties that will be taken up later.

Two other particles of importance in nuclear heat generation are the *positron* and the *neutrino.* The positron is a positively charged electron, or β particle having the same charge and mass as the electron. It has the symbols ${}_{+1}e^0$, e^+ or β^+. The neutrino (little neutron) is a tiny, electrically neutral particle which, because of its size and neutrality, is difficult to observe experimentally. Initial evidence of its existence was based on theoretical considerations. In nuclear reactions where a β particle (of either kind) is emitted or captured, the energy accompanying the reaction (corresponding to the lost mass) was not all accounted for by the energy of the emitted β particle and the recoiling nucleus. This violated the law of the conservation of energy. The neutrino was therefore suggested, first by Wolfgang Pauli in 1934, to be simultaneously ejected in these reactions and to be carrying the balance of the energy. The energy carried by the neutrino is often larger than that carried by the β particle itself. The importance of neutrinos can be judged from the fact that they carry with them some 5 percent of the total energy produced in a fission reactor. This is energy completely lost to us since neutrinos do not react and are not stopped by any practical structural material. The neutrino is usually given the symbol ν.

There are many other atomic subparticles. An example are the *mesons,* unstable positive, negative, or neutral particles having masses intermediate between an electron and a proton, which are exchanged between nucleons and are thought to account for the forces between them. A discussion of these and other particles is, however, beyond the scope of this book.

1-3. THE STRUCTURE OF HEAVY ATOMS

When an atom with a large number of electrons (large Z) is in the ground state, i.e., unexcited (possessing no excitation energy), all its electrons are not located in one orbit. Instead they are distributed over various *shells* represented by quantum numbers 1, 2, 3, 4, ⋯, etc., which are customarily given the designations K, L, M, N, O, ⋯, respectively. When the atom is in the ground state, the number of electrons orbiting in any one shell cannot exceed the value $2n^2$. Thus the maximum number of electrons that can exist in any one shell is given in Table 1-1.

Usually, but not always, lower shells are filled first, Fig. 1-2. As an example, lithium ($Z = 3$), with a total of three electrons, fills the K shell with two of its electrons and places only one in the L shell. Neon ($Z = 10$) has the K and L shells completely filled with two and eight

TABLE 1-1

Maximum Number of Electrons in Different Shells

Quantum Number n	Shell Designation	Maximum Number of Electrons
1	K	2
2	L	8
3	M	18
4	N	32
5	O	50

electrons, respectively. Phosphorus ($Z = 15$) has two, eight, and five, leaving the M shell partially filled, and so on.

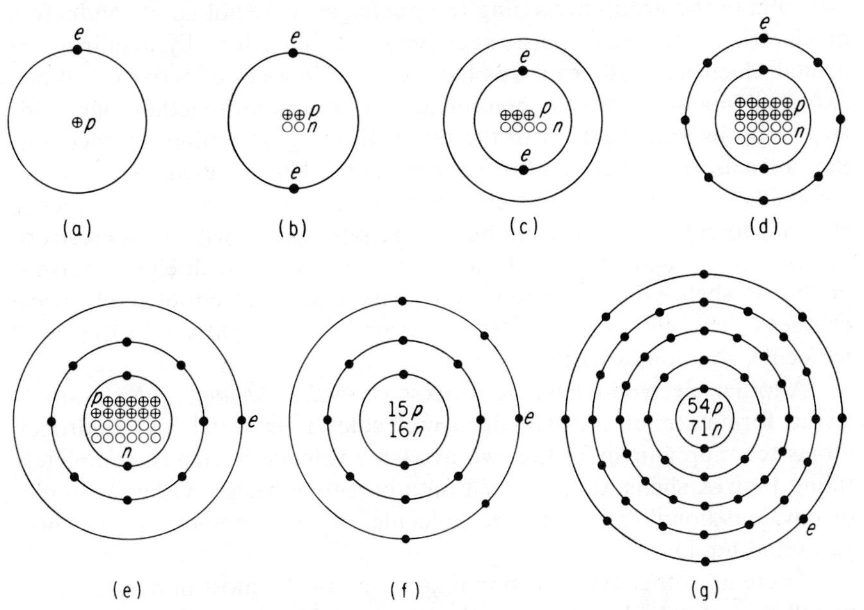

FIG. 1-2. Structure of some atoms (orbit radii not to scale). (a) Hydrogen ($Z = 1$, $A = 1$); (b) helium ($Z = 2$, $A = 4$); (c) lithium ($Z = 3$, $A = 7$); (d) neon ($Z = 10$, $A = 20$); (e) sodium ($Z = 11$, $A = 23$); (f) phosphorus ($Z = 15$, $A = 31$); (g) xenon ($Z = 54$, $A = 125$).

Electrons that orbit in the outermost shell of an atom are called the *valence electrons,* and the outermost shell is called the *valence shell.* Thus hydrogen has one valence electron and its K shell is the valence shell, and so on. Chemical properties of an element are a function of the number of valence electrons.

The so-called noble, or inert, gases $_2He$, $_{10}Ne$, $_{18}A$, $_{36}Kr$, $_{54}Xe$, and $_{86}Rn$ are monatomic gases that show little tendency to enter into chemical combination with other elements or even to combine with their own kind to form multiatomic molecules. There are two valence electrons for He (filling the *K* shell), eight for Ne (filling the *L* shell), and eight for the rest, occupying the *M*, *N*, *O*, and *P* shells, respectively, a number that satisfies these shells. These gases are inert because their valence shells are satisfied and need not give up or acquire electrons.

On the other hand, the elements in groups succeeding and preceding those gases in the *periodic table* of elements are very active chemically. The succeeding group, containing lithium, sodium, copper, etc., have only one valence electron each. They are mostly metallic, are good conductors of electricity, and easily form positive electrolytic ions; i.e., they easily lose their one valence electron. On the other hand, the elements in the group preceding the noble gases are not good conductors of electricity but easily form negative electrolytic ions by acquiring an orbital electron. Other groups have varying degrees of activity. When two elements of the above-mentioned two groups get together, one tends to give up its lone electron to the other, forming oppositely charged ions that attract each other and form a molecule. This is called an *ionic bond.* An example is the NaCl molecule in which sodium loses its lone valence electron to chlorine. Sodium becomes a positive ion with eight electrons in the *L* shell, and chlorine becomes a negative ion with eight electrons in the *M* shell. The two ions thus have equal but opposite electrical charges. The electrostatic attraction binds them tightly into the NaCl molecule, sometimes written Na^+Cl^-.

Another electron-exchange process is *covalent bonding.* An example is the formation of the hydrogen molecule H_2 in which two hydrogen atoms team up and share the two available valence electrons, completely filling their *K* shells, though each on a part-time basis. Other examples of covalent bonding are the Cl_2 molecule, organic compounds and most gases and liquids.

There are other types of bonding. One of the most important is the *metallic bond,* which occurs in the cases of Na, Fe, and Cu. In this, the outer-shell electrons are free to migrate, thus forming a uniform sea of electrons moving freely and easily about a lattice of positive ions of the same element. This type of bonding is characterized by high electrical and thermal conductivity.

It should be mentioned here that in some solid materials the exact type of bonding is not known. Only in a relatively few cases, as with

NaCl, are we sure that one or the other type of bonding exists. Frequently there is a gradual transition between two types. In uranium fuels, for example, uranium metal has a metallic bond, uranium dioxide (UO_2) is both covalent and ionic, while uranium carbide (UC) is covalent and metallic.

A knowledge of the type of bond that a material has is important in nuclear work, since it determines the ability of the material to resist change under nuclear irradiations. An example is the change in molecular structure of organic-reactor coolants [2].

1-4. THE CHEMICAL REACTION

Chemical reactions involve the combination or separation of whole atoms. For example, one whole atom of carbon combines with two whole oxygen atoms (or one oxygen molecule) to form a new molecule, that of carbon dioxide:

$$C + O_2 \longrightarrow CO_2 \tag{1-2}$$

This reaction is accompanied by the release of about 4 electron volts (abbreviated ev) of energy.

An *electron volt* is a unit of energy in common use in nuclear engineering. 1 ev $= 1.519 \times 10^{-22}$ Btu $= 4.44 \times 10^{-26}$ kwhr $= 1.6021 \times 10^{-19}$ watt-seconds (joules), Appendix H. One *million electron volt* (1 Mev) $= 10^6$ ev.

In chemical reactions, each atom participates as a whole and retains its identity after the reaction is completed. The reactions, however, result in molecules different from those entering the reactions. The only effect is a sharing or exchange of valence electrons. The nuclei of the participating atoms are unaffected. The molecules that enter the reaction on the left hand side of the equation are called the *reactants.* The right-hand side includes the resulting molecules, called the *products.* In writing chemical equations it is necessary to see that as many atoms of each participating element show up in the products as did in the reactants. In other words, the number and identity of the atoms are conserved in a chemical reaction.

Chemical equations, such as that in which uranium dioxide (UO_2) is converted into uranium tetrafluoride (UF_4), called green salt, by heating the dioxide in an atmosphere of highly corrosive anhydrous (without water) hydrogen fluoride (HF) and water vapor (H_2O) appears in the products, are balanced by writing

$$UO_2 + \quad HF \longrightarrow \quad H_2O + \quad UF_4 \tag{1-3a}$$

leaving space preceding the symbols. We balance the products. A uranium balance gives us one molecule of UF_4. Oxygen balance gives us two molecules of H_2O. We then proceed to find the number of HF molecules in the reactants. Fluorine balance from the products indicates that we need four molecules. Thus

$$UO_2 + 4HF \longrightarrow 2H_2O + UF_4 \tag{1-3b}$$

In this reaction the resulting water vapor is driven off. Uranium tetrafluoride is used further to prepare uranium hexafloride (UF_6) which is used in the separation of the U^{235} and U^{238} isotopes of uranium by the gaseous diffusion method. (Fluorine has only one isotope, F^{19}, and thus combinations of molecules of uranium and fluorine have molecular masses depending only on the uranium isotope in the molecule.)

It should be pointed out that chemical reactions, like nuclear reactions, are either *exothermic or endothermic.* That is, they either release or absorb energy. Since energy and mass are convertible according to Einstein's law, Eq. 1-1, chemical reactions involving energy *do* undergo changes in mass Δm (a mass decrease in exothermic reactions and an increase in endothermic ones), just as nuclear reactions do. However, the quantities of energy resulting from a chemical reaction are very small compared with those in a nuclear reaction, and the fraction of mass of the reactants that is lost or gained is minutely small. This is why we assume a preservation of mass in chemical reactions–undoubtedly an incorrect assumption but one that is sufficiently accurate for usual engineering calculations.

1-5. NUCLEAR EQUATIONS

In nuclear reactions, the same reactant nuclei do not show up as products. In the products we may find either isotopes of the reactants or completely different ones. In balancing nuclear equations it is necessary to see that the same, or equivalent, nuclear *particles* show up in the products as entered the reaction.

For example, if K, L, M, and N were the symbols of elements of the nuclei or particles participating in a nuclear reaction, the corresponding nuclear equation might look like this:

$$_{Z_1}K^{A_1} + {}_{Z_2}L^{A_2} \longrightarrow {}_{Z_3}M^{A_3} + {}_{Z_4}N^{A_4} \tag{1-4}$$

In order for Eq. 1-4 to balance, the following relationships must be satisfied:

$$Z_1 + Z_2 = Z_3 + Z_4 \tag{1-5a}$$

and

$$A_1 + A_2 = A_3 + A_4 \tag{1-5b}$$

Sometimes the symbols γ or ν are added on the right-hand side to indicate the emission of electromagnetic radiation or a neutrino, respectively. These have no effect on equation balance since both have zero Z and A, but often carry large quantities of the resulting energy.

Although the mass numbers are preserved in a nuclear reaction, the masses of the isotopes on both sides of the equation do not balance. *Exothermic* or *endothermic* energy is obtained when there is a reduction or an increase in mass between reactants and products, respectively.

Example 1-1. One exothermic reaction occurs when common aluminum is bombarded with high-energy α particles (helium-4 nuclei) and is transmuted into Si^{30}, a heavy isotope of silicon whose most abundant isotope has mass number 28. In the reaction, a small particle is emitted. Write the complete reaction and calculate the change in mass.

Solution. The reaction is

$$_{13}Al^{27} + {}_2He^4 \longrightarrow {}_{14}Si^{30} + {}_{Z_4}X^{A_4}$$

where X is a symbol of a yet unknown particle. Balancing gives

$$Z_4 = 13 + 2 - 14 = 1 \quad \text{and} \quad A_4 = 27 + 4 - 30 = 1$$

The only particle satisfying these is a proton. Thus the complete reaction is

$$_{13}Al^{27} + {}_2He^4 \longrightarrow {}_{14}Si^{30} + {}_1H^1 \tag{1-6}$$

Now the isotope masses of the nuclei showing up in this reaction, obtained from the table in Appendix B, are as follows:

Reactants		Products	
Al^{27}	26.98153 amu	Si^{30}	29.97376 amu
He^4	4.00260 amu	H^1	1.00783 amu
Total	30.98413 amu	Total	30.98159 amu

From the above, we see that there is a *decrease* in mass, since $\Delta m = 30.98159 - 30.98143 = -0.00254$ amu. This mass is converted to negative energy, i.e., energy is released or is exothermic.

An example of an endothermic nuclear reaction is

$$_7N^{14} + {}_2He^4 \longrightarrow {}_8O^{17} + {}_1H^1 \tag{1-7}$$

The sum of the masses of these reactants and products (see Appendix B) are 14.00307 + 4.00260 = 18.00567 amu and 16.99914 + 1.00783 = 18.00697 amu respectively. Thus there is a net gain in mass of 0.00130 amu, which means that energy must have been absorbed or that the reaction is endothermic.

In the above two reactions, the positively charged α particles have to be accelerated to high kinetic energies in order to overcome electrical repulsion and bombard the positively charged aluminum or nitrogen nuclei. Thus the reactants in either case possess initial kinetic energy equal to the kinetic energy of the α particle plus the kinetic energy of the nucleus, though the latter is usually negligible. (This process is analogous to raising a fuel-air mixture to its ignition temperature by adding activation energy before combustion can take place in a chemical reaction.) When the reactions are completed, the energy released will be equal to the initial energy of the reactants plus the energy corresponding to the lost mass (or minus the energy corresponding to the gained mass).

This energy shows up in the form of kinetic energy of the resultant particles, in the form of γ energy, and as *excitation energy* of a product nucleus, if any become so excited. The total kinetic energy of the products is divided among the nuclei and particles in such a manner that the lighter particles have higher kinetic energies than the heavier ones.

The isotope masses, obtained from Appendix B and used above, included the masses of the orbital electrons. The nuclear masses can be computed by subtracting the sum of the masses of Z orbital electrons. This for example, makes the mass of the Al^{27} nucleus equal to $26.98153 - 13 \times 0.0005486$ amu, and so on. Such corrections, however, are unnecessary in most cases since the same number or electrons show up on both sides of the equation. For example, in the reaction given by Eq. 1-6, the energy produced corresponds to the change in masses of the nuclei as given by

$$\Delta m = [(M_{\mathrm{Si}} - 14m_e) + (M_{\mathrm{H}} - m_e)] - [(M_{\mathrm{Al}} - 13m_e) + (M_{\mathrm{He}} - 2m_e)]$$

where M denotes the isotope atomic masses and m_e the mass of one electron. It can be seen that the number of electrons balance and that

$$\Delta m = (M_{\mathrm{Si}} + M_{\mathrm{H}}) - (M_{\mathrm{Al}} + M_{\mathrm{He}})$$

which is the relationship used to compute Δm for the reaction given by Eq. 1-6. The principle holds even if neutrons (whose mass, 1.008665 amu, does not include any electrons) are involved in a reaction. In general, then,

$$\Delta m = \Sigma M_{\mathrm{products}} - \Sigma M_{\mathrm{reactants}} \tag{1-8}$$

and the electron masses are neglected. This rule applies even though a negative β particle may appear on either side of the equation. An example is the following reaction:

$$_{16}S^{35} \longrightarrow {_{17}Cl^{35}} + {_{-1}e^{0}} \tag{1-9}$$

In this case

$$\Delta m = [(M_{\text{Cl}} - 17m_e) + m_e] - (M_S - 16m_e)$$
$$= M_{\text{Cl}} - M_{\text{S}}$$

An *exception* to the rule, however, is in reactions involving positrons. This is shown by the reaction

$$_6\text{C}^{11} \longrightarrow {}_5\text{B}^{11} + {}_{+1}e^0 \tag{1-10}$$

In this case,

$$\Delta m = [(M_{\text{B}} - 5m_e) + m_e] - (M_{\text{C}} - 6m_e)$$
$$= M_{\text{B}} - M_{\text{C}} + 2m_e$$

Therefore, *two* electron masses are added if the positron appears on the right-hand side of the equation and are subtracted if it appears on the left-hand side of the equation.

1-6. ENERGY FROM NUCLEAR REACTIONS

The energy corresponding to the change in mass in a nuclear reaction can be calculated from Einstein's law, Eq. 1-1 here repeated

$$\Delta E = \frac{1}{g_c} \Delta mc^2 \qquad [1\text{-}1]$$

where g_c is a conversion factor that has the following values

$$1.0 \text{ g}_m\text{cm}^2/\text{erg sec}^2$$
$$32.2 \text{ lb}_m\text{ft}/\text{lb}_f\text{sec}^2$$
$$4.17 \times 10^8 \text{ lb}_m\text{ft}/\text{lb}_f\text{hr}^2$$
$$0.965 \times 10^{18} \text{ amu cm}^2/\text{Mev sec}^2$$

Thus if Δm is in grams and c in centimeters per second, ΔE will be in ergs. Since $c = 3 \times 10^{10}$ cm/sec, Eq. 1-1 can be written in the form

$$\Delta E \text{ (ergs)} = 9 \times 10^{20} \Delta m \text{ (grams)} \tag{1-11a}$$

But since it is convenient to express the masses of nuclei in atomic mass units and since 1 amu equals 1.66×10^{-24} g_m, Eq. 1-11a may be written as

$$\Delta E \text{ (ergs)} = 1.49 \times 10^{-3} \Delta m \text{ (amu)} \tag{1-11b}$$

In energy-mass relations, it is common to use the electron volt (ev), or the million electron volt (Mev) as units of energy. Using the Mev, Eq. 1-11b becomes

$$\Delta E \text{ (Mev)} = 931 \Delta m \text{ (amu)} \tag{1-12}$$

A useful relationship to remember, therefore, is that 1 amu of mass is equivalent to 931 Mev of energy. The reaction given in Example 1-1 thus produces $-0.00254 \times 931 = -2.365$ Mev of energy. Mass-energy conversion factors based on various other units are given in Appendix G.

1-7. NUCLEAR FUSION AND FISSION

Two classes of nuclear reactions that are of most importance from the point of view of energy production are *fusion* and *fission**. In *fusion,* two or more light nuclei are fused together to form a heavier nucleus. In *fission,* a heavy nucleus is split into two or more lighter nuclei. In both types, there is a net decrease in mass resulting in a net exothermic energy.

Energy from the sun and stars is produced by continuous fusion reactions in which four nuclei of hydrogen fuse and produce one nucleus of helium and two positrons:

$$4\,{}_1H^1 \longrightarrow {}_2He^4 + 2\,{}_{+1}e^0 \tag{1-13}$$

resulting in a decrease in mass by about 0.0276 amu which is equivalent to about 24.7 Mev. Equation 1-13 describes only the net effect of solar and stellar reactions. Actually, there are series, or chains, of reactions involving other particles that continually appear and disappear in the course of the reactions, such as He^3, nitrogen, carbon, and other nuclei.

The commonest stars are composed of hydrogen, helium and traces of heavier elements. The sun and the stars are continuous *fusion* or *hydrogen reactors.* The heat produced in these reactions maintains temperatures of the order of several million degrees in their cores and serves to trigger and sustain succeeding reactions.

On earth, although the utilization of fission preceded that of fusion in both weapons and power generation, the basic fusion reaction was discovered first. In the 1920's, research on particle accelerators produced the first man-made fusion reaction.

To cause fusion, it is necessary to accelerate the positively charged nuclei to high speeds so that collisions between them take place despite electrical repulsive forces. The necessary kinetic energies are produced by raising their temperature to hundreds of millions of degrees and preventing contact with the walls of the container. This is done by confining the plasma within magnetic lines of force in a "magnetic bottle" for a sufficient period of time (of the order of a second), preheating it by passing a large electrical current through it (ohmic heating), and

* A third important class, but one that produces much less energy per reaction, is radioactivity. It is discussed in Sec. 1-10.

heating it to the required temperature by pressurizing it by the magnetic field lines (magnetic pumping). The resulting reaction releases a larger sum of energy. Fusion reactions are called *thermonuclear* because very high temperatures are required to trigger and sustain them [2, 6].

Man-made fusion can be accomplished by having two atoms of deuterium (D or ${}_1H^2$) fuse into one atom of helium. Deuterium shows the greatest promise as fuel for artificially produced fusion reactions, since there is a much greater probability of two particles colliding than of four. The 4-hydrogen reaction requires, on an average, billions of years for completion, whereas the deuterium-deuterium reaction requires a fraction of a second.

Table 1-2 lists some of the possible artificial fusion reactions and the energies produced by them. n, p, D, and T are the symbols for the neutron, proton, deuterium (H^2) and *tritium* (H^3) respectively.

TABLE 1-2

Number	Fusion Reaction		Energy per Reaction, Mev
	Reactants	Products	
1	D + D	T + p	4
2	T + D	$He^4 + n$	17.6
3	D + D	$He^3 + n$	3.2
4	He^3 + D	$He^4 + p$	18.3

Many problems have to be solved before a man-made fusion reactor becomes a reality [2]. The most important of these are the difficulty in generating and maintaining high temperatures and the instabilities in the medium (plasma). There are many other problems of an operational nature.

Unlike fusion, which involves particles of similar electrical charge and therefore requires high kinetic energies to initiate reaction, *fission* can be caused by a neutral particle, the neutron. The neutron can strike and fission a heavy nucleus at high, moderate, or low speeds, without being repulsed. Fission can also be caused by other particles. Neutron bombardment, however, is the only practical way of obtaining a sustained reaction, since two or three neutrons are usually released for each one engaging in fission. These keep the reaction going. There are only a few isotopes that are fissionable. U^{235}, Pu^{239} and U^{233} are fissionable by neutrons of all energies. U^{238}, Th^{232}, and Pu^{240} are fissionable by high-energy neutrons only. An example of a fission reaction shown schematically in Fig. 1-3 is

$$ {}_{92}U^{235} + {}_0n^1 \longrightarrow {}_{54}Xe^{140} + {}_{38}Sr^{94} + 2\,{}_0n^1 \qquad (1\text{-}14)$$

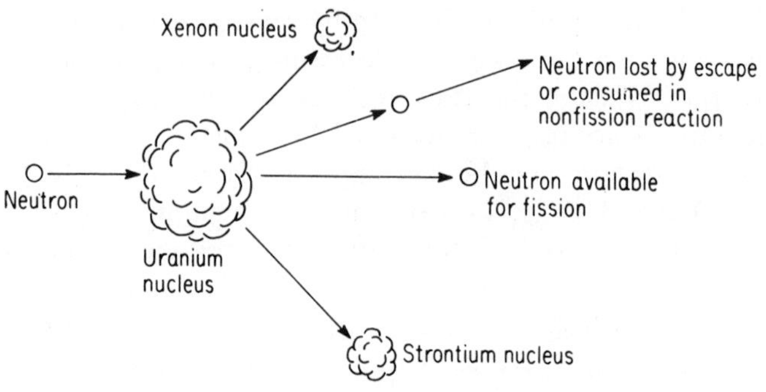

FIG. 1-3. A fission reaction.

The immediate (prompt) products of the fission reaction, such as Xe^{140} and Sr^{94} above, are called *fission fragments.* They, and their decay products (Sec. 1-10) are called *fission products.* Figure 1-4 shows fission

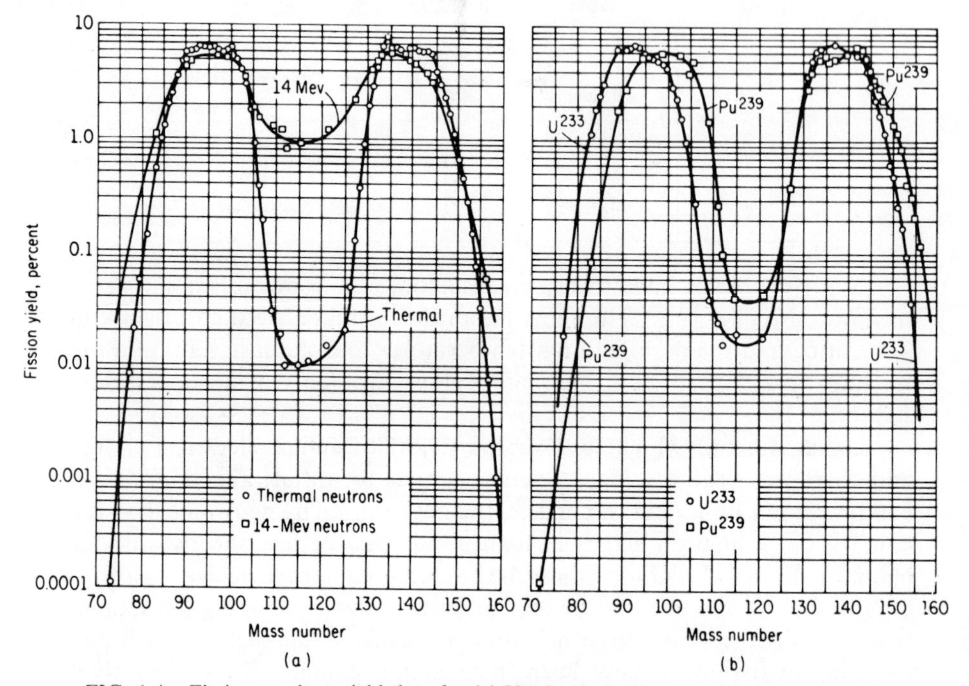

FIG. 1-4. Fission-product yield data for (a) U^{235} by thermal and 14-Mev neutrons and (b) U^{233} and Pu^{239} by thermal neutrons (Ref. 7).

product yield data for U^{235} by thermal and fast neutrons (Secs. 2-2 and 2-4) and for U^{233} and Pu^{239} by thermal neutrons. The fission products are represented by their mass numbers.

1-8. CONVERSION AND BREEDING

The isotopes U^{238} and Th^{232} are called *fertile* because they can be converted into the fissionable isotopes Pu^{239} and U^{233} respectively by nonfission capture of neutrons. The process called *conversion*, comprises a series of reactions (Fig. 1-5) as follows:

$$\begin{aligned} {}_0n^1 + {}_{92}U^{238} &\longrightarrow {}_{92}U^{239} + \gamma \\ {}_{92}U^{239} &\longrightarrow {}_{93}Np^{239} + {}_{-1}e^0 \\ {}_{93}Np^{239} &\longrightarrow {}_{94}Pu^{239} + {}_{-1}e^0 \end{aligned} \tag{1-15}$$

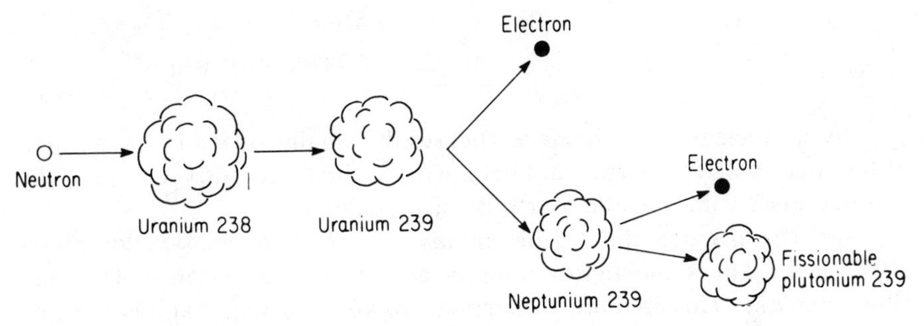

FIG. 1-5. A breeding chain.

where U^{239} is an unstable or radioactive nucleus (Sec. 1-10). Within a very short time after its birth, it emits a β particle and is transmuted into Np^{239}, which likewise is transmuted into long-lived Pu^{239}. Similarly for Th^{232}:

$$\begin{aligned} {}_0n^1 + {}_{90}Th^{232} &\longrightarrow {}_{90}Th^{233} + \gamma \\ {}_{90}Th^{233} &\longrightarrow {}_{91}Pa^{233} + {}_{-1}e^0 \\ {}_{91}Pa^{233} &\longrightarrow {}_{92}U^{233} + {}_{-1}e^0 \end{aligned} \tag{1-16}$$

A reactor in which fissionable nuclei, different from those in the original core loading, are produced is called a *converter*. If more fissionable nuclei are produced than are consumed by fission, or if the same type of nuclei are produced as are consumed, the reactor is called a *breeder*.

1-9. ENERGY FROM FISSION AND FUEL BURNUP

There are many fission reactions which release different energy values. The one in Eq. 1-14, for example, yields 196 Mev. Another is illustrated by the reaction

$${}_{92}U^{235} + {}_0n^1 \longrightarrow {}_{56}Ba^{137} + {}_{36}Kr^{97} + 2{}_0n^1 \tag{1-17}$$

It has the mass balance:

$$235.0439 + 1.00867 \longrightarrow 136.9061 + 96.9212 + 2 \times 1.00867$$

Adding,

$$236.0526 \longrightarrow 235.8446$$

Thus

$$\Delta m = 235.8446 - 236.0526 = -0.2080 \text{ amu}$$

and

$$\Delta E = 931 \times -0.2080 = -193.6 \text{ Mev}$$
$$= 193.6 \times 1.517 \times 10^{-16} = -2.937 \times 10^{-14} \text{ Btu}$$

In a nuclear mass balance the result usually depends on a small difference between large numbers. It is thus necessary to carry the isotope-mass values to the fourth or fifth decimal places.

On the *average* the fission energy of a U^{235} nucleus yields about 193 Mev. The same figure roughly applies to the fission of U^{233} and Pu^{239} nuclei. This amount of energy is *prompt,* meaning that it is released at the time of fission. More energy, however, is produced per fission reaction because of (1) the slow decay of the fission fragments (such as Ba^{137} and Kr^{97} above) into fission products and (2) the nonfission capture of excess neutrons in reactions which also produce energy, although much lower than that produced in fission reactions. The total energy produced *per* fission reaction in a nuclear fuel element therefore is greater than the prompt energy produced in the fission reaction itself. The *average total* energy is about 200 Mev per fission, a useful number to remember.

The complete fission of 1 g_m of U^{235} nuclei in a fuel element thus produces a quantity of energy equal to

$$\frac{\text{Avogadro number}}{U^{235} \text{ isotope mass}} \times 200 \text{ Mev} = \frac{0.60225 \times 10^{24}}{235.0439} \times 200$$
$$= 0.513 \times 10^{24} \text{ Mev}$$
$$= 2.276 \times 10^{4} \text{ kwhr}$$
$$= 948 \text{ kw-day}$$
$$= 0.948 \text{ Mw-day}$$

Another convenient figure to remember, therefore, is that a reactor burning 1 g_m of fissionable material per day generates nearly 1 Mw of energy.

This relates to fuel *burnup*. Maximum theoretical burnup would therefore be about a million Mw-day/ton (metric)* of fuel. This figure applies only if the fuel were entirely composed of fissionable nuclei (such as U^{235}, Pu^{239}, or U^{233}) and if these nuclei were all fissioned. Reactor fuel, however, contains other nonfissionable isotopes of uranium, plutonium or thorium. *Fuel* is defined as all uranium, plutonium, and thorium isotopes. It does not include alloying or other chemical compounds or mixtures. The term *fuel material* is used to refer to fuel plus such other materials.

The fissionable isotopes in the fuel cannot be all fissioned because of the accumulation of fission products that absorb neutrons and eventually stop the chain reaction. Because of this, and owing to metallurgical reasons such as the inability of the fuel material to operate at high temperatures or to retain gaseous fission products (such as Xe and Kr, Eqs. 1-12 and 1-13) in its structure except for limited periods of time, burnup values of reactor fuels are much lower that this figure. They are, however, increased somewhat by the fissioning of some of the new fissionable nuclei, such as Pu^{239}, which are converted from fertile nuclei, such as U^{238}, which were already in the fuel. Depending upon fuel type and *enrichment* (*mass* percent of fissionable fuel in all fuel), they may vary from about 1,000 to 100,000 Mw-day/ton and higher.

In addition to the above-mentioned fission by neutron bombardment, a uranium nucleus may fission by bombardment with other particles. It is also capable of dividing itself into two fragments without the aid of a bombarding particle. This process, called *spontaneous fission,* is quite slow, occurring at the rate of about 3×10^{-4} fission/sec g_m in U^{235} and 10^{-2} fission/sec g_m in U^{238}

1-10. RADIOACTIVITY

Radioactivity is an important source of energy, as well as a source of radiations for use in research, industry, medicine and a wide variety of applications.

Most of the naturally occurring isotopes are *stable*. Those that are not stable, i.e., *radioactive,* are some isotopes of the heavy elements –thallium ($Z = 81$), lead ($Z = 82$), and bismuth ($Z = 83$)– and all the isotopes of the heavier elements beginning with polonium ($Z = 84$) A few lower-mass isotopes are also naturally radioactive, such as K^{40}, Rb^{87}, and In^{115}. In addition, several thousand artificially produced isotopes of all masses are radioactive. Natural and artificial radioactive

* 1 metric ton = 1,000 kg = 2,204.7 lb_m. Other tons in general use: the long ton = 1,016 kg_m = 2,200 lb_m, the short ton = 907.2 kg_m = 2,000 lb_m. (See Appendix H.)

isotopes, also called *radioisotopes,* have similar disintegration rate mechanisms. Figure 1-6 shows a chart of the known isotopes, in the form of horizontal bars on a *Z-N* plot.

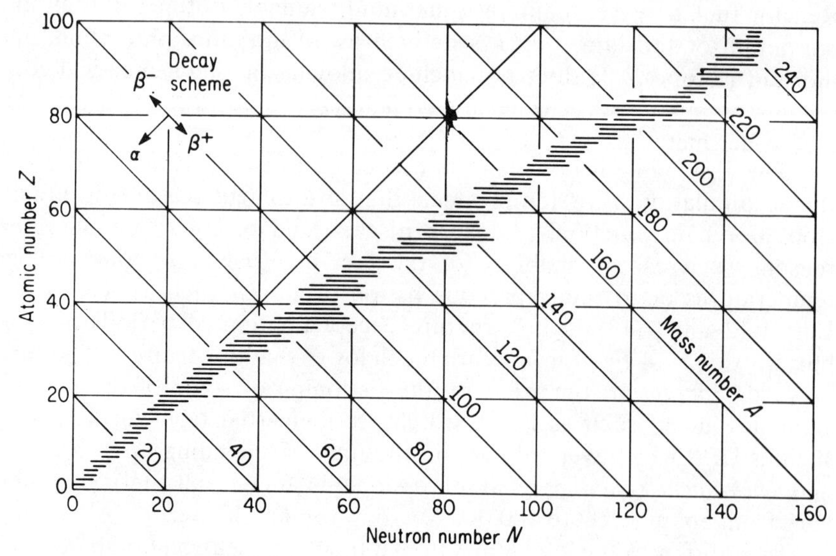

FIG. 1-6. *Z-N* chart of the known isotopes.

Radioactivity means that the radioactive isotope continuously undergoes a spontaneous (i.e., without outside help) disintegration. The process usually involves the emission of one or more of a number of smaller particles from the *parent* nucleus, after which the latter is changed into another, or *daughter,* nucleus. The parent nucleus is said to *decay* into the daughter nucleus. The daughter may not in itself be stable, and several stages of successive decay may then take place before a stable isotope is formed.

An example of radioactivity is the following:

$$_{49}\mathrm{In}^{115} \longrightarrow {}_{50}\mathrm{Sn}^{115} + {}_{-1}e^{0}$$

where In^{115} is a naturally occurring radioisotope and its daughter, Sn^{115}, is stable.

Radioactivity is *always* accompanied by a *decrease* in mass, i.e., by the liberation of energy. The energy thus liberated shows up in the form of kinetic energy of the emitted particles and as electromagnetic radiation. The light particle is ejected at high speed while the heavy particle recoils at a much slower pace in an opposite direction.

Naturally occurring radioisotopes emit one or more of the following three types of particles or radiations: (1) α particles, (2) β particles, and (3) γ radiation. The artificially produced isotopes, in addition to above, emit, or undergo, the following particles or reactions: (4) positron or β^+ particles, (5) orbital electron absorption, called K capture, and (6) neutrons. In addition, neutrino emission accompanies β emission (of either sign) (Sec. 1-2). Figure 1-7 shows the three primary particles (α, β, and γ) when emitted in an electric field in vacuum. The above decay schemes will now be discussed.

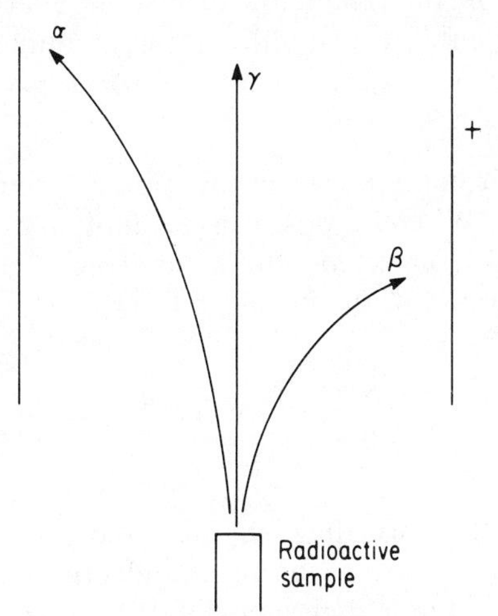

FIG. 1-7. The three primary radiations in vacuum.

Alpha Decay

Alpha particles are positively charged helium nuclei, each consisting of two protons and two neutrons. Alpha particles are commonly emitted by the heavier nuclei and are usually accompanied by γ radiation. An example of α activity is that of the decay of plutonium 239 into fissionable uranium 235:

$$_{94}Pu^{239} \longrightarrow {}_{92}U^{235} + {}_{2}He^{4} \tag{1-18}$$

Beta Decay

An example of β decay is

$$_{82}Pb^{214} \longrightarrow {}_{83}Bi^{214} + {}_{-1}e^{0} + \nu \tag{1-19}$$

where ν is the symbol for the neutrino, usually dropped from the equation. The penetrating power of β-particles is small compared with that of γ-rays but is larger than that of α-particles. β- and α-particle decay are usually accompanied by the emission of γ-radiation.

Gamma Radiation

Gamma radiation is electromagnetic radiation of extremely short wavelength and very high frequency and therefore high energy. Gamma and X-rays are physically similar but differ in their origin and energy levels. Gamma rays originate from the nucleus, and their wavelengths are, on an average, about one-tenth those of X-rays, although the energy ranges overlap to a certain extent. X-rays originate from the atom, owing to orbital electrons changing orbits, or energy levels. Gamma decay does not alter either the atomic or mass numbers.

Positive Beta (Positron) Decay

Positron decay is common when the radioactive nucleus contains an excess of protons. When a positron is emitted from the nucleus, a proton is effectively converted into a neutron. Examples of positive β emission may be found in the decay of nitrogen 13 and phosphorus 30;

$$_{7}N^{13} \longrightarrow {}_{6}C^{13} + {}_{+1}e^{0} \qquad (1\text{-}20)$$

$$_{15}P^{30} \longrightarrow {}_{14}Si^{30} + {}_{+1}e^{0} \qquad (1\text{-}21)$$

Since in positron decay the daughter nucleus has one less proton than the parent nucleus, one of the orbital electrons is released to maintain atom neutrality. Later such an orbital electron meets and combines with an emitted positron according to the equation

$$_{+1}e^{0} + {}_{-1}e^{0} \longrightarrow 2\gamma \qquad (1\text{-}22)$$

The two particles therefore undergo an *annihilation* process, producing γ energy equivalent to the sum of their rest masses (plus their kinetic energies before collision, which are usually minutely small). Since the rest mass of the positron is the same as that of the electron (0.00055 amu) the emitted γ energy equivalent to their rest masses is equal to $2m_ec^2$, or, according to Eq. 1-12, $-(2 \times 0.00055) \times 931 = -1.02$ Mev. Thus positron decay is accompanied by a decrease of at least 1.02 Mev in energy.

The reverse of the annihilation process is called *pair production*. In this, a γ photon of very high energy forms a positron-electron pair. The photon must have an energy threshold of 1.02 Mev, in order for this process to occur. It occurs only in the presence of matter and never in a vacuum. This is an endothermic process, in which the threshold energy is completely converted to mass and excess energy shows up in the form of kinetic energy of the newly born pair. It is important in discussions of the interaction of γ-rays with matter (Sec. 6-2).

K Capture

This is similar to positive β emission in that it takes place when the nucleus has an excess of protons. However, if the nucleus is not at a sufficiently high energy level (1.02 Mev minimum) to permit the emission of a positron, it instead captures an atomic orbital electron, also effectively changing a proton into a neutron. The electron is captured from the orbit or shell nearest to the nucleus, *K* shell; hence the name *K* capture. The vacancy in the *K* shell is filled by another electron falling from a higher orbit. Thus *K* capture is accompanied by X-ray emission from the atom. The process is shown in Fig. 1-8. An example of *K* capture is the reaction

$$_{29}Cu^{64} + {}_{-1}e^{0} \longrightarrow {}_{28}Ni^{64} \tag{1-23}$$

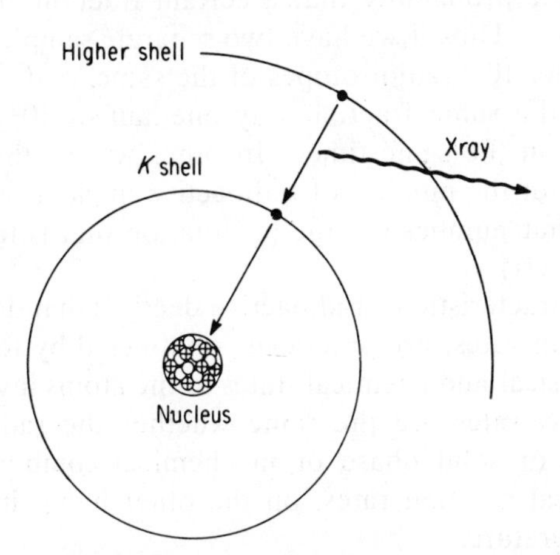

FIG. 1-8. *K* capture.

in which Ni^{64} is a stable nucleus. Because the parent nucleus acquires an electron, the nuclear equation contains the electron symbol on the left-hand side.

Neutron Emission

If a nucleus possesses an extremely high excitation energy, it may emit a neutron. The *binding energy* of a neutron in a nucleus (the energy that would have to be added to a nucleus in order for it to expel a neutron) varies with mass number but is on an average about 8 Mev. Thus if the excitation energy of a nucleus were at least about 8 Mev, it could decay by the emission of a neutron. An example of neutron emission is

$$_{54}Xe^{137} \longrightarrow {}_{54}Xe^{136} + {}_{0}n^{1} \tag{1-24}$$

Neutron decay results in the daughter nucleus being an isotope of the parent. It is usually a rare occurrence. However, it occurs frequently in nuclear reactors and is the source of the so-called *delayed* fission neutrons which are of utmost importance in reactor control. In Eq. 1-24 Xe^{137} is a fission product. It results from the β decay of the fission fragment I^{137}. The latter is called a *precursor*.

1-11. DECAY RATES AND HALF-LIVES

The time of decay of a radioactive nucleus can be predicted only statistically. There can be no indication of the time that it takes any one particular nucleus to decay. However, if there is a very large number of radioactive nuclei of the same kind present in the sample, there is a definite probability that a certain fraction of these will decay in a certain time. Thus if we have two separate samples, one containing 10^{20} and the other 10^{30} radioisotopes of the same kind, we can state with assurance that the same fraction, say one-half or $10^{20}/2$ and $10^{30}/2$, in each will decay in the same time. In other words, the rate of decay is a function only of the number of radioactive nuclei present at any time, provided that that number is large (a situation that is true in most cases of practical interest).

Another characteristic of radioactive decay is that decay rates, unlike chemical reaction rates, are practically unaffected by temperature, pressure, or the physical and chemical states of the atoms involved. In other words, the decay rates are the same whether the radioisotope is in a gaseous, liquid, or solid phase or in chemical combination with other atoms. Chemical reaction rates, on the other hand, increase exponentially with temperature.

Now if N is the number of radioactive nuclei of one species present in a sample at any time θ and if ΔN is the number of parent nuclei decaying in an increment of time $\Delta\theta$, at θ, the rate of decay of the parent is given, in the limit, by $-dN/d\theta$ and is directly proportional to N. This can be expressed by the relationship

$$-\frac{dN}{d\theta} = \lambda N \tag{1-25}$$

where λ is a proportionality factor called the *decay constant*. It has different values for different isotopes and has the dimension $(\text{time})^{-1}$, usually $\sec^{-1}$.

Integrating between an arbitrary zero time, $\theta = 0$, when the initial number of atoms present was N_0, gives

$$-\int_{N_0}^{N} \frac{dN}{N} = \lambda \int_0^{\theta} d\theta$$

Thus

$$-\ln \frac{N}{N_0} = \lambda\theta$$

and

$$N = N_0 e^{-\lambda\theta} \tag{1-26}$$

The rate of decay $-dN/d\theta$ is also called the *activity* of the sample, A, and commonly has the dimension disintegrations per second (dis/sec) or sec^{-1}. The initial activity A_0 is equal to λN_2. Thus

$$A = \lambda N = \lambda N_0 e^{-\lambda\theta}$$

and

$$A = A_0 e^{-\lambda\theta} \tag{1-27}$$

It can be seen that Eqs. 1-26 and 1-27 are similar in form.

A common way of representing decay rates is by the use of the *half-life,* $\theta_{\frac{1}{2}}$, of the radioactive species. This is the time during which one-half of a number of radioactive species decays, or one-half of their activity ceases. Using either of Eqs. 1-26 or 1-27, the half-life is given by

$$\frac{N}{N_0} = \frac{A}{A_0} = \frac{1}{2} = e^{-\lambda\theta_{\frac{1}{2}}}$$

Thus

$$\theta_{\frac{1}{2}} = \frac{\ln 2}{\lambda}$$

or

$$\theta_{\frac{1}{2}} = \frac{0.6931}{\lambda} \tag{1-28}$$

Thus the half-life is inversely proportional to the decay constant. Starting at an arbitrary zero time where $N = N_0$, one-half of N_0 decay after one half-life has elapsed; one-half of the remaining atoms, or one-quarter of N_0, decay during the second half-life; one-eighth of N_0 during the third; and so on. This relationship is shown in Table 1-3 and Fig. 1-9. The fraction of the initial number of parent nuclei or activity remaining after n half-lives is equal to $(\frac{1}{2})^n$.

TABLE 1-3
Activity and Half-life

Number of elapsed half-lives	N/N_0 or A/A_0	
0	1	1.00000
1	$\frac{1}{2}$	0.50000
2	$\frac{1}{4}$	0.25000
3	$\frac{1}{8}$	0.12500
4	$\frac{1}{16}$	0.06250
5	$\frac{1}{32}$	0.031250
6	$\frac{1}{64}$	0.015625
7	$\frac{1}{128}$	0.007813
8	$\frac{1}{256}$	0.003906
9	$\frac{1}{512}$	0.001953
10	$\frac{1}{1,024}$	0.000977

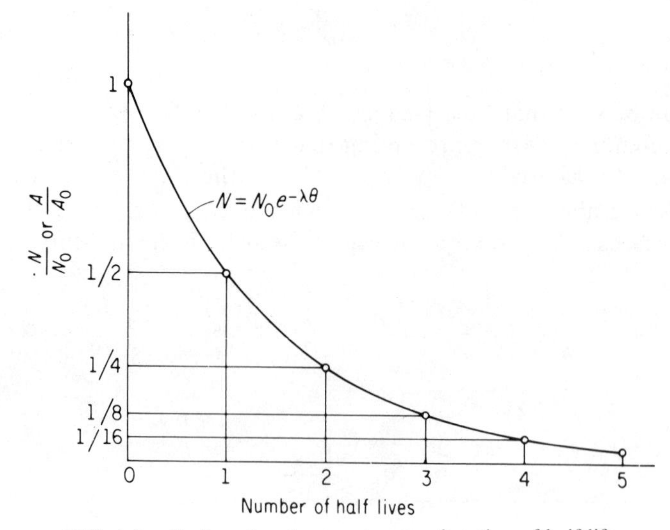

FIG. 1-9. Radioactive-decay rates as a function of half-life.

Theoretically, it takes an infinite time for the activity to become equal to zero. However, a time equivalent to about 10 half-lives reduces the activity to less than one-tenth of one percent of the original—negligible in many practical cases.

Half-lives of the known radioisotopes vary over a wide range, from fractions of a microsecond to billions of years. No two radioisotopes have exactly the same half-lives, and thus half-lives are considered "fingerprints" from which a particular species may be identified. This is done by measuring the change in activity with time and computing λ, say from the slope of the activity curve on a semilog plot, as shown in Fig. 1-10, from which $\theta_{\frac{1}{2}}$ and the unknown specie is identified.

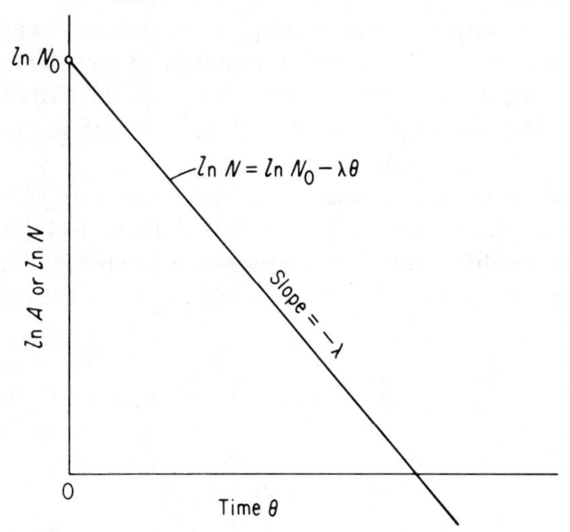

FIG. 1-10. Radioactive-decay curve as affected by the decay constant.

In cases involving two transitions from the parent radioisotope to the stable ground state, there are two decay rates and two characteristic half-lives. In some cases two different half-lives represent one transition.

Table 1-4 gives the half-lives and type of activity (ejected particle) of some important radioactive nuclei. Other half-lives are given in Appendix B.

TABLE 1-4
Half-lives of Some Isotopes

Isotope	$\theta_{\frac{1}{2}}$	Activity	Isotope	$\theta_{\frac{1}{2}}$	Activity
Carbón 14 ...	5,570 yr	β	Thorium 233	22.1 min.	β
Krypton 87 ..	78 min	β	Protactinium 233 .	27.0 days	β and γ
Strontium 90 .	28 yr	β	Uranium 233	1.62×10^5 yr	α and γ
Xenon 135 ...	9.2 hr	β and γ	Uranium 235	7.1×10^8 yr	α and γ
Barium 139 ..	85 min.	β and γ	Uranium 238	4.51×10^9 yr	α and γ
Radium 223 .	11.7 days	α and γ	Neptunium 239 ..	2.36 days	β and γ
Radium 226 .	1,600 yr	α and γ	Plutonium 239 ...	2.43×10^4 yr	α and γ
Thorium 232	1.45×10^{10} yr	α and γ			

It is interesting to note that the readily fissionable isotopes of U^{233}, U^{235} and Pu^{239} have extremely long half-lives, showing that they can be stored practically indefinitely. U^{233} and Pu^{239} are artificially produced (from

Th^{232} and U^{238}, respectively, themselves very long-lived), whereas U^{235} is found in nature.

The phenomenon of radioactivity is of tremendous importance to engineers and scientists. For example, the energy generated by the decaying fission products after reactor shutdown must be known in order to design an adequate coolant system. Also, radioactivity is now utilized in small power-producing devices and as a research tool in an ever-widening field of science and technology.

A less common way of describing the decay rate of a radioisotope is by the use of the *mean life,* θ_m, instead of the half-life, $\theta_{\frac{1}{2}}$. The mean life is the average life span of all the parent nuclei present in a radioactive sample of the same kind. It can be shown that θ_m is the reciprocal of the decay constant λ or

$$\theta_m = \frac{1}{\lambda} \tag{1-29}$$

and

$$\theta_{\frac{1}{2}} = 0.6931\, \theta_m \tag{1-30}$$

Example 1-2. Radium 226 decays into radon gas. Compute (*a*) the decay constant and (*b*) the initial activity of 1 g_m of radium 226.

Solution

(*a*)

$$\begin{aligned} \text{Half-life of } Ra^{226} &= 1{,}600 \text{ yr} \qquad \text{(Table 1-3 or Appendix B)} \\ &= 5.049 \times 10^{10} \text{ sec} \end{aligned}$$

$$\lambda = \frac{0.6931}{5.049 \times 10^{10}} = 1.3727 \times 10^{-11} \text{ sec}^{-1}$$

(*b*)

$$\text{Atomic mass of } Ra^{226} = 226.0245 \qquad \text{(Appendix B)}$$

$$\begin{aligned} \text{No. of atoms per gram} &= \frac{\text{Avogadro's number}}{\text{atomic mass}} \\ &= \frac{0.60225 \times 10^{24}}{226.0245} = 2.6645 \times 10^{21} \end{aligned}$$

$$\begin{aligned} \text{Initial activity } A_0 = \lambda N_0 &= 1.3727 \times 10^{-11} \times 2.6645 \times 10^{21} \\ &= 3.6576 \times 10^{10} \text{ dis/sec} \end{aligned}$$

The above example shows that the initial activity of 1 g_m of Ra^{226} is very small compared to the number of atoms in it. The activity of radium may thus be considered practically constant. This is true for any species with a sufficiently long half-life or small decay constant.

Early measurement of radioactivity indicated that 1 g_m of radium had an activity of 3.7×10^{10} dis/sec instead of the above more correct value. 3.7×10^{10} was adopted as a unit of radioactivity and is called a *curie*.

1-12. RADIOACTIVE CHAINS

Most naturally occurring radioisotopes decay into products themselves radioactive. The latter decay to other products that may also be radioactive, and a number of successive stages of decay occur before a stable final product is formed. They are said to form radioactive chains. (In ore deposits of radioactive material, several related radioisotopes exist in the ore, a proof that the age of the earth is not infinite.)

A similar situation occurs in the case of artificially produced radioisotopes, such as fission fragments which are radioactive and their products also are radioactive and so on. The difference between the two cases is that the naturally occurring radioisotopes are usually much longer-lived than their decay products, while the artificially produced ones often have half-lives not very different from their products. The following examples illustrate the two cases:

1. Radioactive decay of naturally occurring U^{235}:

$$
\begin{gathered}
{}_{92}U^{235} \xrightarrow[7.1 \times 10^{8}\text{ yr}]{\alpha} {}_{90}Th^{231} \xrightarrow[25.6\text{ yr}]{\beta} {}_{91}Pa^{231} \xrightarrow[3.43 \times 10^{4}\text{ yr}]{\alpha} \\
{}_{89}Ac^{227} \xrightarrow[27.7\text{ yr}]{\alpha} {}_{87}Fr^{223} \xrightarrow[21\text{ min}]{\beta} {}_{88}Ra^{223} \xrightarrow[11.2\text{ days}]{\alpha} {}_{86}Rn^{219} \xrightarrow[3.92\text{ sec}]{\alpha} \\
{}_{84}Po^{215} \xrightarrow[1.83 \times 10^{-3}\text{ sec}]{\alpha} {}_{82}Pb^{211} \xrightarrow[3.32\text{ hr}]{\beta} {}_{83}Bi^{211} \xrightarrow[2.16\text{ min}]{\alpha} \\
{}_{81}Tl^{207} \xrightarrow[4.76\text{ min}]{\alpha} {}_{82}Pb^{207} \xrightarrow[0.95\text{ sec}]{\text{isomeric}} {}_{82}Pb^{207}\text{ (stable)}
\end{gathered}
\tag{1-31}
$$

2. Radioactive decay of artificially produced Xe^{140} (a fission fragment):

$$
\begin{gathered}
{}_{54}Xe^{140} \xrightarrow[16\text{ sec}]{\beta} {}_{55}Cs^{140} \xrightarrow[66\text{ sec}]{\beta} {}_{56}Ba^{140} \xrightarrow[12.8\text{ days}]{\beta} \\
{}_{57}La^{140} \xrightarrow[40\text{ hr}]{\beta} {}_{58}Ce^{140}\text{ (stable)}
\end{gathered}
\tag{1-32}
$$

where the type of activity and the half-lives are written above and below the arrows, respectively.

Depending on the relative half-lives of parents and daughters, several modes of equilibrium exist in radioactive chains. A discussion of this is beyond the scope of this text. It suffices here to state that if the parent has a long half-life, as (1) above, a case of *secular* equilibrium exists in which the ratio of the number of nuclei of any two species is proportional to their half-lives and are therefore fixed. In a nuclear reactor, where the radioactive parents are continuously produced by the fission process, such as case (2), above, a state of secular equilibrium also exists provided, of course, the reactor remains in a steady state.

In case the parent has a much shorter half-life than the daughter, *no* equilibrium is possible. In case the half-lives are comparable, a case of *transient* equilibrium is said to exist.

PROBLEMS

1-1. The energy released when carbon (graphite) is burned to carbon dioxide is 14,100 Btu/lb_m of carbon. What is the energy released per atom of carbon?

1-2. Find the number density of U^{235} nuclei (nuclei/cm^3) (*a*) in natural metallic uranium and (*b*) in 93 percent enriched metallic uranium (93 percent U^{235} by mass). Take the uranium density in either case as 19 g_m/cm^3.

1-3. A relatively stationary U^{235} nucleus is bombarded by a 1-Mev neutron, causing it to fission. Two neutrons are ejected, and Xe^{140} appears as one of two fission fragments. Write the complete reaction equation and compute the energy produced per reaction and per pound mass of U^{235} if all fission processes in it were identical to the above.

1-4. A relatively stationary deuterium nucleus may be converted to a hydrogen nucleus by bombarding it with γ radiation, a process called *photodisintegration*. How much minimum γ energy is required for such conversion in Btu per gram of D_2?

1-5. Calculate the percent change in mass of the reactants when carbon burns completely in oxygen to CO_2. The heating value of C is 14,100 Btu/lb_m.

1-6. Consider the following three cases of fuel burnup: (*a*) 1,000 Mw-day/ton with natural metallic uranium, (*b*) 10,000 Mw-day/ton with 3 percent enriched UO_2, and (*c*) 6,000 Mw-day/ton with 47 percent enriched metallic uranium alloyed with Mo (2.5 percent), Ru (1.5 percent), Pd (0.5 percent), Rh (0.3 percent), and Zr (0.2 percent) by mass. For each of these cases compute the percent fuel that has fissioned of all fissionable nuclei, of all fuel nuclei, and of all fuel material in the original loading.

1-7. A reactor contained 1,000 kg_m of fuel material composed of 90 percent enriched UO_2SO_4 in 4 mole percent solution in heavy water. The reactor operated continuously and steadily, producing 10 Mw for 300 days. The entire fuel loading was then removed for reprocessing. Calculate (*a*) the mass of fuel consumed by fission in kilograms and (*b*) the burnup in Mw-day/ton.

1-8. Calculate the rates of decay of equal masses of the three isotopes found in natural uranium, normalizing the lowest to unity.

1-9. Radioactive carbon 14 (radiocarbon) is used to determine the age of materials of organic origin. It is formed by an (*n*,*p*) reaction between atmospheric nitrogen and slow neutrons emitted in cosmic radiation. Organic materials, when living, absorb CO_2, which contains about 0.1 percent $C^{14}O_2$, from the atmosphere. When dead, they absorb no CO_2, and the proportion of C^{14} in them decreases with time because of its decay. The amount of carbon in an old manuscript was determined and the amount of radiocarbon in it was found to be 0.045 percent. What is the age of the manuscript?

1-10. Rutherford once assumed that when the earth was first formed it contained equal amount of U^{235} and U^{238}. From this he was able to determine the age of the earth, and the answer was not very different from that found from astronomical data. Find the Rutherford age of the earth.

1-11. Calculate the activity in curies of (*a*) 1.0 lb_m of natural uranium and (*b*) 1.0 lb_m of 93 percent enriched uranium.

1-12. The human body contains about 0.35 percent by mass of natural potassium. Calculate the radioactivity in microcuries inherent within a 175-lb_m man.

1-13. A radioactive sample is composed of two independently radioactive isotopes. A nuclear counter whose output is directly proportional to the total activity of the sample was used and gave the following counts:

Time, hr	0	2	4	6	8
Counts/min	102,000	50,900	26,300	14,040	7,860
Time, hr	10	12	14	16	18
Counts/min	4,730	3,100	2,210	1,755	1,500
Time, hr	20	22	24	26	28
Counts/min	1,320	1,230	1,135	1,072	1,010

Find the half-life and the decay constant of each of the two isotopes.

1-14. Phosphorous-30 is a radioactive isotope undergoing positron decay. Calculate the energy in Mev/reaction.

1-15. A sample containing 1,000 g_m of N^{13} at an arbitrary time 0 undergoes positron decay. Find (*a*) the nuclear composition of the sample at the end of 20 min and (*b*) the total energy produced in Btu during these 20 minutes.

1-16. Repeat the above problem but the sample is made of C^{11} and the time is 41 min.

1-17. A sample of iodine-126 undergoes β-decay generating 1000 Btu/hr of energy. (*a*) What is the mass of the sample in grams, and (*b*) How many grams of the daughter nucleus are released in the first 26.6 days?

1-18. The U.S. Atomic Energy Commission allows, under license, the manufacture of illuminators for locks, watches, aircraft safety devices (switch plungers, control markers, exit signs, etc.), and other devices. The illuminators contain tritium plus a phosphor in the form of paint sealed in a plastic container. The low-energy β radiation emitted from tritium is too weak to escape the container and no hazard is encountered. However, it acts upon the phosphor to provide luminosity. No more than 15 millicuries may be used in an automobile lock, or 4 curies in an aircraft safety device. In each case find (*a*) the maximum amount of tritium in grams that may be used and (*b*) the percent decrease in luminosity after 5 years of operation.

1-19. Assume that a 1-g_m sample of common salt NaCl has been subjected to neutron bombardment and that all nuclei absorbed neutrons simultaneously. The resulting nuclei were all β emitters, transforming into stable end products independently of each other. Find (*a*) the activity of the sample immediately after neutron bombardment, in curies; (*b*) the activity of the sample after 1 day, in curies; and (*c*) the composition of the sample (nuclei of each species) after 1 day.

1-20. When a fuel rod is removed from a reactor core, it is stored in a pool of water a period of time before it is shipped for reprocessing. This allows the rod to cool, i.e., lose some of its original activity so that it may be handled safely in shipment. For simplicity, let us assume that all the activity in the rod was due to Xe^{133} which β-decays to a stable element with a halflife of 5.27 days. The fuel rod weighs 30 kg_m. When removed, it is assumed to have contained 0.1 percent Xe^{133} by mass.

(*a*) What is the activity of the rod when removed from the core?

(*b*) How much time should that rod be stored so that the activity measures 300 millicuries when shipped?

1-21. A stationary plutonium-239 nucleus is fissioned by a low-energy neutron resulting in the production of 3 fission neutrons, cerium-144 and another fission fragment. The two fission fragments are radioactive. Cerium-144 undergoes 2 stages and the other 5 stages of β-decay to stable isotopes. Identify the final products and calculate the overall energy in Mev produced due to fission and radioactive decay.

1-22. Promethium-147, contained in a phosphorous paint, is used to provide luminosity. The weak β radiation acts upon the phosphors to produce luminosity. Pm^{147} has advantages in brightness level, economy and radiation over both radium and tritium. (Cost of Pm^{147} isotopes per watch is 0.16 ¢, compared to 1.6¢ for Ra and 5¢ for H^3.) Maximum allowable radiations are 0.1 mc for watches, 0.5 for clocks, 2 for lock illuminators and 100 for aircraft safety devices.

In luminosity in a particular application should not vary by more than 2 mc over a 2-year period and starting with a maximum of 100 mc, how can this be accomplished?

1-23. A "food irradiator" contains a cesium-137 source of 170,000 curies. (It is used to irradiate potatoes for sprout control, wheat flour for insect disinfection, etc.) Food is passed by the irradiator source at the rate of 300 lb_m/hr. Cesium-137 is a β emitter of 30-year half-life.

(*a*) What is the approximate mass of the source in g_m?

(*b*) After 5 years of operation, what should be the rate of food processing in lb_m/hr if it were to receive the same dosage per lb_m?

1-24. Derive Eq. 1-29.

1-25. N^{16} (a product of water irradiation in a reactor) is a β emitting radioisotope. Each activity results in the release of 10 Mev. Calculate the isotope mass of N^{16} in amu from known isotope masses of the products.

chapter **2**

Neutrons and Their Interactions

2-1. INTRODUCTION

The heat generation at any point within the fuel in a reactor core is primarily a function of the fission-reaction rate at that point. In turn, this is a function of the neutron *flux* (Sec. 2-6) at that point. The prerequisite for analyzing a reactor core as a source of heat for generating power, therefore, is the establishment of the manner in which the neutron flux varies along the geometrical coordinates of the core.

In order to obtain the neutron-flux distribution in a core, a study of the interactions and conservation of neutrons within media is necessary. The branches of nuclear science that deal with these phenomena are commonly called *neutron physics* and *reactor physics.* They are in themselves the subjects of many textbooks [5, 8-15]. In this chapter we shall cover highlights from the first subject, mainly those dealing with the behavior of neutrons and their interaction with matter. In the next chapter we shall deal with neutron conservation and determine the neutron-flux distributions in reactor cores of different simple geometries. With this as a basis, future chapters will treat the reactor as a heat source for a power plant. While the material presented in this and the next chapter is believed adequate for our purposes, interested readers may decide to do further study of the subjects of neutron and reactor physics.

2-2. NEUTRON ENERGIES

As any other body, the kinetic energy of a neutron, KE_n, is given by

$$KE_n = \frac{1}{g_c} \cdot \frac{1}{2} m_n V^2 \qquad (2\text{-}1)$$

where m_n = mass of neutron
V = speed of neutron
g_c = conversion factor, the same as in Sec. 1-6.

Thus in Eq. 2-1 g_c is numerically equal to 1 if KE_n is in ergs, m_n in g_m and V in cm/sec. However, KE_n is more commonly expressed in electron volts (ev) or million electron volts (Mev). Since $m_n = 1.008665$ amu, 1 amu = 1.6604×10^{-24} grams, and 1 Mev = 1.602×10^{-6} erg, Eq. 2-1 can be written, for the neutron, in the form

$$KE_n = \frac{1}{2} \times 1.008665 \times 1.6604 \times 10^{-24} V^2 \left(\frac{1}{1.602 \times 10^{-6}}\right)$$

or

$$KE_n = 5.227 \times 10^{-19} V^2 \text{ Mev} \tag{2-2a}$$

$$= 5.227 \times 10^{-13} V^2 \text{ ev} \tag{2-2b}$$

where in both cases V is in centimeters per second.

The majority of the newly born fission neutrons have kinetic energies ranging between less than 0.075 and about 17 Mev. As these neutrons travel through matter (e.g., moderator) they are bound to collide with some of the neuclei and be decelerated mainly by the lighter ones in that matter. This process is called *scattering*. In so doing, the neutron gives up some of its kinetic energy and is slowed down with each successive collision.

Neutrons of different kinetic energies are classified into three general categories: fast, intermediate, and slow, as given in Table 2-1. One of many reactor classifications is the energy range of the neutrons primarily causing fission. Thus a fast reactor is one dependent primarily on fast neutrons for fission. A thermal reactor is one utilizing mostly thermal neutrons (Sec. 2-4).

TABLE 2-1
Neutron Classifications

Classification	Neutron Energy, ev	Corresponding Velocity, m/sec
Fast	Greater than 10^5	Greater than 4.4×10^6
Intermediate	1–10^5	1.38×10^4 to 4.4×10^6
Slow	Less than 1	Less than 1.38×10^4

In the discussions that follow, the term kinetic will occasionally be dropped, so that neutron energy means neutron kinetic energy. The symbol KE_n will be simplified to E_n.

2-3. FISSION NEUTRONS

Newly born fission neutrons carry, on an average, about 2 percent of a reactor fission energy, in the form of kinetic energy. Fission neutrons are divided into two categories, prompt and delayed. The former are those released at the time of fission (within about 10^{-14} sec after fission occurs), from fission fragments with a neutron-proton ratio the same as that of the original nucleus but greater than that corresponding to their mass number.

Delayed neutrons are produced in radioactive-decay reactions of the fission products (Sec. 1-10). They constitute only 0.645 percent for the total fission neutrons in the case of U^{235} fission and less for Pu^{239} and U^{233}. Their energies are fairly small compared with those of prompt neutrons but they play a major role in nuclear-reactor control.

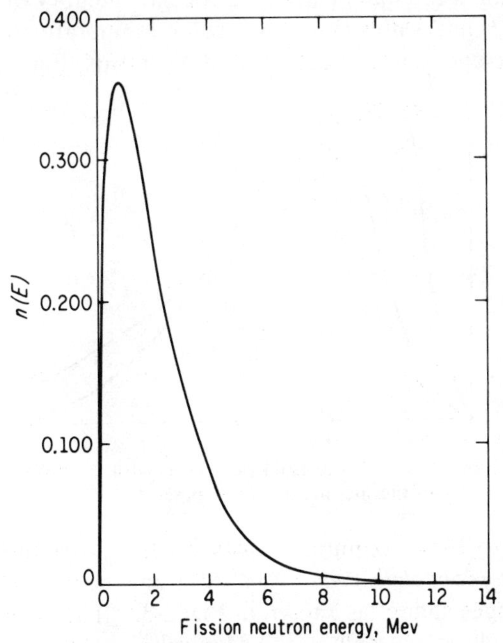

FIG. 2-1. Energy spectrum of fission neutrons.

Prompt neutrons have an energy distribution shown in Fig. 2-1 and quite accurately given (for U^{235} and Pu^{239} fission) by the relationship [16]

$$n(E)dE_n = \sqrt{\frac{2}{\pi e}} \sinh \sqrt{2E_n} e^{-E_n} dE_n \qquad (2\text{-}3)$$

where $n(E)$ is the number of neutrons having kinetic energy E_n per unit energy interval dE_n. Figure 2-1 shows that most of the prompt fission neutrons have energies less than 1 Mev, but an average energy around 2 Mev.

2-4. THERMAL NEUTRONS

Fission neutrons are scattered or slowed down by the materials in the core. An effective scattering medium, called a *moderator* is one which has small nuclei with high neutron scattering cross sections and low neutron absorption cross sections, such as H and D (in H_2O and D_2O), C (graphite), and Be or BeO. Cross sections are discussed in Sec. 2-5.

When neutrons are slowed down in a medium, the lowest energies that they can attain are those that put them in thermal equilibrium with the molecules of that medium. In this state they become thermalized and are called *thermal neutrons,* a special category of slow neutrons.

Particles or molecules at a particular temperature possess a wide range of kinetic energies and corresponding speeds, Fig. 2-2. The speed

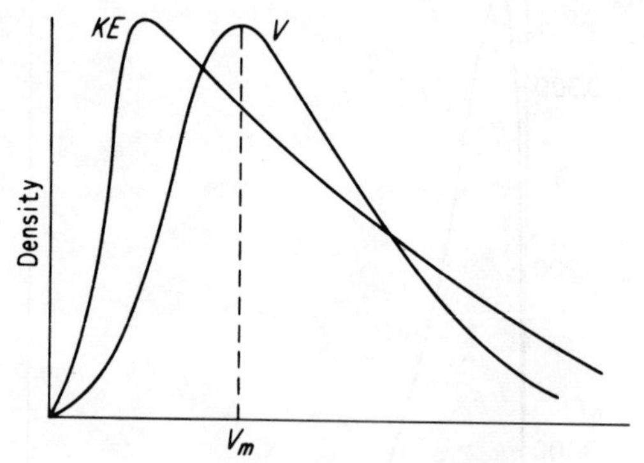

FIG. 2-2. Velocity and kinetic-energy distributions of thermal neutrons at temperature T.

corresponding to the maximum density on the distribution curve V_m is called the *most probable speed.* At a different temperature the distribution curve changes shape, as shown in Fig. 2-3. The distributions shown are mathematically expressed by the *Maxwell distribution law*:

$$n(V)dV = 4\pi n \left(\frac{m}{2\pi kT}\right)^{1.5} V^2 e^{-(mV^2/2kT)} dV \qquad (2\text{-}4)$$

where $n(V)$ = number of particles, present in given volume of medium, with speeds between V and $V + dV$

n = total number of particles in same volume of medium

m = mass of particle

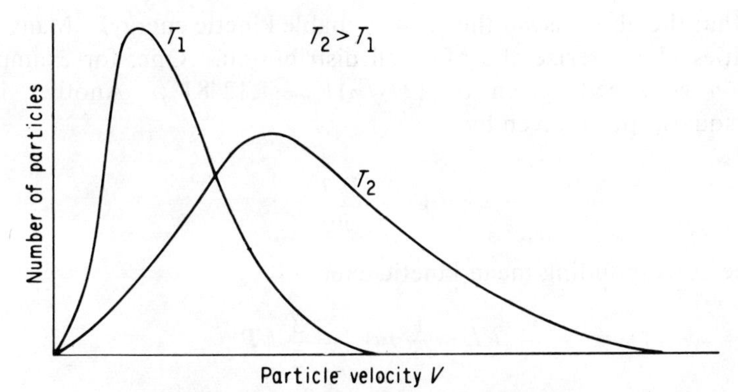

FIG. 2-3. The distribution of particle velocity in thermal equilibrium in two media at different temperatures.

k = Boltzmann's constant (universal gas constant divided by Avogadro number, $= 1.3805 \times 10^{-16}$ erg/°K, or 8.617×10^{-11} Mev/°K)

T = absolute temperature

The most probable speed can be evaluated by differentiating the right-hand side of Eq. 2-4 with respect to V and equating the derivative to zero. Equation 2-4 is first rewritten in the simplified form:

$$n(V) = c_1 V^2 e^{-c_2^2 V^2}$$

where

$$c_1 = 4\pi n \left(\frac{m}{2\pi kT}\right)^{1.5} \quad \text{and} \quad c_2 = \left(\frac{m}{2kT}\right)^{0.5}$$

Thus

$$\frac{dn(V)}{dV} = c_1 \left(2Ve^{-c_2^2 V^2} - 2V^3 c_2^2 e^{-c_2^2 V^2}\right)$$

Equating the right-hand side of this equation to zero and solving for V give a value for the *most probable speed*:

$$V_m = \frac{1}{c_2} = \left(\frac{2kT}{m}\right)^{0.5} \tag{2-5}$$

The *kinetic energy corresponding to the most probable speed* (when g_c is numerically equal to 1) is

$$KE_m = \frac{1}{2} mV_m^2 = \frac{1}{2} m \frac{2kT}{m} = kT \tag{2-6}$$

Note that the above is *not* the most probable kinetic energy. Many other quantities characterize the Maxwell distribution. One, for example, is the *average* speed, given by $(2/\sqrt{\pi})\,V_m = 1.1248\,V_m$. Another is the mean-square speed, given by

$$\overline{V^2} = \frac{3kT}{m} \tag{2-7}$$

and the corresponding mean kinetic energy is

$$\overline{KE} = \frac{1}{2}\,m\overline{V^2} = \frac{3}{2}\,kT \tag{2-8}$$

If species 1, 2, 3, ···, at thermal equilibrium are present in a medium at T (such as thermal neutrons and molecules of a moderator) we can write

$$\frac{1}{2}\,m_1\overline{V_1^2} = \frac{1}{2}\,m_2\overline{V_2^2} = \frac{1}{2}\,m_3\overline{V_3^2} = \cdots = \frac{3}{2}\,kT = \text{const} \tag{2-9}$$

Thus the mean-square speeds of particles in thermal equilibrium are inversely proportional to their respective masses. Equation 2-9 is known as the *law of equipartition of energy.* (The principle is used in the gas diffusion process in which isotopes of the same element are separated.)

Equations 2-6 and 2-8 show that the kinetic energy of a particle is independent of the mass of the particle and is a function only of the absolute temperature of the medium. The independence of particle energy and mass is also true with respect to the shape of the energy-distribution curve. Thus when neutrons become thermalized in a medium, they possess a kinetic-energy distribution identical to that of the molecules of the medium. They also have most probable and mean energies, equal to those of the molecules of the medium. The speeds, however, are dependent on mass and the neutron speed distribution is different from that of the medium molecules because the masses are usually different. (This discussion presupposes that molecules of the medium obey the above gas-distribution laws.)

By using the proper value of neutron mass in grams and Boltzmann's constant (1.38×10^{-16} erg/°K) and the appropriate conversion factor, we can rewrite Eqs. 2-5 and 2-6 as follows:

$$V_m = 1.2839 \times 10^4 T^{0.5} \cdots \text{for a neutron only} \tag{2-10}$$

and

$$KE_m = 8.6164 \times 10^{-5} T \cdots \text{for any particle} \tag{2-11}$$

where the kinetic energy is in electron volts, the speed in centimeters per second, and the absolute temperature in degrees Kelvin. Figure 2-4 shows the dependence of KE_m and V_m on absolute temperature.* Table 2-2 contains some thermal-neutron most probable energies and speeds

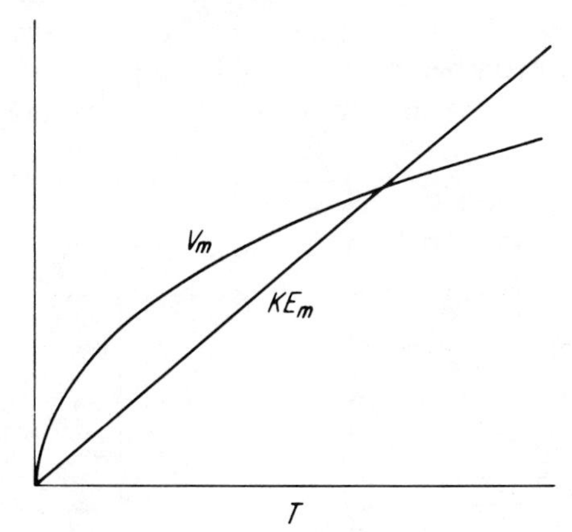

FIG. 2-4. Most probable speed and corresponding kinetic energy of a particle as a function of temperature.

as a function of temperature. At 20°C the neutrons have a most probable speed of 2,200 m/sec and a corresponding energy of 0.0252 ev. At this temperature, the neutron energy and velocity are sometimes said to be "standard." Cross sections (Sec. 2-5) for thermal neutrons are customarily tabulated for 2,200-m/sec neutrons.

TABLE 2-2
Thermal-neutron Speeds and Energies

Temperature		Most Probable Speed, m/sec	Corresponding Energy, ev
°C	°F		
20	68	2,200	0.0252
260	500	2,964	0.0459
537.8	1000	3,656	0.0699
1000	1832	4,580	0.1097

Neutrons having energies greater than thermal, such as those in the process of slowing down in a thermal reactor, are called *epithermal* neutrons.

* The mean speed does not correspond to the mean energy but is given by $\overline{V} = (2/\pi^{0.5}) V_m = 1.4488 \times 10^4 T^{0.5}$.

2-5. NUCLEAR CROSS SECTIONS

Assume a beam of mononergetic neutrons of intensity I_0 neutrons/sec, impinging on a body having a target area A cm² and a nuclear density N nuclei/cm³, as shown in Fig. 2-5. The nuclei of the body have radii roughly 1/1000 those of atoms, and therefore occupy a cross-sectional area, facing the neutron beam, very small compared with the total target area A. Taking one nucleus into consideration, we may liken the situa-

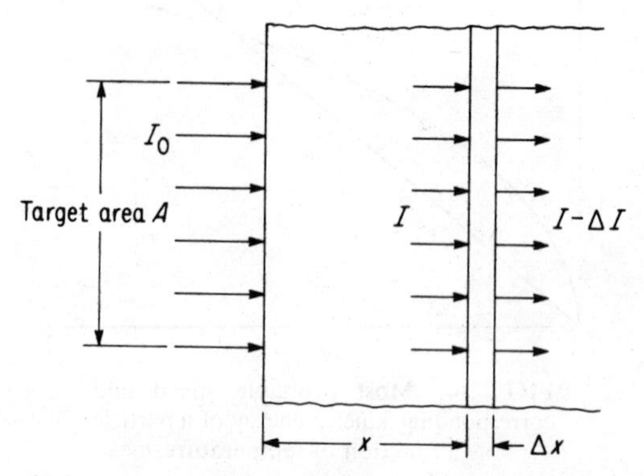

FIG. 2-5. Neutron beam striking target area A.

tion to a case in which a large number of peas are shot at a very large window at the center of which hangs a basketball. The number of peas that will collide with the basketball is proportional to the cross-sectional area of the basketball. However, the fraction of all peas colliding with the basketball, or the *probability* of collision is equal to the crossectional area of the basketball divided by the area of the window. The actual cross-sectional area of a nucleus is obtained from the radius of the nucleus, r_c, which according to numerous experiments [17] is given by

$$r_c = r_o A^{1/3} \tag{2-12}$$

where r_o is a constant varying for different nuclei with an average of 1.4×10^{-13} cm and A is the mass number. The actual cross-sectional areas of nuclei therefore average about 10^{-24} cm².

In nuclear reactions, the degree to which neutrons collide, or interact with nuclei is proportional to an *effective*, rather than actual, cross-sectional area of the nuclei in question. This is called the *microscopic cross section,* or simply the *cross section,* of the reaction and is given by the symbol σ. It varies with the nucleus, type of reaction, and neutron energy.

In Fig. 2-5 the number of nuclei in volume $A\ \Delta x$ is $N\ A\ \Delta x$. As the neutron beam passes through the body, some of the neutrons interact with the nuclei and are therefore removed (by absorption or scatter) from the beam. The fraction of the neutrons removed from the beam in passing through Δx is equal to the ratio of *effective* cross-sectional areas of the nuclei, $\sigma(NA\ \Delta x)$, to the area of the target A. Thus, if at distances x and $x + \Delta x$ the beam intensities become I and $I - \Delta I$, respectively, it follows that, in the limit

$$\frac{-dI}{I} = \frac{\sigma(NA\,dx)}{A} = \sigma N\,dx$$

Thus

$$-\int_{I_0}^{I} \frac{dI}{I} = \sigma N \int_{o}^{x} dx$$

from which

$$I = I_0 e^{-\sigma N x} \tag{2-13a}$$

σ has the units of area. Because nuclear dimensions are small, square centimeters would be too large a unit for σ. Consequently a unit equal to the actual cross-sectional area of an average nucleus, 10^{-24} cm², was taken as the unit of the microscopic cross section and has been given the name *barn*. A unit of millibarn, 10^{-3} barn or 10^{-27} cm², and others are sometimes used. Cross-section values as low as small fractions of a millibarn and as high as several thousand barns are encountered in nuclear reactions. Neutrons have as many microscopic cross sections as there are possible reactions. The most important cross sections are for scattering and absorption. These are

σ_s = microscopic cross section for scattering $= \sigma_{ie} + \sigma_e$

σ_a = microscopic cross section for absorption $= \sigma_c + \sigma_f$

where σ_{ie} = microscopic cross section for inelastic scattering [1]

σ_e = microscopic cross section for elastic scattering

σ_c = microscopic cross section for radiative (nonfission) capture

σ_f = microscopic cross section for fission

The cross sections σ_a and σ_s are given for some of the naturally occurring nuclei and some isotopes in Appendix B for 2,200 m/sec neutrons. Sometimes only a *total* cross section σ_t is given, where $\sigma_t = \sigma_s + \sigma_a +$ any other.

Cross-section-energy plots for some nuclei of interest are shown in Figs. 2-6 through 2-9, 4-5 and 4-6.

The product σN is equal to the total cross sections of all the nuclei present in a unit volume. It is called the *macroscopic cross section* and is given the symbol Σ. It has the unit of (length)$^{-1}$, commonly cm^{-1}. Thus

$$\Sigma = N\sigma \tag{2-14}$$

and Eq. 2-13a can be written in the form

$$I = I_0 e^{-\Sigma x} \tag{2-13b}$$

Macroscopic cross sections are also designated as to the reaction they represent. Thus $\Sigma_f = N\sigma_f$, $\Sigma_e = N\sigma_e$, etc. The reciprocal of macroscopic cross section for any reaction is the *mean free path* for that reaction. It is customarily given the symbol λ, not to be confused with the decay constant in radioactivity (Sec. 1-11).

For an element of atomic mass A_t and density ρ(g_m/cm^3), N (nuclei/cm^3) can be calculated from

$$N = \rho \frac{\text{Avogadro number}}{A_t} \tag{2-15}$$

Note that using the atomic instead of the molecular mass in the denominator makes Eq. 2-15 apply to any element, irrespective of its number of atoms per molecule.

For a mixture of elements 1, 2, 3, ⋯, or a compound, the macroscopic cross section is equal to the sum of the macroscopic cross sections of the different elements present or

$$\Sigma = N_1\sigma_1 + N_2\sigma_2 + N_3\sigma_3 + \cdots \tag{2-16}$$

where N_1, N_2, N_3, ⋯, are the number of nuclei of the different elements per cubic centimeter of mixture or compound.

Example 2-1. Calculate the total macroscopic cross section of the H_2O molecule for 2,200 m/sec neutrons. Assume that the density of H_2O is 1 g_m/cm^3.

Solution

$$\text{Mass of } H_2O \text{ molecule} = 2 \times 1.008 + 16 = 18.016$$

$$\text{Number of molecules/cm}^3 \text{ of } H_2O = \frac{0.60225 \times 10^{24}}{18.016} 1.0$$
$$= 0.03344 \times 10^{24}$$

Thus

$$N_H = 2 \times 0.03344 \times 10^{24} = 0.06688 \times 10^{24} \text{ nuclei/cm}^3$$

and

$$N_O = 1 \times 0.03344 \times 10^{24} = 0.03344 \times 10^{24} \text{ nuclei/cm}^3$$

$$\sigma_t \text{ for hydrogen nucleus} = 0.33 + 38 = 38.33 \text{ barns} \qquad \text{Appendix B}$$

$$\sigma_t \text{ for oxygen nucleus} = 0.0002 + 4.2 = 4.2002 \text{ barns} \qquad \text{Appendix B}$$

$$\Sigma_t \text{ for } H_2O \text{ molecule} = (0.06688 \times 10^{24})(38.33 \times 10^{-24}) + (0.03344 \times 10^{24})(4.2002 \times 10^{-24}) = 2.704 \text{ cm}^{-1}$$

A microscopic cross section for a molecule is sometimes desired and may be evaluated by dividing the macroscopic cross section of the molecule by the number of molecules in a unit volume of medium. In the above example for H_2O, therefore,

$$\sigma_t = \frac{2.704}{0.03344 \times 10^{24}} = 80.86 \times 10^{-24} \text{ cm}^2 = 80.86 \text{ barns}$$

2-6. NEUTRON FLUX AND REACTION RATES

In a beam of neutrons as above the number of neutrons crossing a unit area per unit time in one direction is called the neutron *current* and is proportional to the gradient of neutron *density*. In a reactor core, however, the neutrons travel in all directions. If n is the neutron density, neutrons/cm^3, and V the neutron velocity, cm/sec, the product nV is the number of neutrons crossing a unit area from all directions per unit time and is called the neutron *flux* φ. Thus

$$\varphi = nV \tag{2-17}$$

The neutron flux has the units neutrons/sec cm^2, which are often dropped. Since flux involves all neutrons at a given point, the reaction rate betwen neutrons and nuclei is therefore proportional to it.

Fluxes are dependent upon velocity V, i.e., upon energy, but are usually quoted for broad energy ranges, such as thermal and fast. In a reactor core they vary from maximum (usually at the core geometric center) to minimum (near the edges). In thermal heterogeneous reactors fission neutrons are born in the fuel and thermalized in the moderator. Fast fluxes thus peak above average in the fuel, while thermal fluxes peak in the moderator. Maximum full-power thermal fluxes vary from 10^6 for small training reactors to as high as 10^{15} for power and research reactors.

Now if a medium containing nuclei of density N is subjected to a neutron flux φ, the time rate of interaction, or *reaction rate,* between the nuclei and the neutrons is given by

$$\text{Reaction rate} = (nV)N\sigma = \varphi\Sigma \text{ reactions/sec cm}^3 \tag{2-18}$$

where σ is the microscopic cross section of the particular reaction in question (i.e., absorption, scatter, etc.). Equation 2-18 simply states that the number of neutrons entering a particular reaction (which is the same as the number of reactions) per unit time and volume of the medium is proportional to the total distance traveled by all the neutrons in a unit volume during a unit time (nV) and to the total number of nuclei per unit volume, N. σ, the probability of the reaction, is the proportionality factor. Equation 2-18 thus also states that the rate of reaction is proportional to the neutron flux but inversely proportional to the neutron mean free path of the particular reaction in question. Moreover, since in general N is a fixed quantity in a medium, it can be deduced that the rates of a particular reaction (fixed σ) are directly proportional to the neutron flux, an important deduction that will be used extensively when heat generation in reactor cores is discussed (Chapter 4). It will suffice here to state that heat generation by fission at a given point in a reactor core is proportional to the neutron flux at that point. A knowledge of the neutron-flux distribution in a reactor core is therefore necessary for the study of heat generation and removal in that core.

2-7. THE VARIATION OF NEUTRON CROSS SECTIONS WITH NEUTRON ENERGY

$\sigma - E_n$ plots are usually made on log-log coordinates. In many, but not all, cases, scattering cross sections are so small compared with absorption cross sections that the total cross sections shown may be taken as very nearly equal to the absorption cross sections of these materials. Also, for many nuclei, scattering cross sections vary little with neutron energy.

Examples of the variation of cross section with neutron energy are those for boron, cadmium, and indium (see Fig. 2-6) and for U^{235} and U^{238} (Figs. 2-7 and 2-8 respectively) [18]. In these cases the total, fission, and radiative-capture cross sections are given for all or some of the neutron-energy spectrum.

Cross sections are not, of course, limited to neutrons. Figure 2-9 shows [19] cross sections in barns for four fusion reactions versus the deuteron (deuterium nucleus) energy in Kev. As seen, the fusion cross sections also vary with deuteron energy.

Figures 2-6 through 2-8 show an interesting pattern of variation of neutron cross sections with neutron energy. This pattern is due to the absorption cross sections (fission, radiative capture, etc.) but is reflected in the total cross sections since, as indicated above, scattering cross sections are usually relatively small. It can be seen that absorption-cross-section curves can be divided into three regions which, beginning with low neutron energies, are (1) the *1/V region,* (2) the *resonance region,* and (3) the *fast-neutron region.* These will now be discussed in turn.

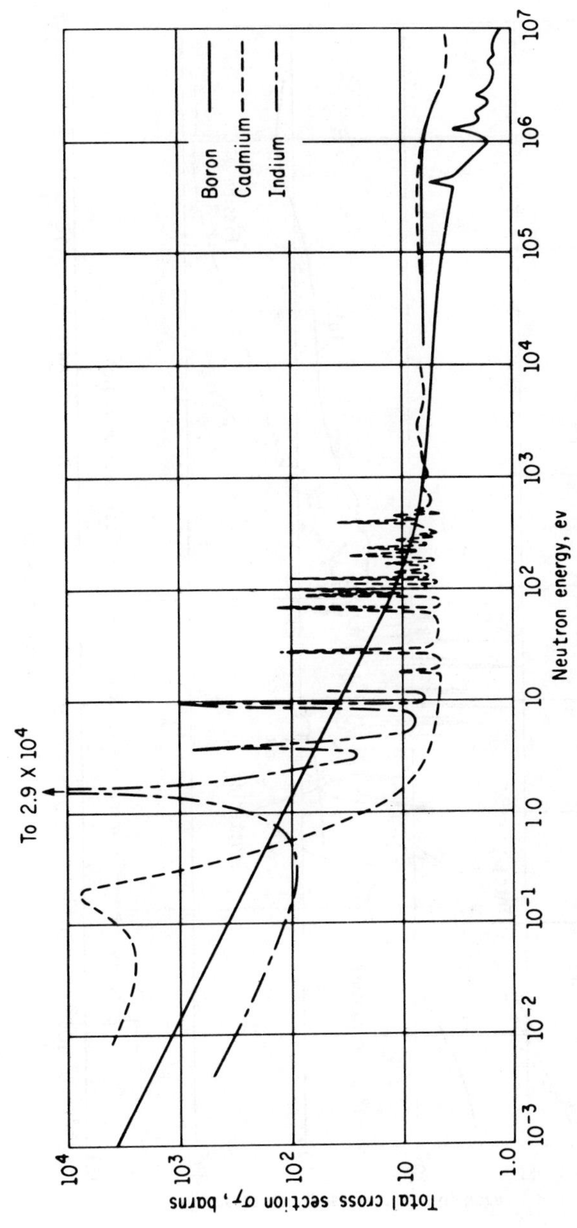

FIG. 2-6. Total neutron cross sections for cadmium, indium, and boron.

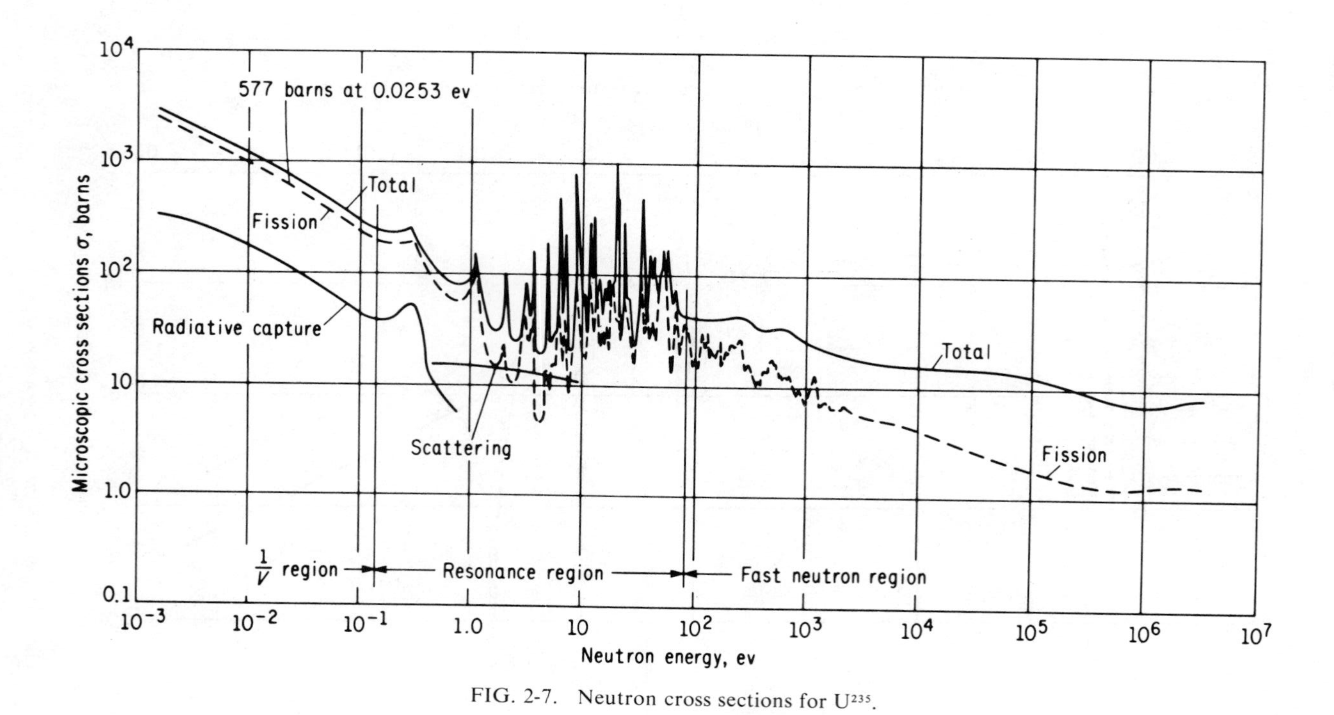

FIG. 2-7. Neutron cross sections for U^{235}.

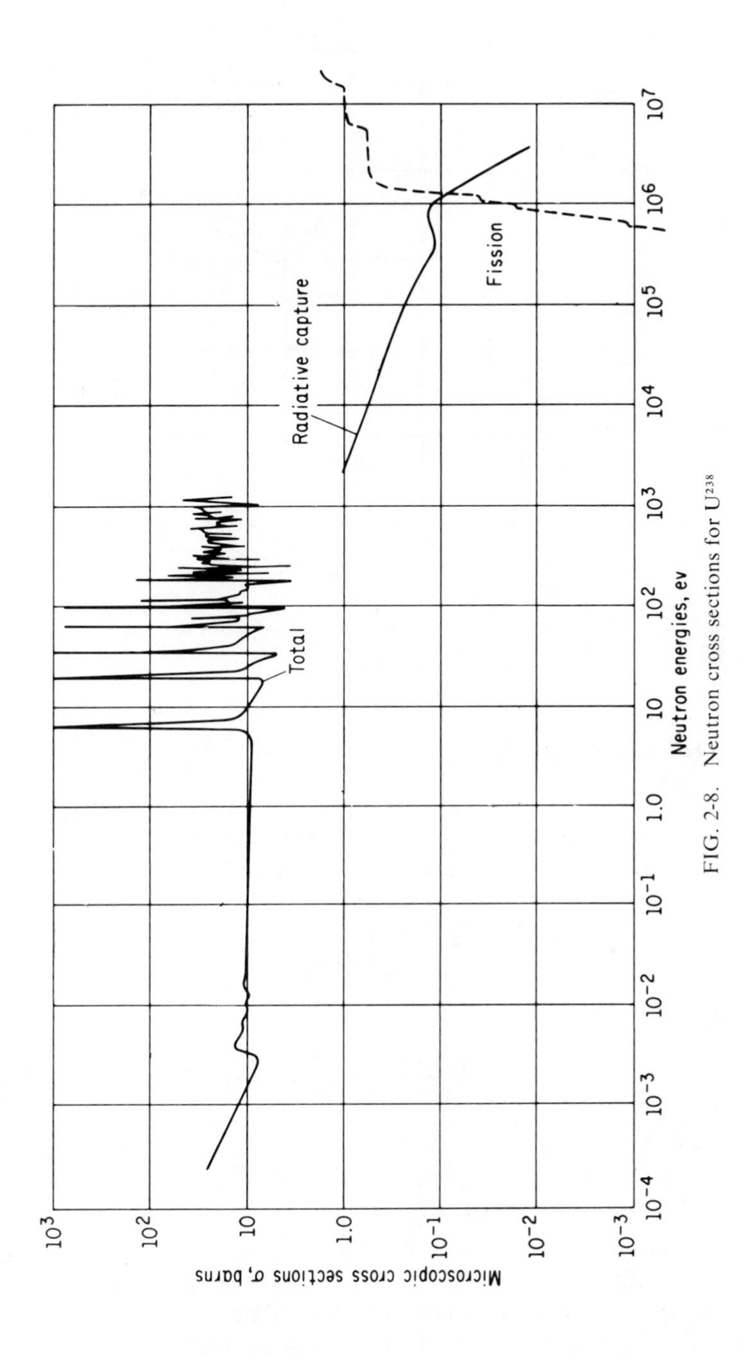

FIG. 2-8. Neutron cross sections for U^{238}

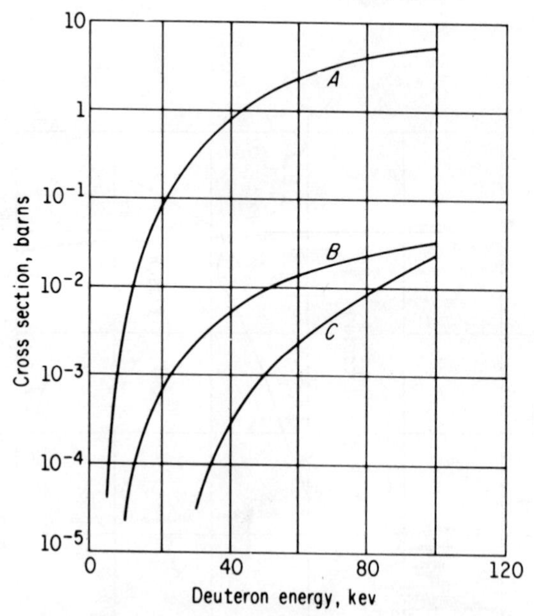

FIG. 2-9. Cross sections for fusion reactions at various deuteron (deuterium nucleus) energies. Reactions: $A, T(d,n)He^4$; $B, D(d,p)T$ and $D(d,n)He^3$; $C, He^3(D,p)He^4$ (Ref. 19).

2-8. THE 1/V REGION

In the low-energy range, the absorption cross sections for many, but not all, nuclei are inversely proportional to the square root of the neutron energy E_n. This can be represented by the equation

$$\sigma_a = C_1 \left(\frac{1}{E_n}\right)^{0.5} \tag{2-19}$$

Thus

$$\sigma_a = C_1 \left[\frac{1}{m_n(V^2/2)}\right]^{0.5} = C_2 \frac{1}{V} \tag{2-20a}$$

where C_1, C_2 = constants
m_n = neutron mass
V = neutron speed

This relationship is known as the $1/V$ *law*. It indicates that the neutron has a higher probability of absorption by a nucleus if it is moving at a lower speed and is thus spending a longer time in the vicinity of that nucleus. The $1/V$ law may also be written in the form

$$\frac{\sigma_{a_1}}{\sigma_{a_2}} = \frac{V_2}{V_1} = \left(\frac{E_{n_2}}{E_{n_1}}\right)^{0.5} \tag{2-20b}$$

where the subscripts 1 and 2 refer to two different neutron energies but within the $1/V$ range. An absorption cross section for monoenergetic neutrons, within the $1/V$ region, may thus be calculated from that at 2,200 m/sec (such as in Appendix B). The evaluation of effective cross sections for neutrons having a range of energies (i.e., nonmonoenergetic) will be discussed in Sec. 4-5.

On the log-log plots of Figs. 2-6 to 2-8, the $1/V$ region is represented by a straight line with a slope of -0.5. The upper energy limit of the $1/V$ region is different for different nuclei, being around 0.3 ev for indium, 0.05 ev for cadmium, 0.2 ev for U^{235}, 150 ev for boron, etc.

2-9. THE RESONANCE REGION

Following the $1/V$ region, most neutron absorbers exhibit a region of one or more absorption-cross-section peaks occurring at definite neutron energies, such as in Figs. 2-6 to 2-8. These are called *resonance peaks.* For medium and heavy nuclei they usually occur in the lower energy levels of the neutron spectrum. They affect neutrons that are in the process of slowing down. Note that indium has but one peak, whereas U^{235} and U^{238} have many. Uranium 238 has very high resonance absorption cross sections, with the highest peak, about 4,000 barns, occurring at about 7-ev neutron energy. This fact affects the design of thermal reactors using natural or low-enriched fuels, in which uranium 238 absorbs neutrons of resonance energy and affects the reactor neutron balance. A theoretical and quantitative formulation of the problem of resonance has been made by Breit and Wigner [20] and is discussed more fully in books on reactor and neutron physics [13, 14]. It will suffice here to present a physical picture of the reasons for the existence of resonance absorption.

Nuclei have discrete quantum energy levels similar to those of atoms, Fig. 2-10. When a nucleus absorbs a neutron the two form a compound nucleus. The excitation energy of the compound nucleus is equal to the kinetic energy of the target nucleus plus the binding energy of the neutron (about 8 Mev), plus a large portion of the kinetic energy of the incident neutron before absorption into the target nucleus. In most cases the target nucleus has relatively little or zero kinetic energy and the excitation energy is very nearly equal to the neutron energy plus the binding energy. An excited nucleus, like an excited atom, can occur only if its energy corresponds to one of the discrete quantum levels as in Fig. 2-10. The probability of neutron absorption, i.e., the cross sec-

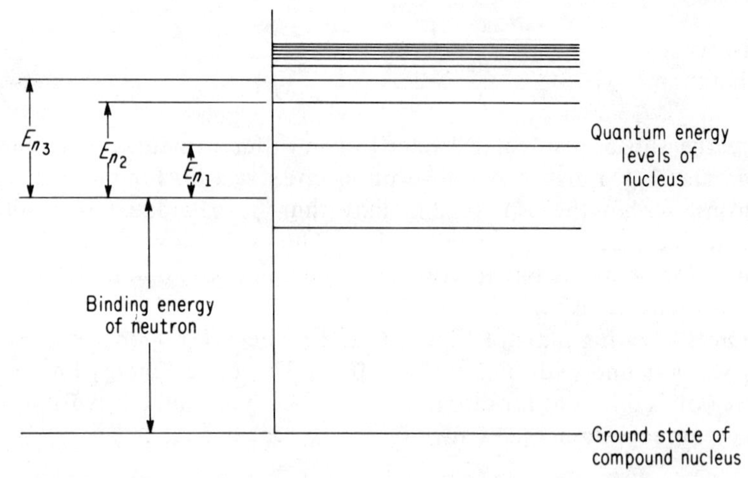

FIG. 2-10. Energy levels of compound nucleus with zero kinetic energy of nucleus.

tion for absorption is therefore greatest when the excitation energy corresponds to one of these discrete energy levels. As can be seen from Fig. 2-10, this can be accomplished only if the neutron possesses the definite kinetic energies E_{n_1}, E_{n_2}, E_{n_3}, etc. These are the *resonance energies*.

Actually, the exact energies are not so precisely defined since each resonance peak has width, usually defined quantitatively by the *half-width,* which is the width in energy units at a cross section equal to one-half that at the peak, Fig. 2-11. The distance between resonance energies,

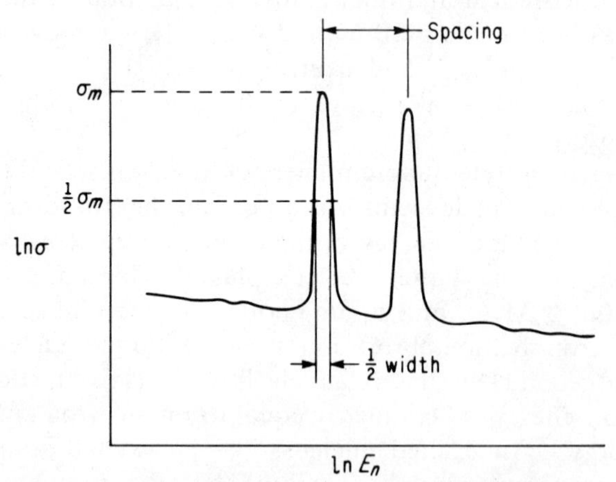

FIG. 2-11. Spacing and half-width of resonance-cross-section peaks.

in energy units, is called the *spacing*. For relatively large nuclei, the spacing is small, being of the order of 1 to 10 ev and the resonance peaks are crowded. In some cases half-widths may be so large, compared with the spacings, that overlap between resonance regions occurs. The width of the resonance peaks increases with temperature. This phenomenon, called *Doppler broadening*, is important from the point of view of reactor control.

It is fortunate that many elements, especially those of low mass numbers, do not exhibit a resonance absorption region. Thus these can be used as reactor construction materials, especially if their absorption cross sections are low.

2-10. THE FAST-NEUTRON REGION

Following the resonance region, cross sections usually undergo a gradual and steady decrease with neutron energies. At very high energies the sum of absorption and inelastic-scattering cross sections is of the same order of magnitude as the elastic-scattering cross section. Each approaches a value equal to the cross-sectional area of the target nucleus. The total cross section therefore approaches a value given by

$$\sigma_t = 2\pi r_c^2 \tag{2-21}$$

where r_c is the radius of the nucleus (Sec. 2-5). Combining Eqs. 2-21 and 2-12 gives

$$\sigma_t = 2\pi r_0^2 A^{\frac{2}{3}}$$

Using the average value of 1.40×10^{-13} cm for r_0 and noting that 1 barn $= 10^{-24}$ cm^2, we obtain

$$\sigma_t = 0.125\ A^{\frac{2}{3}} \text{ barns} \tag{2-22}$$

In the very high neutron-energy range, therefore, absorption plus inelastic- and elastic-scattering cross sections are rather low, usually less than 5 barns each for the largest nuclei. Some nuclei, such as boron, carbon, and beryllium, exhibit some resonance in the high energy range, as shown in Fig. 2-6. The peaks, however, are rather low, and the phenomenon is of little importance.

PROBLEMS

2-1. A number of nuclei belonging to a certain gaseous element are in thermal equilibrium with neutrons so that their most probable speed is 50.2 percent of that of the neutrons. What is the gas?

2-2. A certain nucleus has a cross section of 10 barns for 2,200-m/sec neutrons. Find the cross section if the kinetic energy of the neutrons increases to 0.1 ev. The two neutron energies are within the $1/V$ range of the nucleus.

2-3. Monoenergetic neutrons have a flux of 10^{12} and energy other than standard thermal but within the $1/V$ range of boron. They are absorbed by boron (density 2.3 g_m/cm^3) at the rate of $1.53 \times 10^{13}/cm^3$ sec. Calculate (*a*) the energy of the neutrons and (*b*) the average distance that a neutron travels before absorption.

2-4. A 230-g_m piece of boron absorbs thermal neutrons at the rate of 9.57×10^{13} per second. Boron density = 2.3 g_m/cm^3. Find (*a*) the thermal-neutron flux and (*b*) the average distance that a neutron travels before it is absorbed.

2-5. A parallel beam of 2,200-m/sec neutrons is allowed to escape into the open atmosphere. Personnel safety requires the roping off of a certain length in which the original beam would be attenuated by absorption only to 4.04 percent of its original strength.

(*a*) Calculate the length of the beam that should be roped off, based on atmospheric dry air at 60°F.

(*b*) Will the danger zone increase or decrease if the air is humid? Why?

2-6. Naturally occurring europium is subjected to a neutron flux of 10^{10} (thermal). Its density is 5.24 g_m/cm^3. Find (*a*) the rate of absorption reactions per sec cm^3 and (*b*) the energy produced in absorption if all absorption reactions are of the (n, γ) type, in Mev/sec cm^3.

2-7. A CO_2-cooled reactor has an average neutron flux of 10^{13}. Carbon dioxide spends half of its time in the reactor and half in the primary-coolant loop. The total CO_2 volume is 2,000 ft^3, and it circulates at an average density of 0.013 g_m/cm^3. Calculate the increase in coolant activity, in microcuries per cubic centimeters, due to C^{14} produced in the $C^{13}(n, \gamma)C^{14}$ reaction after one month of operation. The cross section of this reaction is 1 millibarn. The activity-buildup curve of the product nucleus is a mirror image of its decay curve.

2-8. The absorption mean free path for 2,200-m/sec neutrons in a $1/V$ absorber is 1.0 cm. The corresponding reaction rate is 10^{12} sec^{-1} cm^{-3}. The absorber has an atomic mass of 10 and a density of 2.0 g_m/cm^3. Find (*a*) the 2,200-m/sec neutron flux and (*b*) the microscopic absorption cross section of 10-ev neutrons in barns.

2-9. A $1/V$ absorber is in a monoenergetic thermal neutron field of number density n neutrons/cm^3 and temperature T_1. Find the change in absorption rate if it were placed in another monoenergetic thermal neutron field of the same volume density but at another temperature T_2.

2-10. Relative fission rates are sometimes measured by the *catcher foil* technique. A thin uranium foil is sandwiched between two foils of pure aluminum. Fission products have large kinetic energies, and a fraction of them will escape the uranium foil and imbed in the aluminum. Other (n, γ) reactions result in low energy and their products, such as U^{236} and U^{239}, have low kinetic energies and will not transfer to the aluminum. It is assumed that the fraction of all fission products transferring to aluminum is constant. Aluminum activity after a period of irradiation will therefore be a measure of the number of fissions that occurred during that period. (The whole sandwich may be wrapped in cadmium, and by comparison with a bare sandwich, fission rates due to fast and slow neutrons may be obtained.) In a representative run, the following activities (counts per minute) of aluminum were measured from an arbitrary counting time 0:

Time, min	10	20	30	40	60	80	100	200
Activity $\times 10^{-4}$	44	20	10.3	9.8	6	4.4	3.4	1.7

Find an expression for the activity of fission products with time.

2-11. Calculate the absorption macroscopic cross section in cm^{-1} of boron (density 2.3 g_m/cm^3) for (*a*) 2,200 m/sec neutrons, and (*b*) 10 ev neutrons.

2-12. Uranium hexafloride gas is used in the separation of the U^{235} and U^{238} isotopes by the gaseous diffusion method. Calculate the ratio of the root mean square speeds of the molecules of the two isotope hexaflorides when at the same temperature.

chapter 3

Neutron Flux Distribution in Cores

3-1. INTRODUCTION

As is now known, the heat generation at any point in a reactor fuel is primarily a function of the fission-reaction rate at that point. In turn, this is a function of the neutron flux at that point (Sec. 4-2). The prerequisite for analyzing a reactor core as a source of heat for generating power, therefore, is the establishment of the manner in which neutrons distribute themselves in that core.

In order to obtain this flux-distribution pattern, a study of the behavior of neutrons in motion within media is necessary. This is by no means a simple task, since neutrons are born whenever fuel is present in a core, and they travel or diffuse in random fashion and are slowed down at different rates. Thus at any one point in a core, neutrons of all energies, traveling in all directions, are present. At the same point, neutrons may be continuously born and absorbed.

The procedure of determining the neutron flux distribution also results in expressions for the critical dimensions and corresponding critical fuel mass of the core. These are the minimum dimensions and mass necessary for maintaining a chain reaction.

In this chapter we shall first tackle the problem of neutron conservation and diffusion in media as well as the problem of neutron economy and balance in slightly more quantitative detail than we did in Chapter 2. From these, the reactor equations, expressing neutron flux distributions and critical dimensions, will be obtained. The chapter will end with a brief introduction to the unsteady-state behavior of reactors, the knowledge of which is necessary for the evaluation of the response of the heat-transfer mechanism in reactor cores and fuel elements to transient reactor behavior.

3-2. NEUTRON CONSERVATION

It has been shown previously that in order to produce energy by fission a heavy fissionable nucleus such as that of U^{235} has to split into two (sometimes three and rarely four) nuclei. In so doing, the fissioning U^{235} nucleus emits between two and three (2.43, on the average) fission neutrons. If care is taken always to preserve about 40 percent of these neutrons for bombarding other U^{235} nuclei, a continuous process of fission occurs, resulting in a *critical* reactor, one capable of producing energy at a steady rate.

If, on the other hand, more than about 60 percent of the resulting fisson neutrons are lost, either by escape to the outside of the fuel or by absorption in various materials not causing fission, the rate of energy production decreases rapidly to zero and the reactor is *subcritical.* If less than about 60 percent of the fission neutrons are lost, the rate of energy produotion increases and the reactor is *supercritical.*

In a reactor core neutrons are born at all times and in all places containing fissionable material and diffuse in all directions. The understanding of neutron conservation is, however, facilitated by examining an average neutron, or the life cycle of a group of neutrons, assumed all born at the same time, undergo scattering, leakage, absorption, and other reactions, attain the same energy levels, and finally cause fission simultaneously. This hypothetical group is called a *generation,* and the series of events that it undergoes from birth until a new generation is born by fission is called a *life cycle,* or simply a *cycle.*

The life cycle in a thermal reactor of a generation of N fission neutrons is shown in an orderly manner in Fig. 3-1. N first increases by a factor ϵ called the *fast-fission factor* due to fast fission of both U^{235} and U^{238} (the latter is fissionable by fast nuetrons only, Fig. 2-8). The resulting $N\epsilon$ fast neutrons are then decreased by leakage by a factor, P_f, less than 1, called the *fast-neutron nonleakage probability.* $(1 - P_f)N\epsilon$ fast neutrons leak out of the core leaving $N\epsilon P_f$ neutrons. Some of these are absorbed in nonfission reactions in the resonance region, primarily by U^{238}. The fraction of all neutrons that escape resonance capture is called p, the *resonance escape probability.*

The remaining neutrons, $N\epsilon P_f p$, now of thermal energy, suffer a further decrease by leakage according to P_{th}, the *thermal-neutron nonleakage probability.* Now we have $N\epsilon P_f p P_{th}$ thermal neutrons that will remain in the core. They will now be decreased by f, the *thermal utilization factor*, due to nonfission absorptions in all core materials but the fuel (all isotopes). The number of thermal neutrons that will now be absorbed in the fuel is $N\epsilon P_f p P_{th} f$. Some of these will not cause fission, and consequently little or no new neutrons. The number of neutrons produced per neutron *engaged in fission* (2.43 above) is given the symbol ν. The number of neutrons produced per neutron *absorbed* in the fuel is called the *thermal fission factor* η. The number of fission neutrons,

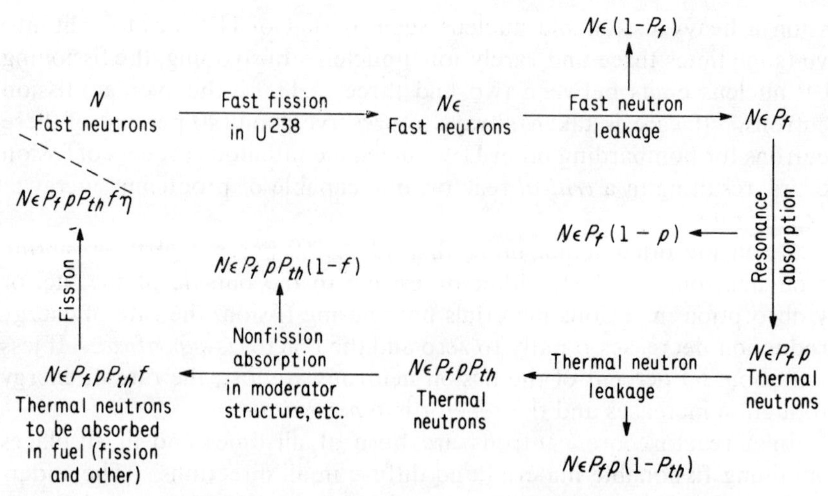

FIG. 3-1. Neutron life cycle in a thermal reactor.

starting a new generation is now $N\epsilon P_f p P_{th} f\eta$. Often $P_f P_{th}$ are combined in one symbol, P, simply called the *nonleakage probability*. The number of new fission neutrons is then $N\epsilon p P f\eta$.

The values of ν and η depend on the fuel and are called *fuel constants*. The factors ϵ, p, and f depend on both fuel and core configuration and materials, and are called *lattice constants*.

The ratio of the number of neutrons at the beginning of a new cycle to the number starting the previous cycle is called the *effective multiplication factor*, k_{eff}. It is given by

$$k_{\text{eff}} = \frac{N\epsilon p P f\eta}{N} = \epsilon p P f\eta \tag{3-1}$$

It is obvious that $k_{\text{eff}} = 1$ for a steady or critical reactor, that $k_{\text{eff}} > 1$ for a divergent or supercritical reactor, and that $k_{\text{eff}} < 1$ for a convergent or subcritical reactor. Equation 3-1 is also written in the form

$$k_{\text{eff}} = k_\infty P \tag{3-2}$$

where k_∞ is called the *infinite multiplication factor*, meaning that it is the factor that would apply if the core in question were made so large by adding more fuel, moderator, structure, etc. with the same lattice arrangement as the original core, that neutron leakage became negligibly small, that is, $P = 1$. The infinite multiplication factor is therefore given by

$$k_\infty = \epsilon p f\eta \tag{3-3}$$

This is the so-called *four-factor formula*.

3-3. THE DIFFUSION OF NEUTRONS IN MEDIA

A rigorous treatment of the problem of conservation of neutrons is done by the *neutron transport theory.* It considers, as a variable, the angular distribution of the velocity vectors* of neutrons. The transport theory, sometimes called the Boltzmann transport theory because of its similarity to Boltzmann's theory of diffusion of gases, is tedious and long.

The mathematical treatment of the problem may be simplified by considering the angular distribution of neutron velocity to be almost isotropic and independent of neutron velocity. The neutron velocity vectors will simply reduce to the scalar neutron flux. Furthermore all neutrons are considered to belong to a set of discrete energy ranges called *groups.* Within these groups the energy variable is eliminated and only an index indicating the energy group is retained. The result is expressions for the variations of neutron flux with geometrical coordinates.

With the above approximations the transport theory reduces to the *neutron diffusion theory* which holds for most of a reactor core. Only in such locations as near a reactor boundary (where there is a predominance of neutrons moving out) or near strong neutron absorbers or sources (where primarily fast neutrons diverge from the source), is the diffusion theory inaccurate and the more rigorous transport theory is resorted to.

We shall primarily use the diffusion theory. First we shall formulate an equation for conservation of neutrons. Such an equation is then transformed with the help of Fick's law into the *neutron diffusion equation.* Its solution results in expressions for the desired neutron flux distribution in the core but also for the *critical* core dimensions and corresponding fuel mass (the minimum dimensions and fuel mass necessary for maintaining a steady fission chain reaction).

In a treatment not too dissimilar to that of heat conduction in solids, Sec. 5-3, a *neutron conservation equation* in a given volume in the core is given by

$$\frac{\partial n}{\partial \theta} = \begin{array}{l} -\text{(leakage rate)} - \text{(elimination rate)} \\ +\text{(production rate)} \end{array} \tag{3-4a}$$

where n is the neutron density, neutrons/cm³ and θ is time, sec. Mathematically, Eq. 3-4a is written as

$$\frac{\partial n}{\partial \theta} = -\nabla . J - \Sigma_{\text{elim}} \varphi + S \tag{3-4b}$$

* The neutron velocity vector is equal to the number of neutrons at a given position per unit volume, traveling at a given speed per unit velocity, in a given direction per unit solid angle. The neutron flux φ is equal to the velocity vector times velocity and solid angle integrated over all solid angles.

and in the steady state

$$-\nabla . J - \Sigma_{\text{elim}} \varphi + S = 0 \tag{3-5}$$

where

J = neutron current, neutrons/sec cm^2
Σ_{elim} = elimination macroscopic cross section
φ = neutron flux
S = neutron source, neutrons/sec cm^3

The above equation is solved by assuming that neutrons belong to a number of groups of distinct energy ranges. *One-* and *two-group models* are simple and approximate, though the latter is often sufficiently accurate for many reactor calculations. The *multigroup* model assumes several energy groups in an attempt to approach the actual case, and is the most accurate. Its use has been facilitated by the availability of computers. Another model, the *Fermi-age* model, assumes that neutrons lose energy in a continuous rather than the actual stepwise fashion. It is accurate for reactors with moderaters of relatively heavy nuclei.

Depending upon the energy group in question, the various parameters in Eq. 3-5 are evaluated. The diffusion term $\Delta . J$, for example, is given in terms of a *diffusion coefficient* D, in the form given by *Fick's law of diffusion* of gases as

$$J = -D \nabla \varphi \tag{3-6}$$

so that the leakage rate becomes $-\nabla . (-D \nabla \varphi)$, and

$$\nabla^2 \varphi - \frac{\Sigma_{\text{elim}}}{D} \varphi + \frac{S}{D} = 0 \tag{3-7}$$

Equation 3-7 is called the *diffusion equation.* D is evaluated in terms of a *slowing-down length* (L_s), or a *thermal diffusion length* (L), Table 3-1. The elimination term is evaluated by assuming that the neutrons are eliminated from one energy group by slowing down to a lower energy group or by absorption. The source term is evaluated from a fission term or slowing down to the group in question from a higher energy group.* For all energy groups, Eq. 3-7 takes the form [1]

$$\nabla^2 \varphi + B^2 \varphi = 0 \tag{3-8}$$

* In the two-group model, for example, the source of fast neutrons is the absorption of thermal neutrons and is given by $f\eta\Sigma_a\varphi_{th}$. The source of thermal neutrons is the slowing down of fast neutrons and is given by $\epsilon p \Sigma_{sl} \varphi_f$. Σ_{elim} is D_f / L_s^2 for fast neutrons and Σ_a is D_{th}/L^2 for thermal neutrons. The subscripts f and th are for the fast and thermal-neutron groups respectively.

B^2 is called the *material buckling,* and sometimes written as B_m^2. It is a number that depends on the properties of the material of a particular reactor core. It has the dimensions (length)$^{-2}$. The name buckling came about because of the similarity of Eq. 3-8 to the well known equation of a loaded column* encountered in strength of materials studies.

TABLE 3-1
Some Properties of Some Pure Moderators at Low Temperature

Moderator	H_2O	D_2O	Meryllium	Beryllium Oxide	Graphite
Density, g_m/cm^3	1	1.1	1.85	2.69	1.8
N, molecules/$cm^3 \times 10^{-24}$	0.03344	0.0332	0.12357	0.06477	0.09030
σ_s, barns	44	11	6	9.8	4.8
σ_a, barns	0.66	0.0026	0.009	0.009	0.0045
Slowing-down power $\xi\Sigma_s$, cm^{-1}	1.53	0.37	0.176	0.125	0.064
Moderating ratio, $\xi\Sigma_s/\Sigma_a$	60	5600	125	170	190
Slowing-down length L_s, cm	5.74	10.93	10	12.22	19.7
Thermal diffusion length L, cm	2.88	116	20.8	29	54.4
Migration length M, cm	6.41	116.5	23.08	31.47	57.86

ξ, the *logarithmic energy decrement,* dimensionless, is the average loss in ln E_n per collision with a moderator nucleus. The average number of collisions to slow a neutron from E_{n_1} to E_{n_2} is therefore given by ln $(E_{n_1}/E_{n_2})/\xi$.

3-4. NEUTRON-FLUX DISTRIBUTIONS IN REACTOR CORES

Equation 3-8 is called the *reactor equation.* It expresses the flux distribution as a function of the space coordinates as independent variables. It is easily solved for various geometries, for homogeneous bare (unreflected) reactor cores. For parallelepiped, spherical and cylindrical geometries, it is written†, respectively, as

$$\frac{\partial^2\varphi}{\partial x^2} + \frac{\partial^2\varphi}{\partial y^2} + \frac{\partial^2\varphi}{\partial^2 z} + B^2\varphi = 0 \tag{3-9}$$

$$\frac{\partial^2\varphi}{\partial r^2} + \frac{2}{r}\frac{\partial\varphi}{\partial r} + B^2\varphi = 0 \tag{3-10}$$

* For a vertical column loaded at the top, the equation is $d^2y/dx^2 + B^2y = 0$ and $B^2 = P/EI$, where y is the transverse deflection, x the vertical distance from the top, P the load, E the modulus of elasticity, and I the moment of inertia of the cross section of the column.

† It is assumed that the flux does not vary with the azimuthal angle in the spherical and cylindrical geometries. For the derivation of the Laplacian ∇^2 in the above geometries, see Sec. 5-3 on heat conduction.

and

$$\frac{\partial^2 \varphi}{\partial r^2} + \frac{1}{r}\frac{\partial \varphi}{\partial r} + \frac{\partial^2 \varphi}{\partial z^2} + B^2 \varphi = 0 \tag{3-11}$$

where r and z are radial and axial distances from the geometrical center of the core in each case. The boundary conditions are:

1. φ positive anywhere in the core and equal to a specified value at the core geometrical center.
2. $\varphi = 0$ outside the physical boundary of the core, because neutron leakage demands a certain positive neutron flux outside the core, see Fig. 3-2. This treatment results in an *extrapolation length,* λ_e, outside the physical boundary at which end the flux effectively goes to zero. For a bare core $\lambda_e = (2/3)\ \lambda_{tr}$ from diffusion theory [1], or 0.71 λ_{tr} from the more exact transport theory. λ_{tr} is a *transport mean free path* (reciprocal of transport macroscopic cross section).

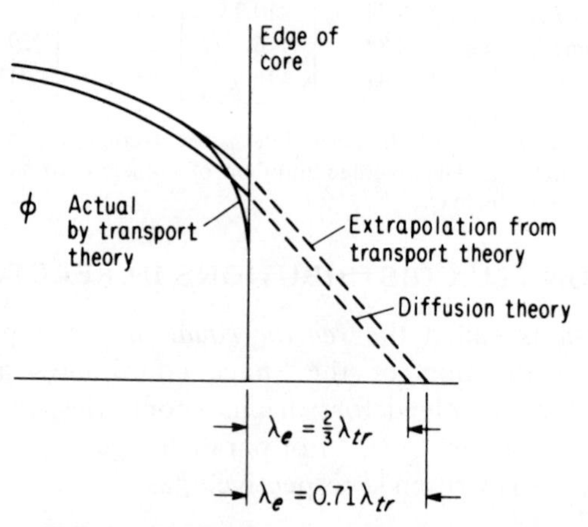

FIG. 3-2. Neutron-flux extrapolation beyond core boundary.

The extended dimensions of a core are called the *extrapolated* dimensions. For example the actual radius and height of cylindrical core R and H, *plus* the extrapolation lengths are called the *extrapolated radius* R_e, and the *extrapolated height* H_e. Thus

$$R_e = R + \lambda_e \tag{3-12}$$

and

$$H_e = H + 2\lambda_e \tag{3-13}$$

and so on. Details of solutions of Eqs. 3-9 to 3-11 are available in many textbooks. They yield values of the buckling. For example, the buckling in a cylindrical core of radius R and height H is

$$B^2 = \left(\frac{\pi}{H_e}\right)^2 + \left(\frac{2.405}{R_e}\right)^2 \tag{3-14}$$

This buckling is a function of core size. Its value decreases as the size increases. It is therefore often referred to as the *geometrical buckling* of the core and sometimes written as B_g^2.

The solutions of the reactor equation also yield the neutron-flux distributions, in terms of spatial coordinates, that are so necessary to evaluate the distribution of energy generation in the core and are therefore of utmost importance in core thermal design. Table 3-2 lists the buckling, minimum core volume and neutron-flux distributions in some simple but important core shapes in terms of φ_{co}, the flux at the geometrical center of each core.

TABLE 3-2
Buckling, Minimum Critical Volume, and Flux Distribution in Some Core Shapes

Geometry of Reactor	Buckling, B^2	Minimum critical Volume	Flux Distribution
Infinite slab (a)	$\left(\frac{\pi}{a_e}\right)^2$	∞	$\varphi_{co} \cos \frac{\pi x}{a_e}$
Parallelepiped (a, b, c; x, y, z)	$\left(\frac{\pi}{a_e}\right)^2 + \left(\frac{\pi}{b_e}\right)^2 + \left(\frac{\pi}{c_e}\right)^2$	$\frac{161}{B^3}$	$\varphi_{co} \cos \frac{\pi x}{a_e} \cos \frac{\pi y}{b_e} \cos \frac{\pi z}{c_e}$
Sphere (R, r)	$\left(\frac{\pi}{R_e}\right)^2$	$\frac{130}{B^3}$	$\frac{\varphi_{co}}{\pi r/R_e} \sin \frac{\pi r}{R_e}$
Finite cylinder (R, r, H)	$\left(\frac{\pi}{H_e}\right)^2 + \left(\frac{2.405}{R_e}\right)^2$	$\frac{148}{B^3}$	$\varphi_{co} \cos \frac{\pi z}{H_e} J_0\left(\frac{2.405r}{R_e}\right)$

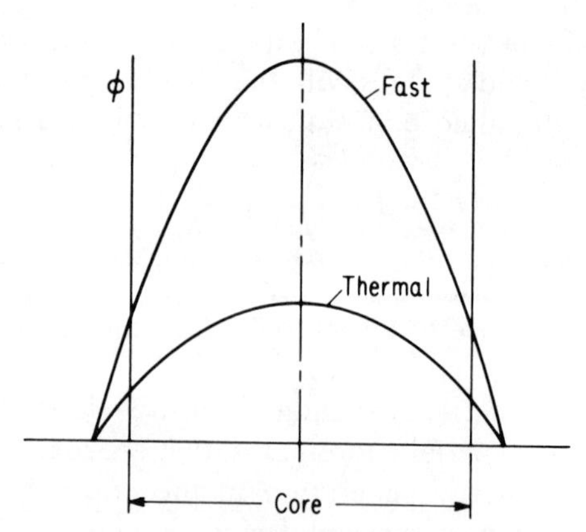

FIG. 3-3. Neutron-flux distributions in a thermal homogeneous unreflected core.

While the solutions have been obtained for homogeneous reactors, Fig. 3-3, they apply with little error to the overall distributions in heterogeneous reactor cores that have a large number of fuel elements, the case of most power reactors. In such heterogeneous cores, the fuel-moderator interfaces represent inner boundaries at which there are sudden changes in material. In that case the smoothed-out flux only averages the actual variable flux pattern through the fuel-moderator lattice, as shown in Fig. 3-4. The actual flux pattern shows a dip in the thermal flux within each fuel element because the fuel acts as a sink for thermal neutrons, (it absorbs them) and as a source for fast neutrons (because of fission). Similarly, the fission neutrons escape into the moderator which thermalizes them and thus acts as a sink for fast neutrons and a source for thermal neutrons.

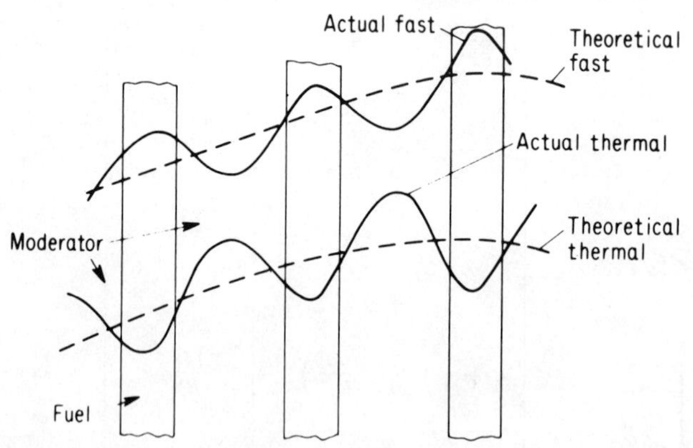

FIG. 3-4. Flux distribution in fuel and moderator in a heterogeneous thermal reactor.

In general, these deviations are small when compared with the absolute magnitude of the flux, so that the heat generated by a particular fuel element, a function of the actual neutron flux in that element, can be obtained with relatively small error from the theoretical flux at the location of that element.

Other deviations from theoretical flux distributions may occur because of inhomogeneities in the lattice structure, which may result from variations from design values in fuel concentration, material dimensions and other parameters caused by manufacturing tolerances. Deviations also arise because the fuel channels may not be uniformly loaded, by having low fuel enrichments in a central region and higher enrichments in outer regions of the core for flux flattening purposes, or where there are strong localized neutron absorbers, such as a control rod.

The neutron-flux distributions also vary with core life. The centermost fuel elements are initially subjected to the highest flux, and more fuel nuclei are consumed there than near the periphery. This results in a flattening of the flux curve with age After a sufficient period of time, the centermost fuel elements may be removed for reprocessing, the outer elements moved in, and new elements put on the outside. This results in further flattening of the flux.

In core thermal design, the deviations in neutron-flux distributions (actually heat flux) from an average value are accounted for by the so-called *hot spot factors,* Sec. 13-6. The ratio of maximum-to-average flux in a normal flux distribution is high for unreflected cores. This results in the peripheral fuel elements producing much less power than the centermost ones which are limited by temperature or burnout considerations (Chapter 13). Flux flattening decreases this ratio of maximum-to-average neutron flux in a core and therefore results in a more uniform power distribution– a great economic advantage.

3-5. REFLECTORS AND THEIR EFFECTS

Reactor cores are often surrounded by *reflectors*. A reflector is a medium of low absorption and high scattering cross sections for neutrons. Thus good moderator materials such as, D_2O, graphite, Be, BeO, and, to a lesser extent, H_2O, are suitable reflectors for thermal reactors. Some of the neutrons leaking out of a reflected core are scattered back into the core by the reflector nuclei. This decreases the *net* neutron leakage (increases neutron nonleakage probability) and reduces the critical dimensions and consequently fuel mass.

A measure of reflector effectiveness is a quantity, less than 1.0, called the *albedo*. It is equal to the ratio of neutron currents leaving and reentering the core. The albedo is larger the smaller the thermal diffusion length L of the reflector material (Table 3-1), and the larger the thickness of the reflector, becoming maximum for an *infinite reflector*. This is a reflector, theoretically of infinite thickness, but practically one with a thickness equal to or greater than $2L$. The albedo also depends upon geometry, being smaller (per unit reflector thickness) the larger the curvature, such as in small spherical cores.

Some cores are normally reflected. For example, boiling or pressurized-water reactor cores are reflected by water because they are normally submerged in a larger pool of water. Because H_2O is not a very effective reflector some low-power water reactors are reflected by graphite. Fast neutrons are best reflected by heavier materials such as natural uranium, although fast reactors are often surrounded by a breeding blanket.

Besides reducing critical dimensions, a reflector improves the neutron-flux distribution in a core by flattening it, or decreasing the ratio of maximum-to-average flux, and thus helps generate more heat in the outer portions of a core. Figure 3-5 shows fast- and thermal-neutron flux distributions in a reflected thermal reactor. Compare this with Fig. 3-3. The reversal of the thermal flux at the core boundaries is caused by fast neutrons which, having escaped into the reflector, become thermalized in it, and upon return to the core cause a flux gradient opposite that in the core itself.

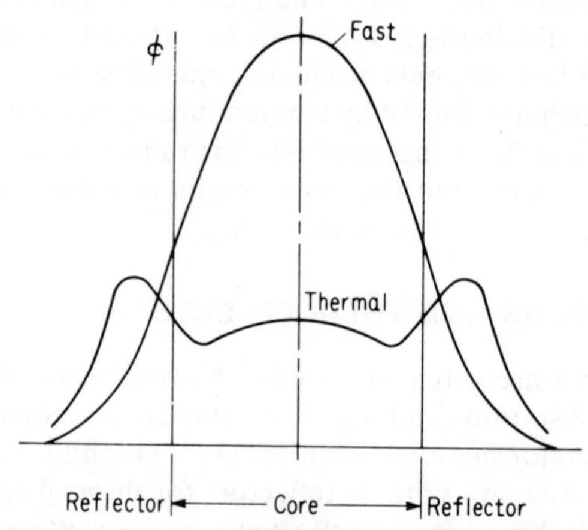

FIG. 3-5. Radial-flux distributions in a reflected core.

The effect of a reflector on neutron-flux distribution is quantitatively evaluated by the concept of the *reflector saving* ψ. The reflector saving is defined as the difference between the dimension of the bare core, measured from the center, $R + \psi$, and the dimensions of a reflected core of the same fuel and lattice structure R, Fig. 3-6. The thickness of the reflector is T. The extrapolated thickness of the reflector is $T_e = T + 0.71\ \lambda_{tr}$. Note that the reflector savings occur at the exterior of the core, and while ψ may be small, it represents a rather substantial saving in volume and mass of fuel.

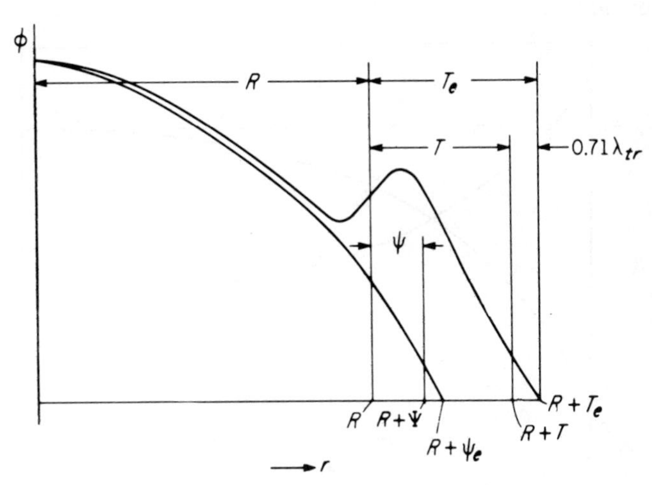

FIG. 3-6. Thermal-flux distribution in bare and reflected reactor cores.

An expression for the reflector savings is necessary for evaluating the extrapolation lengths to be used in the neutron-flux distribution equations, Table 3-2. The extrapolation length of a reflected core is equal to an extrapolated reflector savings ψ_e, so that for a reflected core

$$\lambda_e = \psi_e \tag{3-15}$$

Expressions for ψ_e are obtained by writing diffusion equations for the core and reflector, satisfying the boundary conditions that $\varphi = 0$ at $R + T_e$, that φ is the same at both sides of the interface, and that there is continuity in neutron current at the interface [1]. The resulting expressions are rather complex and can be simplified only for a few cases of interest. One is the case of a larger core where the reflector and core moderator materials are the same. In this case

$$\psi_e = L \tanh \frac{T_e}{L} \tag{3-16}$$

For an infinite reflector (one whose thickness is equal to or greater than $2L$), the hyperbolic tangent is unity, and

$$\psi_e = L \tag{3-17}$$

This state of affairs is shown in Fig. 3-7. Thus a thick graphite reflector reduces graphite-moderated cores by some 50 cm (Table 3-1), and so on. For other than the simplified case above, the reflector savings must be properly evaluated.

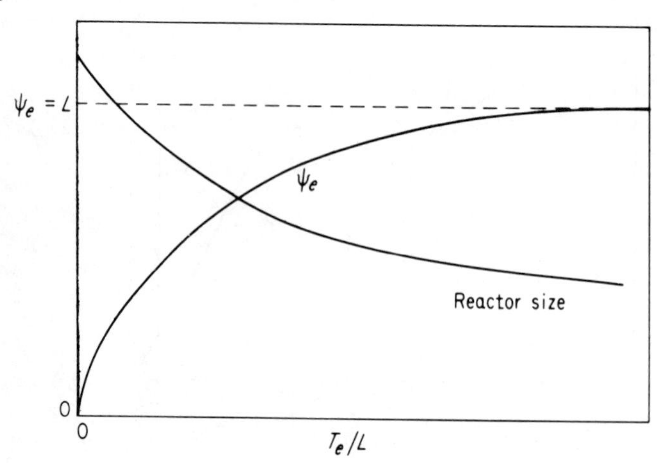

FIG. 3-7. Variation of extrapolated reactor savings and size with $T_e L$.

In summary the neutron-flux distributions are obtained by adding to the various dimensions extrapolation lengths λ_e equal to 0.71 λ_{tr} in bare cores, ψ_e in infinitely reflected cores, or otherwise specified. (Large power reactors that are moderated and reflected by ordinary water, have values of λ_e that are quite small compared to the core dimensions so that the latter may be ignored.) Thus

$$\left.\begin{aligned} a_e &= a + 2\lambda_e \\ b_e &= b + 2\lambda_e \\ c_e &= c + 2\lambda_e \\ R_e &= R + \lambda_e \\ H_e &= H + 2\lambda_e \end{aligned}\right\} \tag{3-18}$$

3-6. THE UNSTEADY STATE AND REACTIVITY

So far neutron-flux distributions in reactor cores were developed in the steady state, i.e., when the flux did not vary with time. In this section we shall attempt to study effects of deviations from steady state. When such deviation takes place, the effective multiplication factor ceases to be unity. The deviation is represented by a quantity called the *reactivity*. Reactivities may be momentary, or short-lived, because of some change in system temperature, pressure, or load. They can develop slowly over a long period of time because of fuel burnup, for example, and the accumulation of fission products in it. If reactor power is to hold constant, some means of compensating for reactivity changes, such as the use of control rod or chemical shim (neutron absorber in coolant or moderator) must be used. In many instances, compensation for momentary reactivity changes are affected by the reactor itself. The study of reactor kinetics is important for our purposes because of its effects on thermal transients in reactor elements (Chapter 8).

It has been shown that a critical reactor has an effective multiplication factor k_{eff} equal to unity. When a reactor deviates from criticality, k_{eff} becomes greater or less than unity, depending on whether the reactor becomes supercritical or subcritical. In such cases the reactor is said to have an *excess multiplication factor* k_{ex}, given by

$$k_{ex} = k_{eff} - 1 \tag{3-19}$$

k_{ex} may be positive or negative. The ratio of excess to effective multiplication factors is called the *excess reactivity* or simply the *reactivity*, ρ, and is given by

$$\rho = \frac{k_{ex}}{k_{eff}} = \frac{k_{eff} - 1}{k_{eff}} \tag{3-20}$$

It is obvious that, for a steady-state or critical reactor, $\rho = 0$. Also, in cases where the deviations from criticality are small, such as occur because of slight changes in temperature or voids within the reactor during normal operation Eq. 3-20 may be simplified to

$$\rho \simeq k_{ex} \simeq k_{eff} - 1 \tag{3-21}$$

Other relationships, easily derived from the above, are

$$k_{eff} = \frac{1}{1 - \rho} \tag{3-22}$$

and

$$k_{ex} = \frac{\rho}{1 - \rho} \tag{3-23}$$

The neutron-flux *distribution* does not materially change when the reactor deviates slightly from criticality. The neutron-flux *level*, and consequently heat generation, may however undergo substantial changes in short periods of time. This can be shown by solving the unsteady-state neutron diffusion equation in terms of an *average neutron lifetime* in the core θ_0, among other parameters. This results [1] in the rate of change of flux,

$$\frac{d\varphi}{d\theta} = \frac{k_{ex}}{\theta_0} \varphi \tag{3-24}$$

where θ is time. This equation integrates, between an arbitrary zero time when the neutron flux was φ_0 to time θ when the flux is φ, as follows:

$$\varphi = \varphi_0 e^{(k_{ex}/\theta_0)\theta} \tag{3-25}$$

The time necessary for the neutron flux to change by a factor of e is called the reactor *period* θ_p (also called the *e folding time*). It follows from Eq. 3-25 that

$$\theta_p = \frac{\theta_0}{k_{ex}} \tag{3-26}$$

Equation 3-26 may now be written in the form

$$\varphi = \varphi_0 e^{\theta/\theta_p} \tag{3-27}$$

Example 3-1. Calculate the rate of increase in flux for a reactor with a neutron average lifetime of 10^{-3} sec after an accidental 0.2 percent step increase in multiplication from criticality occurred.

Solution

$$k_{ex} = 0.002$$

$$\theta_p = \frac{\theta_0}{k_{ex}} = \frac{0.001}{0.002} = 0.5 \text{ sec}$$

$$\frac{\varphi}{\varphi_0} = e^{\theta/\theta_p} = e^{2\theta}$$

Thus the reactor flux, and consequently power, would increase at the rate of e^2 or 7.389 times per second.

A common unit of reactivity ρ is the *inhour* (short for inverse-hour). One inhour is the amount of reactivity that would make the period of a reactor equal to 1 hr or 3,600 sec. In Example 3-1, a 1-hr period would require the value of k_{ex}, and approximately that of ρ, to be 0.2777×10^{-6}.

It was indicated that changes in some operating parameter, such as temperature, with time give rise to reactivity changes. This state of affairs is described by various *coefficients of reactivity* due to the particular parameters that caused the change. Some of the more important reactivity coefficients are discussed briefly in the next three sections.

3-7. THE TEMPERATURE COEFFICIENT OF REACTIVITY

The *temperature coefficient of reactivity,* or simply the *temperature coefficient,* is defined as

$$\text{Temperature coefficient} = \frac{d\rho}{dT} \tag{3-28}$$

where T is temperature. The temperature coefficient may be positive or negative, that is the reactivity may increase or decrease, respectively, with an increase in temperature. However, because an increase in normal reactor temperatures usually denotes an increased heat release beyond normal demand, it could be a prelude to a runaway accident. The temperature coefficient should therefore be negative so that an increase in temperature causes negative reactivity addition; i.e., a decrease in the rate of fission and consequently power, and a return to normal operation. A reactor with a negative temperature coefficient is therefore inherently safe.

Temperature changes in a reactor introduce reactivity primarily in three ways: (1) by altering the mean energy of thermal neutrons in a thermal reactor (fast neutrons are unaffected), (2) by changing reactor core component densities, and (3) by changing the overall size or volume of the core. The temperature coefficient of a reactor is correspondingly separated into three parts: a *Nuclear temperature coefficient,* a *density temperature coefficient,* and a *volume temperature coefficient.* Thus

$$\frac{d\rho}{dT} = \frac{\partial \rho_n}{\partial T} + \frac{\partial \rho_d}{\partial T} + \frac{\partial \rho_v}{\partial T} \tag{3-29}$$

where the subscripts n, d, and v denote nuclear, density, and volume, respectively.

The *nuclear temperature coefficient* arises primarily because of the effect of temperature on the thermal diffusion length L in a thermal reactor. $L = \sqrt{D/\Sigma_a}$ and thus varies with temperature due to the variation of $1/\Sigma_a$ with it. It therefore depends upon the change of σ_a with neutron temperature, and therefore neutron energy, which may well follow a $1/V$ dependence (Sec. 2-8). The effect is an increase in L with temperature. The nuclear temperature coefficient is proportional to L^2 [1] and is usually negative. In a sodium-cooled fast reactor, the expansion of sodium with temperature reduces its slight moderating capability and therefore increases the average neutron energy. This makes the fission factor η and cosequently k_∞ larger, causing the nuclear temperature coefficient to be positive.

The *density temperature coefficient* arises because of the change in the density of core materials with temperature. Density changes introduce reactivity changes by varying the densities of reactor materials, such as the moderator in a thermal reactor, the fuel, etc. An increase in temperature reduces densities and macroscopic scattering and absorption cross sections. This increases the corresponding mean free paths and therefore leakage. The density temperature coefficient is proportional to an average linear coefficient of thermal expansion of the core materials and is usually negative.

The *volume temperature coefficient* occurs due to a change in overall reactor-core size and consequently the core buckling with temperature. It is proportional to an average linear coefficient of thermal expansion of core materials and core vessel. It is usually positive and, numerically, is the smallest of the three components of the temperature coefficient.

The temperature coefficient of reactivity is, in most cases, negative. It is largest for homogeneous reactors where the density effect is greatest, contributing to the nuclear safety of such reactors. Solid-fueled heterogeneous reactors are strongly affected by density changes only if liquid-moderated. The pressurized-water group is in this category. Solid-moderated circulating-liquid-fuel reactors exhibit similar characteristics. Heterogeneous reactors where both the fuel and moderators are in solid form, such as the graphite-moderated gas-cooled and the sodium-cooled groups, are not materially affected by density changes, and their temperature coefficients are rather small.

Under certain design conditions, the density coefficient may be positive. A heterogeneous, light-water-moderated reactor with a large moderator-fuel ratio is an example. In this case, because light water is a neutron absorber, an initial decrease in its density would actually increase k_∞ because the reduction in neutron absorption is greater in effect than the reduction in moderation. A further decrease in density, however, normally reverses this trend.

3-8. THE DOPPLER COEFFICIENT OF REACTIVITY

In much of the discussion on neutron interactions with nuclei, it was assumed that the target nucleus that scatters or absorbs the neutron is at rest. In reality the nucleus is in thermal equilibrium with the environment and consequently possesses a certain energy (vibration) of its own. This energy is small compared to that of the neutron energy and is ignored in most cases. However, the thermal motion is random but has components along the path of the incident neutron. This, in effect, spreads the energy of the neutron about its actual energy. This phenomenon, called the *Doppler broadening* or *effect* is analogous to the Doppler broadening of spectral lines in spectroscopy or to the Doppler shift of the acoustical frequency of a sound source moving with respect to a listener.

A major influence of Doppler broadening is in its effect on resonance cross sections, Sec. 2-9. It broadens the resonance peaks and also lowers their heights. The area under a resonance peak remains unchanged, however. There, therefore, should be no change in reactivity due to the Doppler effect, which may be true if the nuclei were dilute and well dispersed. In most cases of interest, however, a strong self-shielding effect causes a saturation in the cross sections inside the fuel or material in question. The Doppler effect, therefore, results in increasing the rate of interaction (such as resonance absorption) since it broadens resonance without affecting its already saturated height. There will therefore be a *Doppler Coefficient of reactivity* which may be positive due to fissionable fuel such as U^{235}, or negative due to U^{238}, the net effect depending upon composition.

In low-enriched thermal reactors, with a preponderance of U^{238}, Doppler broadening increases resonance absorption by the latter and hence reduces reactivity. These reactors therefore have a negative Doppler coefficient, though its magnitude is rather small, being of the order of $10^{-5}(^{\circ}F)^{-1}$. Its importance, however, lies in the fact that it is prompt, since most fission energy shows up directly in the fuel material nuclei causing an immediate increase in their thermal motion. This is not the case for the coolant or moderator which depend on a slower heat-transport mechanism for their temperature rise.

In fast reactors, the neutron spectrum may be *hard* (narrow distribution, high average energy), or *soft* (wide distribution, lower average energy). Soft spectra are attained if some reactor materials have a moderating influence on neutrons (sodium coolant is an example), or if a small amount of moderator (such as BeO) is deliberately added to the core. A soft spectrum spans more of the resonance region than a hard one. A soft-spectrum fast reactor with U^{238} in the fuel will have a negative Doppler coefficient. A hard-spectrum reactor using highly-enriched fuels, on the other hand, could have a positive Doppler coefficient. This is a serious situation, and design for such reactors must be such that it precludes this from happening.

3-9. VOID AND PRESSURE COEFFICIENTS OF REACTIVITY

In the case of large power surges, a liquid moderator may reach its saturation temperature and begin to boil, i.e., form vapor bubbles. These are called *voids* because they contain very little moderator nuclei. The effects of voids on reactivity are described by the *void coefficient of reactivity,* or simply the *void coefficient,* given by

$$\text{void coefficient} = \frac{d\rho}{d\alpha} \tag{3-30}$$

where α here is the *void fraction,* or the fraction by volume of the moderator that is in void form.

Boiling-water reactors are designed to operate with a certain portion of the water moderator in voids. In a reactor with a normal moderator-to-fuel ratio, an increase in voids exerts a strong negative influence on reactivity, resulting in a negative void coefficient. In an overmoderated reactor, with a high moderator-to-fuel ratio, the void coefficient may, on the other hand, be positive. In sodium-cooled fast reactors, sodium voiding may result in a positive void coefficient [2].

Another important effect on reactivity is due to changes in reactor pressure. Moderators that remain in solid or liquid form are unaffected by such changes (liquid densities are relatively insensitive to pressure changes). A boiling-type reactor, however, is extremely responsive to pressure changes, which may occur because of changes in steam demand by the turbine. For example, if the turbine governor is closed, the reactor pressure rises causing some voids to collapse and thereby increase moderation and reactor power. It can therefore be seen that boiling reactors have a rather large *pressure coefficient of reactivity* which, in most cases, is positive. This pressure coefficient is given by

$$\text{Pressure coefficient} = \frac{d\rho}{dp} \tag{3-31}$$

where p is the system pressure. A large positive pressure coefficient occurs at low system pressures and causes the reactor to be unstable. It is believed to be the major cause of instability observed in low-pressure boiling reactors. However, it decreases rapidly with pressure and becomes relatively small at high system pressures (several hundred psia).

The coefficients of reactivity discussed above make their effects felt rapidly. There are, however, other reactivity effects which are slow and long ranging. Examples are the reactivities due to the change in the composition of the fuel due to fuel burnup (or production in a breeder reactor) which decreases (or increases) fission rates, the accumulation of fission products which consume neutrons and decrease the thermal utilization factor, the temporary buildup of Xe^{135} after reactor shutdown which may affect reactor restarting, and others.

PROBLEMS

3-1. Calculate the thermal utilization factor for 2,200-m/sec neutrons in a homogeneous reactor core containing 20 percent enriched UO_2SO_4 in 20 mole percent solution in heavy water. No structural material is used except the core vessel which may be ignored.

3-2. Calculate the thermal fission factor for a water-cooled and-moderated reactor core. The fuel material is composed of 93.5 percent enriched UO_2 dispersed in stainless steel with a ratio of 23 percent UO_2 by volume. Cladding is made of 0.005-in.-thick stainless steel. The fuel elements are 3 in. wide and 0.5 in. thick in cross section. The fuel elements are spaced so that the water to fuel element area ratio is, on the average, 2:1. Ignore the presence of control rods and assume, for simplicity, that stainless steel is made only of iron. The coolant is saturated at 600°F. The ratio of the average thermal-neutron flux in the moderator to the average thermal-neutron flux in the fuel (called the *thermal disadvantage factor*) is 2:1.

3-3. Develop a relationship between the thermal fission factor η for uranium fuels and ν for U^{235} in terms of the microscopic fission and absorption cross sections and the enrichment.

3-4. Calculate the thermal-fission factor for 2,200-m/sec neutrons for (*a*) natural uranium and for (*b*) 2 percent, (*c*) 20 percent, and (*d*) fully enriched uranium.

3-5. Calculate the infinite multiplication factor for a homogeneous reactor core. The fuel material is a uranium sulfate solution in light water. No structural materials need be considered. The composition in percent by mass of the fuel material-moderator mixture is as follows

U^{235}	U^{238}	S	O	H
5	19.2	2.5	65.9	7.4

The resonance-escape probability is 0.95.

3-6. A natural-uranium, graphite-moderated reactor core is to be built in the shape of a cube. It is proposed that it contain 2,500 vertical fuel rods 1.25 cm in diameter arranged in a square lattice 12 cm on the side. The lattice parameters for such an arrangement are $\epsilon = 1.03$, $p = 0.91$, $f = 0.87$, $\eta = 1.3$. Calculate the size of the core and the masses of natural uranium and U^{235} in it.

3-7. A homogeneous reactor uses a uranium sulfate–heavy water solution for fuel. The core is spherical, 80 cm in diameter, and is reflected by an infinite heavy-water reflector. The maximum flux in the core at full power is 10^{13}.

(*a*) What should the minimum reflector thickness be to qualify as infinite?

(*b*) Calculate the ratio of average to maximum flux in the core.

(*c*) Repeat part *b*, assuming that the core was not reflected.

NOTE: $\int x \sin x \, dx = \sin x - x \cos x$.

3-8. A thermal cylindrical reactor has a height of 5 m and a radius of 2.7 m. It is reflected by an infinite graphite reflector. Operation is at low temperatures. Estimate the ratio of the minimum-to-maximum neutron flux in the core.

3-9. The extrapolation lengths in a cubical reactor core are negligibly small. Find the ratio of average to maximum flux if the flux distribution is:

(*a*) Sinusoidal in the *x*, *y*, and *z* directions.

(*b*) Sinusoidal in the *x* and *y* directions but uniform in the *z* direction.

(*c*) Sinusoidal in the *x* direction only and uniform in the *y* and *z* directions.

3-10. Repeat the above problem but for the case in which the extrapolated sides are 50 percent longer than the actual sides of the cubical core.

3-11. A reactor *period meter* is an instrument that indicates the period of a reactor. A signal proportional to the instantaneous neutron flux is received by the period meter from an ionization chamber. By suitable electronic circuitry, the signal is related to the reactor period. Derive the relationship the electronic circuit should be designed to solve.

3-12. Derive Eqs. 3-25 and 3-27 starting with the relationship $\varphi = \varphi_0(1 + k_{ex})^g$, where k_{ex} is small and g is the number of neutron generations during time θ, measured from an arbitrary zero time when the flux was φ_0.

3-13. A reactor instrumentation system contains a neutron counter whose output is proportional to the average neutron flux. The output is recorded on semilogarithmic chart paper moving at a constant speed of 1.0 in./min. During reactor start-up, it was allowed to increase in power on a fixed period. The record on the chart showed a 10^3 multiplication in average flux as the chart moved 3 in. Show the shape of the trace on the chart and calculate the reactor period. The reactor is uranium-fueled.

3-14. Assuming constant k_∞ and B^2, and a $1/V$ dependence of thermal neutron absorption, show that the nuclear temperature coefficient in a thermal reactor is given by

$$\frac{\partial \rho_n}{\partial T} = -\frac{B^2 L^2}{2k_\infty T}$$

chapter **4**

Reactor Heat Generation

4-1. INTRODUCTION

The importance of the heat-generation and heat-transfer processes in nuclear reactors is probably best emphasized by the fact that the rate of heat release and consequently power generation in a given reactor core is limited by thermal rather than by nuclear considerations. There is no limit to the neutron-flux level attainable in a reactor core, but the heat generated must be removed. The reactor must also be operated at such a power level that, with the available heat-removal system, the temperatures anywhere in the core do not exceed specific safe limits.

In this chapter the processes of heat generation from reactor fuel elements and core will be stressed. Heat transfer from the fuel elements in the steady and unsteady states and from the core as a whole will be discussed in subsequent chapters. In this chapter it will be assumed that the reader possesses a knowledge of the basic principles of neutron and reactor physics. If he does not, a review of Chapters 1, 2, and 3 would be in order.

4-2. HEAT GENERATION RATE IN FUEL

The rate of nuclear heat generation is equal to the rate of reaction producing energy times the energy per reaction. In general, the rate of any reaction between monoenergetic neutrons and the nuclei of a material is given by (Sec. 2-6),

$$R = \Sigma\varphi \qquad [2\text{-}18]$$

where Σ is the macroscopic cross section, cm^{-1}, of the reaction and φ the neutron flux, neutrons/sec cm^2. R therefore has the units reactions/sec cm^3.

The energy generated in a reaction per unit time and volume is called *volumetric thermal source strength, q'''*, given by

$$q''' = GR = G\Sigma\varphi \qquad (4\text{-}1)$$

where G is the energy per reaction, Mev. In the case of energy by fission by neutrons of a given distribution

$$q''' = G_f \bar{\Sigma}_f \varphi \tag{4-2}$$

where the subscript f refers to fission and $\bar{\Sigma}_f$ is an effective macroscopic fission cross section. Recalling that $\Sigma = \sigma N$ (Sec. 2-5),

$$q''' = G_f N_{ff} \bar{\sigma}_f \varphi \tag{4-3}$$

where N_{ff} is the density of fissionable fuel (U^{235}, Pu^{239}, etc.), in nuclei/cm³. The value $\bar{\sigma}_f$ is an effective fission microscopic cross section, cm², for the fissionable fuel used and the energy distribution of the neutrons in the reactor. q''' in Eq. 4-3 has the units Mev/sec cm³. To convert this to Btu/hr ft³, the units commonly used in engineering calculations, multiply by 1.5477×10^{-8}, Table H-11, Appendix H.

It is important to evaluate the volumetric thermal source strength at different positions in a reactor core before evaluating the core temperature distributions, and core heat generation and heat removal. To do this, it is necessary to evaluate the various terms in Eq. 4-3.

Of these terms φ, the neutron flux at any point in the core, is obtained from physics considerations. In Chapter 3 the flux distributions in thermal bare homogeneous cores were obtained. Such flux distributions in a cylindrical core, for example, are sinusoidal in the axial direction and a Bessel function of the radius in the radial direction (Table 3-2). These distributions will be perturbed due to fuel, moderator and other materials in a heterogeneous core, the presence of control rods, reflector, changes in fuel enrichment, etc. The exact flux distributions must, of course, be used to determine heat generation. It is not uncommon, however, for these distributions in turn to be dependent upon temperature distributions of fuel and moderator in the core, so that an iterative solution becomes necessary. In general the determination of the exact flux distribution is not a simple matter. In much of the discussion that follows we will use as examples the homogeneous-core flux distributions. These also apply with little error to heterogeneous cores with a large number of thin fuel elements.

We shall attempt to evaluate the other terms in Eq. 4-3, in the next three sections.

4-3. THE FISSION ENERGY IN REACTORS

It is the intent here to evaluate the numerical value of G_f in Eq. 4-3, in Mev/fission.

The energy resulting from fission in reactors is made up of three types, Table 4-1. It was previously shown (Sec. 1-9) that an average of about 193 Mev of energy is produced in the process of fission itself. This applies to the three fissionable nuclei U^{233}, U^{235}, and Pu^{239}. This energy is made up of types I and II in Table 4-1. Type I includes those processes that produce heat instantaneously at the time of fission of a nucleus. Type II includes the processes that occur after the event of fission takes place. These are mainly due to the several stages of radioactive decay of the fission fragments and products. They continue to generate heat for some time after reactor shutdown. Type III represents those processes that are caused by nonfission absorption (n, γ) of the excess neutrons in the fuel structure, moderator, coolant, cladding, etc., and represent about 7 Mev. Thus a total of about 200 Mev is generated in the core per fission.

The approximate range (last column of Table 4-1) represents the approximate distance that a certain particle travels from its point of origin before its energy is given up and converted to heat. For example, fission fragments are slowed down in a very short distance (less than 0.01 in.) and their kinetic energy can be considered to be converted into

TABLE 4-1
Approximate Distribution of Fission Energy

Type		Process	Percent of total reactor energy	Approximate range
Fission	I Instantaneous energy	Kinetic energy of fission fragments	80.5	Very short
		Kinetic energy of newly born fast neutrons	2.5	Medium
		γ energy released at time of fission	2.5	Long
	II Delayed energy	Kinetic energy of delayed neutrons	0.02	Medium
		β-decay energy of fission products	3.0	Short
		Neutrinos associated with β decay	5.0	Nonrecoverable
		γ energy of fission products	3.0	Long
(n, γ) due to excess neutrons	III Instantaneous and delayed energy	Nonfission reactions due to excess neutrons plus β- and γ-decay energy of (n, γ) products	3.5	Short and long
Total			~ 100.0	

heat in the fuel at the point of fission. Fission neutrons, both prompt and delayed, are slowed down gradually and in successive scattering processes (in a thermal reactor). Their range is medium (from a fraction of an inch to a few feet), and their energy is converted into heat in the various reactor materials with which they interact. Beta particles are also of the short range variety. Gamma energy has long range. A good portion of it escapes the reactor core completely and is absorbed by the reactor vessel shielding material. Of the different particles emitted, the neutrinos do not react with reactor materials, and the energy carried by them, 5 percent of the total, is unrecoverable. The remainder, 95 percent, is recoverable although part of it is long-range. However, even this long-range energy must be removed by cooling shrouds, shields, vessels, etc. However, it is sometimes impractical to cool these by the primary coolant, and the heat generated in them may not be available for power generation.

It is obvious from the above discussion that the amount of heat generated in the fuel per fission is dependent upon the exact range of some of the particles emitted, which in turn is dependent upon the materials used in the reactor as well as its internal configuration. In the absence of more precise information, it can be assumed that about 90 percent of the total energy per fission is produced in the fuel itself, about 4 percent produced in the moderator, about 5 percent carried away by the neutrinos, and the remainder, about 1 percent, produced in various other reactor materials. The amount of heat energy produced in the fuel can then be assumed to have the approximate value of 0.90×200, or 180 Mev per fission, a figure that we may use whenever a more precise one is not available.

It should be noted that, in a new core loading, the amount of energy produced per fission is lower than that given above because the heat given off by the decaying fission products is nonexistent. After a relatively short period of time, the equilibrium value discussed above is reached.

After reactor shutdown, the decay energy continues to be produced by the still radioactive but decaying fission products, although fission energy completely stops a very short time after reactor shutdown. The decay energy is initially some 3 or 4 percent of the energy before shutdown, Sec. 4-10. Heat removal after shutdown is therefore necessary. Since a reactor is "scrammed" in case of power failure, there must be provision for cooling the reactor in such a case. This can be done by having standby power units (say, diesels) to power the coolant pumps or by designing the primary-coolant circuit so that there will be a natural driving head large enough to maintain sufficient coolant flow. Under no circumstances should the coolant be completely dumped out as part of scram in the case of power reactors.

4-4. THE FISSIONABLE FUEL DENSITY

The fissionable fuel density in Eq. 4-3, N_{ff}, fissionable nuclei per cm^3, depends upon the fissionable fuel used and the fuel enrichment, and the fuel material density. Recall that *fuel* is defined as all U, Pu, and Th isotopes only, and that *fuel material* is defined as the entire fuel-bearing material, including chemical (such as UO_2 or UO_2SO_4), alloy or mixture (such as U + Al or U + ZrH) but not cladding or other structural material. The term *fuel element* refers to fuel material and cladding and other structural members of the fuel.

N_{ff} is fixed and independent of position in the core in the case of a fluid-fueled reactor. In a heterogeneous reactor, N_{ff} is usually fixed for any one new fuel element, for an entire new core, or for each of a number (usually three) of concentric zones in a new core loading. These zones are usually chosen so that they contain nearly equal numbers of fuel elements and are more enriched the farther away from core center. This helps flatten the radial neutron-flux and energy distribution. During core life, N_{ff} will vary because of uneven fuel burnup, cycling (replacing a more burned-up inner zone by a less burned-up outer zone), during a fuel change, or because of spiking (scattering new elements in a core).

Care must be taken in evaluating N_{ff} because the enrichment is a *mass* ratio, not a volume or nuclear ratio. In most of the calculations that follow, N_{ff} will be considered constant for simplicity. If the density and enrichment of the fuel are known, N_{ff} can be calculated from

$$N_{ff} = \frac{\text{Av}}{M_{ff}} \rho_{ff} i \tag{4-4}$$

where Av = Avogadro number, 0.60225×10^{24} molecules/g_m mole
M_{ff} = molecular mass of fissionable fuel used
ρ_{ff} = density of fissionable fuel used, g_m/cm^3
i = number of fuel atoms per molecule of fuel

The usual unknown in the above equation is ρ_{ff}. In general it is ρ_f, the density of the fuel (all U, Pu, and Th isotopes), or ρ_{fm}, the density of the fuel material (all isotopes above plus chemical or alloy compounds, but not cladding), that are known. Some of these are given, together with other physical properties of some fuels and fuel materials in Table 4-2. ρ_{ff} is related to ρ_f by the enrichment r:

$$\rho_{ff} = r\rho_f = rf\rho_{fm} \tag{4-5}$$

where r = enrichment or mass ratio of fissionable fuel to total fuel
f = mass fraction of the fuel in the fuel material. It is given by

TABLE 4-2
Some Physical Properties of Some Nuclear Fuels

	Uranium	UO_2	Thorium	ThO_2	Plutonium	UC
Melting point, °F	2,070	4,980	~ 3,180	5,970	1,183	4,200
Boiling point, °F	6,904, (extrapolated from vapor-pressure data)	. . . .	5,480-7,640	. . . . . .	5,840	
Temperature of phase change, °F: α-β	1234	. . . .	2550 (FCC to BCC)	. . . . . .	252	
β-γ	1425				403	
γ-δ					606	
Density, g_m/cm^3	19.05 (77°F) 18.97 (200°F) 18.87 (400°F) 18.76 (600°F) 18.64 (800°F) 18.50 (1000°F) 18.33 (1200°F)	10.5 (10.97 theoretical)	11.72	10.01	19.82 (α,77°F) 17.77 (β,300°F) 17.14 (γ,455°F)	12.97 (13.63 theoretical)
Specific heat, Btu/lb_m °F	0.0278 (200°F) 0.0295 (400°F) 0.0326 (600°F) 0.0366 (800°F) 0.0410 (1000°F) 0.0433 (1100°F) 0.0464 (1200°F)	0.0590 (average between 0 and 2200°F).	0.0282 (100°F) 0.0283 (200°F) 0.0284 (300°F) 0.0285 (400°F)	0.0580 (average between 0 and 1400° F).	0.0375 (100°F) 0.0570 (200°F) 0.0950 (230°F)	0.035 (200°F)

$$f = \frac{\dfrac{r}{r + (1 - r)(M_{ff}/M_{nf})} M_{ff} + \dfrac{(1 - r)}{r(M_{nf}/M_{ff}) + (1 - r)} M_{nf}}{\dfrac{r}{r + (1 - r)(M_{ff}/M_{nf})} M_{ff} + \dfrac{(1 - r)}{r(M_{nf}/M_{ff}) + (1 - r)} M_{nf} + M_{O_2}} \qquad (4\text{-}6a)$$

where M_{nf} and M_{O_2} are the molecular masses of the nonfissionable fuel (such as U^{238}) and of the nonfuel elements that make up the balance of the fuel material (such as O_2 in UO_2). For the case where $M_{ff} \simeq M_{nf}$, the above equation is simplified to

$$f = \frac{rM_{ff} + (1 - r)M_{nf}}{rM_{ff} + (1 - r)M_{nf} + M_{O_2}} \qquad (4\text{-}6b)$$

Combining Eqs. 4-4 and 4-5 gives finally

$$N_{ff} = \frac{\text{Av}}{M_{ff}} r\rho_{fm} f i \qquad (4\text{-}7)$$

Example 4-1. Calculate N_{ff} for 3 percent enriched UO_2.
Solution. For UO_2,

$$\rho_{fm} = 10.5 \; g_m/\text{cm}^3 \qquad \text{(Table 4-2)}$$
$$i = 1.0$$
$$M_{ff} = M_{U^{235}} = 235.0439 \qquad \text{(App. B)}$$
$$M_{nf} = M_{U^{238}} = 238.0508 \qquad \text{(App. B)}$$

Using Eq. 4-6*b* gives

$$f = \frac{0.03 \times 235.0439 + 0.97 \times 238.0508}{0.03 \times 235.0439 + 0.97 \times 238.0508 + 2 \times 15.9944} = 0.8815$$

and

$$N_{ff} = \frac{0.60225 \times 10^{24}}{235.0439} \times 0.03 \times 10.5 \times 0.8815 \times 1 = 7.115 \times 10^{20} \; U^{235} \text{ nuclei/cm}^3$$

4-5. THE FISSION CROSS SECTION IN REACTORS

We shall now attempt to find a value for $\bar{\sigma}_f$ to be used in Eq. 4-3. Since cross sections vary with neutron energy and since neutrons of all energies exist in a thermal reactor, $\bar{\sigma}_f$ must necessarily represent some average or effective microscopic fission cross section.

Thermal Reactors

In a reactor, neutrons are born with large energies and, in a *thermal* reactor, are successively slowed down to thermal energies. In the thermal state they possess a Maxwellian energy distribution dependent upon the temperature of the medium, Eq. 2-4, with an energy spectrum extending from 0 to ∞. The Maxwellian distribution applies to the case of a weakly absorbing medium, and therefore approximately to the case of a thermal reactor with a good low absorbing moderator. Thus in a thermal reactor both thermalized and epithermal neutrons exist. It is obvious that a true and exact value of $\bar{\sigma}_f$ for use in Eq. 4-3 is difficult to come by.

We shall attempt a simplified approach in which it will be assumed that only thermalized neutrons exist, that most of these neutrons fall within the $1/V$ absorption region (Sec. 2-8) of the fuel, and that departure from the $1/V$ region occurs at such high energies that the number of neutrons in the Maxwellian distribution is negligible (see Fig. 4-1). Equation 4-3, is rewritten in the form

$$q'''(E)\,dE_n = G_f N_{ff} \sigma_f(E) \varphi(E)\,dE_n \tag{4-8}$$

where $q'''(E)$ = volumetric thermal source strength per unit energy increment dE_n

$\sigma_f(E)$ = fission cross section at neutron energy E_n

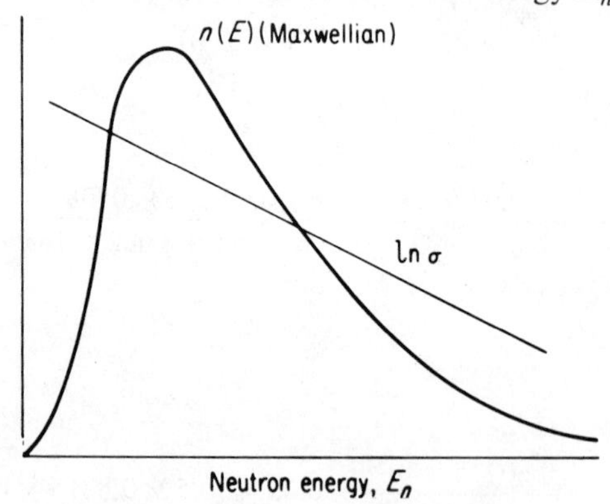

FIG. 4-1. Neutron population and $1/V$ cross sections; one group (thermal).

$\varphi(E)$ = flux of neutrons having energy between E_n and $E_n + dE_n$ per unit energy increment dE_n

Equation 4-8 is integrated to give

$$q''' = G_f N_{ff} \int_0^\infty \sigma_f(E)\varphi(E)\,dE_n \tag{4-9}$$

The flux of all neutrons present, φ, is given by

$$\varphi = \int_0^\infty \varphi(E)\,dE_n \tag{4-10}$$

Equation 4-10 can be combined with Eq. 4-8 to give

$$q''' = G_f N_{ff} \bar{\sigma}_f \int_0^\infty \varphi(E)\,dE_n \tag{4-11}$$

Now Eqs. 4-9 and 4-11 can be combined to give an expression for the desired $\bar{\sigma}_f$:

$$\bar{\sigma}_f = \frac{\int_0^\infty \sigma_f(E)\varphi(E)\,dE_n}{\int_0^\infty \varphi(E)\,dE_n} \tag{4-12a}$$

In the $1/V$ region, σ_f (and this also applies to the absorption and capture cross sections σ_a and σ_c) can be written in terms of a known cross section. A convenient one is at 2,200 m/sec or 0.0235 ev, for which much data are available (Appendix B). If the subscript 0 refers to such cross section and energy,

$$\sigma_f(E) = \sigma_{f_0}\left(\frac{E_0}{E_n}\right)^{0.5}$$

Thus

$$\bar{\sigma}_f = \sigma_{f_0}(E_0)^{0.5} \frac{\int_0^\infty (E_n)^{-0.5}\varphi(E)\,dE_n}{\int_0^\infty \varphi(E)\,dE_n} \tag{4-12b}$$

The Maxwell distribution law, Eq. 2-4, supplies the statistical relationship between $\varphi(E)$ and E_n necessary to integrate Eq. 4-12*b*. It is repeated below in the form

$$n(E)\,dE_n = \frac{2\pi n}{(\pi k T)^{1.5}} E_n^{0.5} e^{-(E_n/kT)}\,dE_n \tag{4-13}$$

where $n(E)$ = number of neutrons per unit volume and energy interval dE_n, having energies between E_n and $E_n + dE_n$

n = total number of neutrons per unit volume

m_n = mass of neutron

k = Boltzmann's constant

T = absolute temperature

Equation 4-13 is obtained from Eq. 2-4 by noting that $E_n = 0.5\, m_n V^2$, $dE_n = m_n V\, dV = (2m_n)^{0.5} E^{0.5}\, dV$, and that $n(E) = n(V)(dV/dE)$, where $V(E)$ is the neutron velocity corresponding to E_n.

Since the neutron flux $\varphi(E) = n(E)V(E)$, it must be proportional to $n(E)E_n^{0.5}$, or, from Eq. 4-13, proportional to the quantity $E_n e^{-(E_n/kT)}$. Equation 4-12*b* can now be rewritten as

$$\bar{\sigma}_f = \sigma_{f_0}(E_0)^{0.5} \frac{\int_0^\infty (E_n)^{0.5} e^{-(E_n/kT)}\, dE_n}{\int_0^\infty E_n e^{-(E_n/kT)}\, dE_n} \tag{4-12c}$$

This equation includes two definite integrals. The solution is

$$\bar{\sigma}_f = \sigma_{f_0}(E_0)^{0.5} \frac{0.5kT(\pi kT)^{0.5}}{(kT)^2}$$

or

$$\bar{\sigma}_f = \frac{\sqrt{\pi}}{2}\, \sigma_{f_0} \left(\frac{E_0}{kT}\right)^{0.5} = 0.8862\, \sigma_{f_0} \left(\frac{E_0}{kT}\right)^{0.5} \tag{4-14}$$

Since E_0 is the energy corresponding to the most probable speed at temperature T_0, it can be equated to the quantity kT_0. Equation 4-14 can thus be written in the form

$$\bar{\sigma}_f = 0.8862\, \sigma_{f_0} \left(\frac{T_0}{T}\right)^{0.5} \tag{4-15}$$

where σ_{f_0} = fission cross section for 0.0253-ev or 2200-m/sec neutrons, Appendix B.

T = effective neutron absolute temperature, $^\circ K$ or $^\circ R$

T_0 = 20°C + 273 = 293°K, or 528°R.

The effective neutron temperature T is that of the fuel-moderator mixture in a homogeneous reactor. In a thermal heterogeneous power reactor, as will be seen in the next chapter, the fuel is normally at a higher temperature than the moderator.

However, since neutrons are thermalized in the moderator, in a well-thermalized system where the scattering macroscopic cross sections are greater than the absorption macroscopic cross sections, T can be taken, with little error, to be the same as the temperature of the moderator. In many cases, such as pressurized- and boiling-water reactors, the moderator and coolant are the same.

Equation 4-15 shows that the average fission cross section for thermal neutrons in a medium at absolute temperature T is less than the fission cross section at the most probable neutron energy corresponding to that temperature by $1 - 0.8862$, or by about 11.4 percent. Remember that the same deduction also applies to absorption cross sections (for which the $1/V$ rule applies).

Now in a thermal reactor the actual neutron spectrum includes not only thermal neutrons but also fast neutrons just born and having a distribution given [16] by Eq. 2-3, Sec. 2-3, here repeated:

$$n(E) = \left(\frac{2}{\pi e}\right)^{0.5} e^{-E_n} \sinh (2E_n)^{0.5} \tag{4-16}$$

as well as neutrons in the process of slowing down to thermal energies. The latter neutrons have a distribution inversely proportional to E_n down to an energy equivalent to $5kT$. Below $5kT$ the epithermal flux rapidly drops to zero because of the conversion of neutrons to thermal energies. Figure 4-2 gives a theoretical composite picture of the neutron density at various energies in a thermal reactor. The actual flux pattern would probably be modified by absorptions in different materials (such as resonance absorption in U^{238}) which depend upon reactor-core internal design. In any case, it can be seen that a true average neutron cross section would have to be based on a modified neutron-flux distribution $\varphi'(E)$ given by

$$\varphi'(E) = \varphi(E) + \varphi_e(E) \tag{4-17}$$

where $\varphi(E)$ is the Maxwellian distribution and $\varphi_e(E)$ is an epithermal distribution.

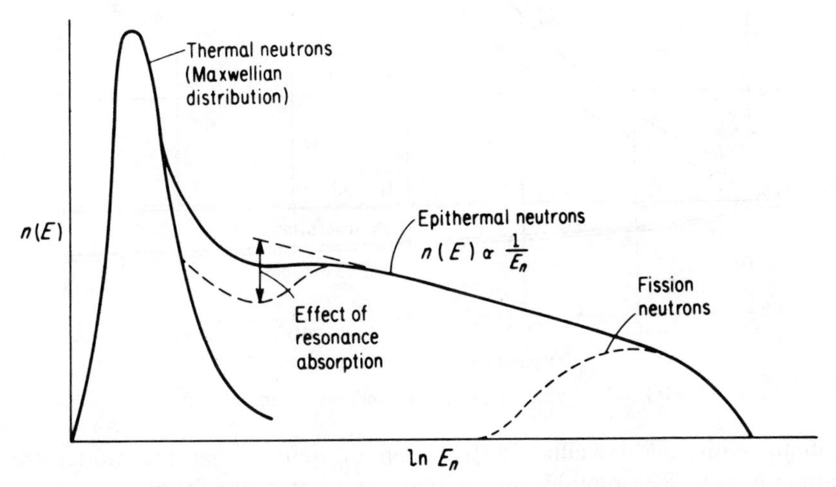

FIG. 4-2. Neutron population versus energy in a thermal reactor.

Furthermore, the fission cross section of the fissionable fuel ceases to follow the $1/V$ law and becomes a complex function of neutron energy at intermediate energies (resonance). Beyond that it is fairly flat in the fast-neutron range (see Fig. 2-7).

The determination of a true effective fission (or absorption) cross section over such a complicated cross-section-neutron-spectrum relationship is obviously a very complex problem indeed.

A partial correction for the cross sections as given by Eq 4-15, which applies to non-$1/V$ absorbers but to neutrons of thermal distribution only, has been attempted [21,22]; in it an empirical correction factor $f(T)$ is to be multiplied by the above cross section. This correction factor depends upon the effective neutron temperature T. It is given for the fuels U^{235}, U^{238}, and Pu^{239} in Fig. 4-3. It applies strictly to well-thermalized

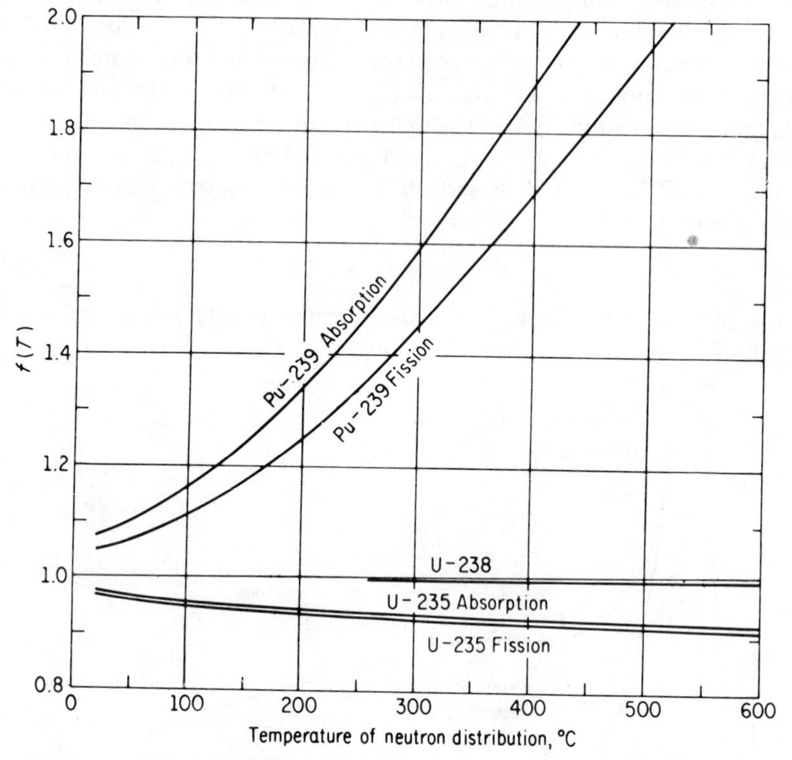

FIG. 4-3. Values of $f(T)$ as a function of temperature.

systems with a Maxwellian distribution of neutrons at the moderator temperature. Reasonable approximations to this are found in reactors moderated by D_2O, graphite Be and BeO (but not ordinary water or other light hydrogen-bearing moderators). Thus for these moderators, Eq. 4-15 is modified to

$$\bar{\sigma}_f = 0.8862 f(T) \sigma_{f_0} \left(\frac{T_0}{T} \right)^{0.5} \tag{4-18}$$

In the case of ordinary water or other light hydrogen-bearing moderators, it is necessary to correct for the departure from Maxwellian distributions and $1/V$ absorption. These departures result in the so-called Wigner-Wilkins spectra, and depend on the fuel-to-moderator ratio. Figure 4-4 [23] shows effective absorption cross sections for U^{235} in an ordinary-water-moderated reactor as a function of the atomic ratio of U^{235}/H in the core. The upper curve assumes no other neutron

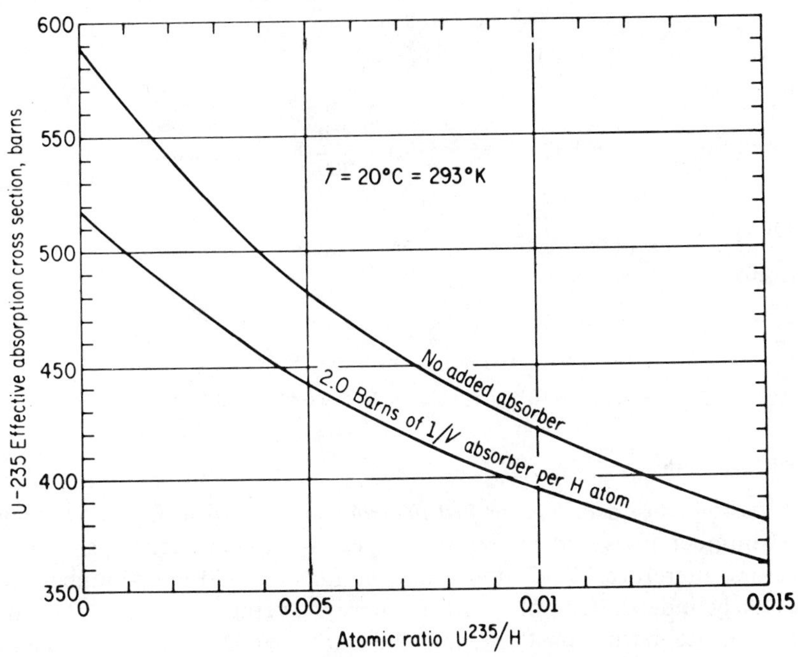

FIG. 4-4. Thermal absorption cross sections of uranium-235 at 293°K averaged over Wigner-Wilkins spectra (Ref. 23).

absorbers. The lower curve assumes an added absorber (cladding, structural material, etc.) contributing an additional 2 barns per H atom present. Both curves are for a temperature of 293°K = 528°R = 68°F. At other temperatures, it is sufficient to assume that the cross sections vary as $1/T^{0.5}$. Since not all neutrons absorbed in a fissionable fuel cause fission, the fission cross sections may be obtained from Fig. 4-4 by multiplying by $1/(1 + \alpha)$. α is the ratio of neutrons asorbed in the fissionable fuel but *not* causing fission to the number of neutrons absorbed and causing fission. $1/(1 + \alpha)$ therefore is the fraction of all neutrons absorbed in the fissionable fuel that cause fission. For U^{235} and thermal neutrons, α may be taken as 0.175, so that $1/(1 + \alpha)$ is 0.851.

Example 4-2. Calculate the volumetric thermal source strength at a position in a reactor core in which the neutron flux is 10^{13}. The core contains the same fuel as in Example 4-1. The moderator temperature at the same core position is 500°F. Assume that neutrons are all thermalized and that the $1/V$ region applies over the bulk of the neutron spectrum. Moderator is heavy water.

Solution

$$N_{ff} = 7115 \times 10^{20}, \text{ from Example 4-1}$$

$$\sigma_{f_0} = 577.1 \text{ barns, Appendix B}$$

$$T = 500°\text{F} = 960°\text{R}$$

$$f(T) = 0.93 \quad \text{(Fig. 4-3)}$$

Therefore

$$\bar{\sigma}_f = 0.8862 \times 0.93 \times 577.1 \left(\frac{528}{960}\right)^{0.5} = 352.7 \text{ barn}$$

$$= 352.7 \times 10^{-24} \text{ cm}^2$$

Assume

$$G = 180 \text{ Mev/fission}$$

Therefore

$$q''' = 180 \times 7.115 \times 10^{20} \times 352.7 \times 10^{-24} \times 10^{13}$$

$$= 4.517 \times 10^{14} \text{ Mev/sec cm}^3$$

$$= 4.517 \times 10^{14} \times 1.5477 \times 10^{-8} = 6.991 \times 10^{6} \text{ Btu/hr ft}^3$$

Fast Reactors

Fission cross sections for *fast neutrons* are shown in Fig. 4-5 for odd-mass-number fuels and in Fig. 4-6 for even-mass-number fuels. In fast reactors, therefore, the fission cross sections for odd-mass-number fuels are fairly independent of neutron energy, averaging about 1.25 barns for U^{235}, 1.8 barns for Pu^{239}, and 2.0 barns for U^{233}. The problem of selecting an effective fission cross section in fast reactors using these fuels in thus much simpler than in thermal reactors. The even-mass-number fuels, such as U^{238} and Th^{232} (the fertile fuels, Sec. 1-8), while nonfissionable by thermal neutrons, are fissionable by high-energy neutrons, Fig. 4-6. Some fast reactors use fuel mixtures such as Pu^{239} and U^{238}, both in oxide form. U^{238} in a fast reactor contributes to breeding, increases the fraction of delayed neutrons and therefore makes the reactor more manageable from a control standpoint (by increasing the average neutron lifetime θ_0, Sec. 3-6), and generally makes the reactor safer by making a negative contribution to the Doppler temperature coefficient of reactivity. When used in fast reactors, U^{238} contributes between 10 and 20 percent of the total fission energy. Its fission cross section is a strong function of neutron energy, making the determination of an effective fission cross section a complex matter.

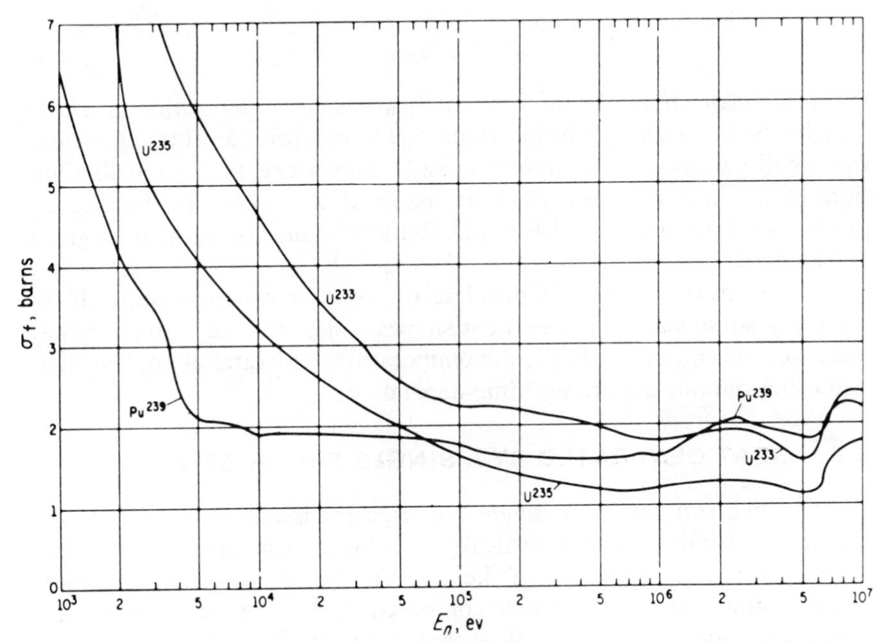

FIG. 4-5. Fast-neutron-fission cross sections for odd-mass-number nuclides (Ref. 24).

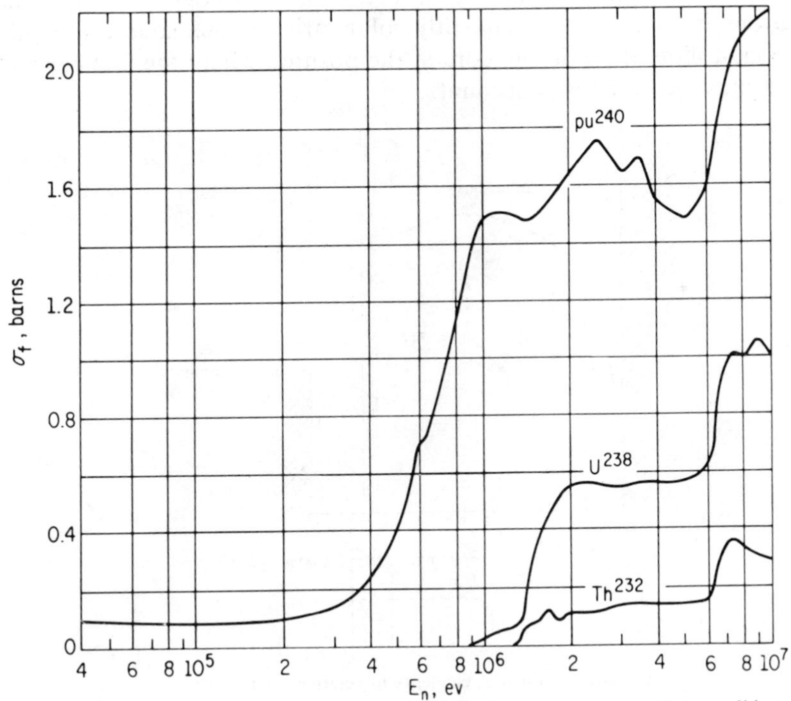

FIG. 4-6. Fast-neutron-fission cross sections for even-mass-number nuclides (Ref. 24).

In Eq. 4-3, here repeated,

$$q''' = G_f N_{ff} \sigma_f \varphi \qquad [4\text{-}3]$$

the volumetric thermal source strength q''' at any one position in a core may be assumed directly proportional to the neutron flux at that position. For small variations in composition and temperature, the spatial distribution of q''' in a core may also be assumed to follow and be directly proportional to the spatial-flux distribution in that core, such as given in Table 3-2.

In the next chapter the problem of *steady-state* temperature distribution in some simple fuel element shapes will be treated. In the steady state, φ, and consequently q''', the temperatures, the rates of coolant flow, and other parameters are not time-dependent.

4-6. HEAT GENERATED BY A SINGLE FUEL ELEMENT

The neutron flux in a single fuel element is not constant. A fuel element is usually situated vertically in a heterogeneous core such that its length equals the height of the core H, Fig. 4-7. The fuel element cross-sectional area is so small compared to that in the core that the radial variation in flux, other than that due to the fuel element acting as a heat sink of thermal neutrons, Fig. 3-4, is negligible. The axial variations in flux, and consequently volumetric thermal source strength, however, follow those in the core at the position where the fuel element is, and must be taken into account.

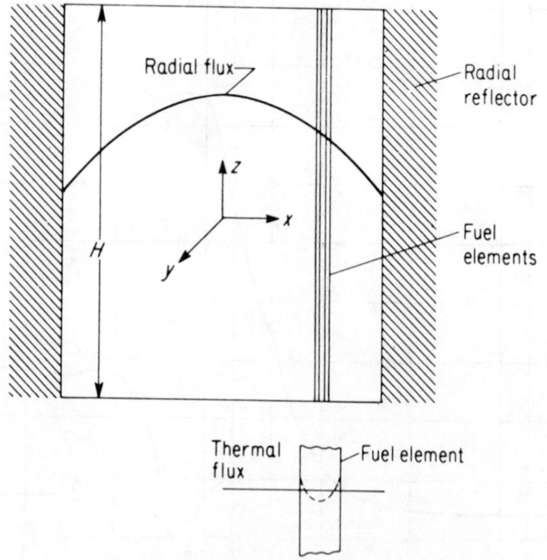

FIG. 4-7. Thin fuel elements in relatively large core.

From considerations of reactor physics (Chapter 3), the neutron flux drops to zero at some extrapolated height H_e, that is, $\varphi = 0$ at $z = \pm H_e/2$, where $z = 0$ corresponds to the core (and fuel element) center plane. If the variation of φ in the axial direction is a pure cosine function of z, the maximum value of φ and q''' occur in a single fuel element at its center and will be designated φ_c and q_c'''. These, of course, are the maximum values for that element only. Other fuel elements closer to the center of the core normally have higher values of φ_c and q_c'''. It should be indicated here that when the reactivity varies with z, because of a large axial temperature rise in a water-moderated core or because of change in phase, such as in the boiling-water reactor, or in fuel enrichment because of uneven burnup, or because of partially inserted control rods, the axial flux may deviate appreciably from the cosine function. In such cases some other relationship between q''' and z should be used. If the q''' variation is complicated, the problem may be solved by separately treating segments of the fuel element where the q''' variation can be approximated by a simple or linearized functions of z.

The total heat generated by one fuel element, called q_t, in Btu/hr, can be obtained as follows:

$$q_t = \int_{-H/2}^{H/2} q'''(z)\, A_s dz \tag{4-19}$$

where $q'''(z)$ is the volumetric thermal source strength at z, and A_s is the cross-sectional area of the fuel element (without cladding).

An expression for $q'''(z)$ is needed to evaluate Eq. 4-19. For the case of sinusoidal flux $q'''(z) = q_c''' \cos (\pi z/H_e)$, so that

$$q_t = \int_{-H/2}^{H/2} q_c''' A_s \cos \frac{\pi z}{H_e}\, dz$$

or

$$q_t = \frac{2}{\pi}\, q_c''' A_s H_e \sin \frac{\pi H}{2H_e} \tag{4-20}$$

Where the extrapolation lengths may be neglected, this equation reduces to

$$q_t = \frac{2}{\pi}\, q_c''' A_s H \tag{4-21}$$

where $(2/\pi)q_c'''$ is the average height of a cosine function having an amplitude of q_c'''.

The permissible value of q_c''' (and consequently the corresponding φ_c) is dictated, of course, by the maximum permissible fuel or cladding temperatures for fixed coolant flow and heat-transfer coefficient h. This will become more apparent in the next chapter.

4-7. THE TOTAL HEAT GENERATED IN CORE–GENERAL

In the previous section, the heat generated by one fuel element of known composition and dimensions and neutron-flux distribution was evaluated. We shall now evaluate the heat generated and removed by all the fuel in a reactor core in which the neutron flux may vary appreciably in the radial as well as the axial direction.

The individual fuel elements in a heterogeneous core are subjected to different values of the neutron flux φ_c and consequently volumetric thermal source strengths q_c'''. In a homogeneous core, there are no individual fuel elements. However, the two cases can be treated in a like manner, provided that (1) the number of fuel elements in the heterogeneous core is sufficiently large, so that negligible variation in radial flux within any one fuel element takes place, and (2) the fuel type and enrichment do not vary in the heterogeneous reactor–or they vary in a manner that can be easily accounted for analytically or by numerical techniques.

It should be remembered here that the total heat generated within the core is composed of the total heat generated by the fuel in the core plus the heat generated by structural components, coolant, moderator, and other materials due to the absorption of radiations. In a thermal reactor, additional heat is generated by the moderator due to the slowing down of neutrons. The latter accounts for about 4 to 6 percent of the total heat generated within the core.

In order to evaluate the heat generated by the fuel in any core, the flux distribution throughout the core should be known. Normal distributions for bare homogeneous cores were presented in Table 3-2 in which the flux φ is given in all cases as a function of φ_{co}, the maximum flux in the geometrical center of the core. a, b, c, R, and H are the actual dimensions, and a_e, b_e, c_e, R_e, and H_e are the extrapolated dimension of each core.

The two cases of homogeneous and heterogeneous reactors will next be treated separately.

4-8. THE CASE OF THE HOMOGENEOUS CORE

In a homogeneous core, where the type and concentration of the fissionable material are independent of location, the volumetric thermal source strength at any point in the core is given by Eq. 4-3, here repeated:

$$q''' = G_f N_{ff} \bar{\sigma}_f \varphi \qquad [4\text{-}3]$$

The value of G_f in a homogeneous core includes the heat generated in the moderator which averages about 5 percent of the total heat. In the absence of more precise information, therefore, a value of 190 Mev/fission (rather than the 180 in the fuel of a heterogeneous core) is used.

To obtain the total heat generated in a given core, Q_t, the proper flux distribution such as from Table 3-2, should be substituted for φ and the equation modified to obtain the heat generated in an infinitesimal volume. The required total heat generated is then obtained by integration over the core volume. This can best be shown by an example.

Example 4-3. Derive the equation for the total heat generated, Q_t (Btu/hr), in an unreflected homogeneous reactor core of spherical shape and radius R (ft). Neglect the extrapolation length (that is, $R = R_e$).

Solution

$$q''' = G_f N_{ff} \bar{\sigma}_f \varphi_{co} \frac{\sin(\pi r/R)}{\pi r/R}$$

For a spherical shell of radius r and thickness Δr, as shown in Fig. 4-8, the heat

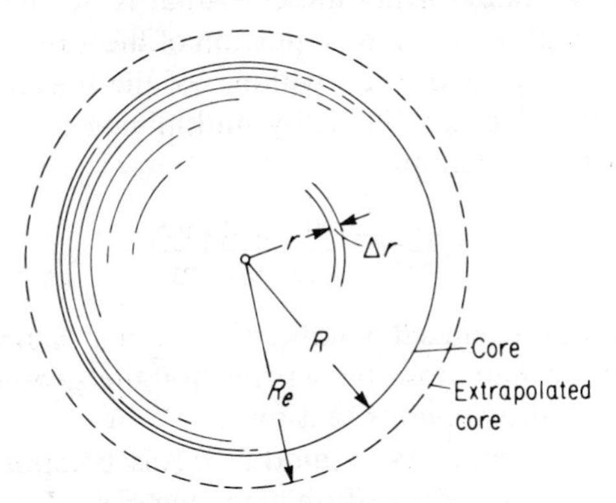

FIG. 4-8. A spherical reactor core.

generated in that shell, ΔQ_r, is equal to the above q''' multiplied by the volume of the shell, $4\pi r^2 \Delta r$. Thus

$$\Delta Q_r = G_f N_{ff} \bar{\sigma}_f \varphi_{co} 4\pi r^2 \frac{\sin(\pi r/R)}{\pi r/R} \Delta r$$

using $G_f = 190$ Mev/fission and the conversion factor 1.5477×10^{-8} to convert Mev/sec cm³ to Btu/hr ft³ result in

$$\Delta Q_r = 1.175 \times 10^{-5} N_{ff} \bar{\sigma}_f R \varphi_{co} r \sin \frac{\pi r}{R} \Delta r \tag{4-22}$$

The total heat generated in the core, Q_t, in Btu/hr, is now

$$Q_t = 1.175 \times 10^{-5} N_{ff} \bar{\sigma}_f R \varphi_{co} \int_0^R r \sin \frac{\pi r}{R} dr$$

This equation integrates as follows:

$$Q_t = 1.175 \times 10^{-5} N \bar{\sigma}_f R \varphi_{co} \left[\left(\frac{R}{\pi}\right)^2 \sin \frac{\pi r}{R} - \frac{R}{\pi} r \cos \frac{\pi r}{R} \right]_0^R$$

$$= 3.75 \times 10^{-6} N_{ff} \bar{\sigma}_f R^3 \varphi_{co} \tag{4-23}$$

In this example the extrapolation length was assumed to be negligible. If it is not or if the core is reflected, the extrapolated dimension should be used. The limits of integration, however, remain as 0 and R.

4-9. THE CASE OF THE HETEROGENEOUS REACTOR WITH A LARGE NUMBER OF FUEL ELEMENTS

In this case, the fuel elements are usually of equal height and are as high as the reactor core itself. In the case of a large number of fuel elements, as is usually the case in a power reactor, Q_t can be obtained, with little error, by *homogenizing* the core–that is by modifying the value of q''' so that it would be based on a portion of the core volume, occupied by one fuel element, instead of the volume of the fuel element itself. In other words q''' would be modified by multiplying it by the ratio of the fuel volume to the core volume:

$$q_h''' = q''' \frac{\text{fuel volume}}{\text{core volume}} \tag{4-24}$$

where q_h''' is the homogenized volumetric thermal source strength. A differential equation can now be written for the geometry of the core and integrated as in the previous section.

An alternate procedure is to modify q_t, as computed by Eqs. 4-19 or 4-21, so that is would be based on a portion of the cross-sectional area of the core occupied by a single fuel element. This last procedure will be explained by an example.

Example 4-4. Derive the equation for the total heat generated, Q_t(Btu/hr), in a cylindrical reactor core of radius R and height H containing n vertical fuel elements. Assume a normal neutron-flux distribution.

Solution. The flux distribution in a cylindrical core is given by

$$\varphi = \varphi_{co} \cos \frac{\pi z}{H_e} J_0\left(\frac{2.4048 r}{R_e}\right) \qquad \text{Table 3-2}$$

where J_0 is the Bessel function of the first kind, zero order (Appendix C). $z = 0$ represents the middle plane of the core where all fuel elements have their maximum neutron flux φ_c. The above equation can be modified to give φ_c for all fuel elements, as a function of r, by putting $z = 0$. Thus

$$\varphi_c = \varphi_{co} J_0\left(\frac{2.4048 r}{R_e}\right) \tag{4-25a}$$

Similarly,

$$q_c''' = q_{co}''' J_0\left(\frac{2.4048r}{R_e}\right) \tag{4-25b}$$

where φ_c and q_c''' are the maximum (midpoint) neutron flux and volumetric thermal source strength for a fuel element at a radial distance r from the core center line. φ_{co} and q_{co}''' are the neutron flux and volumetric thermal source strength at the geometrical center of the core, respectively.

An area A' = reactor cross-sectional area per fuel element will now be defined:

$$A' = \frac{\pi R^2}{n} \tag{4-26}$$

where n is the total number of fuel elements in the core.

In Eq. 4-21, q_t will now be modified to give q_t', the heat generated per unit area A':

$$q_t' = \frac{q_t}{A'} = \frac{n}{\pi R^2}\, q_t \tag{4-27a}$$

Thus

$$q_t' = \frac{2n}{\pi^2 R^2} A_s H_e \sin\frac{\pi H}{2H_e}\, q_c''' \tag{4-27b}$$

or, where the extrapolation lengths may be neglected,

$$q_t' = \frac{2n}{\pi^2 R^2} A_s H q_c''' \tag{4-27c}$$

These equations, in which all components are constant except q_c''', can be combined with Eq. 4-25b to give q_t' as a function of r. Thus

$$q_t'(r) = \frac{2n}{\pi^2 R^2} A_s H_e \sin\frac{\pi H}{2H_e}\, q_{co}''' J_0\left(\frac{2.4048r}{R_e}\right) \tag{4-28}$$

Taking a differential cylindrical element of width Δr at radius r from the axis of the core (Fig. 4-9), the total heat generated in the core, Q_t(Btu/hr), can be computed from the equation

$$\begin{aligned} Q_t &= \int_0^R q_t'(r) 2\pi r\, dr \\ &= \frac{4n}{\pi R^2} A_s H_e \sin\frac{\pi H}{2H_e}\, q_{co}''' \int_0^R r J_0\left(\frac{2.4048r}{R_e}\right) dr \end{aligned} \tag{4-29a}$$

or where the extrapolation lengths may be neglected,

$$Q_t = \frac{4n}{\pi R^2} A_s H q_{co}''' \int_0^R r J_0\left(\frac{2.4048r}{R}\right) dr \tag{4-29b}$$

The terms inside the integral sign integrate as follows:

$$\int_0^R r J_0\left(\frac{2.4048r}{R_e}\right) dr = \frac{R_e}{2.4048}\left[r J_1\left(\frac{2.4048r}{R_e}\right)\right]_0^R$$

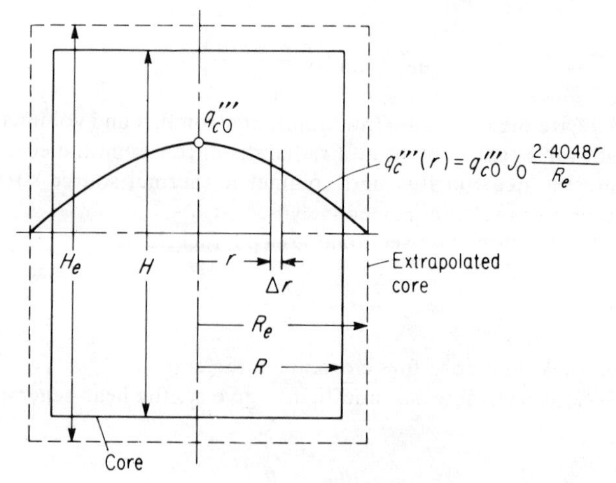

FIG. 4-9. A cylindrical reactor core.

where J_1 is the Bessel function of the first kind, first order. Since $J_1(0) = 0$ and $J_1(2.4048) = 0.519$ (Table C-2, Appendix), the above integral for the case where $R = R_e$ equals $0.2158\ R^2$. This quantity can then be substituted in Eq. 4-29b to give

$$Q_t = 0.275nA_sHq'''_{co} \tag{4-30a}$$

The above Q_t represents the total heat generated in the solid fuel alone. To obtain the total heat generated in the core, allowances must be made for the heat generated in the moderator and other reactor materials. A reasonable approach is to multiply the above Q_t by 1.05 giving

$$Q_t = 0.289nA_sHq'''_{co} \tag{4-30b}$$

There now remains the case of a small number of relatively large fuel elements such as in research or training reactor (which may include a hydrogenous moderator with the fuel, such as uranium in mixture with polyethylene, zirconium hydride, or graphite). In such a case the total heat generated in the core has to be found by calculating and summing the heat generated in the individual elements.

4-10. REACTOR SHUTDOWN HEAT GENERATION

In reactor shutdown, the reactor power does not immediately drop to zero but falls off rapidly according to a negative period, eventually determined by the half-life of the longest-lived delayed neutron group (delayed neutrons are those emitted in the neutron decay of some fission products). Even then the fission fragments and fission products existing in the fuel continue to decay (β and γ), at decreasing rates of course, for long periods of time.

After shutdown, a reactor therefore continues to generate power, P_s, a function of time. The amount of such power generation depends on (a) the level of power before shutdown, P_0 and (b) the length of time, θ_0, it operated at such a level; since both these factors determine the amount of fission products present. Figure 4-10 [25] shows the fraction of decay power to power before shutdown, for various reactor operating

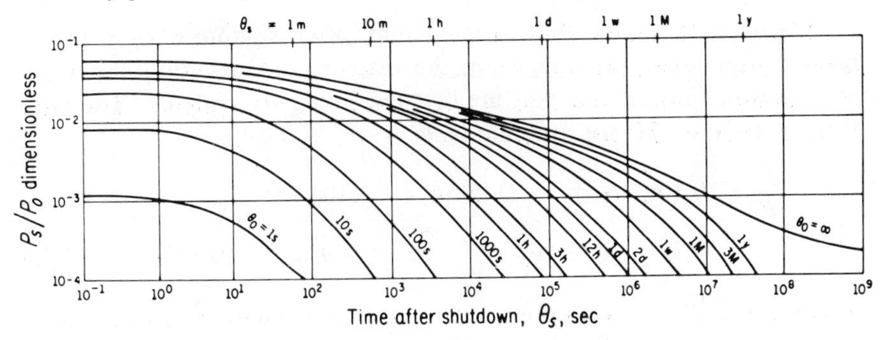

FIG. 4-10. Ratio of power after, to power before shutdown for various operation times before shutdown: s,sec; m,min; h,hour; d,day; w,week; M,month; y,year (Ref. 25).

times before shutdown, θ_0, ranging from 1 second to infinite (very long) operation. The ordinate is dimensionless, (Btu/hr)/(Btu/hr), Mw/Mw, etc. The abscissa is time θ_s after shutdown. The data is for uranium fuels (figures for natural uranium and U^{235} are nearly identical). Figure 4-10, based on 1955 data, is instructive, but has since been found to overestimate the ratio of power production by a factor of about 2. More recent information [26], more accurate especially for power reactor work, is given in Fig. 4-11 for the case of $\theta_0 = \infty$ only. $\theta_0 = \infty$ here can be construed as one year or more of operation.

Because of such decay heat, it is essential to provide *shutdown cooling* or *decay cooling* for the fuel of a reactor. If this is not done, say due to power failure to the coolant pumps which may have caused an emergency shutdown in the first place, the fuel and cladding temperature will rise to a peak before cooling again. In small training or research reactors, the reactor power levels are low and sufficient cooling after shutdown may be provided by the natural convective flow of coolant or air. In power reactors the peak temperatures would almost invariably be in excess of safe temperatures resulting in structural damage to the fuel elements, radioactivity release, etc. In such cases, either a natural-circulation cooling capability of the normal coolant is built into the system, or an emergency cooling system is incorporated into the design. Natural-circulation flow is covered in more detail in Chapter 14.

The ratio of the volumetric thermal source strength after shutdown q_s''' to that before shutdown q_0''' could be taken as the same as the ratio of the respective power, or

$$\frac{q_s'''}{q_0'''} = \frac{P_s}{P_0} \tag{4-31}$$

Figure 4-11 shows that, immediately after shutdown of a power reactor after long time operation, the reactor produces about 3 percent of its original power, and that the first few hours are critical. The ratio P_s/P_0 is given in [25] for uranium fuels as

$$\frac{P_s}{P_0} = [0.1(\theta_s + 10)^{-0.2} - 0.087(\theta_s + 2 \times 10^7)^{-0.2}]$$
$$- [0.1(\theta_s + \theta_0 + 10)^{-0.2} - 0.087(\theta_s + \theta_0 + 2 \times 10^7)^{-0.2}] \tag{4-32}$$

where θ and θ_0 are in seconds. The more up-to-date data of Fig. 4-11 appear to correlate well, for $\theta_s > 200$ seconds and $\theta_0 = \infty$, by

$$\frac{P_s}{P_0} = 0.095\theta_s^{-0.26} \tag{4-33}$$

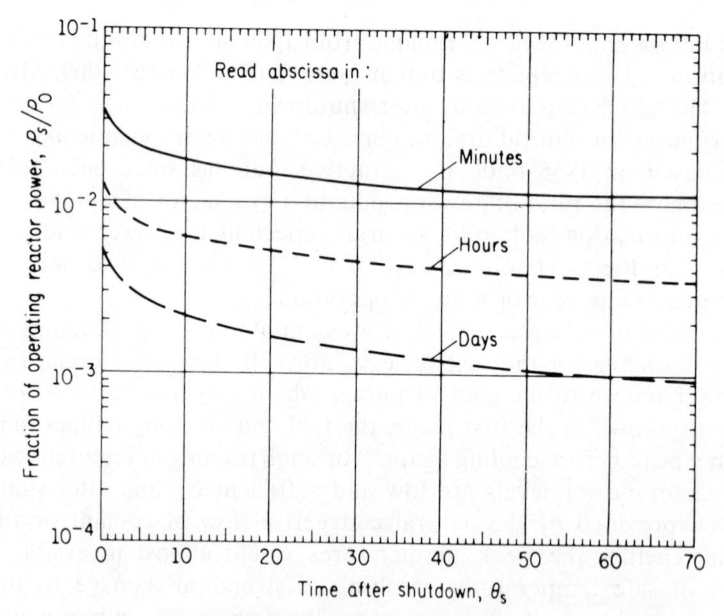

FIG. 4-11. Fission product power vs. time after shutdown following infinite operation (Ref. 26)

and, for finite θ_0, the right-hand side of the above equation may be multiplied by a factor $f(\theta_0)$ given by

$$f(\theta_0) = \left[1 - \left(1 + \frac{\theta_0}{\theta_s}\right)^{-0.2}\right] \tag{4-34}$$

Of interest is the total energy release as a function of time after shutdown, E_s. This is obtained by integrating Eq. 4-33 to give

$$E_s = \int_0^{\theta_s} \left(\frac{P_s}{P_0}\right) d\theta_s = 0.128\theta_s^{0.74} \tag{4-35}$$

where the units are energy per unit power level before shutdown, (Btu/hr) seconds/(Btu/hr), or simply time in seconds, etc. Figure 4-12 shows the value of E_s in units of minutes, hours and days, vs. time in corresponding units.

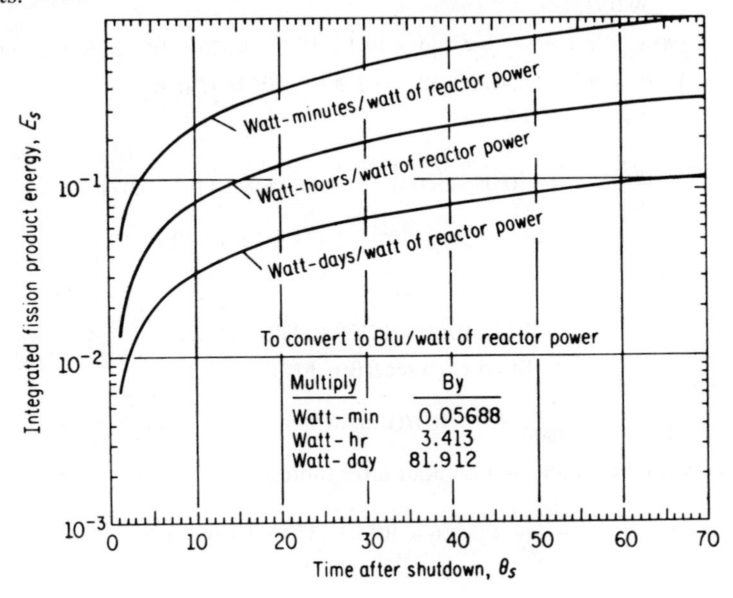

FIG. 4-12. Integrated fission product energy generated vs. time after shutdown following infinite operation (Ref. 26).

Equations 4-33, 4-34, and 4-35 and Figs. 4-10 and 4-11 may be assumed to apply to fission by all fuels.

Example 4-5. A fast reactor uses fuel material of density $10 g_m/cm^3$ and composed of 80 percent depleted UO_2 and 20 percent PuO_2 by mass. At the center of the core, the fast neutron flux is 10^{16}, and the effective fission cross section for Pu^{239} is 1.78 barns. Ignoring the U^{235} content of the fuel and assuming that plutonium is 80 percent enriched in fissionable Pu^{239} (the rest being essentially nonfissionable Pu^{240}) calculate (a) the volumetric thermal source strength at the center of the core, (b) the total energy produced by the same fuel in Btu/ft³, one hour after shutdown if the reactor operated at the above flux for more than a year, and (c) the volumetric thermal source strength after one hour of shutdown if the reactor operated 6 weeks at the above flux.

Solution. Using data from Appendix B, f, the mass fraction of Pu^{239} in the fuel element

$$= 0.2 \times \frac{0.8 \times 239.0522 + 0.2 \times 240.0540}{0.8 \times 239.0522 + 0.2 \times 240.0540 + 2 \times 15.9994}$$

$$= 0.1766$$

$$N_{ff} = \frac{Av}{Mff} r\rho_{fm} f i \qquad \text{(Eq. 4-7)}$$

$$= \frac{0.60225 \times 10^{24}}{239.0522} + 1 \times 10 \times 0.1766 \times 1 = 4.449 \times 10^{21} \text{ nuclei/cm}^3$$

$$q_0''' = 180 \times 1.78 \times 10^{-24} \times 4.449 \times 10^{21} \times 10^{16} = 1.426 \times 10^{16} \text{ Mev/sec cm}^3$$

$$= 1.426 \times 10^{16} \times 1.5477 \times 10^{-8} = 2.206 \times 10^{8} \text{ Btu/hr ft}^3$$

(a)

$$\frac{q_s'''}{q_0'''} = \frac{P_s}{P_0} = 0.095\,(3600)^{-0.26} = 0.095 \times 0.119 = 0.0113$$

$$q_s''' = 0.0113 \times 2.206 \times 10^{8} = 2.493 \times 10^{6} \text{ Btu/hr ft}^3$$

(b)

$$E_s = 0.128\,(3600)^{0.74} = 0.128 \times 430 = 55 \text{ sec.}$$

$$= 55 \text{ (Btu/hr ft}^3\text{) sec/(Btu/hr ft}^3\text{)}$$

$$= \frac{55}{3600} \text{ (Btu/ft}^3\text{)/(Btu/hr ft}^3\text{)}$$

Thus energy produced during first hour after shutdown

$$= \frac{55}{3600} \times 2.206 \times 10^{8} = 3.370 \times 10^{6} \text{Btu/ft}^3$$

(c)

$$f(\theta_0) = 1 - \left(1 + \frac{6 \times 24}{1}\right)^{-0.2} = (1 - 0.37) = 0.63$$

$$q_s''' = 2.493 \times 10^{6} \times 0.63 \times 1.571 \times 10^{6} \text{Btu/hrft}^3$$

4-11. HEAT GENERATION BY RADIOISOTOPES

Radioisotopic fuels are widely used in small power devices to generate heat. This heat is usually then directly converted to electrical energy in compact power devices by *thermoelectric* or *thermionic* means [2].

In the United States, such systems are called *Systems for Nuclear Auxiliary Propulsion,* or *SNAP.* These SNAP devices are given odd numbers. For example, SNAP 17, used to power communications satellites, uses strontium-90 as fuel. The device weighs 30 lb_m, produces 30 watts of electricity, and has a design life of 3–5 years. Another, SNAP 27, a plutonium-238-fueled thermoelectric generator that produces 67 watts after one year of operation, was deployed on the lunar surface by the Apollo 12 astronauts in November 1969. It powers the Apollo Lunar Surface Experiment Package (ALSEP) as well as enables the experiments to transmit data back to earth during both the lunar day and long (350 hr) lunar night. Four more SNAP 27-ALSEP units are planned for deployment. (Another entirely different category of compact power sources, using small fission reactors rather than radioisotopes for heat generation, are also referred to as SNAP but are given even numbers.)

Isotopic power devices are compact, lightweight, and have no moving parts. They are however, very costly (fuel prices currently vary between one hundred to several thousand dollars per watt output) and, depending upon the activity and half life of the fuel used, produce low power densities for relatively short periods of time. There are generally two types of fuel used: (1) those processed from spent reactor fuels and (2) α emitters prepared by irradiation in a reactor. Table 4-3 lists some of the radioisotopes suitable as fuels and their important characteristics. Type 1 fuels are generally longer lived, produce lower power densities, and are also expected to be lower in cost than fuels of type 2. They also present a greater hazard of X and γ radiation and must therefore be incapsuled more carefully in suitable shielding material.

TABLE 4-3
Radioisotopic Fuels

Type	Radioisotope	Activity	Half-life	Probable Fuel Material	Fuel Material Density, g_m/cm^3	Power Density $w(t)/cm^3$
1	Sr^{90}	β	28 yr	$Sr\,Ti\,O_3$	4.8	0.54
	Cs^{137}	β	33 yr	Cs Cl	3.9	1.27
	Ce^{144}	β	285 d	$Ce\,O_2$	6.4	12.5
	Pm^{147}	β	2.5 yr		6.6	1.1
2	Po^{210}	α	138.4 d	Po	9.3.	132.0
	Pu^{238}	α	86 yr	Pu C	12.5	6.9
	Cm^{242}	α	163 yr	Cm C	11.75	1,169

Heat generation in a radioisotope is due to the exothermic decay reactions and is uniform throughout a fuel element. The volumetric thermal source strength q''' is obtained by evaluating the mass defect per decay reaction and the energy associated with it. This is best illustrated by an example.

Example 4-6. A SNAP generator is fueled with 475g_m of $Pu^{238}C$, 100 percent enriched in Pu^{238}. Calculate the volumetric thermal source strength in Btu/hr ft^3, and the total thermal power generated in watts. Density of PuC = 12.5g_m/cm^3.

Solution. Using data from Appendix B and Table 4-3, and conversion factors from Appendix H:

Since Pu^{238} is an α emitter,

$$_{94}Pu^{238} \longrightarrow {}_{92}U^{234} + {}_2He^4$$

where U^{234} has a 2.47×10^5 yr half-life, Appendix B, and can therefore be considered stable. The mass defect is

$$\begin{aligned} \Delta m &= \text{mass of products} - \text{mass of reactants} \\ &= (234.0409 + 4.0026) - (238.0495) \\ &= -0.0060 \text{ amu} \\ \Delta E\,(\text{Mev}) &= 931\ \Delta m(\text{amu}) \qquad (\text{Sec. 1-6}) \\ &= 931 \times -0.0060 = -5.586 \text{ Mev/reaction} \end{aligned}$$

where the negative sign indicates an exothermic reaction. Rate of reaction $R = \lambda N$, where λ = decay constant = $0.6931/\theta_{\frac{1}{2}}$, $\theta_{\frac{1}{2}}$ = half-life, and N is the nuclear density of the radioisotope. Thus

$$\theta_{\frac{1}{2}} = 86 \text{ yr} = 86 \times 3.1557 \times 10^7 = 2.7156 \times 10^9 \text{ sec.}$$

$$N = \frac{\text{Av}}{\text{M}_{\text{PuC}}}\,\rho = \frac{0.60225 \times 10^{24}}{238.04 + 12.01} \times 12.5 = 3.01 \times 10^{22} \text{ nuclei/cm}^3$$

$$R = \frac{0.6931}{2.715 \times 10^9} \times 3.01 \times 10^{22} = 7.682 \times 10^{12} \text{ reaction/sec cm}^3$$

$$\begin{aligned} q''' &= R\Delta E = 7.682 \times 10^{12} \times 5.586 = 4.291 \times 10^{13} \text{ Mev/sec cm}^3 \\ &= 4.291 \times 10^{13} \times 1.5477 \times 10^{-8} = 6.641 \times 10^5 \text{ Btu/hr ft}^3 \\ &= 4.291 \times 10^{13} \times 3600 \times 4.44 \times 10^{-17} = 6.86 \text{ w(t)/cm}^3 \end{aligned}$$

$$Q_t = q''' \times \text{volume} = 6.641 \times 10^5 \times \frac{475}{12.5}\,(0.03281)^3 = 891.3 \text{ Btu/hr}$$

$$= 891.3 \times 2.931 \times 10^{-1} = 261 \text{ w(t)}$$

The electrical output will be far less than Q_t because the efficiency of thermoelectric devices is rather low (5-10 percent).

PROBLEMS

4-1. Calculate the number density of U^{235}, nuclei/cm^3, in a 10 percent enriched uranyl sulphate (UO_2SO_4) in a 50 percent by mass aqueous solution at 500°F Assume density of solution to be equal to that of water.

4-2. A fast-reactor fuel material is composed of 20 percent by mass of Pu^{239} O_2 and 80 percent by mass UO_2. The U is depleted uranium and may be

considered to be all U^{238}. The density of the fuel material is 10 g_m/cm^3. Calculate
(a) the nuclear densities, in nuclei/cm^3, of Pu^{239} and U^{238}
(b) The percent fissions by Pu^{239} and by U^{238} if the neutrons in the reactor were all assumed to have an energy of 1 Mev
(c) Repeat (b) but for a neutron energy of 2 Mev

4-3. A graphite-moderated, uranium-fueled reactor generates 2,000 Mw(t) in the core. The helium coolant mass flow rate is 2.07×10^7 lb_m/hr. It enters the core at 800°F. Estimate the effective thermal-neutron fission cross section in the core.

4-4. Calculate the effective fission cross section of 3 percent enriched UO_2 (as in Example 4-1) if it is used in an ordinary water-moderated thermal reactor. The core lattice arrangement is such that the fuel/moderator ratio is 1:1.9 by volume. Assume the moderator is at 500°F and 2,000 psia, and that the other reactor-core materials contribute 2.0 barns of $1/V$ absorption per atom of hydrogen.

4-5. Calculate the volumetric thermal source strength, Btu/hr ft^3 at a point 49.9 percent of the radial distance and halfway above the centerplane of a cylindrical, homogeneous bare reactor core containing enriched UO_2SO_4 in solution in H_2O. The density of the *fuel* is 0.255 g_m/cm^3. The temperature of the solution is 500°F The enrichment is 10 percent. The maximum neutron flux in the core is 10^{14}

4-6. A heterogeneous cylindrical reactor core contains 10,000 fuel rods. The maximum thermal-neutron flux is 10^{13} and the fuel enrichment is 3 percent. The average moderator temperature is 500°F. The fuel pellets are 0.6 in. in diameter. The core is 8 ft in diameter and 20 ft high. Neglecting the extrapolation lengths, calculate the total heat generated in the core in kw(t). Fuel is UO_2 and moderator is heavy water.

4-7. A homogeneous thermal cylindrical reactor core is 4 ft diameter and 4 ft high. The extrapolation lengths in all directions are 0.5 ft. The maximum neutron flux is 10^{13}. Find the volumetric thermal source strength at a point 0.832 ft radial distance and 1.667 ft above the center of the core. The fissionable fuel density is 6×10^{20}. The average moderator temperature is 500°F.

4-8. A cylindrical reactor core is 4 ft in diameter and 4.8 ft high. The maximum neutron flux is 10^{13}. The extrapolation lengths are 0.186 ft in the radial direction and 0.3 ft in the axial direction. The fuel is 20 percent -enriched UO_2. $\bar{\sigma}_f = 500$ b. Determine (*a*) the neutron flux at the upper and lower rims, and (*b*) the maximum heat generated in the fuel in Mev/sec cm^3 and Btu/hr ft^3.

4-9. A thermal homogeneous circulating-fuel reactor has a spherical core 200 cm in diameter. The fuel is a uranium salt solution in heavy water and the U^{235} concentration is 10^{20} nuclei/cm^3. The core is surrounded by a graphite reflector. The extrapolated radius is 104.3 cm. The maximum thermal neutron flux is 10^{13}. Calculate the heat produced in the core in Btu/hr if the effective moderator temperature is 550°F.

4-10. A pressurized-water reactor has a cylindrical core 6.25 ft in diameter and 8.5 ft high. The total heat generated in the fuel within the core is 53 Mw(t). The core contains 20,000 vertical cylindrical rods containing 3 percent enriched UO_2 cylindrical pellets 0.3 in. in diameter. 4×10^6 lb_m/hr of coolant enters the core at 460°F. Estimate the maximum thermal-neutron flux in the core. Neglect the extrapolation lengths.

4-11. A thermal heterogeneous reactor core is in the form of a cube 4 ft on the side. It contains 4,900 vertical fuel pins made of 20 percent enriched uranium metal of 0.45 in. diameter. The total heat generated in the fuel is 5×10^8 Btu/hr. The effective graphite moderator temperature is 500°F. Assume that the neutron-flux distribution is undistorted by coolant or other reactor materials, and neglect the extrapolation lengths. Find (*a*) the average thermal-neutron flux in the central horizontal plane of the core and (*b*) the maximum core thermal-neutron flux.

4-12. A fast homogeneous fluid-fueled reactor uses uranium in solution in molten bismuth so that the fuel concentration is 10^{20} U^{235} nuclei/cm³. The reactor core is 5 ft in diameter and 8 ft high. The extrapolated height is 12 ft. Because of strong radial reflection, the neutron flux may be considered uniform in the radial direction. The fast-neutron flux at the core entrance (bottom) is 10^{15}. Take σ_f for fast neutrons as 5.0 barns. Calculate the total heat generated in the core.

4-13. A cylindrical reactor core 4 ft in diameter and 4 ft high contains 1,000 vertical plate-type fuel elements 3.5 in. wide and 0.2 in. thick, clad in 0.005 in.-thick 304 L stainless steel. A fuel element 0.5 ft from the core center line has $\varphi_c = 3.057 \times 10^{13}$ and $q_c''' = 2.221 \times 10^7$ Btu/hr ft³. The fuel elements may be considered equally spaced within the core. The core is reflected only in the axial direction by 2 ft of light water above and below the core. Neglect extrapolation length in the radial direction. Calculate (*a*) the maximum neutron flux in the core and (*b*) the total heat generated in the fuel in Btu/hr.

4-14. A reactor generates 2,000 mw(t) for a year or more. Find the amount of decay heat in Btu that must be removed from the core during (a) the first hour, and (b) the second hour, following shutdown.

4-15. A heavy-water-moderated thermal reactor operated at full load for more than a year. The maximum volumetric thermal source strength was 5×10^7 Btu/hr ft³. The core is cylindrical, 10 ft diameter and 10 ft high. All extrapolated dimensions are 50 percent greater than the respective physical dimensions. The fuel material constituted 30 percent of the core by volume. Estimate the total energy in Btu that must be removed during the first day following shutdown. Compare it to the total energy generated in an equivalent period of time during normal operation.

4-16. A reactor generates 2,000 Mw(t) for one month. Find the amount of decay heat in Btu that must be removed from the core during (a) the first hour, and (b) the second hour, following shutdown.

4-17. For the radiosotopic fuel materials of Table 4-3 calculate the specific power Btu/hr lb_m, assuming 100 percent enrichment of the material (a) during initial operation, and (b) one year afterward.

chapter **5**

Heat Conduction in Reactor Elements

I. General and One-Dimensional Steady-State Cases

5-1. INTRODUCTION

Solid reactor and radioisotopic fuel elements (and such others as electrical resistance heaters) are devices in which heat is both generated and conducted. The problems of temperature distribution in and heat removal from these elements are important in evaluating reactor-core thermal performance. Elements of different geometrical shapes in both steady- and unsteady-state conduction need be treated.

In the previous chapter the heat generation in fuel elements was evaluated. In this chapter the general heat-conduction mechanism will first be discussed, after which equations will be derived for the temperature distribution in fuel elements of simple geometries, their cladding and coolant. Only *one-dimensional steady-state* heat flow with *uniform* heat generation is considered in this chapter. More complex geometries and situations and the unsteady-state are covered in subsequent chapters.

5-2. GENERAL ASSUMPTIONS

In treating the problems of heat removal from an individual fuel element in a heterogeneous reactor, several simplifying assumptions are made. Some of these are given below.

It will generally be assumed that the thermal conductivities of the fuel k_f and the cladding k_c, as well as the physical properties of the coolant (density, viscosity, specific heat) are constant and independent of temperature. The heat-transfer coefficient h between solid and coolant will therefore be considered constant also. The errors involved in these assumptions depend of course on the severity of the temperature gradients and the dependence of the above properties on temperature. These properties may be evaluated at the mean temperatures existing in the fuel, cladding, and other elements. This method is accurate if they are linear functions of temperature as will be shown in Sec. 6-13. If the functions are more complicated, absolute accuracy necessitates the integration of the exact function over the temperature range.

Tables 5-1 and 5-2 list thermal conductivities and other physical properties of fuel and cladding materials commonly encountered in reactor work.

TABLE 5-1
Thermal Conductivity k_f of Some Fuel Materials*
(Btu/hr ft °F)

Temperature, °F	Uranium	UO_2	UC	PuO_2	Thorium	ThO_2
200	15.80	4.5	14.77	3.60	21.75	7.29
300	16.40	...	14.07	.....	22.18	6.25
400	17.00	3.5	13.48	.....	22.60	5.34
500	17.50	...	13.02	.....	23.00	4.61
600	18.10	2.8	12.67	.....	23.45	4.03
700	18.62	...	12.39	.....	23.90	3.59
800	19.20	2.5	12.19	.....	24.30	3.21
900	19.70	...	12.02	.....	24.65	2.91
1000	20.25	2.2	11.91	.....	25.75	2.68
1100	20.75	...	.....	.....	25.60	2.47
1200	21.20	2.0	11.82	.....	26.13	2.30
1300	21.60	...	.....	.....	.....	2.17
1400	22.00	1.6	11.76	1.57	.....	2.07
1600	.....	1.5	11.70	.....	.....	1.90
1800	.....	1.4	11.67	.....	.....	1.80
2000	.....	1.3	11.57	.....	.....	1.70
2200	.....	1.2	.....	.....	.....	1.69
2400	.....	1.1	.....	.....	.....	1.68
2600	.....	1.1	.....	.....	.....	....
2800	.....	1.1	.....	.....	.....	....
3000	.....	1.1	.....	.....	.....	....
3200	.....	1.1	.....	.....	.....	....

* Values given are for unirradiated materials and usually decrease on irradiation. The percent decrease is a function of both irradiation temperature and burnup. The ceramics (UO_2, ThO_2, and UC) in particular suffer a large decrease. For example, k_f for UO_2 decreases by some 60 percent on irradiation at 200°F and after 4,000 Mw-day/ton burnup. For UC the decrease is about 58 percent on irradiation at 1000–1500°F and after 7,500-10,000 Mw-day/ton burnup.

It will also be assumed that no heat is generated except in the fuel; i.e., none is generated in cladding or coolant. The problem of heat generation and conduction in the special case of finned cladding will be taken up in Chapter 6. The case of heat generation and transport in a moving fluid (such as a molten-salt fuel) will be taken up in Chapter 10.

TABLE 5-2
Properties of Some Cladding Materials

Material	Atomic Number	ρ, Density lb_m/ft^3	N, no. of Nuclei/cm^3	Microscopic Absorption Cross Section σ_a, Barns (thermal)	Macroscopic Absorption Cross Section Σ_a, cm^{-1} (thermal)	k_c, Thermal Conductivity, Btu/ft hr °F	Specific Heat, c_p Btu/lb_m °F	Melting Point, °F
Beryllium	4	115 (77°F)	0.12×10^{24}	0.010	0.00123	128 (77°F) 82.2 (200°F)	0.506 (261°F)	2330
Magnesium	12	109	0.0431×10^{24}	0.059	0.00254	97 (200°F) 94 (390°F) 91 (570°F)	......	1205
Aluminum	13	169 (68°F)	0.0603×10^{24}	0.215	0.0130	132 (68°F) 131 (390°F) 131 (750°F)	0.230 (212°F) 0.245 (572°F)	1040
Zirconium	40	406 (75°F)	0.0423×10^{24}	0.18 ± 0.02	0.00765	12.1 (120°F) 11.8 (200°F) 11.5 (300°F) 11.0 (500°F) 10.6 (750°F)	0.0739 (260°F) 0.0815 (800°F)	3370
Zircaloy 2 (1.2–1.7% Sn. 0.07–0.2% Fe, 0.05–0.15% Chr, 0.03–0.06% Ni), also Zircaloy 4	...	409	....	....	....	Less than 10	....	3310
Types 304 and 304L stainless steel, 18–20% Cr, 8–12% Ni, 2% max. Mn, 0.08% max. C, 1% max. Si, 0.03% max. C (L)	...	501	....	....	....	9.4 (200°F) 12.4 (930°F)	0.12 (32–212°F)	2550–2650
Type 347 stainless steel, 17–19% Cr, 9–12% Ni, 0.8% Cb, 2% max. Mn, 1% max. Si, 0.08% max. C	...	....	....	....	....	9.3 (200°F) 12.8 (930°F)	0.12 (32–212°F)	2550–2600

The resistance to heat transfer in the contact areas between solid materials, such as between fuel and cladding, will either be neglected or taken into account via a conductance term. Ignoring the resistance is not a very severe assumption in view of the good contact engineered in the fuel-element fabrication. In the case of uranium oxide fuel, a helium-filled gap exists between fuel and cladding and its resistance must be included in computations.

Other assumptions will be outlined when specific cases are treated. In the next section we shall derive the general heat conduction equations that are applicable to nuclear (and other) elements with heat generation.

5-3. THE HEAT-CONDUCTION EQUATIONS

Figure 5-1 shows a stationary cartesian volume $\Delta x \Delta y \Delta z$ within a heat

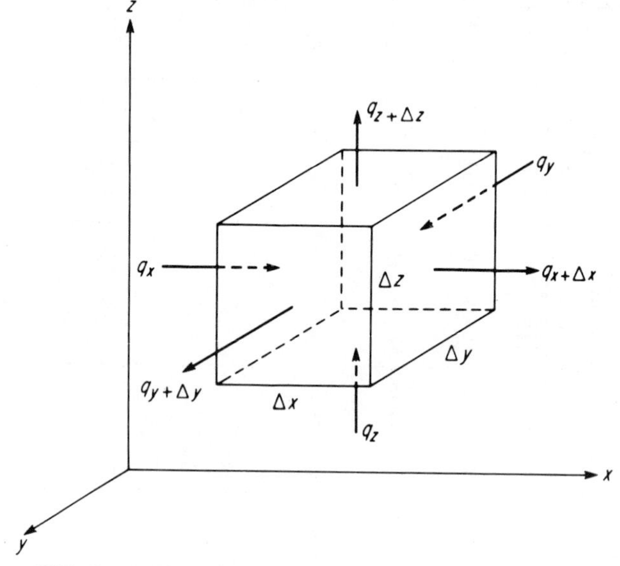

FIG. 5-1. Three-dimensional heat conduction in cartesian coordinates.

generating and conductive element. A heat balance on this volume during an increment of time $\Delta\theta$ can be written as*:

$$\begin{pmatrix}\text{Change in internal energy of}\\ \text{material within volume during } \Delta\theta\end{pmatrix} = \begin{pmatrix}\text{Heat conducted into}\\ \text{volume during } \Delta\theta\end{pmatrix} - \begin{pmatrix}\text{Heat conduced}\\ \text{out during } \Delta\theta\end{pmatrix} + \begin{pmatrix}\text{Heat generated}\\ \text{during } \Delta\theta\end{pmatrix}$$

* Note the similarity to the derivation of the neutron conservation equation, Sec. 3-3.

or

$$(U^{\theta+\Delta\theta} - U^{\theta}) = (q_x + q_y + q_z)\,\Delta\theta - (q_{x+\Delta x} + q_{y+\Delta y} + q_{z+\Delta z})\,\Delta\theta + \Delta x\,\Delta y\,\Delta z\,q'''\,\Delta\theta \tag{5-1}$$

The symbols U^{θ} and $U^{\theta+\Delta\theta}$ are the internal energies of the material within the volume at time θ and $\theta+\Delta\theta$ respectively, and q''' is the volumetric thermal source strength, Btu/hr ft³. The internal energies can be written as

$$U^{\theta} = \Delta x\,\Delta y\,\Delta z\,\rho u^{\theta} \tag{5-2}$$

and

$$U^{\theta+\Delta\theta} = \Delta x\,\Delta y\,\Delta z\,\rho u^{\theta+\Delta\theta} = \Delta x\,\Delta y\,\Delta z\left(\rho u^{\theta} + \frac{\partial\,\rho u^{\theta}}{\partial\theta}\,\Delta\theta\right) \tag{5-3}$$

where ρ is the density of the material of the element, lb_m/ft^3 and u is the specific internal energy Btu/lb_m. Noting that $\partial u = c\partial t$, where c is the specific heat of the element, Btu/lb_m °F, the above two equations can be combined, for constant ρ and c into the form

$$U^{\theta+\Delta\theta} - U^{\theta} = \Delta x\,\Delta y\,\Delta z\,\rho c\left(\frac{\partial t}{\partial\theta}\right)\Delta\theta \tag{5-4}$$

The symbols q_x and $q_{x+\Delta x}$ are the rates of heat conducted in the x direction, perpendicular to area $\Delta y\Delta z$, at faces x and $x+\Delta x$ respectively, Btu/hr. They can be written, with the help of the *Fourier* equation

$$q_x = -\,kA\,\frac{\partial t}{\partial x} \tag{5-5}$$

as

$$q_x = -\,k\,\Delta y\,\Delta z\,\frac{\partial t}{\partial x} \tag{5-6}$$

and

$$q_{x+\Delta x} = q_x + \frac{\partial q_x}{\partial x}\,\Delta x = -\,k\,\Delta y\,\Delta z\,\frac{\partial t}{\partial x} - \Delta y\,\Delta z\,\frac{\partial}{\partial x}\left(k\,\frac{\partial t}{\partial x}\right)\Delta x \tag{5-7}$$

and for constant thermal conductivity k

$$= -\,k\,\Delta y\,\Delta z\,\frac{\partial t}{\partial x} - k\,\Delta y\,\Delta z\left(\frac{\partial^2 t}{\partial x^2}\right)\Delta x \tag{5-8}$$

Thus

$$q_x - q_{x+\Delta x} = \Delta x\,\Delta y\,\Delta z\,k\left(\frac{\partial^2 t}{\partial x^2}\right) \tag{5-9}$$

Writing similar expressions for the heat conducted in the y and z directions, substituting with Eq. 5-4 into Eq. 5-1 and rearranging give

$$\Delta x\,\Delta y\,\Delta z k\left(\frac{\partial^2 t}{\partial x^2}+\frac{\partial^2 t}{\partial y^2}+\frac{\partial^2 t}{\partial z^2}\right)\Delta\theta+\Delta x\,\Delta y\,\Delta z\, q'''\,\Delta\theta$$
$$=\Delta x\,\Delta y\,\Delta z \rho c\left(\frac{\partial t}{\partial \theta}\right)\Delta\theta \tag{5-10}$$

Dividing the entire equation by $\Delta x \Delta y \Delta z \Delta\theta$ and noting that $k/\rho c=\alpha$, the *thermal diffusivity*, ft²/hr, the above equation reduces to the *general heat-conduction equation in cartesian coordinates:*

$$\left(\frac{\partial^2 t}{\partial x^2}+\frac{\partial^2 t}{\partial y^2}+\frac{\partial^2 t}{\partial z^2}\right)+\frac{q'''}{k}=\frac{1}{\alpha}\frac{\partial t}{\partial \theta} \tag{5-11}$$

Putting

$$\nabla^2 t=\frac{\partial^2 t}{\partial x^2}+\frac{\partial^2 t}{\partial y^2}+\frac{\partial^2 t}{\partial z^2} \tag{5-12}$$

where $\nabla^2 t$ above is the *Laplacian operator* of temperature in *cartesian coordinates.* Substituting into Eq. 5-11 results in the *general heat-conduction* equation:

$$\nabla^2 t+\frac{q'''}{k}=\frac{1}{\alpha}\frac{\partial t}{\partial \theta} \tag{5-13}$$

The special case of steady-state heat conduction, where $\partial t/\partial\theta=0$, the general equation becomes the *Poisson* equation:

$$\nabla^2 t+\frac{q'''}{k}=0 \tag{5-14}$$

The special case of no heat generation, where $q'''=0$, gives the *Fourier equation**

$$\nabla^2 t=\frac{1}{\alpha}\frac{\partial t}{\partial \theta} \tag{5-15}$$

The special case of steady state and no heat generation gives the *Laplace* equation

$$\nabla^2 t=0 \tag{5-16}$$

Another equation of interest is the *Helmholtz* equation†:

$$\nabla^2 t+B^2 t=0 \tag{5-17}$$

* Equations 5-5 and 5-15 share the name of Fourier.
† Recall the reactor equation, Sec. 3-3.

where B^2 is a constant. This is a form of the Poisson equation in which the constant in it (the heat generation term) is replaced by a linear function of the dependent variable t. A solution of this special case will be given in Sec. 6-12.

Because most fuel elements are cylindrical, the cylindrical coordinate system, Fig. 5-2, is particularly important. Only the Laplacian in the

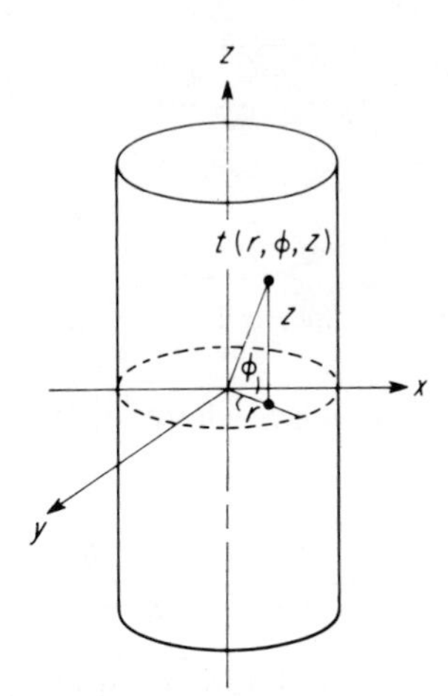

FIG. 5-2. Cylindrical coordinate system.

above five basic equations changes. In the cylindrical case, x becomes $r \cos \varphi$, y becomes $r \sin \varphi$, z remains as is, and Eq. 5-12 transforms into the *Laplacian operator* of temperature in *cylindrical* or *polar coordinates*:

$$\nabla^2 t = \left(\frac{\partial^2 t}{\partial r^2} + \frac{1}{r} \frac{\partial t}{\partial r} + \frac{1}{r^2} \frac{\partial^2 t}{\partial \varphi^2} + \frac{\partial^2 t}{\partial z^2} \right) \tag{5-18}$$

In the one-dimensional case, where $\partial t / \partial z = 0$ and $\partial t / \partial \varphi = 0$, the above equation results in

$$\nabla^2 t = \left(\frac{d^2 t}{dr^2} + \frac{1}{r} \frac{dt}{dr} \right) \tag{5-19}$$

Spherical fuel elements as used in pebble-bed reactors, are also of interest. In the spherical coordinate system, Fig. 5-3, x becomes $r \sin \psi \cos \varphi$, y becomes $r \sin \psi \sin \varphi$ and z becomes $r \cos \psi$. Equation 5-12 transforms into the *Laplacian operator* of temperature in *spherical coordinates*:

$$\nabla^2 t = \left(\frac{\partial^2 t}{\partial r^2} + \frac{2}{r} \frac{\partial t}{\partial r} + \frac{1}{r^2 \tan \psi} \frac{\partial t}{\partial \psi} + \frac{1}{r^2} \frac{\partial^2 t}{\partial \psi^2} + \frac{1}{r^2 \sin^2 \psi} \frac{\partial^2 t}{\partial \varphi^2} \right) \tag{5-20}$$

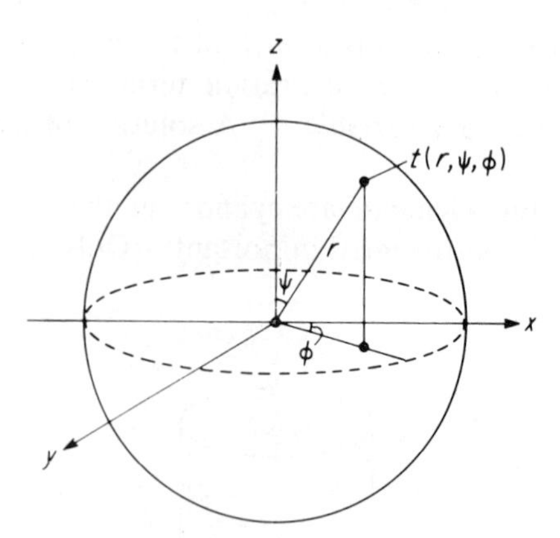

FIG. 5-3. Spherical coordinate system.

In the one-dimensional case where $\partial t/\partial\psi = 0$ and $\partial t/\partial\varphi = 0$, the above equation results in

$$\nabla^2 t = \left(\frac{d^2t}{dr^2} + \frac{2}{r}\frac{dt}{dr}\right) \tag{5-21}$$

The general-conduction, Poisson, Fourier, Laplace, and Helmholtz equations can now be obtained for cylindrical and spherical as well as for cartesian coordinates. Tables 5-3a and 5-3b list the above derived equations.

The most important equations in nuclear work are the Poisson equation (in the steady state) and the general conduction equation (in the unsteady state). They will be solved for various cases of interest. In the remainder of this chapter, simple one-dimensional steady-state cases will be solved. In subsequent chapters, more complicated two-dimensional steady-state and unsteady-state cases will be solved.

In this chapter short sections of fuel elements of simple shapes will be treated. The height of a fuel element is generally equal to the height of the active core. The thickness and width, however, are small compared with the horizontal dimensions of the core, as shown in Fig. 4-7. Thus while the axial neutron flux distribution (in the z direction) in the fuel element varies in the same manner as that in the core, the radial distributions (in the x and y directions) in the fuel element show relatively small change. The small change is due to both the overall core-flux radial gradient and the fuel element acting as a sink for thermal neutrons, causing a slight dip in flux, Fig. 4-7. These two effects, however, are small and the neutron flux φ, and consequently the volumetric thermal

TABLE 5-3a
Differential Equations of Temperature in Heat Conduction

Equation name	Conduction Mode	Equation
General conduction	Transient with heat generation	$\nabla^2 t + \frac{q'''}{k} = \frac{1}{\alpha}\frac{\partial t}{\partial \theta}$
Poisson	Steady state with heat generation	$\nabla^2 t + \frac{q'''}{k} = 0$
Fourier	Transient with no heat generation	$\nabla^2 t = \frac{1}{\alpha}\frac{\partial t}{\partial \theta}$
Laplace	Steady state with no heat generation	$\nabla^2 t = 0$
Helmholtz	Steady state with linear function of temperature term	$\nabla^2 t + B^2 t = 0$

TABLE 5-3b
The Laplacian of Temperature, $\nabla^2 t$

Coordinates	Cartesian	Cylindrical	Spherical
3-dimensional	$\frac{\partial^2 t}{\partial x^2} + \frac{\partial^2 t}{\partial y^2} + \frac{\partial^2 t}{\partial z^2}$	$\frac{\partial^2 t}{\partial r^2} + \frac{1}{r}\frac{\partial t}{\partial r} + \frac{1}{r^2}\frac{\partial^2 t}{\partial \varphi^2} + \frac{\partial^2 t}{\partial z^2}$	$\frac{\partial^2 t}{\partial r^2} + \frac{2}{r}\frac{\partial t}{\partial r} + \frac{1}{r^2 \tan \psi}\frac{\partial t}{\partial \psi} + \frac{1}{r^2}\frac{\partial^2 t}{\partial \psi^2} + \frac{1}{r^2 \sin^2 \psi}\frac{\partial^2 t}{\partial \varphi^2}$
1-dimensional (in x or r)	$\frac{d^2 t}{dx^2}$	$\frac{d^2 t}{dr^2} + \frac{1}{r}\frac{dt}{dr}$	$\frac{d^2 t}{dr^2} + \frac{2}{r}\frac{dt}{dr}$

source strength q''', shall be considered constant in any one cross section of a fuel element.

When short sections of fuel elements (in the z direction) compared with the core height are considered, there will be no appreciable change in φ or q''' in the z direction in such sections. Consideration of axial φ and q''' change in the entire length of a fuel element will be taken up in Chapter 13.

5-4. HEAT FLOW OUT OF SOLID-PLATE-TYPE FUEL ELEMENTS

Figure 5-4a shows a thin, bare (unclad) plate-type fuel element of thermal conductivity k_f and constant cross-sectional area. The dimensions of the element are large in the y and z directions compared with that in the x direction, so that heat flow can be considered one-dimen-

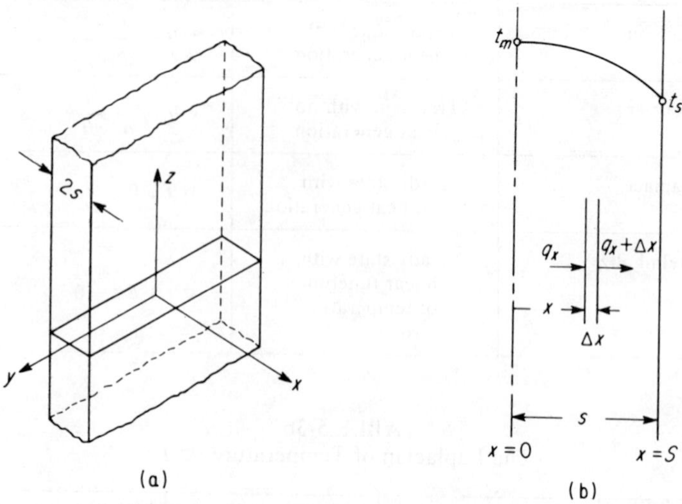

FIG. 5-4. Unclad thin plate-type fuel element.

sional, i.e., in the x direction only. Two-dimensional heat flow in such a geometry will be treated in Chapter 7.

Because φ and q''' are constant over the element cross section, heat is conducted equally in the $+x$ and $-x$ directions, and the midplane $x = 0$ (Fig. 5-4b) is the plane of highest temperature. We shall then simply treat heat flow only in one half of the element, the $+x$ direction.

The heat-conduction equation for this case is the one-dimensional Poisson equation

$$\frac{d^2t}{dx^2} + \frac{q'''}{k_f} = 0 \tag{5-22}$$

The same equation can be obtained by a heat balance. It is instructive to do this. Considering a thin layer of thickness Δx at distance x from the midplane, Fig. 5-4b, a heat balance may be written for the case of steady-state heat transfer as follows:

$$\begin{pmatrix}\text{Heat crossing}\\ \text{plane } x + \Delta x\end{pmatrix} = \begin{pmatrix}\text{Heat crossing}\\ \text{plane } x\end{pmatrix} + \begin{pmatrix}\text{Heat generated}\\ \text{in layer}\end{pmatrix}$$

Thus

$$q_{x+\Delta x} = q_x + q''' A\,\Delta x \tag{5-23}$$

But
$$q_x = -k_f A \frac{dt}{dx}$$

and
$$q_{x+\Delta x} = q_x + \frac{dq_x}{dx} \Delta x$$

$$= \left(-k_f A \frac{dt}{dx}\right) + \left(-k_f A \frac{d^2 t}{dx^2} \Delta x\right)$$

where A is the area of the layer in the yz plane, perpendicular to the direction of heat flow. Equation 5-23 can now be written as

$$q''' A \,\Delta x = -k_f A \frac{d^2 t}{dx^2} \Delta x$$

which reduces to

$$\frac{d^2 t}{dx^2} = -\frac{q'''}{k_f} \tag{5-22}$$

Equation 5-22 indicates that the temperature profile in x-direction has constant curvature, i.e., it is parabolic, and since q''' and k_f are considered constant and are positive, it is convex. Equation 5-22 now integrates twice to give

$$\frac{dt}{dx} = -\frac{q'''}{k_f} x + C_1 \tag{a}$$

and
$$t = -\frac{q'''}{2k_f} x^2 + C_1 x + C_2 \tag{b}$$

where C_1 and C_2 are the constants of integration. Because of symmetry around the midplane and because there is equal and opposite and consequently no net heat flow at the midplane, t_m, the temperature at the midpoint, is the maximum temperature in the section. The boundary conditions are therefore

$$\frac{dt}{dx} = 0 \qquad \text{at } x = 0 \tag{5-24a}$$

and
$$t = t_m \qquad \text{at } x = 0 \tag{5-24b}$$

Thus $C_1 = 0$ and $C_2 = t_m$. Substituting the values for C_1 and C_2 into Eqs. (a) and (b) gives

$$\frac{dt(x)}{dx} = -\frac{q'''}{k_f} x \tag{5-25}$$

and
$$t(x) = t_m - \frac{q'''}{2k_f} x^2 \tag{5-26}$$

The temperature at the surface, t_s, can be obtained by putting $x = s$, where s is equal to half the element thickness in the x direction. Thus

$$t_s = t_m - \frac{q'''}{2k_f} s^2 \qquad (5\text{-}27)$$

q_x, the heat conducted past any plane x, is equal to the total heat generated between $x = 0$ and x. Thus

$$q_x = q''' A x \qquad (5\text{-}28)$$

The same expression can be derived from $-k_f A \, dt/dx|_x$, where dt/dx is given by Eq. 5-25.

q_s, the heat conducted out of one surface ($x = s$), equal to the heat generated from one-half of the element, is given by

$$q_s = q''' A s \qquad (5\text{-}29a)$$

Another expression of q_s in terms of the boundary temperatures can be obtained by combining Eqs. 5-27 and 5-29a and rearranging to give

$$q_s = 2k_f A \frac{t_m - t_s}{s} \qquad (5\text{-}29b)$$

showing that the heat transfer by conduction in a system with uniform internal heat generation is twice that without heat generation for the same path length s and overall temperature drop. This is because heat, produced uniformly throughout the element, has, on an average, to travel only one-half the distance between $x = 0$ and $x = s$.

The total heat generated in the two halves of the element q_{2s} (Btu/hr) is twice that given by Eq. 5-29b. If A is replaced by the total surface area of the fuel element A_s (neglecting the edges), where $A_s = 2A$, it is given by

$$q_{2s} = 2k_f A_s \frac{t_m - t_s}{s} \qquad (5\text{-}30)$$

Effect of Cladding

Figure 5-5 shows a system containing a plate-type fuel element of half thickness s with cladding of thickness c and thermal conductivity k_c. The equations for the temperature variation and heat conduction within the fuel element, i.e., where $x \leqslant s$, are the same as those given above. Because of the presence of the cladding, an extra, though usually small, resistance to heat conduction results in the decrease of q_s for the same overall temperature drop (i.e., if $t_m - t_c$ were the same as $t_m - t_s$ in the previous case).

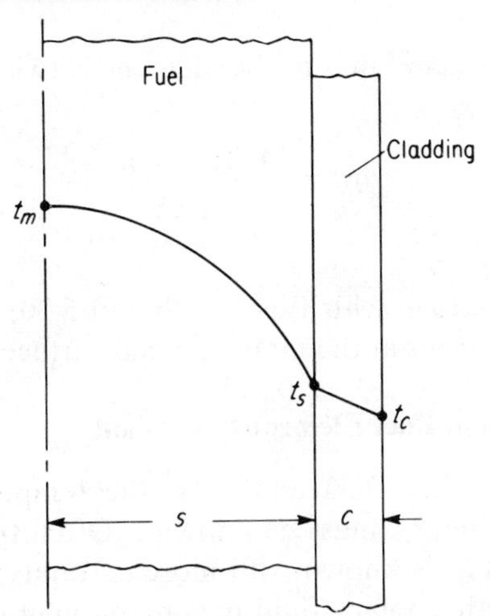

FIG. 5-5. Temperature profile in clad fuel plate ($k_f < k_c$).

In the steady state and with the assumption of no heat generated in the cladding material, the amount of heat leaving surface s must be the same as that leaving surface c. For a constant k_c, dt/dx through the cladding is constant. (The ratio of dt/dx in the fuel at $x = s$ to dt/dx in the cladding is inversely proportional to the ratio of their thermal conductivities.) Also neglecting resistance to heat flow at the fuel-cladding interface, we can write

$$q_s = q'''As = 2k_f A \frac{t_m - t_s}{s} = -k_c A \left.\frac{dt}{dx}\right|_{\text{clad.}} = k_c A \frac{t_s - t_c}{c} \tag{5-31}$$

Solving for the temperature differences,

$$t_m - t_s = \frac{q_s s}{2k_f A} = \frac{q''' s^2}{2k_f}$$

and

$$t_s - t_c = \frac{q_s c}{k_c A} = \frac{q''' sc}{k_c}$$

Adding,

$$t_m - t_c = \frac{q_s}{A}\left(\frac{s}{2k_f} + \frac{c}{k_c}\right) = \frac{q''' s^2}{2k_f} + \frac{q''' sc}{k_c} \tag{5-32}$$

Rearranging gives q_s in terms of the temperatures at the boundaries of the system, that is, t_m and t_c. Thus

$$q_s = \frac{A(t_m - t_c)}{\dfrac{s}{2k_f} + \dfrac{c}{k_c}} \tag{5-33}$$

Again the total heat generated in the element is twice the above amount or is equal to

$$q_{2s} = \frac{A_s(t_m - t_c)}{\dfrac{s}{2k_f} + \dfrac{c}{k_c}} \tag{5-34}$$

Compare these equations with Eqs. 5-29b and 5-30, where q_s is given in terms of the temperatures at the midplane and surface of the fuel element.

Heat Transfer from Fuel Element to Coolant

In order to use Eqs. 5-33 and 5-34, the temperature at the outer surface of the cladding, t_c, must be known. Often the bulk temperature of the coolant fluid, t_f, is known with more certainty. Figure 5-6 shows a clad fuel plate with coolant fluid passing parallel to it. Again, in the

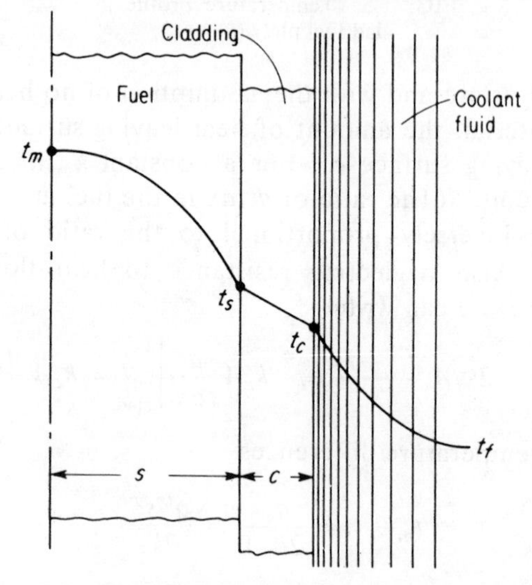

FIG. 5-6. Temperature profile in clad plate-type fuel element with coolant.

steady state, with no heat produced in cladding or coolant, the heat passing out of the fuel-plate surface is the same as that passing out of the cladding's outer surface and into the coolant.

The heat passing from the cladding to the coolant is by convection. This may be natural or forced convection. In either case, the coefficient of heat transfer is given the symbol h and commonly has the units of Btu/hr ft² °F. Recommended methods for the evaluation of h in different systems will be presented in Chapters 9 and 10.

If q_s is again the heat leaving the fuel element (see Fig. 5-6), the following relationships hold:

$$q_s = q''' As = 2k_f A \frac{t_m - t_s}{s} = k_c A \frac{t_s - t_c}{c} = hA(t_c - t_f)$$

Solving for the temperature differences,

$$t_m - t_s = \frac{q_s s}{2k_f A} = \frac{q''' s^2}{2k_f} \tag{5-35}$$

$$t_s - t_c = \frac{q_s c}{k_c A} = \frac{q''' sc}{k_c} \tag{5-36}$$

and

$$t_c - t_f = \frac{q_s}{hA} = \frac{q''' s}{h} \tag{5-37}$$

Adding,

$$t_m - t_f = \frac{q_s}{A}\left(\frac{s}{2k_f} + \frac{c}{k_c} + \frac{1}{h}\right) = \frac{q''' s^2}{2k_f} + q''' s\left(\frac{c}{k_c} + \frac{1}{h}\right) \tag{5-38a}$$

which may be written as

$$q_s = \frac{t_m - t_f}{\dfrac{s}{2k_f A} + \dfrac{c}{k_c A} + \dfrac{1}{hA}} \tag{5-38b}$$

Using familiar electrical analogy, Eq. 5-38b shows that the heat flow in Btu/hr (analogous to current in amperes or coulomb/sec) is equal to a temperature difference in °F (analogous to voltage difference in volts), divided by the sum of resistances to heat flow in hr°F/Btu (analogous to resistances in ohms) in series. If there are other resistances to heat flow, such as that due to a helium gas layer between the fuel material and cladding (common in rod-type fuel elements containing UO_2), or due to the bonding between metallic fuel and cladding, they must be added to the denominator. The latter is usually given in the form $1/h_b A$, where h_b is a bonding conductance (Btu/hrft^2°F) which is a complex function of the type of bonding and the materials bonded. In most cases, good bonding exists between fuel and cladding, so that h_b is large and the resistance $1/h_b A$ is negligible by comparison to the other resistances.

In Eqs. 5-38, if only t_m and t_f are known or assumed, t_s and t_c can be obtained by calculating q_s and then substituting into Eqs. 5-35, 5-36 or 5-37. If k_f and k_c are strong functions of temperature, t_s and t_c may have to be first assumed and an iterative solution becomes necessary.

5-5. THE INTERDEPENDENCE OF TEMPERATURE, HEAT TRANSFER AND HEAT (OR NEUTRON) FLUX

Equations 5-38 may be used to explain some of the limitations on heat generation in nuclear reactors. For any constant value of q''', if t_f is to be kept as high as possible for good plant thermal efficiency, h has to be increased materially (since it affects only part of the equation) to keep the maximum fuel temperature t_m or the maximum cladding temperature t_s from becoming excessively high. On the other hand, if for metallurgical or other reasons these temperatures are limited to certain values, h again has to be increased materially to allow a high value of q'''. The value of q''' may be regulated by changing the neutron flux with the help of the control rods. In essence, there is no limit to the quantity of heat that a nuclear reactor is capable of generating, so long as adequate cooling is provided to keep the temperatures in the system from exceeding their safe limits.

The interdependence, shown for a particular case in Fig. 5-7, can best be illustrated by a numerical example.

Example 5-1. A plate-type fuel element is made of 1.5 per cent enriched uranium metal. The element is 4 ft long, 3.5 in. wide, and 0.2 in. thick. It is clad in Type 304L stainless steel, 0.005 in. thick. The effective thermal-neutron flux at the point of maximum temperature (slightly above the center of the core; see Sec. 13-4) is 3.0×10^{13} neutrons/sec cm^2. For good plant efficiency, the coolant bulk temperature at that point should be no lower than 600°F. Find (*a*) the heat transferred per unit area of element at that point, q''; (*b*) the minimum value of h if the fuel temperature should not exceed 700°F; (*c*) the maximum cladding and coolant temperatures; (*d*) the minimum value of h if the fuel temperature should not exceed 700°F, but for neutron fluxes of 2×10^{13} and 4×10^{13}. The effective fission cross section $\bar{\sigma}_f$ is 364 b.

Solution. The physical properties of the fuel and cladding are evaluated at their respective mean temperatures. Since these are not known, the boundary temperatures are assumed as follows:

$$t_m = 700°\text{F} \qquad \text{given}$$
$$t_s = 660°\text{F} \qquad \text{assumed}$$
$$t_c = 650°\text{F} \qquad \text{assumed}$$
$$t_f = 600°\text{F} \qquad \text{given}$$

Thus mean fuel and cladding temperatures are 680 and 655°F, respectively.

ρ, density of uranium metal at 680°F = 18.71 g_m/cm^3 (Table 4-2)

k_f, conductivity at same temperature = 18.5 Btu/hr ft °F (Table 5-1)

k_c of 304L stainless steel at 655°F = 11.36 Btu/hr ft °F (Table 5-2)

Therefore

$$N_{ff} = \frac{0.60225 \times 10^{24}}{235.044} \times 18.71 \times 0.015 = 7.19 \times 10^{20} \text{ U}^{235} \text{ nuclei/cm}^3$$

(*a*) Using Eq. (4-3),

$$q''' = 180 \times 7.19 \times 10^{20} \times 364 \times 10^{-24} \times 3 \times 10^{13} = 1.413 \times 10^{15} \text{ Mev/sec cm}^3$$
$$= 1.413 \times 10^{15} \times 1.5477 \times 10^{-8} = 2.185 \times 10^{7} \text{ Btu/hr ft}^3$$

$$\text{Fuel-element surface area per unit fuel volume} = \frac{(3.5 + 2 \times 0.005)2 \times 12}{3.5 \times 0.2} = 120.34 \text{ ft}^2/\text{ft}^3$$

$$q'' = \text{heat transferred per unit of fuel-element surface area} = \frac{2.185 \times 10^7}{120.34} = 181{,}570 \text{ Btu/hr ft}^2$$

(*b*) Using Eq. (5-38),

$$700 - 600 = \frac{2.185 \times 10^7 \times (0.1/12)^2}{2 \times 18.5} + 2.185 \times 10^7 \times \frac{0.1}{12}\left(\frac{0.005}{12 \times 11.36} + \frac{1}{h}\right)$$

from which

$$h\,(\text{for } \varphi = 3 \times 10^{13}) = 3480 \text{ Btu/hr ft}^2 \text{ °F}$$

(*c*) By Eq. (5-35),

$$700 - t_s = \frac{2.185 \times 10^7 \times (0.1/12)^2}{2 \times 18.5} = 41.3\text{°F}$$

Thus

$$t_s = 700 - 41.3 = 658.7\text{°F}$$

By Eq. (5-37),

$$t_c - 600 = \frac{2.185 \times 10^7 \times 0.1/12}{3480} = 52.3\text{°F}$$

Thus

$$t_c = 600 + 52.3 = 652.3\text{°F}$$

The temperature t_c is the highest coolant temperature at the section in question. If boiling is to be avoided, the coolant pressure at that section should be higher than the saturation pressure corresponding to that temperature. We shall see later that, even if this condition is satisfied, local hot spots may develop in the fuel and cladding, causing some surface boiling to take place.

At this point these calculated boundary temperatures are compared with the ones assumed at the beginning of the solution. If the differences had been large enough to alter materially the physical properties used in the calculation, a second iteration would be necessary. In Example 5-1, the differences are sufficiently small, and the answers are therefore sufficiently accurate.

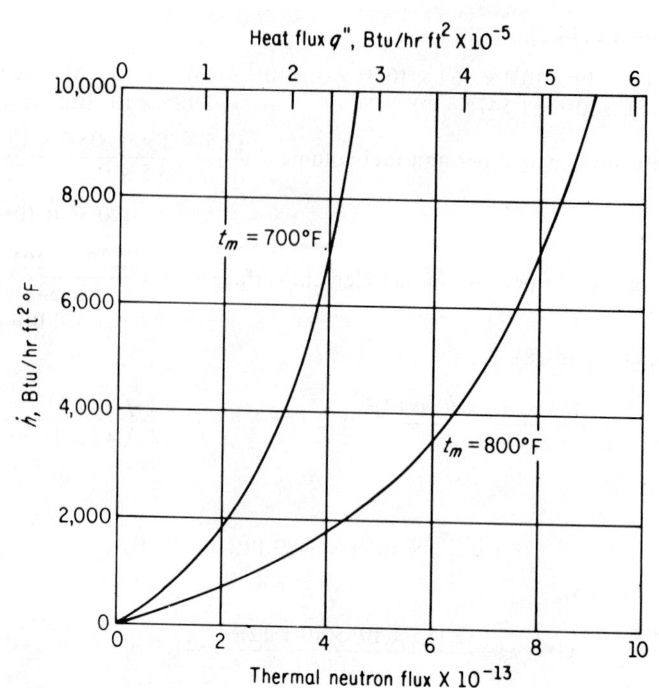

FIG. 5-7. Variation of h with φ and q''' for fuel element and coolant temperatures of Examples 5-1 and 5-2.

(*d*) By the same method used in part (*b*),

$$h(\text{for } \varphi = 2 \times 10^{13}) = 1776 \text{ Btu/hr ft}^2 \text{ °F}$$

and

$$h(\text{for } \varphi = 4 \times 10^{13}) = 6662 \text{ Btu/hr ft}^2 \text{ °F}$$

It can be seen that these values of h, the minimum necessary to keep the maximum fuel temperature within 700°F, increase rapidly with neutron flux φ. If φ is increased beyond a certain point, however, it is impossible to maintain the specified maximum fuel temperature (unless the coolant temperature is lowered), no matter what value of heat-transfer coefficient is designed into the system. This theoretical maximum value of flux, φ_{max}, is obtained by putting $h = \infty$ in Eq. 5-38a and evaluating a corresponding value of volumetric thermal source strength, q'''_{max}, to give

$$q'''_{max} = \frac{t_m - t_f}{\dfrac{s^2}{2k_f} + \dfrac{sc}{k_c}} \tag{5-39}$$

(In effect, this makes the temperature of the coolant equal to the temperature of the cladding's outer surface.)

φ_{max} can then be evaluated from q'''_{max} by the use of Eq. 4-3. Equation 5-39 shows that, to increase φ beyond φ_{max} for a given fuel and cladding configuration, it is necessary to lower t_f (with its undesirable effect on plant thermodynamic efficiency) or to increase t_m (at the expense of reduced fuel burnup or otherwise by fuel alloying or using ceramic fuels), or both.

Example 5-2. Calculate the maximum possible neutron flux corresponding to the data of Example 5-1.

Solution. Using Eq. 5-39,

$$q'''_{max} = \frac{700 - 600}{\dfrac{(0.1/12)^2}{2 \times 18.5} + \dfrac{0.1/12 \times 0.005/12}{11.36}} = 4.59 \times 10^7 \text{ Btu/hr ft}^3$$

$$= \frac{4.59 \times 10^7}{1.5477 \times 10^{-8}} = 2.97 \times 10^{15} \text{ Mev/sec cm}^3$$

$$\varphi_{max} = \frac{q'''_{max}}{180 N_{ff} \bar{\sigma}_f} = \frac{2.97 \times 10^{15}}{180 \times 7.19 \times 10^{20} \times 364 \times 10^{-24}} = 6.30 \times 10^{13}$$

Figure 5-7 shows the variation of h with φ and q''' for the fuel element and coolant temperature of Examples 5-1 and 5-2, and for $t_m = 700°F$ and $t_m = 800°F$. It can be seen that h rises sharply beyond a certain value of φ. Recognizing the fact that large values of h are produced via high coolant velocities V(h proportional to $V^{0.8}$, other things being constant) and that pumping losses increase with V^2, we find that a point of diminishing returns is approached much before the maximum theoretical flux is reached.

5-6. HEAT FLOW OUT OF SOLID CYLINDRICAL FUEL ELEMENTS

Here again a short section of an unclad cylindrical fuel element, or *fuel rod* or *pin*, of a diameter small compared with that of the reactor core will be treated. The neutron flux will be assumed not to change appreciably in either the axial or radial directions. The heat flow out of the fuel element will substantially be radial and will be equal in all directions. The pertinent equation here is the one-dimensional Poisson equation in cylindrical coordinates, given by Eqs. 5-14 and 5-19. This equation can also be obtained by a heat balance on a cylindrical shell within the fuel element.

Figure 5-8 shows the cross section of an unclad element having a radius R and length (in the axial direction, not shown) L. Consider a thin cylindrical shell at r of thickness Δr. A heat balance may be written for the case of steady-state heat transfer as follows:

$$q''' 2\pi r \Delta r L = q_{r+\Delta r} - q_r \tag{5-40}$$

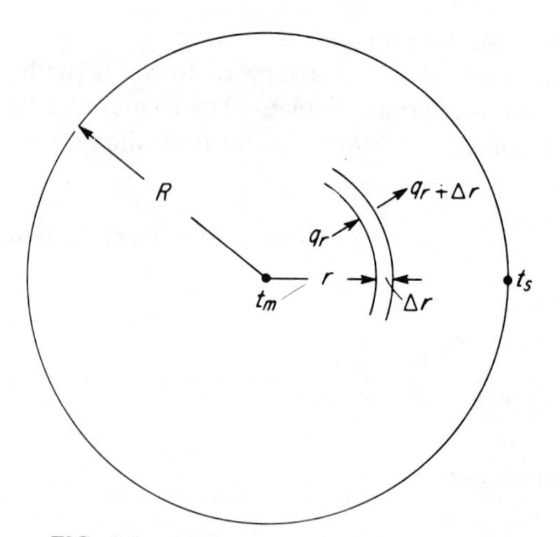

FIG. 5-8. Cross section of solid cylindrical fuel element.

where

$$q_r = -k_f A \frac{dt}{dr} = -2\pi L k_f r \frac{dt}{dr}$$

and

$$q_{r+\Delta r} = q_r + \frac{dq_r}{dr}\Delta r$$

$$= -2\pi L k_f r \frac{dt}{dr} - 2\pi L k_f \left(r \frac{d^2t}{dr^2} + \frac{dt}{dr}\right)\Delta r$$

Thus

$$q''' 2\pi r \Delta r L = -2\pi L k_f \left(r \frac{d^2t}{dr^2} + \frac{dt}{dr}\right)\Delta r$$

This reduces to

$$\frac{d^2t}{dr^2} + \frac{1}{r}\frac{dt}{dr} + \frac{q'''}{k_f} = 0 \tag{5-41}$$

The solution of this differential equation is

$$t = -q''' \frac{r^2}{4k_f} + C_1 \ln r + C_2 \tag{5-42}$$

C_1 and C_2, the constants of the double integration, are evaluated from the boundary conditions of the system, which are

$$\frac{dt}{dr} = 0 \qquad \text{at } r = 0 \tag{5-43a}$$

and

$$t = t_m \qquad \text{at } r = 0 \tag{5-43b}$$

so that $C_1 = 0$ and $C_2 = t_m$

and the solution expressing t in terms of r is

$$t(r) = t_m - \frac{q'''r^2}{4k_f} \tag{5-44}$$

The temperature distribution here, as in the flat plate, is parabolic. The maximum temperature difference in the system may be obtained by putting $r = R$ and $t = t_s$ in Eq. 5-44, so that

$$t_m - t_s = \frac{q'''R^2}{4k_f} \tag{5-45}$$

The heat flow $q(r)$ at any radius r is equal to the total heat generated by the element within that radius. Thus

$$q(r) = \pi r^2 L q''' \tag{5-46}$$

The heat conducted out of the periphery of the element q_s is equal to the heat generated in the entire element and is given by

$$q_s = \pi R^2 L q''' \tag{5-47}$$

Another expression of q_s in terms of the boundary temperatures can be obtained by combining Eqs. 5-45 and 5-47 and rearranging to give

$$q_s = 4\pi k_f L(t_m - t_s) \tag{5-48}$$

Equation 5-48 shows that in a fuel rod, the total heat generated per unit length is only a function of the maximum temperature drop $(t_m - t_s)$ and is independent of the volumetric thermal source strength q''' or the radius of the rod R. Putting $2\pi RL = A_s$, the peripheral area of the element through which the total heat generated by the element must pass, and rearranging give

$$q_s = 2k_f A_s \frac{t_m - t_s}{R} \tag{5-49}$$

Note the similarity in form between Eqs. 5-30 and 5-49 for the total heat transfer of a plate-type and a cylindrical fuel element, respectively.

Effect of Cladding and Coolant

Figure 5-9 shows the cross section of a cylindrical fuel element of radius R and axial length L, having a center temperature t_m and surrounded by cladding of radial thickness c and coolant fluid having bulk temperature t_f.

In the steady state with no heat generated in cladding or coolant, and

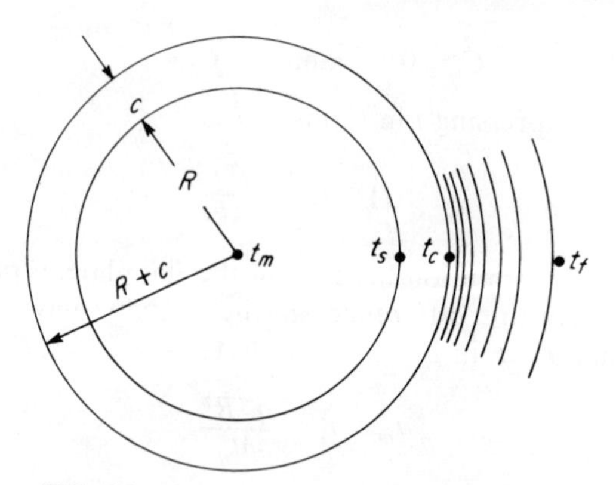

FIG. 5-9. Cross section of cylindrical fuel element with cladding and coolant.

with negligible resistance to heat flow at the fuel-cladding interface, the heat passing out of the fuel surface is the same as that passing through the cladding and into the coolant fluid. (The expression for heat flow through the cladding is that for heat flow by conduction through a hollow cylinder.) Thus

$$q_s = \pi R^2 L q''' = 4\pi k_f L(t_m - t_s) = \frac{2\pi k_c L(t_s - t_c)}{\ln|(R+c)/R|}$$
$$= 2\pi(R+c)Lh(t_c - t_f) \qquad (5\text{-}50)$$

Solving for the temperature differences and adding, in the same fashion as was done in the case of the plate-type fuel element, give

$$t_m - t_f = \frac{q''' R^2}{4k_f} + \frac{q''' R^2}{2}\left[\frac{1}{k_c}\ln\frac{R+c}{R} + \frac{1}{h(R+c)}\right] \qquad (5\text{-}51\text{a})$$

which may be written as

$$q_s = \frac{t_m - t_f}{\dfrac{R}{2k_f A_R} + \dfrac{c}{k_c A_m} + \dfrac{1}{hA_{R+c}}} \qquad (5\text{-}51\text{b})$$

where $A_R = 2\pi RL$, A_m is the logarithmic mean area of the cladding, given by $2\pi cL/\ln[(R+c)/c] = (A_{R+c} - A_R)/\ln(A_{R+c}/A_R)$, and $A_{R+c} = 2\pi(R+c)L$. In the usual case, where c is very small compared to R, A_m may be taken, with little error, to be the arithmetic mean area of the cladding $(A_R + A_{R+c})/2$, or $[2\pi R + 2\pi(R+c)]L/2$.

Compare this equation with Eq. 5-38b for a plate-type element. The same remarks regarding resistances to heat flow apply here, and any additional resistances must be taken into account. A notable example is when ceramic fuels are used. Unbonded fuel-cladding elements are built and the space between the fuel material (UO_2) and cladding is filled with an inert gas such as helium (during the jacket-welding process) or by a liquid metal. The space presents an extra resistance to heat flow. Because the thickness of this space is quite small, no convection effects, even in the case of a gas, need be taken into account, and heat transfer through it may be assumed to occur totally by conduction. In the case of the gas layer, thermal conductivity of the gas changes with fuel burnup because of the liberation of fission gases which mix with it. In the case of helium the effect is to decrease the gas thermal conductivity [27].

An expression for the maximum theoretical flux attainable with fixed configuration and given fuel and coolant temperatures can be obtained by a procedure similar to that for a plate-type fuel element.

5-7. HEAT FLOW OUT OF HOLLOW THICK CYLINDRICAL FUEL ELEMENTS

A hollow thin fuel element may be treated, with little error, as a flat plate, Sec. 5-4. The case of a hollow thick cylindrical element having an inner radius r_i and an outer radius r_o, Figs. 5-10 and 5-11, will be treated in this section. In the one-dimensional case, where heat flow is zero in the axial directions and equal in all radial directions, Eq. 5-42 applies. The boundary conditions, however depend upon whether or not the inner or outer surfaces are insulated, i.e., whether all heat generated flows out of the outer or inner surface, or both.

Heat Flow out of Outer Surface

In this case, the boundary conditions are

$$\frac{dt}{dr} = 0 \text{ at } r = r_i \tag{5-52a}$$

and

$$t = t_i \text{ at } r = r_i \tag{5-52b}$$

so that

$$C_1 = \frac{q'''r_i^2}{2k_f}$$

and

$$C_2 = t_i - \frac{q'''r_i^2}{2k_f}\left(\ln r_i - \frac{1}{2}\right)$$

and the solution is

$$t(r) = t_i - \frac{q'''r_i^2}{4k_f}\left[\left(\frac{r}{r_i}\right)^2 - 2\ln\left(\frac{r}{r_i}\right) - 1\right] \tag{5-53}$$

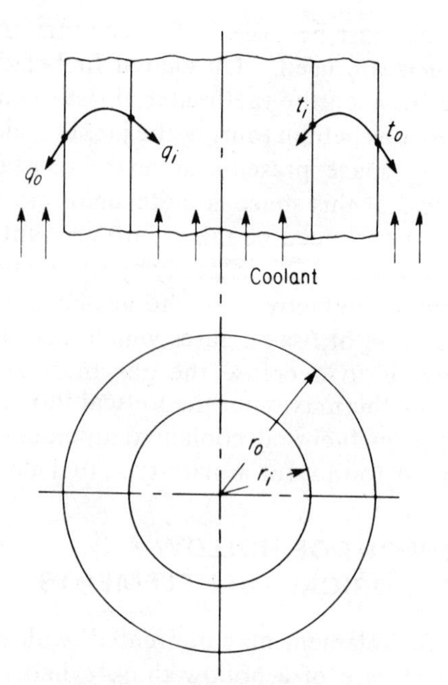

FIG. 5-10. An unclad hollow thick cylindrical fuel element with heat flow from inner and outer surfaces.

The value of t_0 can be obtained by putting $r = r_o$ in the above equation, giving

$$t_i - t_0 = \frac{q''' r_i^2}{4k_f} \left[\left(\frac{r_o}{r_i} \right)^2 - 2 \ln \left(\frac{r_o}{r_i} \right) - 1 \right] \tag{5-54}$$

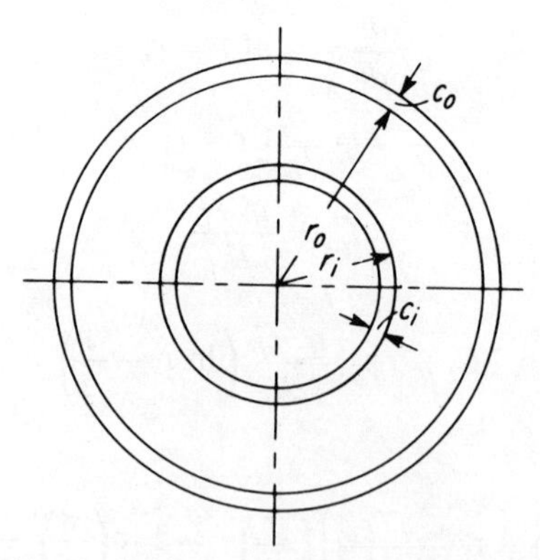

FIG. 5-11. A clad hollow thick cylindrical fuel element

q_{s_o}, the heat flow out of the outer surface of the element, is equal to the heat generated in the element or

$$q_{s_o} = \pi(r_o^2 - r_i^2)Lq''' \tag{5-55}$$

or by combining Eq. 5-54 and 5-55 and rearranging to give

$$q_{s_o} = 4\pi k_f L(t_i - t_0) \frac{\left(\frac{r_o}{r_i}\right)^2 - 1}{\left(\frac{r_o}{r_i}\right)^2 - 2\ln\left(\frac{r_o}{r_i}\right) - 1} \tag{5-56}$$

With cladding on the outer surface of thickness c_o and coolant at t_{f_o},

$$t_i - t_{f_o} = \frac{q'''r_i^2}{4k_f}\left[\left(\frac{r_o}{r_i}\right)^2 - 2\ln\left(\frac{r_o}{r_i}\right) - 1\right] + \frac{q'''r_i^2}{2}\left\{\left[\left(\frac{r_o}{r_i}\right)^2 - 1\right]\left[\frac{1}{k_c}\ln\frac{r_o + c_o}{r_o} + \frac{1}{h(r_o + c_o)}\right]\right\} \tag{5-57}$$

Heat Flow out of Inner Surface

In this case the boundary conditions are

$$\frac{dt}{dr} = 0 \text{ at } r = r_o \tag{5-58a}$$

and

$$t = t_o \text{ at } r = r_o \tag{5-58b}$$

and the solutions are

$$t(r) = t_0 - \frac{q'''r_o^2}{4k_f}\left[\left(\frac{r}{r_o}\right)^2 - 2\ln\left(\frac{r}{r_o}\right) - 1\right] \tag{5-59}$$

$$t_0 - t_i = \frac{q'''r_o^2}{4k_f}\left[\left(\frac{r_i}{r_o}\right)^2 - 2\ln\left(\frac{r_i}{r_o}\right) - 1\right] \tag{5-60}$$

and

$$q_{s_i} = 4\pi k_f L(t_0 - t_i) \frac{1 - \left(\frac{r_i}{r_o}\right)^2}{\left(\frac{r_i}{r_o}\right)^2 - 2\ln\frac{r_i}{r_o} - 1} \tag{5-61}$$

With cladding on the inner surface of thickness c_i and coolant at t_{f_i},

$$t_0 - t_{f_i} = \frac{q'''r_o^2}{4k_f}\left[\left(\frac{r_i}{r_o}\right)^2 - 2\ln\left(\frac{r_i}{r_o}\right) - 1\right] + \frac{q'''r_o^2}{2}\left\{\left[1 - \left(\frac{r_i}{r_o}\right)^2\right]\left[\frac{1}{k_c}\ln\frac{r_i}{r_i - c_i} + \frac{1}{h(r_i - c_i)}\right]\right\} \tag{5-62}$$

Heat Flow out of Both Surfaces

If the surface temperatures t_i at r_i and t_0 at r_o are known, the solution for the temperature would be

$$t(r) = t_i - \frac{q'''(r^2 - r_i^2)}{4k_f} + \left[(t_i - t_0) - \frac{q'''}{4k_f} (r_o^2 - r_i^2) \right] \left(\ln \frac{r_o}{r_i} \right) \left(\ln \frac{r}{r_i} \right) \tag{5-63}$$

5-8. HEAT FLOW OUT OF SPHERICALLY SHAPED FUEL

A heat-balance treatment analogous to that for cylindrical fuel elements for the case of a bare (unclad) sphere of radius R, shown in Fig. 5-12, and having a uniform volumetric thermal source strength q'''

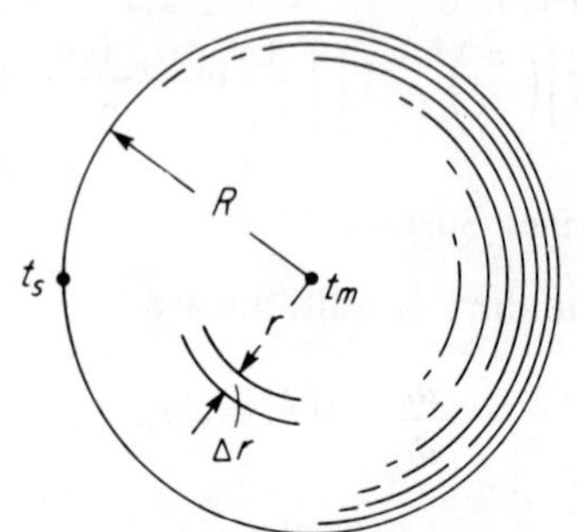

FIG. 5-12. Spherical fuel element.

yields the one-dimensional Poisson equation in spherical coordinates, also obtained by combining Eqs. 5-14 and 5-21 to give

$$\frac{d^2t}{dr^2} + \frac{2}{r}\frac{dt}{dr} + \frac{q'''}{k_f} = 0 \tag{5-64}$$

Integration of this equation for the boundary conditions $t(r)$ at r, $t = t_m$ at $r = 0$ and $t = t_s$ at $r = R$ gives

$$t(r) = t_m - \frac{q''' r^2}{6k_f} \tag{5-65}$$

and

$$t_m - t_s = \frac{q''' R^2}{6k_f} \tag{5-66}$$

The total heat generated, q_s, is given by the expression

$$q_s = \frac{4}{3}\pi R^3 q''' \tag{5-67}$$

Combining Eqs. 5-66 and 5-67 gives

$$q_s = 8\pi R k_f (t_m - t_s) \tag{5-68}$$

Putting $4\pi R^2 = A_s$, the total peripheral area of the spherical element, and rearranging give

$$q_s = 2k_f A_s \frac{t_m - t_s}{R} \tag{5-69}$$

Again note the similarity in form between this expression and those given by Eqs. 5-30 and 5-49.

PROBLEMS

5-1. A heavy-water-cooled research reactor is fueled with natural-uranium metal rods 0.9 in. in diameter clad with 0.05-in.-thick aluminum. The maximum temperature in the fuel at a certain cross section in the rod is 700°F. The coolant bulk temperature at the cross section is 180°F. The heat-transfer coefficient is 5000 Btu/hr ft² °F. Determine for the above cross section (*a*) the neutron flux (thermal); (*b*) the specific power, kw/kg_m fuel; (*c*) the maximum cladding temperature; (*d*) the maximum coolant temperature; (*c*) the minimum coolant pressure to avoid boiling; and (*f*) the theoretical maximum flux beyond which the above maximum fuel temperature cannot be maintained.

5-2. A 0.5-in.-diameter fuel element is made of 3 percent enriched UO_2. It is surrounded by a 0.003-in.-thick helium layer and a 0.03-in.-thick Zircaloy 2 cladding. A certain section of the element operates in boiling light water at 1,000 psia. The boiling heat-transfer coefficient is 10,000 Btu/hr ft² °F, and the temperature drop in the boiling film at the section is 30.4°F. Calculate the maximum fuel temperature at that section. $k_{He} = 0.16$, $k_{clad} = 8$ Btu/hr ft °F.

5-3. A thin cylindrical-shell fuel element is used in the superheater region of a boiling-water reactor. The element is 0.2 in. thick with an inside diameter of 4 in. Because of its thinness, however, it may be treated as a flat slab. At a particular section q''' is 50×10^6 Btu/hr ft³, $k_f = 10$ Btu/hr ft °F, the temperature of the steam vapor is 700°F on both sides, and the heat-transfer coefficients are 400 on one side and 280 Btu/hr ft² °F on the other (because of differences in geometry). Neglect cladding. Determine (*a*) the position within the element at which the maximum temperature occurs and (*b*) the temperature at both surfaces.

5-4. A UO_2-fuel manufacturer made fuel pellets 0.05 in. in diameter and 0.60 in. long placed end to end in a 0.032-in.-thick Zircaloy 2 can, 0.57 in. OD. The gap between fuel and cladding was filled with helium gas during the can welding process. The fuel was 1.5 percent enriched. The pellets were to be used in a 50-Mw(t) boiling-water reactor where the average core heat flux is 74,444 Btu/hr ft² of can OD. Four percent of the thermal output of the core is produced in the moderator. Because of manufacturing considerations, the manufacturer decided to reduce all diameters by 0.12 in. (cladding thickness remains constant). What should be the new average core heat flux (based on the same maximum fuel temperature and the same number of pellets)? The average coolant bulk temperature is 545°F, and the average boiling heat-transfer coefficient is 10,000 Btu/hr ft² °F. Use thermal conductivities of Zircaloy 2, fuel, and helium of 8.0, 1.15, and 0.135 Btu/hr ft °F, respectively.

5-5. A spherical reactor core is 5 ft in diameter. It contains 20 percent enriched UO_2 spherical fuel pellets 1 in. in diameter each. In one of these pellets, the temperature drop from center to edge of fuel is 3000°F. The maximum flux in the core is 10^{13}. What is the radial position of this pellet? Neglect extrapolation lengths. Take $k_f = 1.1$ Btu/hr ft °F, fuel density $= 11\ g_m/cm^3$, $\bar{\sigma}_f = 500$ b.

5-6. A SNAP (Systems for Nuclear Auxiliary Power) type electrical generator is fueled with 560 g_m of Pu^{238} C. The fuel is in the form of a sphere 0.0720 ft radius, has a density of 12.5 g_m/cm^3, a thermal conductivity of 12 Btu/hr ft °F, and is surrounded by a shperical shell containing thermoelectric elements and maintained at 235°F. The radioisotope in the fuel decays by the expulsion of alpha particles. The energy thus generated is transferred to the shell by thermal radiation with a heat transfer coefficient $h = 10$ Btu/hr ft² °F. Find (*a*) the energy generated by the fuel in Btu/hr ft³ (volumetric thermal source strength), and (*b*) the maximum fuel temperature in °F. Neglect fuel cladding.

5-7. Malfunctioning in a uranium oxide compacting machine resulted in the manufacture of a 0.2-in. thick plate-type fuel element with a discontinuity in it. At a particular cross section, this discontinuity is approximated by assuming that the plate is made of two distinct regions. The midsection, 0.1-in. thick has a thermal conductivity of 1 Btu/hr ft °F and a U^{235} concentration of 10^{21} nuclei/cm³. The remaining outer regions have 0.9 and 0.8×10^{21} respectively. Neglect cladding and find the maximum temperature at that cross section if the thermal neutron flux is 10^{14}, $h = 5000$ Btu/hr ft² °F, and the coolant-moderator temperature is 500°F.

5-8. A pressurized-water pebble-bed reactor contains spherical fuel elements 1 in. in diameter. The reactor pressure, chosen to completely suppress subcooled boiling is 2,500 psia. Find the maximum temperature in the core center if $q'''_{co} = 10^7$ Btu/hr ft³ and $k_f = 10$ Btu/hr ft °F. Ignore cladding.

5-9. Derive Eq. 5-64 by taking a spherical shell of thickness Δr at r and writing a heat balance.

5-10. Derive expressions for (a) the position r of maximum temperature, and (b) the heat transfer through the inner and outer surfaces of an unclad hollow thick cylindrical fuel element with uniform heat generation and known surface temperatures if heat is allowed to flow through both surfaces.

chapter 6

Heat Conduction in Reactor Elements

II. Some Special One-Dimensional Steady-State Cases

6-1. INTRODUCTION

In the previous two chapters the heat generation in fission and radioactivity, and the heat conduction in some simple one-dimensional steady-state cases with uniform energy generation were investigated. Reactors contain elements that do not fit into these categories. Some bodies such as core vessels, supports, shields, and pressure vessels are subjected to, and absorb, core radiations resulting in nonuniform energy generation. Fins or extended surfaces, on fuel-element cladding of gas-cooled reactors absorb radiation but dissipate energy at their surfaces, resulting in unique temperature distributions and heat-transfer characteristics. Also, there are cases where the heat-generation term q''' may not be uniform, but a function of temperature, and where the material properties, such as thermal conductivity, may be strongly dependent on temperature or pressure. These cases will be covered in this chapter.

6-2. THE ABSORPTION OF CORE RADIATIONS

Strong γ, neutron, and other radiations emanate from active reactor cores. These radiations encounter and are absorbed by structural materials, pressure vessels, shields, and the like. The absorbed radiation is converted into heat that should be removed from these bodies.

In this section we shall discuss the mechanism of absorption of neutron and γ radiations which are the largest contributors to radiation heating. In the absorption process, the neutron, and γ photon beams are attenuated as they penetrate thicknesses of bodies. The attenuation, in this case a function of distance, is analogous to the decay process of

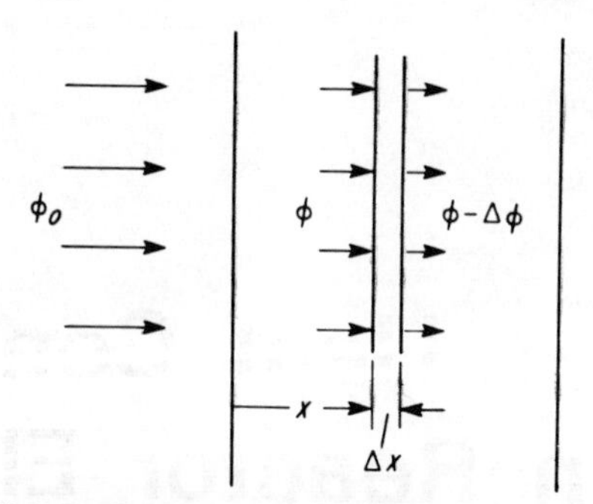

FIG. 6-1. Radiation-beam attenuation in a body.

radioactive materials, which is a function of time (Sec. 1-11). The number of particles absorbed by the body at a certain location is proportional to the total number of particles at that location. Thus if φ is the flux of a parallel beam in photons, or neutrons, per/sec cm² at x, the change in flux with distance (Fig. 6-1) is given, in the limit, by

$$-\frac{d\varphi}{dx} = \mu\varphi \tag{6-1}$$

where μ is a constant of proportionality which depends on the material of the body and the energy of the particles in question. μ is called the *absorption coefficient*. It has the unit (length)$^{-1}$, commonly cm^{-1}. Equation 6-1 integrates to

$$\varphi = \varphi_0 e^{-\mu x} \tag{6-2}$$

where φ_0 is the flux incident on the body, at $x = 0$.

In the case of neutrons, μ is equal to the neutron macroscopic cross section Σ of the material of the body for the particular reaction in question (absorption or scattering). For γ, μ is a complex function of the material of the body and the γ energy (see below).

The intensity of beams I (Mev/sec cm²) is equal to the product of flux and energy of the particles. For neutrons, $I = \varphi\Delta E$, where ΔE is the energy per neutron reaction. For γ, $I = \varphi(h\nu)$, where $h\nu$ is the energy per photon, h is Planck's constant (4.1358×10^{-21} Mev sec) and ν is the frequency of γ radiation (sec^{-1}). In the cases of monoenergetic photons or neutrons, I is directly proportional to φ so that Eq. 6-2 may be written in the form

$$I = I_0 e^{-\mu x} \tag{6-3}$$

where I_0 is the intensity at $x = 0$.

The attenuation may also be expressed in terms of the *half-thickness* of the body, $x_{\frac{1}{2}}$. This is the thickness (analogous to half-life in radioactive decay) through which the intensity of the beam is reduced by one-half, and it is obtained by putting $I/I_0 = 0.5$ in Eq. 6-3 to give

$$x_{\frac{1}{2}} = \frac{0.693}{\mu} \tag{6-4a}$$

This equation can also be written in the form

$$\frac{\mu}{\rho} = \frac{0.693}{x_{\frac{1}{2}}\rho} \tag{6-4b}$$

where ρ is the density of the material of the body. It is found that for γ, μ/ρ and consequently $x_{\frac{1}{2}}\rho$ vary slowly from element to element (i.e., vary slowly with the atomic number Z). μ/ρ is called the *mass-absorption coefficient*. It commonly has the units square centimeters per gram.

It follows from the above that it takes approximately the same *mass* of material to attenuate γ beams to equal degrees. In other words, thicknesses of denser materials (such as lead) are more effective in stopping γ radiation than equal thicknesses of lighter ones.

The absorption of individual γ-ray photons is accomplished mainly by three processes. As shown in Fig. 6-2 for lead, these processes are (a) *photoelectric absorption,* (b) *Compton scattering,* and (c) *pair production.* The absorption coefficient μ is equal to the sum of the absorption coefficients due to the three processes.

In the photoelectric effect, the entire energy of a γ photon, $h\nu$, is transferred to an orbital electron, belonging to one of the atoms of the

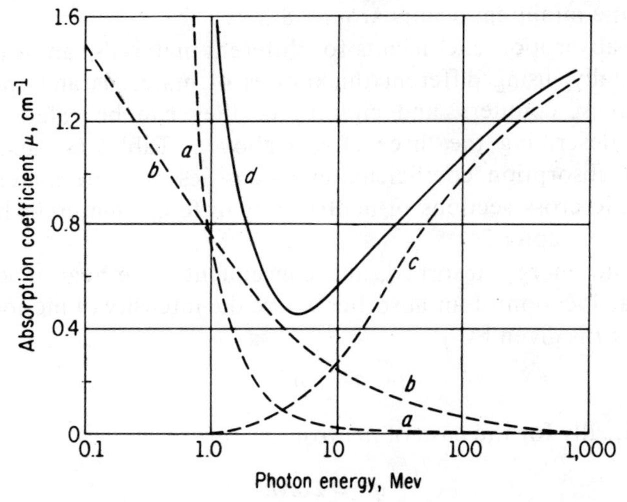

FIG. 6-2. Absorption coefficient of lead. a = photoelectric effect, b = Compton scattering, c = pair production, d = total.

body. This electron is knocked out of its atom but is quickly reabsorbed by another, with a consequent release of energy in the form of an equivalent quantity of heat. At low photon energies (below 0.5 Mev for lead), the photoelectric effect is the chief contributor to the absorption coefficient μ.

In Compton scattering, the incident photon encounters an orbital electron, loses only part of its energy in freeing that electron from its atom, and is scattered away from the beam with less energy than it initially had. It can thus be reasoned that Compton scattering becomes important at higher photon energies than the photoelectric effect. For lead, Compton scattering becomes the dominant factor between about 2 and 8 Mev.

In pair production, the absorption of a photon results in the creation of an electron-positron pair (Sec. 1-10). The minimum energy of a photon engaging in pair production should be $2m_e c^2$, where m_e is the rest mass of the electron (or positron) and c is the speed of light. This quantity is equivalent to 1.02 Mev. For lead, pair production is the chief contributor to the absorption coefficient at photon energies beyond 20 Mev (Fig. 6-2). The balance between the incoming photon energy $h\nu$ and 1.02 Mev is added to the electron-positron pair as kinetic energy.

There are other processes contributing to the absorption coefficient besides those mentioned above. Their effect is, however, minor. As can be seen from Fig. 6-3, the absorption coefficient μ shows a minimum value at some photon energy level, depending upon the distribution of the three effects. For lead, this minimum occurs between 3 and 5 Mev. For iron, the minimum occurs around 8 Mev.

The γ-absorption coefficients for different materials can be measured experimentally, using different thicknesses of materials and appropriate γ-ray sources, counters, and circuits, or they can be calculated from equations describing the three effects above. Tables 6-1 and 6-2 give calculated absorption coefficients and densities of some materials [28]. Macroscopic cross sections of neutrons by different materials have been discussed in Chapter 2.

Now the energy absorbed, and consequently the heat generated, in a particular location of an absorber where the intensity of monoenergetic radiation is I is given by

$$q''' = \mu I \tag{6-5}$$

and, specifically for the case of neutrons,

$$q''' = \Sigma \varphi \Delta E \tag{6-6}$$

where q''' is a volumetric thermal source strength having the units Mev/sec cm^3. (Multiply Mev/sec cm^3 by 1.5477×10^{-8} to get Btu/hr ft^3.) This

TABLE 6-1
Calculated γ Mass-absorption Coefficients*
(In square centimeters per gram)

Photon Energy Mev	Water	Aluminum	Iron	Lead
0.1	0.171	0.169	0.370	5.46
0.15	0.151	0.138	0.196	1.92
0.2	0.137	0.122	0.146	0.942
0.3	0.119	0.104	0.110	0.378
0.4	0.106	0.0927	0.0939	0.220
0.5	0.0967	0.0844	0.0840	0.152
0.6	0.0894	0.0779	0.0769	0.119
0.8	0.0786	0.0683	0.0668	0.0866
1.0	0.0706	0.0614	0.0598	0.0703
1.5	0.0576	0.0500	0.0484	0.0523
2.0	0.0493	0.0431	0.0422	0.0446
3.0	0.0396	0.0353	0.0359	0.0413
4.0	0.0330	0.0310	0.0330	0.0416
5.0	0.0302	0.0284	0.0314	0.0430
6.0	0.0277	0.0266	0.0305	0.0445
8.0	0.0242	0.0243	0.0298	0.0471
10.0	0.0221	0.0232	0.0300	0.0503
20.0	0.0180	0.0217	0.0321	0.0625
50.0	0.0167	0.0230	0.0384	0.0817

* From Ref. 28.

TABLE 6-2
Calculated Absorption Coefficients of Gamma Rays for Various Materials Used in Shielding*

Material	Density ρ, g_m/cm^3	Absorption Coefficient μ, cm^{-1}		
		1 mev	3 mev	6 mev
Air	0.001294	0.0000766	0.0000430	0.0000304
Aluminum	2.7	0.166	0.0953	0.0718
Ammonia (liquid)	0.771	0.0612	0.0322	0.0221
Beryllium	1.85	0.104	0.0579	0.0392
Beryllium carbide	1.9	0.112	0.0627	0.0429
Beryllium oxide (hot-pressed blocks)	2.3	0.140	0.0789	0.0552
Bismuth	9.80	0.700	0.409	0.440
Boral	2.53	0.153	0.0865	0.0678
Boron (amorphous)	2.45	0.144	0.791	0.0679
Boron carbide (hot-pressed)	2.5	0.150	0.0825	0.0675
Bricks:				
Fire clay	2.05	0.129	0.0738	0.0543
Kaolin	2.1	0.132	0.0750	0.0552
Silica	1.78	0.113	0.0646	0.0473
Carbon	2.25	0.143	0.0801	0.0554
Clay	2.2	0.130	0.0801	0.0590
Clements:				
Colemanite borated	1.95	0.128	0.0725	0.0528
Plain (1 portland cement: 3 sand mixture)	2.07	0.133	0.0760	0.0559

TABLE 6-2 *(continued)*

Material	Density ρ, g_m/cm^3	Absorption Coefficient μ, cm^{-1}		
		1 mev	3 mev	6 mev
Concretes:				
Barytes	3.5	0.213	0.127	0.110
Barytes-boron frits	3.25	0.199	0.119	0.101
Barytes-limonite	3.25	0.200	0.119	0.0991
Barytes-lumite-colemanite	3.1	0.189	0.112	0.0939
Iron-portland	6.0	0.364	0.215	0.181
MO (ORNL mixture)	5.8	0.374	0.222	0.184
Portland (1 cement:	2.2	0.141	0.0805	0.0592
2 sand: 4 gravel mixture)	2.4	0.154	0.0878	0.0646
Flesh	1	0.0699	0.0393	0.0274
Fuel oil (medium weight)	0.89	0.0716	0.0350	0.0239
Gasoline	0.739	0.0537	0.0299	0.0203
Glass:				
Borosilicate	2.23	0.141	0.0805	0.0591
Lead (Hi-D)	6.4	0.439	0.257	0.257
Plate (av.)	2.4	0.152	0.0862	0.0629
Iron	7.86	0.470	0.282	0.240
Lead	11.34	0.797	0.468	0.505
Lithium hydride (pressed powder)	0.70	0.0444	0.0239	0.0172
Lucite (polymethyl methacrylate)	1.19	0.0816	0.0457	0.0317
Paraffin	0.89	0.0646	0.0360	0.0246
Rocks:				
Granite	2.45	0.155	0.0887	0.0654
Limestone	2.91	0.187	0.109	0.0824
Sandstone	2.40	0.152	0.0871	0.0641
Rubber:				
Butenediene copolymer	0.915	0.0662	0.0370	0.0254
Natural	0.92	0.0652	0.0364	0.0248
Neoprene	1.23	0.0813	0.0462	0.0333
Sand	2.2	0.140	0.0825	0.0587
Stainless steel, Type 347	7.8	0.462	0.279	0.236
Steel (1% carbon)	7.83	0.460	0.276	0.234
Uranium	18.7	1.46	0.813	0.881
Uranium hybride	11.5	0.903	0.504	0.542
Water	1.0	0.0706	0.0396	0.0277
Wood:				
Ash	0.51	0.0345	0.0193	0.0134
Oak	0.77	0.0521	0.0293	0.0203
White pine	0.67	0.0452	0.0253	0.0175

* From Ref. 28.

volumetric thermal source strength differs from that encountered in fuel elements in that it is not constant at any one cross section but varies as I (or φ). It is thus an exponential function of x, as can be ascertained by combining Eqs. 6-3 and 6-5 to give

$$q''' = q_0''' e^{-\mu x} \tag{6-7}$$

where q_0''' is the volumetric thermal source strength at $x = 0$, that is, at the surface of the body facing the radiation.

6-3. HEAT REMOVAL IN SLABS SUBJECTED TO RADIATION

The case of a flat slab will be treated. Although most pressure vessels and shields are cylindrical, their thickness-radius ratio is usually so small that the problem of heat generation and removal in them can be treated as if occurring in a flat slab. Figure 6-3 shows a slab of thickness L subjected to radiation from one side only, resulting in an exponential volumetric source strength distribution. If heat flow is assumed to be

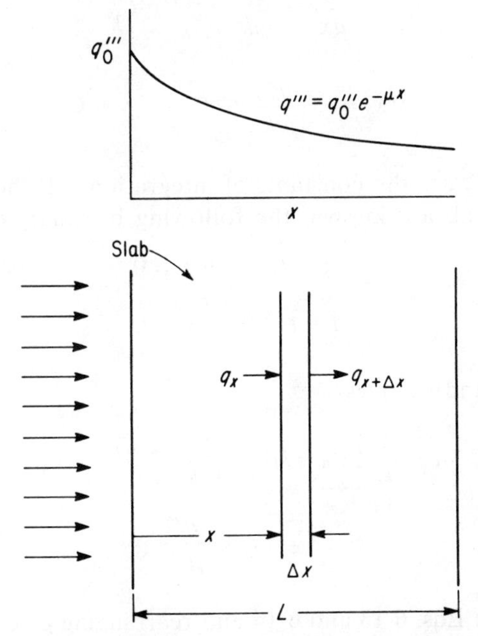

FIG. 6-3. Heat transfer and heat generation in a body subjected to radiation on one side.

positive when in the $+x$ direction, the heats entering and leaving an infinitesimal thickness Δx are given by

$$q_x = -kA\frac{dt}{dx} \tag{6-8}$$

and

$$q_{x+\Delta x} = q_x + \frac{dq_x}{dx}\Delta x = -kA\frac{dt}{dx} - kA\frac{d^2t}{dx^2}\Delta x \tag{6-9}$$

where t is the temperature at x and k is the thermal conductivity of the material of the slab, assumed constant. Thus the heat generated in Δx is

$$q_{x+\Delta x} - q_x = -kA\frac{d^2t}{dx^2}\Delta x \tag{6-10}$$

and

$$q_{x+\Delta x} - q_x = q'''A\,\Delta x \tag{6-11}$$

Combining these two equations with Eq. 6-7 gives

$$\frac{d^2t}{dx^2} = -\frac{q'''}{k} = -\frac{q_0'''}{k}\,e^{-\mu x} \tag{6-12}$$

Integrating gives

$$\frac{dt}{dx} = \frac{q_0'''}{\mu k}\,e^{-\mu x} + C_1 \tag{6-13}$$

and

$$t(x) = -\frac{q_0'''}{\mu^2 k}\,e^{-\mu x} + C_1 x + C_2 \tag{6-14}$$

where C_1 and C_2 are the constants of integration. If the surface temperatures of the slab are known, the following boundary conditions apply:

$$t = t_i \qquad \text{at } x = 0 \tag{6-15a}$$

and

$$t = t_0 \qquad \text{at } x = L \tag{6-15b}$$

Solving for C_1 and C_2 gives

$$C_1 = \frac{1}{L}\left[(t_0 - t_i) + \frac{q_0'''}{\mu^2 k}\,(e^{-\mu L} - 1)\right]$$

and

$$C_2 = t_i + \frac{q_0'''}{\mu^2 k}$$

Combining with Eqs. 6-13 and 6-14 and rearranging give

$$\frac{dt(x)}{dx} = (t_0 - t_i)\,\frac{1}{L} + \frac{q_0'''}{\mu k}\left[\frac{1}{\mu L}\,(e^{-\mu L} - 1) + e^{-\mu x}\right] \tag{6-16}$$

and

$$t(x) = t_i + (t_0 - t_i)\,\frac{x}{L} + \frac{q_0'''}{\mu^2 k}\left[\frac{x}{L}\,(e^{-\mu L} - 1) - (e^{-\mu x} - 1)\right] \tag{6-17}$$

On the right-hand side of Eq. 6-17, $(t_0 - t_i)(x/L)$ is the straight-line temperature gradient in bodies of constant cross section, constant thermal conductivity, and no heat sources. The remaining part represents the effect of the exponential heat source.

Under most conditions, when cooling is provided on both sides of the slab, the temperature t within the slab rises to a maximum t_m somewhere in its interior (Fig. 6-4). Physically, this may be explained by the fact that the temperature profile must slope in opposite directions at $x = 0$

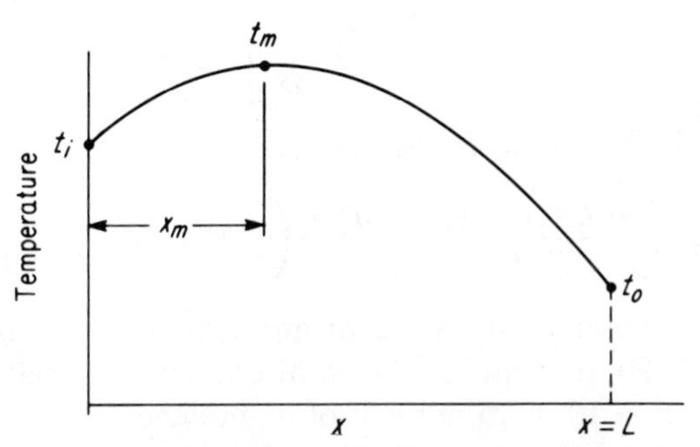

FIG. 6-4. Temperature distribution in a body subjected to radiation and cooled on both sides.

and $x = L$ (that is, dt/dx is respectively positive and negative, but not necessarily equal numerically) if cooling is to be provided on both sides of the slab. x_m, the distance at which t_m occurs, can be found by equating $dt(x)/dx$ to zero and solving for x. Equation 6-16 is now equated to zero and solved for x_m, to give

$$x_m = -\frac{1}{\mu} \ln \left[\frac{\mu k}{q_0''' L} (t_i - t_0) + \frac{1}{\mu L} (1 - e^{-\mu L}) \right] \tag{6-18}$$

The value of the maximum temperature t_m can be obtained by substituting x_m obtained from Eq. 6-18 in Eq. 6-17.

To cool the slab, heat is usually removed by passing coolant parallel to, and on both sides of, the slab. The total heat removed must then be equal to the total heat generated in the slab, q_t (Btu/hr), which is given by

$$q_t = \int_0^L q''' A \, dx = \int_0^L q_0''' A e^{-\mu x} \, dx$$

Integrating gives

$$q_t = \frac{q_0''' A}{\mu} (1 - e^{-\mu L}) \tag{6-19}$$

The quantity of heat removed on each side of the slab can be obtained as follows:

1. Side facing radiation ($x = 0$):

$$q_{x=0} = -kA \left. \frac{dt}{dx} \right]_{x=0}$$

Combining with Eq. 6-15, putting $x = 0$, and rearranging give

$$q_{x=0} = \frac{kA(t_i - t_0)}{L} - \frac{q_0''' A}{\mu} \left(1 + \frac{e^{-\mu L} - 1}{\mu L} \right) \tag{6-20}$$

2. Other side of the slab ($x = L$):

$$q_{x=L} = -kA \left.\frac{dt}{dx}\right]_{x=L}$$

In a treatment similar to the above, we get

$$q_{x=L} = \frac{kA(t_i - t_0)}{L} - \frac{q_0''' A}{\mu}\left(e^{-\mu L} + \frac{e^{-\mu L} - 1}{\mu L}\right) \tag{6-21}$$

Because of the exponential nature of absorption, it can be shown that between 80 and 90 percent of the total energy absorbed by a body is absorbed in the first 10 or 15 percent of its thickness.

Example 6-1. A large iron plate of 6-in. thickness is subjected to pure γ radiation of 10^{14} photons/sec cm² flux and 5.0 Mev/photon strength. The two sides of the plate are cooled by forced convection so that the side facing the radiation is at 700°F and the other side is at 500°F.

Calculate (*a*) the maximum temperature within the plate and (*b*) the heat carried by the coolant on each side in Btu per hour per square foot of plate surface area.

Solution. For the plate material:

$\frac{\mu}{\rho} = 0.0314$ cm²/g$_m$ (Table 6-1 at 5.0 Mev)

$\rho = 7.8$ g$_m$/cm³ for steel at estimated average slab temperature

$\mu = 0.0314 \times 7.8 = 0.245 \text{ cm}^{-1} = 7.5 \text{ ft}^{-1}$

$k = 28$ Btu/hr ft °F

$q_0''' = \mu\varphi_{\gamma 0}(h\nu) = 0.245 \times 10^{14} \times 5.0 = 1.225 \times 10^{14}$ Mev/sec cm³

$= 1.225 \times 10^{14} \times 1.5477 \times 10^{-8} = 1.894 \times 10^{6}$ Btu/hr ft³

Using Eq. 6-18,

$$x_m = -\frac{1}{7.5}\ln\left[\frac{7.5 \times 28}{1.894 \times 10^6 \times 0.5}(700 - 500) + \frac{1}{7.5 \times 0.5}(1 - e^{-7.5\times 0.5})\right]$$

$$= -\frac{1}{7.5}\ln 0.30475 = -\frac{1}{7.5}(-1.185) = 0.158 \text{ ft}$$

Substituting this value of x in Eq. 6-17,

$$t_m = 700 + (500 - 700)\frac{0.158}{0.5} + \frac{1.894 \times 10^6}{7.5^2 \times 28}\left[\frac{0.158}{0.5}(e^{-7.5\times 0.5} - 1) - (e^{-7.5\times 0.158} - 1)\right]$$

$$= 1101.4°\text{F}$$

Cooling on inner surface (Eq. 6-20):

$$q_{x=0} = \frac{28 \times 1 \times (700 - 500)}{0.5} - \frac{1.894 \times 10^6 \times 1}{7.5}\left(1 + \frac{e^{-7.5\times 0.5} - 1}{7.5 \times 0.5}\right)$$

$$= -1.755 \times 10^5 \text{ Btu/hr ft}^2$$

(The negative sign indicates that heat is in the $-x$ direction, i.e., toward the radiation source.)

Cooling on the outer surface (Eq. 6-21):

$$q_{x=L} = \frac{28 \times 1 \times (700 - 500)}{0.5} - \frac{1.894 \times 10^6 \times 1}{7.5}\left(e^{-7.5\times 0.5} + \frac{e^{-7.5\times 0.5} - 1}{7.5 \times 0.5}\right)$$

$$= +0.711 \times 10^5 \text{ Btu/hr ft}^2 \quad \text{in the } +x \text{ direction}$$

The sum of the absolute values of these two heat quantities is 2.466×10^5 Btu/hr ft². This is the total heat generated per square foot of plate. The same figure can be obtained from Eq. 6-19.

Results of computations for $t(x)$ as a function of (x/L) for three materials, lead, iron and aluminum for the same data as in Example 6-1, except that $t_i = t_0$, are shown in Fig. 6-5 for comparative purposes. It can be seen that lead, having the highest absorption coefficients, attains the highest temperatures with a peak closest to the side facing the radiation $(x/L = 0)$, followed by iron and aluminum in that order.

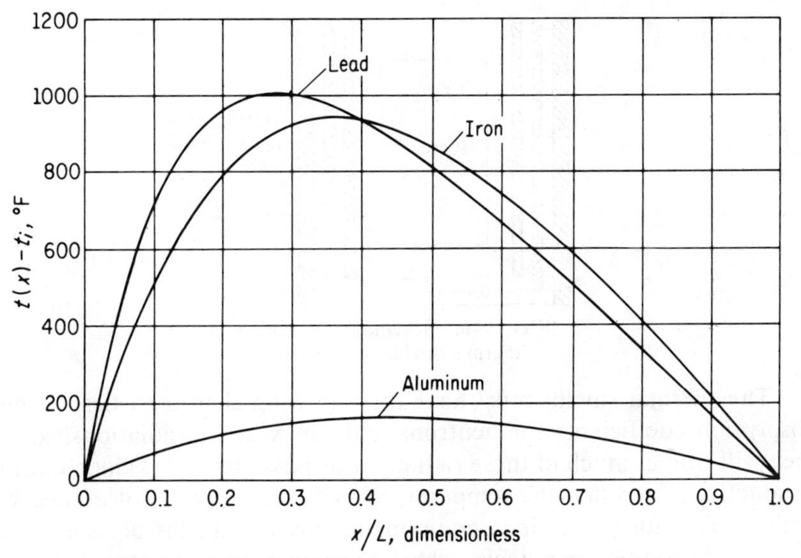

FIG. 6-5. $t(x) - t_i$ vs x/L where $t_i = t_0$ for data of Example 6-1.

6-4. THERMAL SHIELDS

Although Example 6-1 neglects the added radiations such as neutron, β, and X and other effects of γ radiations, it indicates the considerable magnitude of heat generated in bodies subjected to reactor radiations. It also shows that large temperatures and temperature gradients can exist in such bodies.

A concrete biological shield or a steel pressure vessel under 2,000-

psia pressure (as in a pressurized-water reactor) is under large mechanical stresses. Also, the nuclear radiations they are subjected to may have detrimental effects on their mechanical properties. Such bodies must either be designed to meet, or be protected against, these severe thermal conditions. They may be protected by placing one or more *thermal shields* between the core and the vessel. The thermal shields must be cooled. The coolant may or may not be the same as that used in cooling the reactor core itself. A common coolant is often used for both core and thermal shield; passed in series, first by the shield and then into the core. Thermal shields are usually in the form of thin cylindrical shells surrounding the core shroud, Fig. 6-6, but sometimes in the form of closely packed rods as in the Enrico Fermi reactor. Since they are placed inside the reactor pressure vessel, no pressure differential exists between their faces, and they are better able to withstand the heating to which they are subjected.

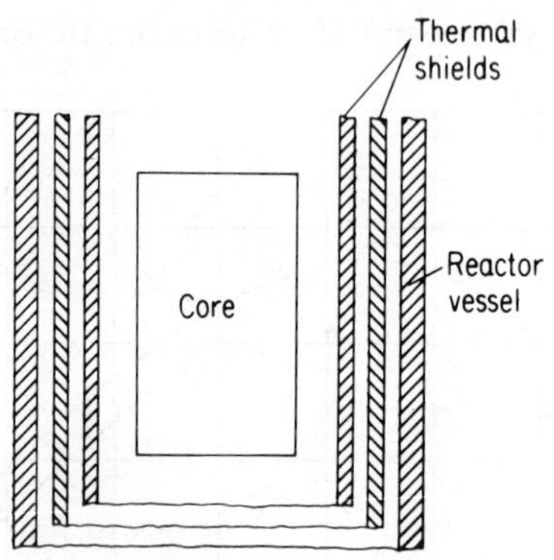

FIG. 6-6. Schematic of core, thermal shields, and vessel.

The thermal shields must have the following characteristics: (1) high absorption coefficients for neutrons and for X and γ radiations, so that they will trap as much of these radiations as possible, and (2) high thermal conductivity k so that the temperatures within them will not exceed safe limits. To satisfy the first requirement, thermal shields are sometimes made of boron-bearing steels, where the boron helps to absorb thermal neutrons and the steel has a high absorption coefficient for γ radiations, as well as good inelastic-scattering cross sections for fast neutrons. Calculations for thermal shields are carried out in the manner presented in Sec. 6-3.

Example 6-2. If the 6-in. steel plate of Example 6-1 were protected by two 2-in.-thick steel thermal shields, (*a*) what would be the γ flux reaching the plate's inner surface and (*b*) how much heat should be carried away from each of these two thermal shields and the plate? Neglect the attenuation of the γ radiations due to the coolant between and around the thermal shields.

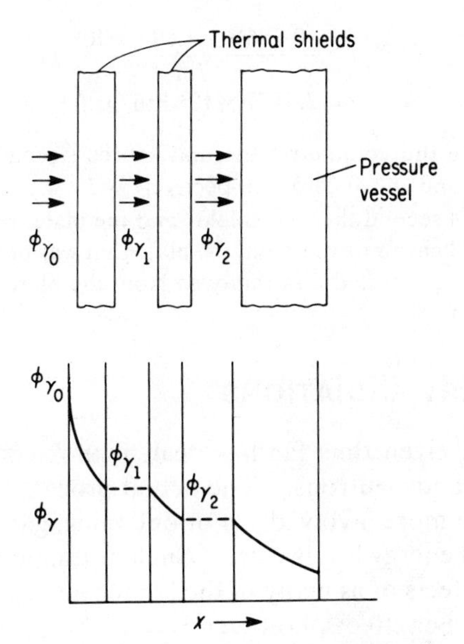

FIG. 6-7. Schematic arrangement of Example 6-2.

Solution. Assume that $\varphi_{\gamma 0}$ is the initial γ flux reaching the first thermal shield (10^{14} photons/sec cm²) and $\varphi_{\gamma 1}$ and $\varphi_{\gamma 2}$ are the γ fluxes reaching the second thermal shield and the plate, respectively (Fig. 6-7).

Since $\varphi_\gamma = \varphi_{\gamma 0} e^{-\mu x}$ and the two thermal shields are of equal thickness, and assuming the same absorption coefficient for thermal shields and plate,

$$\frac{\varphi_{\gamma 1}}{\varphi_{\gamma 0}} = \frac{\varphi_{\gamma 2}}{\varphi_{\gamma 1}} = e^{-7.5 \times \frac{2}{12}} = 0.286$$

$$\varphi_{\gamma 0} = 1 \times 10^{14} \quad \text{photons/sec cm}^2$$

$$\varphi_{\gamma 1} = 2.86 \times 10^{13} \quad \text{photons/sec cm}^2$$

$$\varphi_{\gamma 2} = 8.18 \times 10^{12} \quad \text{photons/sec cm}^2$$

Thus the radiation intensity reaching the plate has been attenuated to a little over 8 percent of its original value. The total heat that must be removed from the first thermal shield is obtained by using Eq. 6-19:

$$\frac{q_0'''A}{\mu}(1 - e^{-\mu L_1}) = \frac{1.894 \times 10^6}{7.5}(1 - e^{-7.5 \times \frac{2}{12}})$$

$$= 1.803 \times 10^5 \text{ Btu/hr ft}^2$$

The heat to be removed from the second thermal shield is 0.286 that in the first, or

$$\frac{q_1'''A}{\mu}(1 - e^{-\mu L_2}) = 5.157 \times 10^4 \text{ Btu/hr ft}^2$$

The heat to be removed from the plate is

$$\frac{q_2'''A}{\mu}(1 - e^{-\mu L_3}) = \frac{(0.286)^2 \times 1.894 \times 10^6}{7.5}(1 - e^{-7.5 \times 0.5})$$
$$= 2.0177 \times 10^4 \text{ Btu/hr ft}^2$$

where q_1''' and q_2''' are the volumetric thermal source strengths at faces 1 and 2, equal to q_0''' (0.286) and q_0''' $(0.286)^2$, respectively. L_1, L_2 and L_3 are the thicknesses of the first and second thermal shields and the plate, respectively.

The quantity of heat removed from the plate that was obtained above is some 8 percent of the heat generated and removed from the plate without the thermal shields (Example 6-1).

6-5. SECONDARY RADIATIONS

The treatment given thus far has dealt with fixed fluxes and energy levels of gammas and neutrons. The actual process in nuclear reactors is, of course, much more involved. For one thing, gammas and neutrons of many fluxes and energy levels exist. Such a situation may be dealt with by summing the effects of as many of the higher-intensity beams as can be accounted for. Other effects, however, render this procedure inaccurate. One is the fact that *secondary radiations* of various intensities are created by the interaction of the original or primary radiations with matter. Another is the so-called *geometrical attenuation,* which is a function of the body thickness and material. In the case of gamma radiation, secondary gammas are obtained in Comptom scattering and in pair production (in the photoelectric effect the primary photons are eliminated, producing β radiation of limited range). In Compton scattering, a new photon of reduced energy results, so that it can be assumed that the primary photon has simply been deenergized. In pair production, the positron eventually collides with an electron, resulting in two new photons of energy (to conserve momentum) equivalent to the sum of the rest masses of the new pair, plus their kinetic energies at the time of collision. Thus the primary photon has effectively created two photons of lower energy. If new photons continue to be created within the body, not near its boundary where they may easily escape, successive stages of lower-energy photons resulting from Compton scattering and pair production will be produced. This continues until the energy of the photons is in the photoelectric range (below 1 Mev for lead; see Fig. 6-2). There the absorption coefficient is high, and they are finally eliminated.

In the case of neutrons, interaction with materials also results in secondary radiations. Four cases are of interest: (n,p), (n,α), (n,n), and (n,γ). The first two are reactions of extremely low cross sections, i.e., probabilities. They result in protons and α particles, which are of extremely short range, so that the neutrons can be assumed to be effectively removed at the time and place of absorption. The (n,n), or scattering, reaction results in lower-energy neutrons. The (n,γ), or radioactive-capture, reaction usually results in very-high-energy gammas at the point of absorption, which may be anywhere in the body in question. The effects of the (n,n) and the (n,γ) reactions are therefore of great importance.

It can be seen that secondary radiations of many energies are created from primary neutron and gamma radiations within a body. Primary and secondary radiations may also be scattered and change direction as they travel through the body. This is called *geometric attenuation*. This effect depends upon the body thickness and the mean free path of the radiation. The relationships given in the previous three sections are therefore inadequate in actual shielding problems. Because of the apparent difficulty of expressing such situations with exact relationships, an empirical *buildup factor* is applied to the attenuation relationships, such as Eqs. 6-12, 6-13, and 6-17. Thus Eq. 6-13, for example, may be written in the form

$$I = (1 + B)I_0 e^{-\mu x} \tag{6-22}$$

where B [or sometimes $(1 + B)$] is the empirical buildup factor.

The problem is important in reactor-shielding studies. It is dealt with in more detail in other texts [13] and in texts on reactor shielding [29-31] but is beyond the scope of this book.

6-6. FINS IN REACTORS

If the heat transfer characteristics of a coolant are poor, the reactor may have to be limited in specific power (power per unit mass of fuel) and volumetric thermal source strength, or else fuel element temperatures may exceed safe limits. This, in particular, is the case of gases or organic liquids. One method of alleviating this problem is the use of finned cladding or, when reactor physics permit, fuel elements of small diameter. Fins are also used in heat exchangers. In gas-cooled-plant steam generators, for example fins are put on the gas side of the tubes, outside for ease of manufacture, while water and steam are made to flow inside the tubes. Fins are recommended whenever there exists a large difference in heat-transfer coefficient between the two sides of a heat-transfer surface, and are placed on the side of the low coefficient.

In nuclear and other applications where it is important to minimize resistance to heat flow, and consequently operating temperatures, the fins are usually made of the same material and are an integral part of the cladding. This eliminates the possibility of separation of the fin from the parent surface due to vibration, repeated expansions and contractions, or other causes. Separation results in increased resistance to heat transfer and in the possibility of electrolytic reaction between fin and parent materials.

Fins may take many forms (Fig. 6-8). Some are (1) longitudinal or axial, (2) circumferential or transverse (not shown), (3) helical, of different pitch and height, and (4) pin and strip type, in line or staggered.

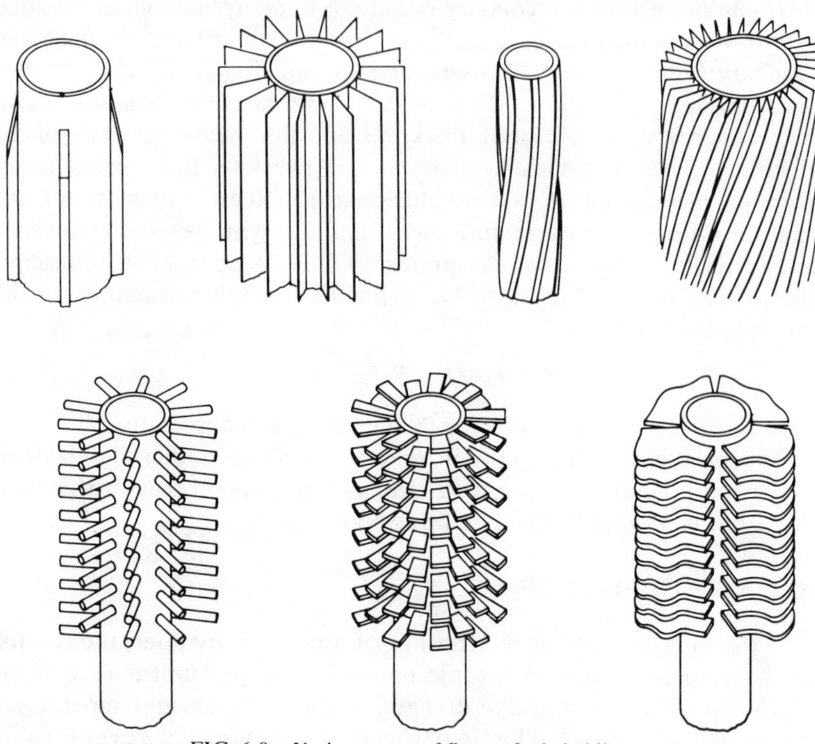

FIG. 6-8. Various types of fins on fuel cladding.

Fins on nuclear fuel elements, while transferring fission heat to the coolant, generate some heat of their own because of the absorption of nuclear radiations. Unlike pressure vessels, thermal shields, and other bodies subjected to γ and neutron radiations emanating from the core, where heat generation is exponential with distance (Sec. 6-2), fins on fuel elements are subjected to many radiations from all directions. They can consequently be assumed, with little error, to have uniform volumetric thermal source strengths q''' Btu/hr ft^3. q''' for a fin is, of course, far less than q''' within a fuel element, being of the order of a hundredth to a thousandth of the latter.

The choice of fin material for a fuel element is a complex function of many variables. Because of the greater added surface, nuclear compatibility is even more important than in the case of plain, unfinned cladding.

In other words, low neutron absorption cross sections are a primary requisite. Good thermal conductivity k, however, is also necessary for good heat dissipation per given volume of fin or, alternatively, for the least amount of fin material per given amount of heat dissipation. A compromise here is possible. For example, magnesium has a lower k than aluminum but also lower macroscopic neutron absorption cross section, so that a large fin may be used. Also, as will be seen later, the lower the internal heat generation within the fin, the better the heat transfer (fuel to coolant) of that fin. In this respect, materials of low γ and fast-neutron absorption cross sections are preferred. Table 6-3 shows that, from this standpoint, beryllium is the best and iron is the worst of the materials listed. Aluminum, which sits in the middle of the scale, may, however, be as good a fin material as beryllium, because of its higher k. Note also that iron and zirconium have poor k. Cost and manufacturing problems are other factors to be taken into account in choosing fin materials.

TABLE 6-3
Some Properties of Some Fin Materials

Material	Relaxation length,* cm		Σ_a, cm^{-1} (thermal)	k, Btu/hr ft °F
	Fast neutrons	γ rays		
Beryllium	~9	18	0.00123	68 (600°F)
Aluminum	~10	13	0.01300	131 (600°F)
Magnesium			0.00254	91 (572°F)
Iron	~6	3.7	0.1900	27 (600°F)
Zirconium			0.00765	11 (480°F)

* A term used in shielding work; equal to the thickness of material in which the intensity of radiation is reduced by a factor of $1/e$.

In designing a finned surface, the heat dissipated from the parent surface by a single fin, q_0, may be found by dividing the total heat to be dissipated by that surface by the number of fins that that surface can accommodate. (The heat transferred by the surface between the fins is usually ignored.) Once q_0, the coolant conditions, and the space available from the lattice chosen are known, the fins may be designed. There is a lower limit on the distance between individual fins because of the reduction in heat transfer whenever the boundary layers of two adjacent surfaces overlap. The spacing between two fins should therefore not be smaller than twice the boundary-layer thickness. This is smaller, the faster the coolant. No exact estimates can be given here but, for speeds encountered in reactor work, the minimum boundary-layer thickness may be taken roughly as 0.1 in., making the spacing between fins 0.2 in. Generally the number of fins required on a fuel rod is such that the spacing exceeds this value. The ratio of the total finned surface to the inside surface of a tube varies from 3 or 4, for short fins, to as much as 10 or 15 for long fins. Integral long fins are manufactured by a process similar to thread rolling. Strips and pins are usually welded to the parent metal.

6-7. HEAT TRANSFER FROM FINS OF LOW BIOT NUMBERS

We shall first treat the general case of a fin of variable cross section, in which heat is generated uniformly. The fin, Fig. 6-9, is surrounded by a coolant fluid of constant temperature t_f with a constant heat-transfer coefficient h around the entire surface of the fin. These assumptions, though conventional in fin theory, are not exact. Variations in t_f and h are, however, usually included implicity in experimentally determined values of h.

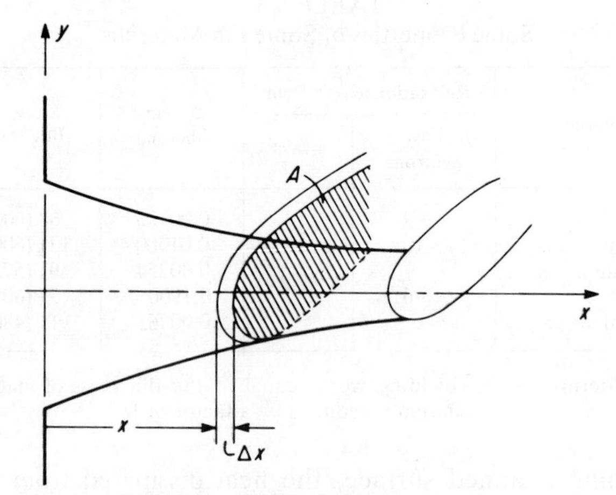

FIG. 6-9. Fin of variable cross section.

It will be assumed further that the fin is made of homogeneous material of constant thermal conductivity k, that the temperature in the fin varies in the x direction only and that heat conduction within the fin will therefore be in that direction only. The criterion for this to be true is that the fin-coolant system has a low *Biot* number. The Biot number is a measure of the ratio of internal (conductive) to external (convective) resistances to heat transfer. The mere thinness of the fin, a condition often mentioned for this treatment, is not a sufficient criterion [32].

If we consider a differential element at x of thickness Δx, cross-sectional area A, and temperature t, the amount of heat entering plane x is given by

$$q_x = -kA\,\frac{dt}{dx} \tag{6-23}$$

The heat leaving plane $x + \Delta x$ is given by

$$\begin{aligned} q_{x+\Delta x} &= q_x + \frac{dq_x}{dx}\,\Delta x \\ &= -kA\,\frac{dt}{dx} - k\left(A\,\frac{d^2t}{dx^2} + \frac{dt}{dx}\,\frac{dA}{dx}\right)\Delta x \end{aligned} \tag{6-24}$$

The difference between q_x and $q_{x+\Delta x}$ plus the heat generated within the element must equal the heat given off by convection at the peripheral surface of the element.

$$q_x - q_{x+\Delta x} + q'''A\,\Delta x = C\Delta x h(t - t_f) \tag{6-25}$$

where C is the peripheral length at x. Since the thickness of a fin everywhere is usually small compared with its length, the effect of the slope of the surface on convection will be ignored.

Introducing $\theta = t - t_f$, the *excess* temperature at x over the assumed constant coolant-fluid temperature, and noting that $dt/dx = d\theta/dx$ and $d^2t/dx^2 = d^2\theta/dx^2$, we combine Eqs. 6-23, 6-24, and 6-25 into the general form

$$\frac{d^2\theta}{dx^2} + \frac{1}{A}\,\frac{d\theta}{dx}\,\frac{dA}{dx} - m^2\theta = -\frac{q'''}{k} \tag{6-26}$$

where $m^2 = Ch/Ak$, a variable having the dimension (length)$^{-2}$.

This equation will now be solved for some cases of interest.

6-8. THE CASE OF FINS OF CONSTANT CROSS SECTIONS

If the fin has constant cross-sectional area and constant periphery, so that $A = A_0$ and $C = C_0$, where A_0 and C_0 are the area and peripheral length at the base of the fin where $x = 0$, Eq. 6-26 becomes

$$\frac{d^2\theta}{dx^2} - m_0^2\theta = -\frac{q'''}{k} \tag{6-27}$$

where $m_0^2 = C_0h/A_0k$, a constant.

Fin geometries that belong to this case are

1. Rectangular or helical of wide pitch (Fig. 6-10), having length L,

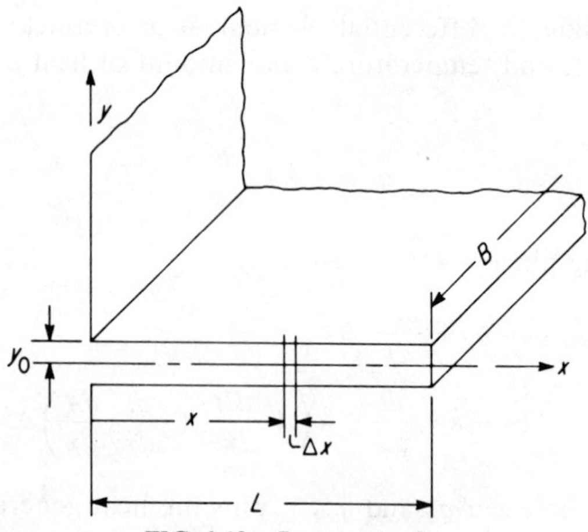

FIG. 6-10. Rectangular fin.

width B, and constant thickness $2y_0$. In this case, $m_0^2 = (2B + 4y_0)h/2y_0Bk$. Where the thickness is small compared with the width, $m_0^2 = h/y_0k$.

2. Circular or pin type, of constant radius r_0 and length L (Fig. 6-11), for which $m_0^2 = 2\pi r_0 h/\pi r_0^2 k = 2h/r_0 k$.

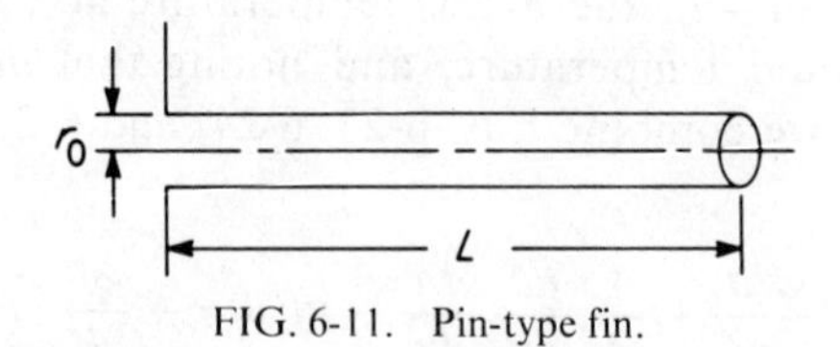

FIG. 6-11. Pin-type fin.

Values of m_0^2 for other geometries of constant cross section can be evaluated with equal ease.

The solution of Eq. 6-27 is

$$\theta(x) = ae^{-m_0 x} + be^{+m_0 x} + \frac{q'''}{m_0^2 k} \tag{6-28a}$$

where a and b are the constants of integration. This equation may be written in the form

$$\theta(x) = ae^{-m_0 x} + be^{+m_0 x} + R_g \theta_0 \tag{6-28b}$$

and therefore

$$\frac{d\theta(x)}{dx} = -am_0 e^{-m_0 x} + bm_0 e^{+m_0 x} \tag{6-29}$$

where

$$R_g = q'''/m_0^2 k\theta_0 \tag{6-30}$$

R_g is a dimensionless number called the *generation ratio*. It is equal to the ratio of the heat generated within the fin to the heat dissipated by the fin by convection, if the fin were all at the base temperature t_0 (where $\theta = \theta_0$). For a rectangular fin, $R_g = q'''y_0/h\theta_0$, and for a circular fin, $R_g = q'''r_0/2h\theta_0$.

a and b are determined from the boundary conditions of the system, one of which is

$$\theta = \theta_0 \qquad \text{at } x = 0 \tag{6-31a}$$

The other boundary condition could be given by $\theta = 0$ at $x = L$, if the fin were so long that no excess temperature, and consequently no heat transfer, occurred at $x = L$. In reality, however, most fins are not overly long, and some heat is transferred from the tip of the fin. If h is constant over the entire fin surface, as previously assumed, one may imagine the fin to be of length $L' = L + \epsilon$ but with an insulated tip (Fig. 6-12). For identical heat transfer as before, the added length ϵ replaces the previously exposed tip area by an equal peripheral area.

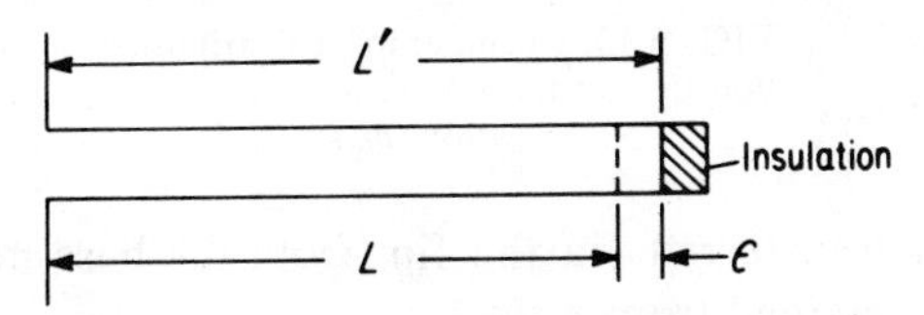

FIG. 6-12. Fin with insulated tip.

Thus, for the rectangular fin, $2\epsilon B = 2y_0 B$ and $\epsilon = y_0$. Similarly, for the circular fin, $\epsilon = r_0/2$. The second boundary condition now becomes

$$\frac{d\theta}{dx} = 0 \qquad \text{at } x = L' \tag{6-31b}$$

Using the two boundary conditions in Eqs. 6-28b and 6-29 gives a and b. Substituting in Eq. 6-28b and rearranging give the temperature profile along the fin

$$\theta(x) = R_g\theta_0 + (1 - R_g)\theta_0 \left(\frac{e^{-m_0 x}}{1 + e^{-2m_0 L'}} + \frac{e^{+m_0 x}}{1 + e^{+2m_0 L'}} \right) \tag{6-32a}$$

Multiplying and dividing the first quantity in the second parentheses by $e^{+m_0 L'}$ and the second quantity by $e^{-m_0 L'}$ and noting that $\cosh x = 0.5$ $(e^{+x} + e^{-x})$, Eq. 6-32a is written in the more convenient form

$$\frac{\theta(x)}{\theta_0} = R_g + (1 - R_g)\frac{\cosh m_0(L' - x)}{\cosh m_0 L'} \tag{6-32b}$$

Figure 6-13 shows $\theta(x)/\theta_0$ lines of constant values of the generation ratio plotted against x/L', all for $m_0 L' = 2.5$.

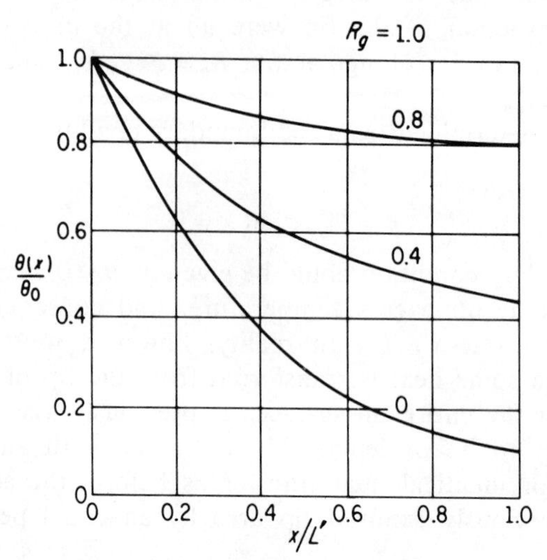

FIG. 6-13. Temperature distribution in fin of constant cross section and internal heat generation, $m_0 L = 2.5$.

The heat transferred via the fin (from the base material), q_0 Btu/hr, may now be obtained from

$$q_0 = -kA_0 \left.\frac{d\theta}{dx}\right]_{x=0}$$

Substituting for a and b in Eq. 6-29, putting $x = 0$, and multiplying by $-kA_0$ give

$$q_0 = -kA_0\left[m_0(1 - R_g)\theta_0\left(-\frac{1}{1 + e^{-2m_0L'}} + \frac{1}{1 - e^{+2m_0L'}}\right)\right] \tag{6-33a}$$

Rearranging as with Eq. 6-32a and noting that $\tanh x = (\sinh x)/(\cosh x) = (e^{+x} - e^{-x})/(e^{+x} + e^{-x})$ give

$$q_0 = (1 - R_g)m_0 kA_0\theta_0 \tanh m_0 L' \tag{6-33b}$$

$\theta(x)/\theta_0$ and q_0 for the special case of no internal heat generation are obtained by putting $R_g = 0$. Figure 6-14 is a plot of the quantity $q_0/m_0kA_0\theta_0$ versus m_0L' for various values of R_g. The ordinate can be expressed as $(q_0/A_0h\theta_0)\sqrt{hA_0/kC_0}$, where $q_0/A_0h\theta_0$ is known as the *removal number.* It is equal to the ratio of the heat dissipated by the fin to the heat that would be dissipated by convection from the base area of the fin if the fin were not present and if the same h applied.

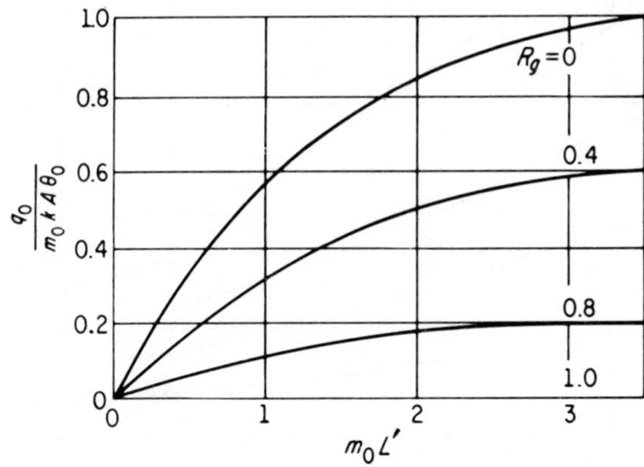

FIG. 6-14. Heat-transfer characteristics of fin of constant cross section and internal heat.

For a fin to be worthwhile, the removal number must be greater than 1.0. However; to offset such things as cost, complexity, and increased neutron absorption, the removal number should be much greater, possibly 5 or more. A removal number less than unity means, of course, that the fin acts as an insulator.

The abscissa of Fig. 6-14 is a function of the aspect ratio of the fin. For a rectangular fin, for example, $m_0 L' = (L'/\sqrt{y_0})\sqrt{h/k}$. Note, then, that as the aspect ratio is increased beyond a certain value, no great improvement in removal number results. Note also that this occurs earlier, the higher the generation ratio.

Another parameter of interest in fin work is the fin *effectiveness* η. This is the ratio of the heat dissipated by the fin to the heat that would be dissipated if the entire fin were at the base temperature t_0, that is, if it had infinite thermal conductivity. Thus for a fin of constant cross section

$$\eta = \frac{q_0}{hC_0 L' \theta_0} = \frac{q_0}{m^2 k A_0 L' \theta_0} \tag{6-34a}$$

Combining with Eq. 6-33b gives

$$\eta = (1 - R_g) \frac{\tanh m_0 L'}{m_0 L'} \tag{6-34b}$$

Figure 6-15 is a plot of this relationship. Note that the effectiveness decreases with $m_0 L'$, approaching $(1 - R_g)/m_0 L'$ for very long fins, and

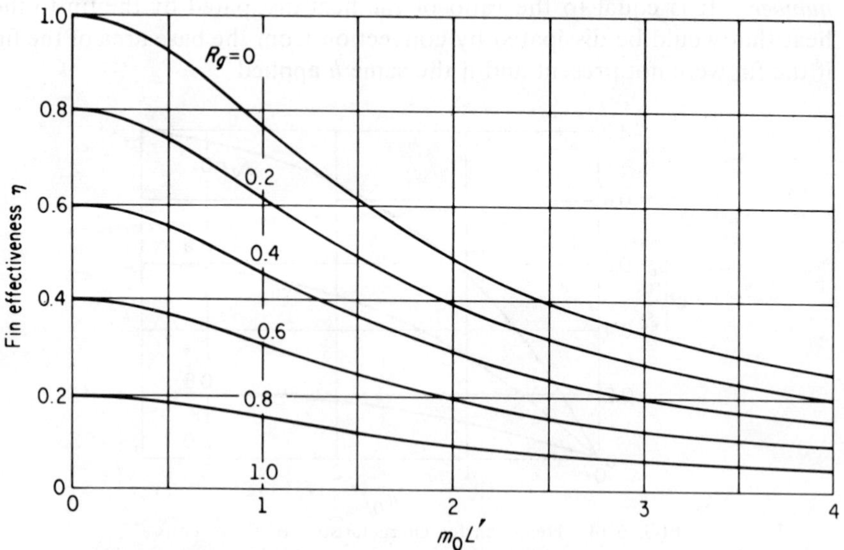

FIG. 6-15. Effectiveness of fin of constant cross section and internal heat generation.

that it decreases with R_g, becoming zero at $R_g = 1.0$. It can also be seen by examining Eq. 6-34b that, for a constant R_g, η is a strong function of m_0, decreasing as $\sqrt{h/k}$ increases (a logical thing) and as $\sqrt{C_0/A_0}$ increases, i.e., as the fin becomes thinner.

6-9. OPTIMUM AND TRIANGULAR FINS

Minkler and Rouleau [33] obtained relationships for straight thin fins of optimum design. An *optimum fin* is one which has a minimum volume of material for a given amount of heat dissipation. It is one in which $d\theta/dx$ is constant. The relationships obtained are functions of the generation ratio and have the boundary condition $y = 0$ at $x = L$ and $\epsilon = 0$.

With no heat generation, the optimum fin is parabolic and concave (Fig. 6-16a). As q''' increases, the fin becomes thicker and eventually convex (Fig. 6-16b). Such optimum fins are difficult to fabricate and, when concave, weak structurally. As with no heat generation, an optimum fin may be approximated by a triangular fin (Fig. 6-16c). The relationships for such a shape may be obtained by putting $A = 2By = 2By_0(L - x)/L$, where B is the width of the fin. The resulting relationships [33] are

$$\frac{\theta(x)}{\theta_0} = \left[1 - R_g - \frac{R_g}{(m_0L)^2}\right] \frac{I_0(2m_0L\sqrt{x/L})}{I_0(2m_0L)} + \frac{x}{L} R_g + \frac{1}{(m_0L)^2} \tag{6-35}$$

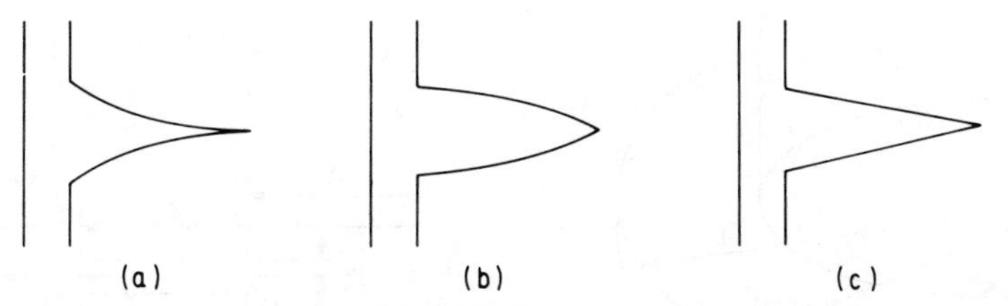

FIG. 6-16. (*a*) Optimum fin for small q'''; (*b*) optimum fin for large q'''; (*c*) triangular fin.

and
$$q_0 \doteq \left[1 - R_g - \frac{R_g}{(m_0 L)^2}\right] m_0 k A_0 \theta_0 \frac{I_1(2m_0 L)}{I_0(2m_0 L)} + \frac{R_g}{m_0 L} \tag{6-36}$$

where I_0 and I_1 are the modified Bessel functions of the first kind, zero and first order, respectively (Appendix C).

If a straight, or nearly straight, fin of fixed cross section is preferred, the optimum ratio between length L and thickness, which would result in the maximum heat dissipation for a given volume of material for that shape, can be obtained by writing L' in terms of the thickness and the given volume, differentiating q_0 in Eq. 6-33b with respect to the thickness, and equating to zero. For a rectangular fin the result is given implicitly [33] by

$$\frac{\sinh 2m_0 L'}{2m_0 L'} = \frac{1 - R_g}{\frac{1}{3} - R_g} \tag{6-37}$$

6-10. THE CASE OF CIRCUMFERENTIAL FINS OF CONSTANT THICKNESS

In the case of circumferential (or transverse) fins of constant cross section, on small rods or tubes, the cross-sectional area depends strongly on the radius. If the thickness is $2y_0$ (Fig. 6-17), the heat balance on a differential element of radial thickness Δr may be written by putting $A = (2y_0)(2\pi r)$ and $dA/dr = 4\pi y_0$ into Eq. 6-26 and modifying to

$$\frac{d^2\theta}{dr^2} + \frac{1}{r}\frac{d\theta}{dr} - m_0^2\theta = -\frac{q'''}{k} \tag{6-38}$$

where here $m_0^2 = h/ky_0$. This is a Bessel equation whose solution is

$$\theta(r) = aI_0(m_0 r) + bK_0(m_0 r) + \frac{q'''}{m_0^2 k} \tag{6-39}$$

where I_0 and K_0 are the modified Bessel functions of zero order, first and second kind, respectively (Appendix C). a and b are found from the boundary conditions:

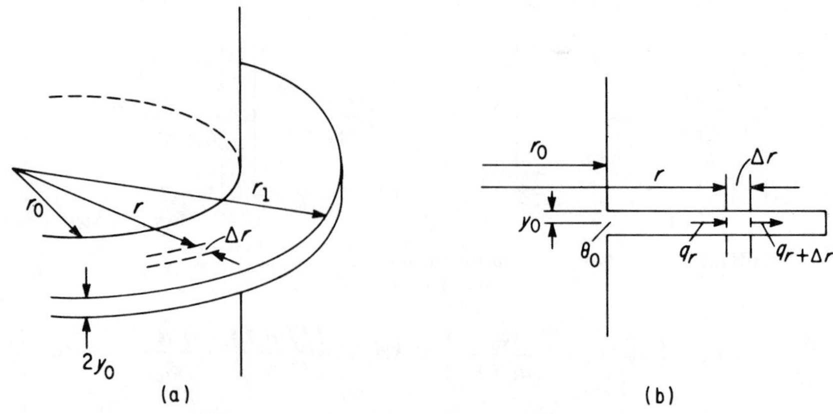

FIG. 6-17. Circumferential fin of constant thickness.

$$\theta = \theta_0 \qquad \text{at } r = r_0 \tag{6-40a}$$

and

$$\frac{d\theta}{dr} = 0 \qquad \text{at } r = r_1 + y_0 = r' \tag{6-40b}$$

where r_0 and r_1 are the radii of the fin at its base and tip, respectively. The second boundary condition is similar to that for straight fins, provided that y_0 is small compared with r_1. r' is analogous to L'. The solution is

$$\frac{\theta}{\theta_0} = R_g + (1 - R_g) \frac{K_1(m_0 r')I_0(m_0 r) + I_1(m_0 r')K_0(m_0 r)}{K_1(m_0 r')I_0(m_0 r_0) + I_1(m_0 r')K_0(m_0 r_0)} \tag{6-41}$$

and

$$q_0 = (1 - R_g) m_0 k A_0 \theta_0 \frac{I_1(m_0 r')K_1(m_0 r_0) - K_1(m_0 r')I_1(m_0 r_0)}{I_1(m_0 r')K_0(m_0 r_0) + K_1(m_0 r')I_0(m_0 r_0)} \tag{6-42}$$

where A_0 is the area at the base of the fin $4\pi r_0 y_0$. I_1 and K_1 are the modified Bessel functions of the first order, first and second kind, respectively.

Performance graphs of fins of various shapes, without heat generation, can be found in the literature and texts on heat transfer [34-36].

6-11. HEAT TRANSFER IN POROUS ELEMENTS

As previously discussed, Sec. 6-7, gases usually have low heat transfer coefficients when used to cool surfaces. Fuel elements that are surface cooled by a gas, therefore, are limited in specific powers (power per unit mass of fuel) and volumetric thermal source strengths, so that their temperatures would not exceed safe limits. Fins on such elements are used to partially alleviate the problem (Sec. 6-6). That problem is particularly acute in the case of fast reactors where specific powers are high. Such reactors have been proposed with gases (helium, superheated steam) as coolants.

A proposal that would materially reduce gas-cooled fuel element temperatures for a given power density has been made by El-Wakil and coworkers [37-39]. In this it is proposed to construct the fuel elements in porous form which may then be cooled by passing the gaseous coolant *through* them. Figure 6-18 shows two possible configurations of such a system.

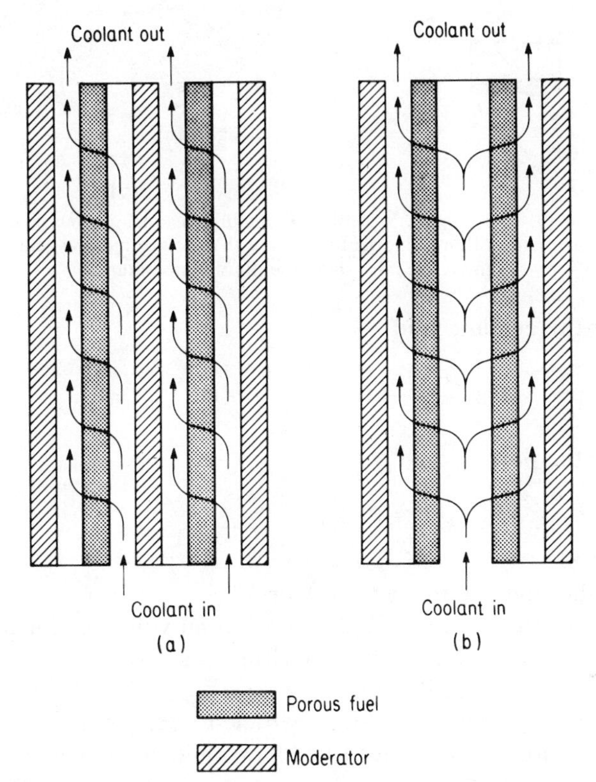

FIG. 6-18. Possible flow patterns in a porous-fuel reactor core subassembly. (*a*) Flat-plate fuel, (*b*) hollow-cylindrical fuel.

Figure 6-19a shows the coordinate system of a one-dimensional porous element which may be a plate-type or a thin hollow cylinder (one in which the inner radius is half or more of the outer radius). For such a system, the governing equations in the steady state and for uniform heat generation are:

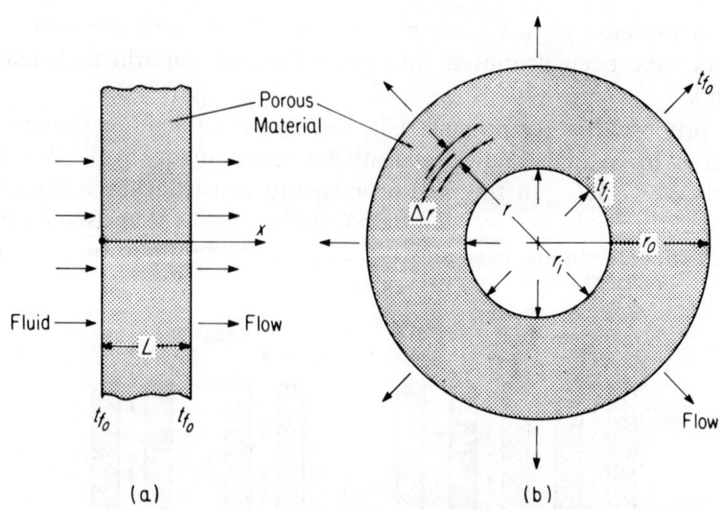

FIG. 6-19. Coordinate systems of one-dimensional porous systems. (a) Flat plate or thin cylindrical geometry, (b) thick cylindrical geometry. t_{f_i} and t_{f_o} are inlet and outlet fluid temperatures.

(a) for the coolant fluid

$$k_f \varphi \frac{d^2 t_f}{dx^2} - G \varphi c_f \frac{dt_f}{dx} - h_v(t_f - t_s) = 0 \tag{6-43}$$

(b) for the porous solid

$$k_s \frac{d^2 t_s}{dx^2} + h_v(t_f - t_s) + q''' = 0 \tag{6-44}$$

where k = thermal conductivity Btu/hr ft °F
φ = porosity of fuel = void volume/total volume, dimensionless
G = mass velocity, lb_m/hr ft^2 of free area
c = specific heat, Btu/lb_m °F
h_v = *volumetric* heat transfer coefficient, Btu/hr ft^3 °F
and the subscripts f and s refer to coolant fluid and porous solid material respectively.

Figure 6-19b shows the coordinate system of a thick cylindrical element in which the coolant flows radially. The one-dimensional (in r) governing equations in the steady state and uniform heat generation in this system are

(a) for the coolant fluid

$$k_f \varphi \left(\frac{d^2 t_f}{dr^2} + \frac{1}{r} \frac{dt_f}{dr} \right) - \frac{m_f c_f}{2\pi H} \left(\frac{1}{r} \frac{dt_f}{dr} \right) - h_v(t_f - t_s) = 0 \tag{6-45}$$

(b) for the solid

$$k_s\left(\frac{d^2t_s}{dr^2} + \frac{1}{r}\frac{dt_s}{dr}\right) + h_v(t_f - t_s) + q''' = 0 \tag{6-46}$$

where $\dot{m}_f$ = mass-flow rate of coolant fluid lb_m/hr, (used here instead of G because of the variable flow areas)

H = height of element, ft

The above equations have been solved for the boundary conditions:

$$\left.\begin{array}{llll} t_f = t_{f_0} & \text{at } x = 0 & \text{or} & r = r_i \\ dt_f/dx = 0 & \text{at } x = 0, & dt_f/dr = 0 & \text{at } r = r_i \\ dt_s/dx = 0 & \text{at } x = 0, & dt_s/dr = 0 & \text{at } r = r_i \\ dt_s/dx = 0 & \text{at } x = L, & dt_s/dr = 0 & \text{at } r = r_o \end{array}\right\} \tag{6-47}$$

These boundary conditions have been justified after analyzing their effects on the solutions within the body of the element, and the effects of possible surface losses [38].

The heat-transfer coefficient in the above equations is necessarily based on volume rather than area because of the near impossibility of assigning a heat-transfer surface area and temperature difference between the surface and the fluid in a porous system. This volumetric heat-transfer coefficient h_v, is experimentally obtained by the so-called *single-blow technique* in which the exit fluid temperature is recorded with time in a transient experiment, avoiding the need for knowing the temperature differences between the solid and fluid within the body of the porous material [38]. A volumetric Nusselt number, Nu_v, is obtained for individual porous systems and is related to a Reynolds number by a relationship of the form (the Prandtl number varies little for different gases)

$$Nu_v = C\,Re^n \tag{6-48}$$

where C and n are constants. Both the Nusselt ($h_v l^2/k_f$) and Reynolds ($V_f \rho_f l/\mu_f$) numbers are based on a *characteristic length* l which characterizes the intricacies of structure and flow in a porous material. It is numerically equal to an *inertial coefficient* b having the dimension ft^{-1}, divided by a *viscous coefficient* a having the dimension ft^{-2}. a and b are obtained from experimental pressure drop data and the modified Darcy law

$$-g_c\frac{dp}{dx} = a\mu_f V_f + b\rho_f V_f^2 \tag{6-49}$$

where p = pressure, lb_f/ft^2

V_f = fluid velocity, ft/hr

μ_f = fluid viscosity, lb_m/ft hr

ρ_f = fluid density, lb_m/ft^3

g_c = conversion factor, 4.17×10^8 lb_m ft/lb_f hr^2

Porous materials made by sintering aggregates of stainless steel fibres were experimentally investigated [37]. Nuclear heat generation was simulated by passing electrical currents through them. These materials, having porosities of 0.6 and higher and thermal conductivities, k_s (including voids), between 0.1 and 0.3 Btu/hr ft °F, resulted in C and n so that Eq. 6-48, for all porous materials, becomes

$$\mathrm{Nu}_v = \left[\frac{l(1-\varphi)}{0.00377}\right]^{1.33} \mathrm{Re}^{0.65} \tag{6-50}$$

where l is in feet. The value of l for these particular materials were found to vary between 2.6×10^{-6} and 2.04×10^{-5} ft. Representative values of h_v for these materials varied between 5000 and 8000 Btu/hr ft³ °F for mass velocities between 870 and 1,500 lb_m/hr ft² of frontal area respectively.

The governing heat-transfer Equations 6-43 through 6-46 have been numerically solved for flat plates and thin hollow cylinders [38] and for thick hollow cylinders [39] using the porosity and characteristic length of 0.6 and 2.0375×10^{-5} ft of one of the materials referred to above, and for helium as a coolant. The solutions yielded temperature profiles for the fluid and solid phases. These temperature profiles were then compared to temperature profiles obtained with conventional surface cooling of solid-type fuel elements using relationships developed in Secs. 5-4 to 5-7. The comparisons were made on the basis of the same total heat generated (same total fissionable nuclei and same neutron flux), same coolant mass flow rate and therefore same inlet and exit coolant temperatures. Comparisons were made on the basis of one-dimensional heat flow (in x or r) and for uniform heat generation in the radial direction (x or r) but for two cases of (1) uniform heat generation and (b) sinusoidal heat generation, in the axial (z) direction and for helium as the coolant. It should be noted here that the solid temperature profiles change from one cross section to another along the axis (because of coolant heating), being maximum at the exit for uniform axial heat generation and at a position z_m between the midplane and exit of the element for sinusoidal axial heat generation (Sec. 13-4).

Figures 6-20 and 6-21 show such comparisons for a hollow cylindrical fuel element of 1-in. inner radius and 2-in. outer radius. The calculations were made for the data indicated in the figure captions. While the heat-transfer correlation, Eq. 6-50, was obtained in experiments on other than nuclear fuel materials, it can be shown that the heat-transfer coefficient h is not a strong function of l, the characteristic length, being proportional to $l^{-0.2}$, so that Eq. 6-50 may be used for a wide variety of porous materials, and was used in the above comparison. The 150 fps coolant surface

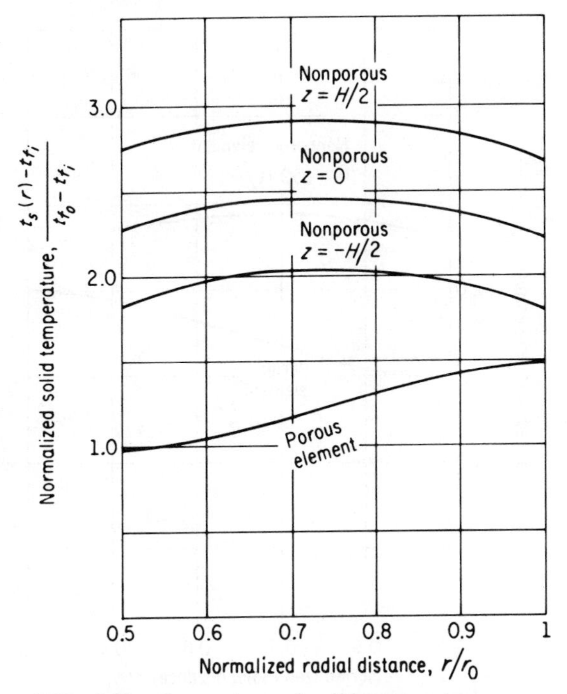

FIG. 6-20. Comparison of solid temperature profiles in nonporous and porous hollow cylindrical fuel elements with uniform energy sources. For both elements: $r_i = 1$ in., $r_o = 2$ in., $H = 20$ ft, $q''' = 2 \times 10^6$ Btu /hr ft³ (based on total volume), $\dot{m} = 4{,}200$ lb_m/hr, $t_{f_i} = 500°\text{F}$, $t_{f_o} = 1000°\text{F}$. For the solid element: $V_f = 150$ fps.

velocity chosen for the nonporous fuel element is high and therefore results in high heat transfer and conservatively low solid temperatures. Nevertheless, as the figures indicate, all calculations show that the porous elements attain much lower solid temperatures than nonporous ones. Conversely a higher volumetric thermal source strength, may be used for the same maximum temperatures.

Porous materials may also be used for bodies, such as thermal shields, Sec. 6-4, (and solar collectors), which are subjected to radiations and are normally surface-cooled. The volumetric thermal source strength in such bodies is usually an exponential function of the distance through the body, Eq. 6-7. A comparison between porous and nonporous shields in which the radiation intensity does not vary in the z direction is shown in Fig. 6-22. The calculations were made for a flat plate (all thermal shields are very thin hollow cylinders) and were based on a porosity of 0.6. The absorption coefficient for the porous material is therefore 40 percent

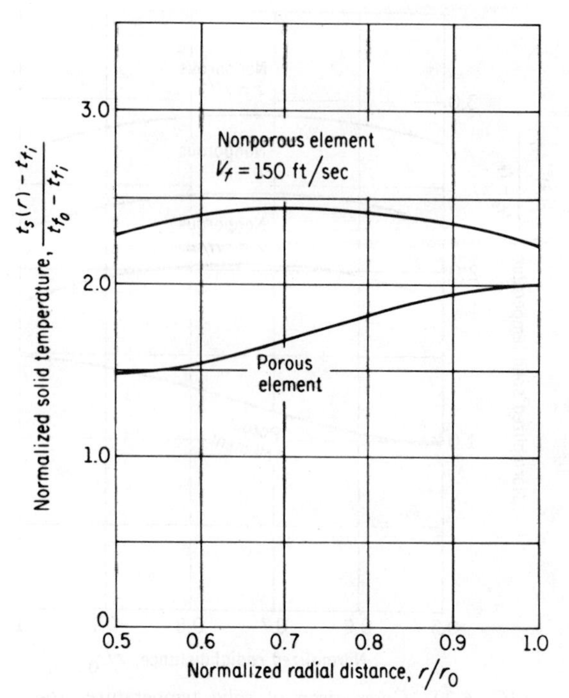

FIG. 6-21. Comparison of solid temperature profiles in nonporous and porous hollow cylindrical fuel element with sinusoidal axial energy sources. Same data as in Fig. 6-20 except $q_0''' = 2 \times 10^6$ Btu/hrft^3 and $\dot{m} = 3{,}000$ lb$_m$/hr.

of the solid material. To make the total energy absorbed the same, the thickness of the porous shield is therefore 2.5 that of the nonporous shield. Again, as Fig. 6-22 shows, porous materials can operate at much lower temperatures than solid materials when used as thermal shields. Conversely, a lower mass of total material may be used for the same maximum temperatures.

The foregoing discussion has been mainly on the heat transfer mechanism of porous materials. Further research work must be done on such materials before they become feasible for nuclear applications. One area is the necessity of investigating the dependence of the characteristic length l on the internal structure of the porous medium. While h is fairly independent of l, the pressure drop is not. Preliminary work has shown that the pressure drop through porous fuel is not expected to pose limitations on core design. Another area concerns fission gases, such as Xe and Kr, Eqs. 1-14 and 1-17, that emanate in fission, and the degree of their retention by the porous material. It is interesting to note that some current fast-reactor designs envisage purging the fission gases into the coolant and trapping them outside the fuel elements (to increase burnout), while others contemplate fission gas retention within the fuel. It is believed possible to construct the porous material with particles having permeable or nonpermeable coatings to meet either of the above design requirements.

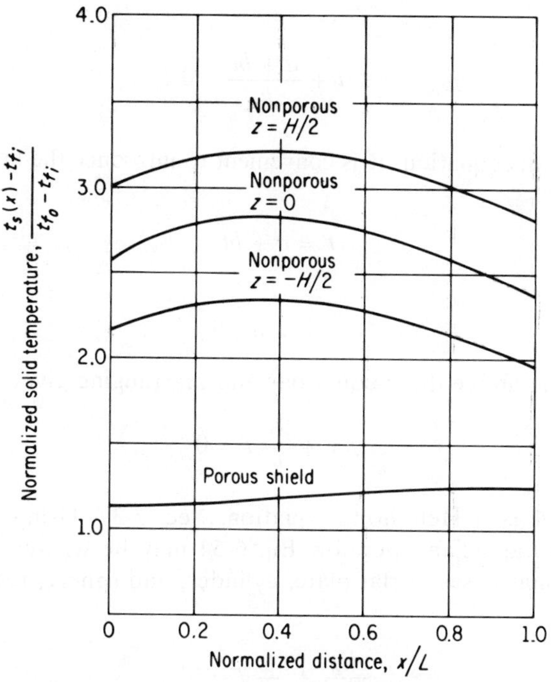

FIG. 6-22. Comparison of solid temperature profiles in nonporous and porous thermal shields. $H = 20$ ft, $q_0''' = 2 \times 10^6$ Btu/hrft^3 (based on total volume), $t_{f_i} =$ 500°F, $t_{f_o} = 1000$°F. For nonporous shield: $L = 2$ in., $\mu = 7.5$ ft^{-1}. For porous shield: $L = 5$ in., $\mu = 3.0$ ft^{-1}.

6-12. THE SPECIAL CASE OF THE POISSON EQUATION WITH $q''' = f(t)$

In previous analyses, the heat generation was assumed uniform in a fuel element. It is, however, sometimes a function of the temperature of the fuel. Such cases occur when the reaction rate between neutrons and fuel nuclei shifts appreciably with temperature (the *Doppler* effect), a case which affects the reactivity of reactors (Sec. 3-8). We shall here take up the case of the heat generation being related to temperature by a linear function of the form

$$q''' = a + bt \tag{6-51}$$

where a and b are constants. The Poisson equation (Sec. 5-3) may now be written as

$$\nabla^2 t + \frac{a + bt}{k_f} = 0 \tag{6-52}$$

To solve this equation, it is convenient to introduce the linear transformation

$$t' = a + bt \tag{6-53}$$

Thus

$$\nabla^2 t' = b \nabla^2 t$$

Combining the above three equations and rearranging give

$$\nabla^2 t' + \frac{b}{k_f} t' = 0 \tag{6-54}$$

Equation 6-54 is a Helmholtz equation, Sec. 5-3. Using the different forms of the Laplacian operator, Eq. 6-54 may be written for the three one-dimensional cases of flat plate, cylinder, and sphere, respectively, as follows:

$$\frac{d^2 t'}{dx^2} + \frac{b}{k_f} t' = 0, \tag{6-55a}$$

$$\frac{d^2 t'}{dr^2} + \frac{1}{r}\frac{dt'}{dr} + \frac{b}{k_f} t' = 0 \tag{6-56a}$$

and

$$\frac{d^2 t'}{dr^2} + \frac{2}{r}\frac{dt'}{dr} + \frac{b}{k_f} t' = 0 \tag{6-57a}$$

Equation 6-56a is a Bessel equation of zero order, Appendix C. The above equations, respectively, have the following general solutions:

$$t' = A \cos\left(\sqrt{\frac{b}{k_f}}\, x\right) + B \sin\left(\sqrt{\frac{b}{k_f}}\, x\right) \tag{6-55b}$$

$$t' = A\, J_0\left(\sqrt{\frac{b}{k_f}}\, r\right) + B\, Y_0\left(\sqrt{\frac{b}{k_f}}\, r\right) \tag{6-56b}$$

and

$$t' = A \frac{\cos\left(\sqrt{\frac{b}{k_f}}\, r\right)}{\left(\sqrt{\frac{b}{k_f}}\, r\right)} + B \frac{\sin\left(\sqrt{\frac{b}{k_f}}\, r\right)}{\left(\sqrt{\frac{b}{k_f}}\, r\right)} \tag{6-57b}$$

where A and B are constants, to be evaluated in each case from the boundary conditions. J_0 and Y_0 in the solution for the cylinder are the Bessel functions of zero order, first, and second kinds respectively.

The case of the one-dimensional (infinitely long) cylinder, Eqs. 6-56, will now be solved for an unclad fuel rod of radius R, cooled by a fluid at t_f with a heat-transfer coefficient h. The boundary conditions for this case are

$$t = \text{finite at } r = 0 \tag{6-58a}$$

and

$$-k_f \frac{dt}{dr}\bigg|_{r=R} = h(t_s - t_f) \tag{6-58b}$$

where t_s is the fuel-rod surface temperature. Introducing

$$t_s' = a + bt_s$$

and

$$t_f' = a + bt_f$$

the second boundary condition becomes

$$-k_f \frac{dt'}{dr}\bigg|_{r=R} = h(t_s' - t_f') \tag{6-58c}$$

Introducing the first boundary condition into Eq. 6-56b gives $B = 0$, since $Y_0(0) = -\infty$, Table C-3, Appendix C. The second boundary condition, after some manipulation, results in

$$A = \frac{h(a + bt_f)}{hJ_0(\sqrt{b/k_f}\, R) - k_f\sqrt{b/k_f} \cdot J_1(\sqrt{b/k_f} R)}$$

where J_1 is the Bessel function of the first order, first kind. Substituting A in Eq. 6-56b and transforming back to t, result finally in

$$t = \frac{\left(\frac{a}{b} + t_f\right) J_0\left(\sqrt{\frac{b}{k_f}}\, r\right)}{J_0\left(\sqrt{\frac{b}{k_f}}\, R\right) - \left(\frac{k_f}{h}\right)\sqrt{\frac{b}{k_f}} \cdot J_1\left(\sqrt{\frac{b}{k_f}}\, R\right)} - \frac{a}{b} \tag{6-59}$$

The above and other cases relating to other geometries have been extensively treated by Jakob [34].

6-13. THE DEPENDENCE OF MATERIAL PROPERTIES ON TEMPERATURE AND PRESSURE

In our solutions of the general conduction and Poisson equations, the material properties k, ρ, and c have been (and will be) considered constant and independent of temperature. This is a reasonably good approximation, especially if these properties vary little with temperature within the temperature range the material is subjected to, and if care is taken in choosing the temperature at which these properties are evaluated. Otherwise the general conduction equation should be written as

$$\nabla \cdot k(t) \nabla t + q''' = \rho(t) c(t) \frac{\partial t}{\partial \theta} \tag{6-60}$$

which is a nonlinear equation. Solutions of nonlinear equations are accomplished by many techniques which are essentially approximate, rather tedious and outside the scope of this discussion. Of the three properties mentioned above, ρ of most solid materials varies little with temperature because of their relatively low coefficients of thermal expansion, and can therefore be considered constant, and c can be treated in a like manner; as in Tables 4-2 and 5-2. Only k can vary greatly. For example, between t_m and t_s in a typical fuel element under full load, k for UO_2 may vary between 1.1 and 2.5 Btu/hr ft °F, an increase of more than 120 percent (Table 5-1). Furthermore, only k appears in the steady-state Poisson and Fourier equations.

The conduction heat flux due to a temperature gradient (one-dimensional), q''_x, is given by the Fourier equation

$$q''_x = -k(t) \frac{dt}{dx} \tag{6-61}$$

With no heat generation, this equation integrates for the case of a flat plate between $t = t_1$ at $x = 0$, and $t = t_2$ at $x = L$, the thickness of the plate, to give

$$q''_x = -\frac{\int_{t_1}^{t_2} k(t)\, dt}{L} \tag{6-62}$$

If the variation of k with temperature can be approximated, within the temperature range of interest, by a linear relationship of the form

$$k(t) = a + bt \tag{6-63}$$

where a and b are constants, the analytical treatment, pointing the way to a choice of an average value of k in the equations, is simplified. In such a case, the heat flux would be

$$q_x'' = -\frac{a(t_2 - t_1) + \frac{b}{2}(t_2^2 - t_1^2)}{L} \tag{6-64}$$

and in terms of the needed average k

$$q_x'' = -k\frac{(t_2 - t_1)}{L} \tag{6-65}$$

Comparing the above two equations results in

$$k = a + \frac{b}{2}(t_2 + t_1) \tag{6-66}$$

which is the value of k at the *arithmetic mean temperature* between t_1 and t_2. The same result applies for cylindrical and spherical geometries. Often it is one of the temperatures that is known prior to the solution of the problem. In such cases an estimate of the mean temperature, and therefore k, is first made, then checked against the result, and recalculation made if necessary.

Pressure usually has very little effect on k, ρ, and c in solids. The effect can become noticeable, however, only if very large temperature gradients exist in the element in question, giving rise to very large local thermal stresses which may accentuate the pressure-dependence. There is little information in the literature on pressure effects in solids and, being small and difficult to account for, are usually neglected.

PROBLEMS

6-1. A 3-in.-thick flat-plate iron shield is subjected at one face to 3.0-Mev/photon gamma radiation of uniform 10^{13} flux. If the temperature of both sides of the plate were to remain equal, find (*a*) the position within the plate at which the maximum temperature occurs; (*b*) the surface cooling on each side of the plate, in Btu/hr ft², necessary to maintain the above conditions; and (*c*) the difference between maximum and surface temperatures.

6-2. Two parallel iron shields are subjected to uniform γ radiation of 3.0 Mev/photon energy and 10^{14} flux, coming in from one side only. The shields are cooled on all sides with air. Determine the thickness of each plate if the total attenuation of γ radiation in both plates is 90% and the amount of heat removed is the same from each plate.

6-3. Two parallel aluminum plates each 6 in. thick and 1 ft apart are subjected on one side to a uniform 10^{12} γ flux of 2.0 Mev/photon energy, measured at the side of one plate. The plates are water-cooled on all sides. The plate surface temperatures, beginning with the surface receiving radiation, are 400, 335, 285, and 250°F, respectively. Calculate the amount of heat, in Btu/hr ft^2 of plate surface carried away by the water between the plates.

6-4. A 6-in.-thick iron plate is subjected on one side to uniform 10^{13} γ flux of 3.0 Mev/photon energy. It is insulated completely against thermal losses on the other side. Determine (*a*) the temperature of the insulated side of the plate if the exposed side is at 500°F and (*b*) the amount of cooling required on the exposed side in Btu/hr ft^2.

6-5. A 3-in.-thick slab of pure nickel 60 (density 8.9 g_m/cm^3, specific heat 0.103) is subjected on one side to a 10^{11} flux of 2,200 m/sec neutrons and to γ radiation. Find the average increase in slab temperature of °F/hr due to neutron absorption alone, if the slab were not cooled.

6-6. A parallel beam of thermalized neutrons strike a flat plate 6 in. thick. The neutrons are assumed to be thermalized to 500°F. They interact with the material of the plate by absorption only. This absorption is assumed to occur in the $1/V$ range. The energy released per absorption reaction is 8.5 Mev. The neutron flux in the beam is 10^{11}. The microscopic absorption cross section for monoenergetic neutrons at 68°F is 2.6 barns. The nuclear density of the material of the plate is 10^{23} nuclei/cm^3. Find the total energy generated in the plate in Btu/hr ft^2.

6-7. A 4-in.-thick iron thermal shield receives 5 Mev gamma rays on one side. The total heat generated in the shield is 200,000 Btu/hr ft^2. What is the incident gamma flux? Ignore secondary radiations.

6-8. A wall is composed of an 8-in.- thick aluminum plate lined by 2 in. of type 347 stainless steel. The wall is subjected, on the liner side, to gamma radiation of 4×10^{12} flux and 3-Mev energy. The buildup factors in the plate and liner maybe taken as 1.5 each. The wall is cooled by 400°F water with a convective heat-transfer coefficient of 200 Btu/hr ft^2 °F on each side. Neglecting any contact resistance between plate and liner, find (*a*) the temperatures of the two outside wall surfaces and of the plate-liner interface, (*b*) the heat generated in the plate, and (*c*) the heat generated in the liner, in Btu /hr ft^2.

6-9. A 3-inch thick iron slab is subjected on one side to 3-Mev gamma rays. The total heat generated in the slab is 10^6 Btu/hr ft^2. Find the gamma flux in photons/sec cm^2.

6-10. A lead slab receives a gamma beam of 1 Mev-energy photons from one side only. It absorbs 90% of the beam. What is the thickness of the slab in inches? Assume a build-up factor of 3.0.

6-11. A reactor pressure vessel made of 6-in. thick steel plate is subjected on one side only to pure γ-radiation of 10^{14} photons/sec cm^2 and 5 mev/photon strength. The two sides are cooled by forced convection with equal heat transfer coefficients of 1000 Btu/hr ft^2 °F, by water at 300°F. Calculate: (*a*) the two surface temperatures, and (*b*) the maximum temperature within the plate.

6-12. In designing a pressure vessel, the maximum thermal stress had to be reduced by 90% by the use of two thermal shields made of iron plate. Assuming only pure γ-radiation of 5 Mev energy and 10^{14} flux, find the amount of heat in Btu/hr

ft^2 that must be removed from each shield if they were made of equal thicknesses. Assume no buildup factor and that thermal stresses are proportional to q'''.

6-13. A pressure vessel is protected by a thermal shield. Assuming for simplicity pure gamma radiation of 5.0 Mev/photon and 10^{14} photon/sec cm^2 flux reaching the vessel, calculate (*a*) the thickness (inches) of an iron thermal shield that would reduce the above flux by 90%, and (*b*) the heat generated in the thermal shield in Btu/hr ft^2. Ignore secondary radiations and curvature of the shield.

6-14. A large iron plate 1-foot thick is subjected to pure gamma radiations from both sides. On one side it is subjected to 10^{14} γ-flux of 3.0 Mev/photon energy; on the other to 10^{13} flux and 1 Mev/photon energy. Both surfaces of the plate are held at 500°F. Assuming no interaction between the radiations and no buildup factor; find (*a*) the maximum temperature within the slab; (*b*) the position of this maximum temperature; and (*c*) the amount of heat that must be removed from each side in Btu/hr ft^2 in order to maintain the above temperatures.

6-15. Derive expressions for the temperature distribution in, and heat conducted at the base of, an infinitely long fin of constant cross sectional area and known generation ratio. Assume that the convective heat transfer coefficient between fin surface and fluid is constant.

6-16. An axial fin of constant cross section is 0.1 in. thick and 0.5 in. long. It is subjected to a thermal neutron flux of 10^{14}. The temperature at the base of the fin is 300°F above the coolant temperature, and the heat-transfer coefficient is 100 Btu/hr ft^2 °F. Find the percent change in heat transfer from the fuel via the fin if the fin material were changed from steel to beryllium. Assume for simplicity that heat generation within the fin is solely due to thermal neutron absorption and that the steel fin material is Fe^{56}.

6-17. A natural uranium fuel rod 1 in. in diameter and 3 ft long is used in a large gas-cooled reactor. It has magnesium cladding 0.030 in. thick plus integral circumferential fins. The element, which is in a vertical channel, is short enough compared to the height of the core so that it may be assumed subjected to a constant neutron flux of 1.66×10^{12} neutrons/sec cm^2. The fins are of constant 0.2 in. thickness. The gap between adjacent fins is 0.2 in. Determine the outside diameter of the fins necessary for heat removal. Take $h = 37.50$ Btu/hr ft^2 °F, gas temperature = 700°F, temperature of the cladding at the base = 740°F, and k for magnesium = 88 Btu/hr ft °F. Neglect heat generation in the fin.

6-18. It is desired to compare porous and nonporous thermal shields operating under the conditions of Fig. 6-22. For each calculate (*a*) the minimum solid temperature, (*b*) the maximum solid temperature, (*c*) the total heat generated per unit frontal area, and (*d*) the coolant mass flow rate per unit width of shield.

6-19. Assume a fuel material operates such that the volumetric thermal source strength increases linearly with its temperature from 4.9×10^6 to 5×10^6 Btu/hr ft^3 when the temperature changes from 3000 to 4000°F. A cylindrical fuel element of 0.1-ft radius using the above material is cooled at a particular cross section by a coolant at 500°F with a heat-transfer coefficient of 2000 Btu/hr ft^2 °F. Neglect cladding and take $k_f = 1.0$ Btu/hr ft°F. Find (*a*) the maximum, (*b*) the minimum temperatures at the above cross section, and (*c*) the heat generated per foot of height, Btu/hr ft.

6-20. A cylindrical fuel element, 0.06 ft in diameter using UO_2 as fuel material, generates 5×10^6 Btu/hr ft^3 at its surface when that surface is at 2000°F. The thermal conductivity may be considered constant at 1.25 Btu/hr ft°F. How much hotter would the center of this element operate if the heat generation increases by 1 percent for each 1000°F temperature rise above surface conditions instead of being uniform at the above value?

chapter 7

Heat Conduction in Reactor Elements

III. Two-Dimensional Steady-State Cases

7-1. INTRODUCTION

In the previous two chapters, exact solutions for the temperature distributions in some simple elements with one-dimensional heat conduction and internal heat generation were obtained analytically by integrating the pertinent Poisson equations and applying appropriate boundary conditions. Exact solutions are available only for some multi-dimensional and transient cases. An example of an exact analytical solution will be presented for a two-dimensional transient case in Sec. 8-10. In many cases with complicated geometries or boundary conditions, however such solutions are either complex or nonexistent, and other methods of solution are adopted. These may be classified into *experimental, numerical, approximate analytical,* and *analogical* (including *graphical*) methods.

In the *experimental* method of solution, a model of the element in question is constructed and heat is generated in it electrically, nuclearly or chemically. The temperature distribution in the element is then measured and recorded (necessary in transient cases) at key locations. The method is usually time consuming and expensive and the other methods mentioned above have found wider use.

This chapter deals primarily with two-dimensional steady-state cases. The next chapter discusses transient cases. Numerical methods will first be emphasized, an approximate analytical method of solution will then be illustrated, and, finally, a survey of analogical methods will be given in this chapter.

7-2. FINITE-DIFFERENCE TECHNIQUES–GENERAL

Numerical methods are based on *finite-difference* techniques. These techniques approximate the differential equations to be solved by finite-difference equations. The principles involved are relatively simple, and the technique can be adopted, in the simpler cases, to a desk calculator. In more complicated cases, or where more accuracy is sought by using finer differences, the size of the job would have made the technique unattractive except that, fortunately, digital computers have become widey available. Many permanent computer programs for calculating temperature distributions are available. In some cases, even though exact analytical solutions may exist a finite-difference technique might prove more economical and convenient.

The derivatives in a differential equation describing a continuous temperature distribution $t(x)$ as that shown in Fig. 7-1, are approximated

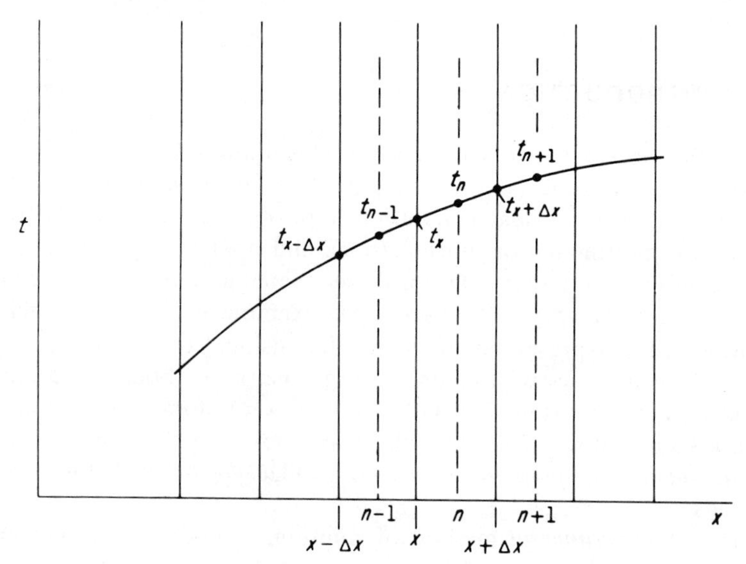

FIG. 7-1. Nodes in a one-dimensional system.

by finite differences by dividing the independent variable x into a finite number of increments of equal width Δx. At n, a coordinate halfway between x and $x + \Delta x$, the first derivative is given by

$$\left(\frac{\partial t}{\partial x}\right)_n = \lim_{\Delta x \to 0} \frac{t_{x+\Delta x} - t_x}{\Delta x} \tag{7-1}$$

If Δx is small enough, then *approximately*

$$\left(\frac{\partial t}{\partial x}\right)_n = \frac{t_{x+\Delta x} - t_x}{\Delta x} \tag{7-2}$$

The second derivative at n is obtained from first derivatives in finite difference form as

$$\left(\frac{\partial^2 t}{\partial x^2}\right)_n = \frac{\left.\frac{\partial t}{\partial x}\right|_{x+\Delta x} - \left.\frac{\partial t}{\partial x}\right|_x}{\Delta x} = \frac{\frac{t_{n+\Delta x} - t_n}{\Delta x} - \frac{t_n - t_{n-\Delta x}}{\Delta x}}{\Delta x}$$

or

$$\left(\frac{\partial^2 t}{\partial x^2}\right)_n = \frac{(t_{n+\Delta x} - t_n) - (t_n - t_{n-\Delta x})}{(\Delta x)^2} \tag{7-3}$$

n, $n + \Delta x$, $n - \Delta x$, etc. are called *nodal planes* or *nodes*.

At this stage it is instructive to evaluate the *errors* in the first and second derivatives due to the above finite-difference approximation. These errors are found by expanding the temperature function $t(x)$ in a Taylor series about n as

$$t(x) = t_n + (x-n)\left(\frac{\partial t}{\partial x}\right)_n + \frac{(x-n)^2}{2!}\left(\frac{\partial^2 t}{\partial x^2}\right)_n + \frac{(x-n)^3}{3!}\left(\frac{\partial^3 t}{\partial x^3}\right)_n + \frac{(x-n)^4}{4!}\left(\frac{\partial^4 t}{\partial x^4}\right)_n + \cdots \tag{7-4}$$

From which, at $x = n - \Delta x$ and $x = n + \Delta x$:

$$t_{n-\Delta x} = t_n - \Delta x\left(\frac{\partial t}{\partial x}\right)_n + \frac{(\Delta x)^2}{2}\left(\frac{\partial^2 t}{\partial x^2}\right)_n - \frac{(\Delta x)^3}{6}\left(\frac{\partial^3 t}{\partial x^3}\right)_n + \frac{(\Delta x)^4}{24}\left(\frac{\partial^4 t}{\partial x^4}\right)_n - \cdots \tag{7-5}$$

and

$$t_{n+\Delta x} = t_n + \Delta x\left(\frac{\partial t}{\partial x}\right)_n + \frac{(\Delta x)^2}{2}\left(\frac{\partial^2 t}{\partial x^2}\right)_n + \frac{(\Delta x)^3}{6}\left(\frac{\partial^3 t}{\partial x^3}\right)_n + \frac{(\Delta x)^4}{24}\left(\frac{\partial^4 t}{\partial x^4}\right)_n + \cdots \tag{7-6}$$

Subtracting Eq. 7-5 from 7-6 and rearranging give

$$\left(\frac{\partial t}{\partial x}\right)_n = \frac{t_{n+\Delta x} - t_{n-\Delta x}}{2\Delta x} - \frac{(\Delta x)^2}{6}\left(\frac{\partial^3 t}{\partial x^3}\right)_n - \cdots \tag{7-7}$$

Comparing Eqs. 7-2 and 7-7, noting that $t_{n+\Delta x} - t_{n-\Delta x} \simeq 2(t_{x+\Delta x} - t_x)$, and ignoring derivatives of order 5 or higher, then

$$\text{Error in first derivative} = \frac{(\Delta x)^2}{6}\left(\frac{\partial^3 t}{\partial x^3}\right)_n \tag{7-8}$$

Adding Eqs. 7-5 and 7-6 and rearranging give

$$\left(\frac{\partial^2 t}{\partial x^2}\right)_n = \frac{(t_{n+\Delta x} - t_n) - (t_n - t_{n-\Delta x})}{(\Delta x)^2} - \frac{(\Delta x)^2}{12}\left(\frac{\partial^4 t}{\partial x^4}\right)_n - \cdots \tag{7-9}$$

Comparing Eqs. 7-3 and 7-9, and ignoring derivatives of order 6 and higher, then

$$\text{Error in second derivative} = \frac{(\Delta x)^2}{12}\left(\frac{\partial^4 t}{\partial x^4}\right)_n \tag{7-10}$$

and the errors are all functions of $(\Delta x)^2$. Finer divisions, therefore, are conducive to greater accuracies.

7-3. TRANSFORMATION OF THE POISSON EQUATION

The One-Dimensional Case

In the steady-state case, the basic Poisson differential equation (Sec. 5-3) is written in the form

$$\frac{d^2 t}{dx^2} + \frac{q'''}{k} = 0 \tag{7-11}$$

where q''' is the volumetric thermal source strength (Sec. 4-2) and k is the thermal conductivity. The above equation is converted to a finite difference form with the help of Eq. 7-3, as

$$\frac{(t_{n+\Delta x} - t_n) - (t_n - t_{n-\Delta x})}{(\Delta x)^2} + \frac{q'''(n)}{k} = 0 \tag{7-12a}$$

where q''' is assumed a function of position n. Multiplying by $(\Delta x)^2/2$, and rearranging give

$$t_n = \frac{t_{n+\Delta x} + t_{n-\Delta x}}{2} + \frac{(\Delta x)^2 q'''(n)}{2k} \tag{7-12b}$$

The last term has the units of temperature. Defining

$$\Delta t_g(n) = \frac{(\Delta x)^2 q'''(n)}{2k} \tag{7-13a}$$

where $\Delta t_g(n)$ is a *temperature increment due to heat generation*, gives

$$t_n = \frac{t_{n+\Delta x} + t_{n-\Delta x}}{2} + \Delta t_g(n) \tag{7-12c}$$

If heat generation is uniform, $\Delta t_g(n)$ is a constant given by

$$\Delta t_g = \frac{(\Delta x)^2 q'''}{2k} \tag{7-13b}$$

and

$$t_n = \frac{t_{n+\Delta x} + t_{n-\Delta x}}{2} + \Delta t_g \tag{7-12d}$$

The Two-Dimensional Case

The temperature is expressed as a function of the two independent variables, x and y, in rectangular coordinates. In the steady-state case the pertinent equation is the Poisson equation,

$$\left(\frac{\partial^2 t}{\partial x^2} + \frac{\partial^2 t}{\partial y^2}\right)_n + \frac{q'''(n)}{k} = 0 \tag{7-14}$$

The field is divided into a fine square grid of widths $\Delta x = \Delta y$, Fig. 7-2.

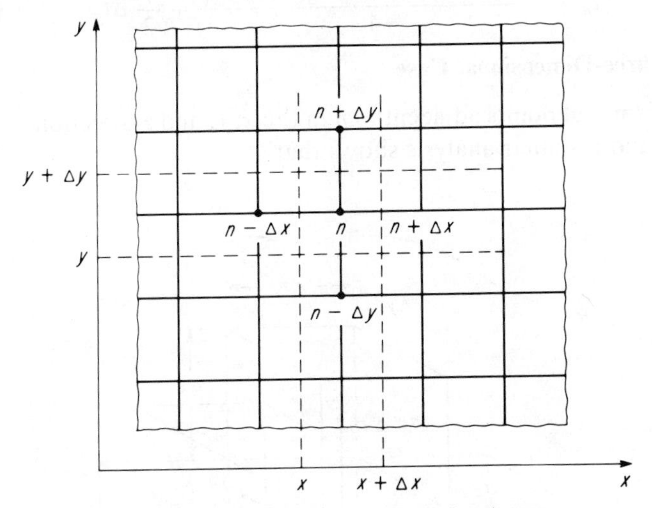

FIG. 7-2. Nodes in a two-dimensional system.

The Laplacian at n is now obtained in terms of the temperatures at the *nodal points* shown, with the help of Eq. 7-3, as

$$\left(\frac{\partial^2 t}{\partial x^2} + \frac{\partial^2 t}{\partial y^2}\right)_n = \frac{(t_{n+\Delta x} - t_n) - (t_n - t_{n-\Delta x})}{(\Delta x)^2} + \frac{(t_{n+\Delta y} - t_n) - (t_n - t_{n-\Delta y})}{(\Delta y)^2}$$

$$= \frac{(t_{n+\Delta x} + t_{n-\Delta x} + t_{n+\Delta y} + t_{n-\Delta y}) - 4t_n}{(\Delta x)^2} \tag{7-15}$$

Combining the above two equations and rearranging give

$$(t_{n+\Delta x} + t_{n-\Delta x} + t_{n+\Delta y} + t_{n-\Delta y}) - 4t_n + \frac{(\Delta x)^2 q'''(n)}{k} = 0$$

or

$$t_n = \frac{t_{n+\Delta y} + t_{n+\Delta x} + t_{n-\Delta y} + t_{n-\Delta x}}{4} + \frac{(\Delta x)^2 q'''(n)}{4k} \tag{7-16a}$$

Noting that

$$\frac{(\Delta x)^2 q'''(n)}{4k} = \frac{1}{2} \Delta t_g(n)$$

$$t_n = \frac{t_{n+\Delta y} + t_{n+\Delta x} + t_{n-\Delta y} + t_{n-\Delta x}}{4} + \frac{1}{2} \Delta t_g(n) \tag{7-16b}$$

In the case of uniform heat generation, q''' is not a function of position, and

$$t_n = \frac{t_{n+\Delta y} + t_{n+\Delta x} + t_{n-\Delta y} + t_{n-\Delta x}}{4} + \frac{1}{2} \Delta t_g \tag{7-16c}$$

The Three-Dimensional Case

The six nodal points adjacent to n in the x, y, and z directions are used (Fig. 7-3) and a similar analysis shows that

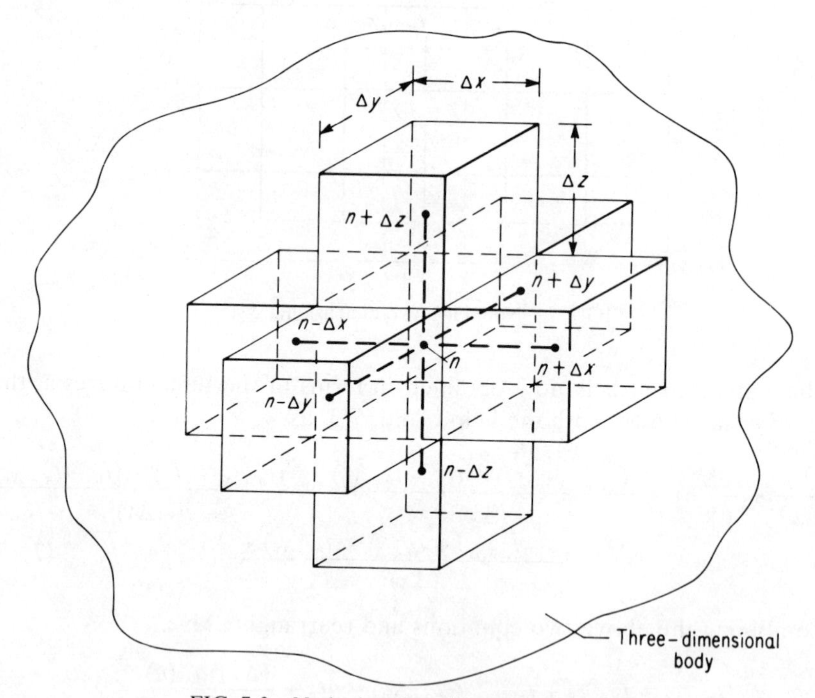

FIG. 7-3. Nodes in a three-dimensional system.

$$t_n = \frac{t_{n+\Delta z} + t_{n+\Delta y} + t_{n+\Delta x} + t_{n-\Delta z} + t_{n-\Delta y} + t_{n-\Delta x}}{6} + \frac{1}{3}\,\Delta t_g(n) \tag{7-17}$$

The temperature at a nodal point is therefore equal to the average of the temperatures at the surrounding nodal points, plus an additional quantity, due to internal heat generation, equal to $\Delta t_g(n)$ in the one-dimensional case, $1/2\ \Delta t_g(n)$ in the two-dimensional case and $1/3\ \Delta t_g(n)$ in the three-dimensional case. In the case of no heat generation the temperature at a nodal point is simply the average of the temperatures of the surrounding adjacent nodal points.

It is seen that the Poisson differential equation with constant properties becomes, in finite difference form, a linear algebraic equation. The technique is now to write out similar equations for all nodal points where the temperature is not known, and to simultaneously solve the resulting set of equations with the help of known boundary conditions.

7-4. METHODS OF SOLUTIONS

There are several methods for the solution of a system of simultaneous linear equations. The simplest, of course, is the familiar method of *algebraic elimination.* This method is suited to cases where there are only a small number of unknowns, such as two or three. The method of *matrix algebra,* usually used for a moderately large number of unknowns, and the method of *iteration,* usually used for a large number of unknowns, are both suitable for digital-computer use. Finally the method of *relaxation* is suitable for manual operation where the number of unknowns is moderately large.

These methods will now be illustrated for the case of an infinitely long cruciform-shaped element, shown in cross section in Fig. 7-4 as an example. The following data is assumed: Heat generation is uniform at 10^7 Btu/hr ft^3, all surfaces held at 600°F, and $k = 19.84$ Btu/hr ft°F. Since the element is infinitely long, heat flows in two dimensions only. It is required to obtain the temperature distribution within it. An accurate temperature distribution would require a fine mesh. However since it is only intended to illustrate the method here, and to conserve time, the coarse mesh shown with $\Delta x = \Delta y = 0.2$ in. is chosen. Also, because of symmetry, it will only be necessary to compute the temperatures at nodal points marked 1 through 6. For this case,

$$\Delta t_g = \frac{(0.2/12)^2 \times 10^7}{2 \times 19.84} = 70°\text{F}$$

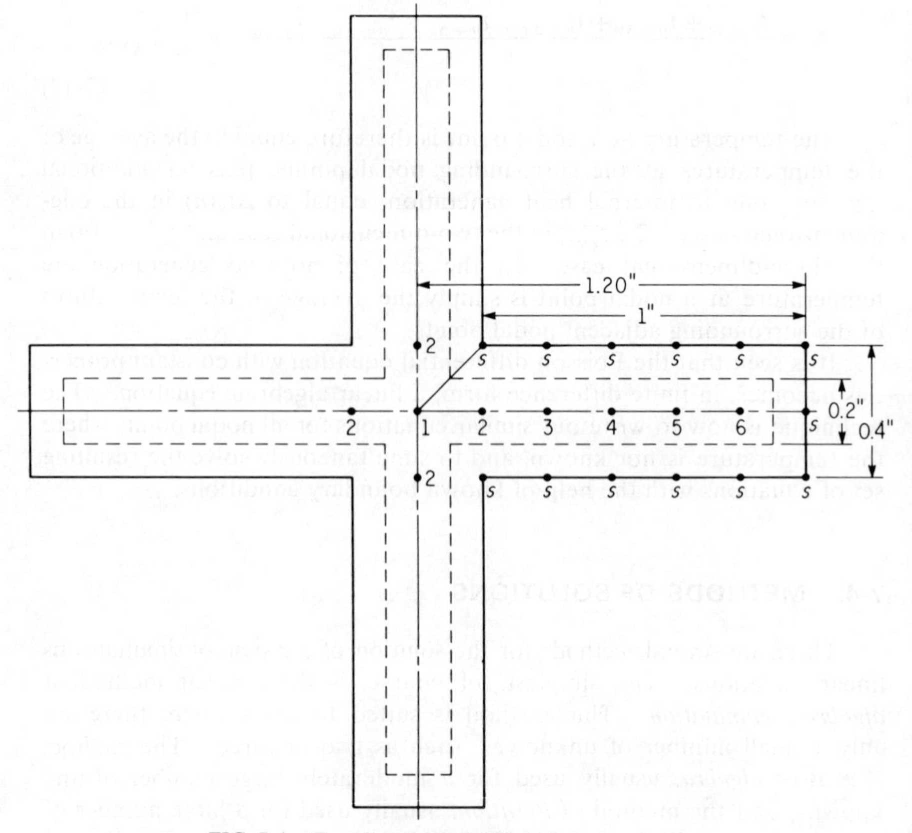

FIG. 7-4. Two-dimensional (infinitely long) cruciform.

Noting symmetry, the temperature at nodal point 1, t_1, is given with the help of Eq. 7-16c by

$$t_1 = \frac{t_2 + t_2 + t_2 + t_2}{4} + \frac{1}{2}\,\Delta t_g$$
$$= t_2 + 35$$

Likewise

$$t_2 = \frac{t_1 + 600 + t_3 + 600}{4} + \frac{1}{2}\,\Delta t_g$$
$$= \frac{1}{4}\,t_1 + \frac{1}{4}\,t_3 + 335$$

and so on for the rest of the nodal points. The equations are now written in sequence in the following manner

$$
\begin{aligned}
t_1 - t_2 &= +35 \\
\frac{1}{4}t_1 - t_2 + \frac{1}{4}t_3 &= -335 \\
\frac{1}{4}t_2 - t_3 + \frac{1}{4}t_4 &= -335 \\
\frac{1}{4}t_3 - t_4 + \frac{1}{4}t_5 &= -335 \\
\frac{1}{4}t_4 - t_5 + \frac{1}{4}t_6 &= -335 \\
\frac{1}{4}t_5 - t_6 &= -485
\end{aligned}
$$

7-5. SOLUTION BY MATRIX ALGEBRA

The order of the equations must be such that the leading coefficient in the first row is not zero. The coefficients and constants are then grouped in *augmented matrix* form as

$$
\begin{bmatrix}
1 & -1 & 0 & 0 & 0 & 0 & +35 \\
\frac{1}{4} & -1 & +\frac{1}{4} & 0 & 0 & 0 & -335 \\
0 & \frac{1}{4} & -1 & +\frac{1}{4} & 0 & 0 & -335 \\
0 & 0 & \frac{1}{4} & -1 & +\frac{1}{4} & 0 & -335 \\
0 & 0 & 0 & \frac{1}{4} & -1 & +\frac{1}{4} & -335 \\
0 & 0 & 0 & 0 & \frac{1}{4} & -1 & -485
\end{bmatrix}
$$

The values of the temperatures are obtained by row manipulation. The leading coefficient in each row is made equal to 1 (if not already so) by dividing that row by its value. The matrix now becomes

$$
\begin{bmatrix}
1 & -1 & 0 & 0 & 0 & 0 & +35 \\
1 & -4 & +1 & 0 & 0 & 0 & -1{,}340 \\
0 & 1 & -4 & +1 & 0 & 0 & -1{,}340 \\
0 & 0 & 1 & -4 & +1 & 0 & -1{,}340 \\
0 & 0 & 0 & 1 & -4 & +1 & -1{,}340 \\
0 & 0 & 0 & 0 & 1 & -4 & -1{,}940
\end{bmatrix}
$$

Since all equations (rows) are interdependent, they can be added and subtracted at will. For example, the second row is now subtracted from the first row, so that its leading coefficient disappears and a new leading coefficient (+.3) shows up in its second column. The new second row is used together with the third row (by multiplying the latter by 3 and subtracting) so that a new leading coefficient for the third row (+11) falls in the third column, and so on. The resulting matrix is

$$\begin{bmatrix} 1 & -1 & 0 & 0 & 0 & 0 & +\ 35 \\ 0 & +3 & -1 & 0 & 0 & 0 & +\ 1{,}375 \\ 0 & 0 & +11 & -3 & 0 & 0 & +\ 5{,}395 \\ 0 & 0 & 0 & +41 & -11 & 0 & +\ 20{,}135 \\ 0 & 0 & 0 & 0 & +153 & -41 & +\ 75{,}075 \\ 0 & 0 & 0 & 0 & 0 & +571 & +371{,}895 \end{bmatrix}$$

The rows are now divided by their leading terms to give

$$\begin{bmatrix} +1 & -1 & & & & & +\ 35 \\ & +1 & -\frac{1}{3} & & & & +458.333 \\ & & +1 & -\frac{3}{11} & & & +490.454 \\ & & & +1 & -\frac{11}{41} & & +491.097 \\ & & & & +1 & -\frac{41}{153} & +490.686 \\ & & & & & +1 & +651.304 \end{bmatrix}$$

where the zero terms have been omitted for simplicity. Note that in the above matrix the value of t_6 already appears as 651.304. Proceeding now with the last two rows, the sixth row is multiplied by the last term of the fifth row (41/153) and added to the fifth row to eliminate the latter's last term. The result leaves the leading (and only) term of the fifth row as 1.0 and gives the value of t_5. This manipulation is repeated for the fourth row and so on. This eliminates all but diagonal terms of unity and results in

$$\begin{bmatrix} 1 & & & & & & 717.687 \\ & 1 & & & & & 682.687 \\ & & 1 & & & & 673.064 \\ & & & 1 & & & 669.570 \\ & & & & 1 & & 665.218 \\ & & & & & 1 & 651.304 \end{bmatrix}$$

The last column of the above matrix gives the temperatures t_1 through t_6 in that order.

7-6. SOLUTION BY ITERATION

This method is lengthy and tedious, and lends itself to manual operation only in the simplest cases. It is, however, well suited to digital computers. It consists of successive steps. The first is to estimate initial values for the unknown temperatures at all nodal points. In the next step the temperature at one nodal point is recalculated using, for the two-dimensional case, Eq. 7-16c. The temperature at the next nodal point is now recalculated using the new adjacent temperature and the previous three adjoining temperatures, and so on. The sequence is stopped when further iteration does not result in significant changes in the temperatures.

The highest temperature in the mesh, Fig. 7-4, will be t_1. Let us assume initial values shown by superscript 1 as follows: $t_1^1 = 980$°F, $t_2^1 = 850$°F, $t_3^1 = 750$°F, $t_4^1 = 700$°F, $t_5^1 = 650$°F, and $t_6^1 = 620$°F. The nodal points at the surface will always be at $t_s = 600$°F.

The second step in computation shown by superscript 2, starting with t_1 gives

$$t_1^2 = \frac{t_2^1 + t_2^1 + t_2^1 + t_2^1}{4} + \frac{1}{2}\,\Delta t_g = \frac{850 + 850 + 850 + 850}{4} + 35 = 885\text{°F}$$

$$t_2^2 = \frac{t_1^2 + t_s + t_3^1 + t_s}{4} + \frac{1}{2}\,\Delta t_g = \frac{885 + 600 + 750 + 600}{4} + 35 = 743.75\text{°F}$$

$$t_3^2 = \frac{t_2^2 + t_s + t_4^1 + t_s}{4} + \frac{1}{2}\,\Delta t_g = \frac{743.75 + 600 + 700 + 600}{4} + 35 = 695.94\text{°F}$$

$$t_4^2 = \frac{t_3^2 + t_s + t_5^1 + t_s}{4} + \frac{1}{2}\,\Delta t_g = \frac{694.94 + 600 + 650 + 600}{4} + 35 = 671.49\text{°F}$$

$$t_5^2 = \frac{t_4^2 + t_s + t_6^1 + t_s}{4} + \frac{1}{2}\,\Delta t_g = \frac{671.49 + 600 + 620 + 600}{4} + 35 = 657.87\text{°F}$$

$$t_6^2 = \frac{t_5^2 + t_s + t_s + t_s}{4} + \frac{1}{2}\,\Delta t_g = \frac{657.87 + 600 + 600 + 600}{4} + 35 = 649.47\text{°F}$$

Note that t_1^1 need not have been estimated in step 1. The third step in computation gives

$$t_1^3 = \frac{t_2^2 + t_2^2 + t_2^2 + t_2^2}{4} + \frac{1}{2}\,\Delta t_g = \frac{743.75 + 743.75 + 743.75 + 743.75}{4} + 35$$
$$= 778.75\text{°F}$$

$$t_2^3 = \frac{t_1^3 + t_s + t_3^2 + t_s}{4} + \frac{1}{2}\,\Delta t_g = \frac{778.75 + 600 + 695.94 + 600}{4} + 35$$
$$= 703.67\text{°F}$$

and so on. The solution up to 10 steps, is given in Table 7-1. An advantage of this method is that an error made in one iterative step only lengthens the solution time and is corrected in later steps. Compare the results of this solution with the more exact one obtained by matrix algebra.

TABLE 7-1
Iterative Solution For Cruciform of Fig. 7-4

Step	t_1	t_2	t_3	t_4	t_5	t_6
1	980	850	750	700	650	620
2	885.00	743.75	695.94	671.49	657.87	649.47
3	778.75	703.67	678.79	669.17	664.66	651.17
4	738.67	689.37	674.64	669.83	665.25	651.31
5	724.37	684.75	673.65	669.73	665.26	651.32
6	719.75	683.35	673.27	669.63	665.24	651.31
7	718.35	682.91	673.14	669.60	665.23	651.31
8	717.91	682.76	673.09	669.58	665.22	651.31
9	717.76	682.71	673.07	669.57	665.22	651.31
10	717.71	682.70	673.07	669.57	665.22	651.31

The element may be divided into a finer mesh with Δx cut in half to 0.1 in., so that Δt_g is reduced to a quarter of its value above, or 17.5°F. The procedure is, of course, much lengthier, but the solution, shown in Fig. 7-5, is much more accurate.

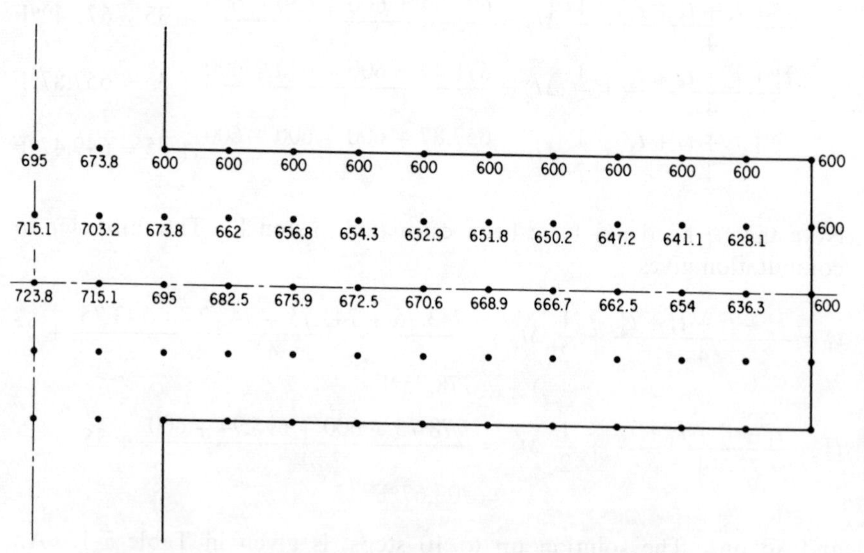

FIG. 7-5. Solution of cruciform of Fig. 7-4 with a 0.1-in. mesh.

7-7. SOLUTION BY THE RELAXATION METHOD

The method of relaxation, widely used in various fields of engineering such as heat conduction, fluid flow, stress analysis and vibrations, is a method of successive approximations. It is flexible, rapid, and suitable for manual operation.

As in the case of the iterative method, the first step here is to make estimates of unknown temperatures at various nodal points. These temperatures, if correct, would satisfy, in the two-dimensional case, all equations of the form of Eq. 7-16c for all nodal points.

The method is to evaluate the errors in the first assumption and then successively reduce them until they nearly vanish. This is done by first evaluating an initial *residual* R_n at nodal point n, given with the help of Eq. 7-16c as

$$R_n^1 = (t_{n+\Delta y}^1 + t_{n+\Delta x}^1 + t_{n-\Delta y}^1 + t_{n-\Delta x}^1) + 2\Delta t_g - 4t_n^1 \qquad (7\text{-}18)$$

for all nodal points. The method is expedited by first eliminating, or *relaxing*, only one residual, the one that has the largest numerical value. This is done by adding one quarter of the residual to t_n. The method is best illustrated by an example. The same problem of the cruciform and the course mesh of Fig. 7-4, above, will be solved using the same initial values assumed in the previous section for nodal points 1 through 6. The initial residuals are computed in the first step as follows:

$$\begin{aligned} R_1^1 &= (t_2^1 + t_2^1 + t_2^1 + t_2^1) + 2\Delta t_g - 4t_1^1 = (850 + 850 + 850 + 850) \\ &\qquad + 140 - 4 \times 980 = -380^\circ\text{F} \\ R_2^1 &= (t_s + t_3^1 + t_s + t_1^1) + 2\Delta t_g - 4t_2^1 = (600 + 750 + 600 + 980) \\ &\qquad + 140 - 4 \times 850 = -330^\circ\text{F} \\ R_3^1 &= (t_s + t_4^1 + t_s + t_2^1) + 2\Delta t_g - 4t_3^1 = (600 + 700 + 600 + 850) \\ &\qquad + 140 - 4 \times 750 = -110^\circ\text{F} \\ R_4^1 &= (t_s + t_5^1 + t_s + t_3^1) + 2\Delta t_g - 4t_4^1 = (600 + 650 + 600 + 750) \\ &\qquad + 140 - 4 \times 700 = -60^\circ\text{F} \\ R_5^1 &= (t_s + t_6^1 + t_s + t_4^1) + 2\Delta t_g - 4t_5^1 = (600 + 620 + 600 + 700) \\ &\qquad + 140 - 4 \times 650 = +60^\circ\text{F} \\ R_6^1 &= (t_s + t_s + t_s + t_5^1) + 2\Delta t_g - 4t_6^1 = (600 + 600 + 600 + 650) \\ &\qquad + 140 - 4 \times 620 = +110^\circ\text{F} \end{aligned}$$

In the second step, the residual with the largest numerical value, is noted. This happens to be R_1^1 at nodal point 1. One quarter of its value or -95 is added to the temperature at that nodal point, t_1^1, resulting in a new value $t_1^2 = t_1^1 + (1/4)\,R_1^1 = 980 - 95 = 885^\circ$F, and leaving the other

TABLE 7-2
Relaxation Solution of Example 8-2, Fig. 8-3. *(Residuals in italic type)*

Step	t_1	R_1	t_2	R_2	t_3	R_3	t_4	R_4	t_5	R_5	t_6	R_6
1	980.00	*−380.00*	850.00	*−330.00*	750.00	*−110.00*	700.00	*−60.00*	650.00	*+60.00*	620.00	*−110.00*
2	885.00			*−425.00*								
3		*−425.00*	743.75	*0*		*−216.25*						
4	778.75			*−106.25*								
5				*−160.31*	695.94	*0*		*−114.00*				
6		*−160.32*	703.67	*0*		*−40.00*						
7	738.67			*−40.00*								
8						*−68.59*	671.50	*0*		*+31.50*		
9										*+59.60*	647.50	
10				*−57.22*	678.79	*0*		*−17.21*				*0*
11								*−2.46*	664.75	*0*		*+14.75*
12		*−57.20*	689.37	*0*		*−14.29*						
13	724.37	*0*		*−14.30*								
14										*+3.69*	651.19	*0*
15		*−14.28*	685.80	*0*		*−17.86*						
16				*−4.50*	674.33	*0*		*−6.92*				
17	720.80	*0*		*−8.07*								
18		*−8.08*	638.78	*0*		*−2.04*						
19	718.78	*0*		*−2.01*								
20						*−3.77*	669.77	*0*		*+1.96*		
21				*−2.95*	673.39	*0*		*−0.94*				
22		*−2.96*	683.04	*0*		*−0.75*						
23	718.04	*0*		*−0.73*								
24								*−0.45*	665.24	*0*		*+0.48*
25				*−0.92*	673.20	*0*		*−0.64*				
26		*−0.92*	682.81	*0*		*−0.22*						
27	717.82	*0*		*−0.23*								
28						*−0.38*	669.61	*0*		*−0.16*		
29										*−0.04*	651.31	*0*
30				*−0.32*	673.11	*0*		*−0.09*				
31		*−0.32*	682.73	*0*		*−0.10*						
32	717.73	*0*		*−0.08*								
33				*−0.10*	673.09	*0*		*−0.11*				
34						*−0.05*	669.58	*0*		*−0.07*		
35		*−0.08*	682.71	*0*		*−0.07*						
	717.71	*0*	682.71	*−0.04*	673.09	*−0.07*	669.58	*0*	665.24	*−0.07*	651.31	*0*

temperatures unchanged. New residuals are now computed. Note that R_1^2 should now be 0, R_2^1 will change to $R_2^2 = -425$, but the other residuals will remain unchanged.

In the third step, the process is repeated, beginning by adjusting the residual with the largest numerical value, now R_2^2, to zero and so on until all residuals nearly vanish. The solution is tabulated in Table 7-2. The last row in that table shows final residuals having numerical values less than 0.1°F, sufficiently negligible for the problem at hand.

Like the iterative method, errors in intermediate steps only lengthen the process and are corrected in subsequent steps. There exist some techniques to shorten the process. For example, if residuals in one area of the field are predominantly of one sign, two techniques are available. One is *overrelaxation*, in which the largest residual in that area is relaxed to the opposite sign instead of zero. This reduces the adjacent residuals and speeds up the process. The second is known as *block relaxation.* A block-relaxation region is selected, but the coefficients of the temperatures are altered. For example, in the simple case of two nodal points in a block, Fig. 7-6a, their coefficients become −3 each since the block is connected to 6 nodal points outside the block. In Fig. 7-6, the coefficients are shown next to the nodal points and the block is shown by the dashed boundary. Block relaxation is advantageous, however, in

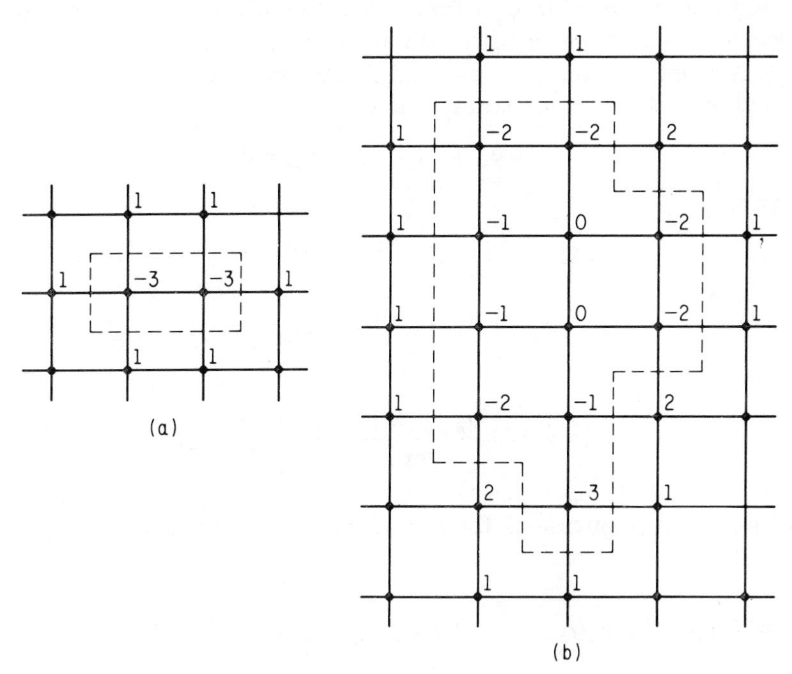

FIG. 7-6. Block relaxation.

the case of a larger mesh. While the coefficient of the temperature whose residual is to be relaxed in ordinary relaxation is -4 (Eq. 7-16), the coefficient in block relaxation is -4 plus the number of lines connecting this nodal point to other nodal points within the chosen block. The coefficient of temperatures of nodal points outside the block is equal to the number of lines connecting them to nodal points inside the block. This is shown in Fig. 7-6b.

Numerical methods will also be used in solving transient problems, Chapter 8. The iteration method is more suitable in transient cases, however.

7-8. HEAT CONDUCTION IN FINITE DIFFERENCE SYSTEMS

In the preceding discussions the temperature at nodal points was obtained. In this section the heat transfer by conduction when finite difference solutions are available will be evaluated. Consider again the two-dimensional grid of Fig. 7-2. The nodal point n is located at the center of an element of unit thickness (perpendicular to the paper) and of sides $\Delta x = \Delta y$, bounded by the dashed lines at x, $x + \Delta x$, y and $y + \Delta y$. Heat generated in the square about n is assumed to be generated at n and conducted only along the four rods interconnecting at n. That this physical model is true in the finite difference system can be ascertained by a heat balance where the heat generated at n is called q_g, the heats conducted into n in the x and y directions are called q_{xi}, and q_{xo}, q_{yi}, and q_{yo}. Thus (in the steady state)

$$q_g + q_{xi} + q_{yi} + q_{xo} + q_{yo} = 0 \tag{7-19}$$

where

$$q_g = q'''(n)\Delta x \Delta y \, 1 = q'''(n)(\Delta x)^2 \tag{7-20}$$

$$q_{xi} = k\Delta y \, 1 \, \frac{t_{n-\Delta x} - t_n}{\Delta x} = k(t_{n-\Delta x} - t_n) \tag{7-21a}$$

and

$$q_{xo} = k\Delta y \, 1 \, \frac{t_{n+\Delta x} - t_n}{\Delta x} = k(t_{n+\Delta x} - t_n) \tag{7-21b}$$

Writing similar expressions for q_{yi} and q_{yo} and substituting into Eq. 7-19 give

$$q'''(n)(\Delta x)^2 + k[(t_{n-\Delta x} - t_n) + (t_{n+\Delta y} - t_n)(t_{n+\Delta x} - t_n) + (t_{n-\Delta y} - t_n)] = 0 \tag{7-22}$$

Rearranging gives

$$t_n = \frac{t_{n+\Delta y} + t_{n+\Delta x} + t_{n-\Delta y} + t_{n-\Delta x}}{4} + \frac{(\Delta x)^2 q'''(n)}{4k}$$

which is identical to Eq. 7-16a.

The heat conducted per foot length of the cruciform element above, is found by using the course grid of Fig. 7-4, and noting the symmetry of the element. The total heat conducted out, Q_k, is given by the heat conducted via rods to nodal points on the boundaries plus heat generated at these nodal points.

$$\begin{aligned}
Q_k &= k\{8[(t_2 - t_s) + (t_3 - t_s) + (t_4 - t_s) + (t_5 - t_s)] \\
&\quad + [12(t_6 - t_s)]\} + q''' \times \text{area between boundary and dashed line} \times 1 \\
&= k[8(t_2 + t_3 + t_4 + t_5) + 12t_6 - 44t_s] + q''' \times \frac{0.92}{144} \\
&= 19.84\,[8(682.69 + 673.06 + 669.57 + 665.22) + 12(651.30) \\
&\qquad - 44 \times 600] + 10^7 \times \frac{0.92}{144} \\
&= 19.84 \times 2{,}940 + 10^7 \times \frac{0.92}{144} = 58{,}328.8 + 63{,}888.9 \\
&= 122{,}217.7 \text{ Btu/hr ft}
\end{aligned}$$

The total heat generated, Q_g, which should be equal to Q_k, is

$$\begin{aligned}
Q_g &= q''' \times \text{element volume per foot length} \\
&= 10^7 \times \frac{4 \times 0.4 \times 1 + 0.4 \times 0.4}{144} \times 1 = 122{,}222.2 \text{ Btu/hr ft}
\end{aligned}$$

7-9. NODAL POINTS NEAR CURVED BOUNDARIES

In the example solved above, the field was divided in such a way that, in the resulting mesh, nodal points fell on straight boundaries. The symmetry of the element was used to advantage and only a one-eighth portion of the element was studied, Figs. 7-4 and 7-5. Thus a line of symmetry is used as boundary with nodal points lying on it.

There are cases, however, when the geometry of an element is such that nodal points on a boundary would not constitute a square spacing with inside nodal points. In such cases, the fractions of the distances along the x and y directions from internal nodal points is noted and used. In the general case (in two dimensions) of a curved boundary, Fig. 7-7, let points a and b fall on the boundary such that

$$\frac{a - n}{\Delta x} = f_x \tag{7-23a}$$

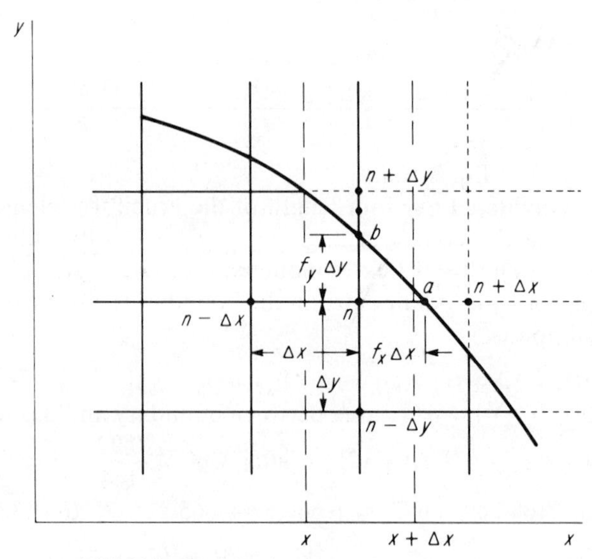

FIG. 7-7. Nodal points near curved boundaries.

and

$$\frac{b-n}{\Delta y}=f_y \tag{7-23b}$$

Equation 7-3 for the second derivative with respect to x of the temperature at n is now rewritten in the form

$$\left(\frac{\partial^2 t}{\partial x^2}\right)_n = \frac{\dfrac{t_a - t_n}{f_x \Delta x} - \dfrac{t_n - t_{n-\Delta x}}{\Delta x}}{\dfrac{f_x \Delta x + \Delta x}{2}}$$

$$= \frac{2}{(\Delta x)^2(1+f_x)}\left[\frac{t_a}{f_x} + t_{n-\Delta x} - t_n\left(1+\frac{1}{f_x}\right)\right] \tag{7-24}$$

A similar expression for $(\partial^2 t/\partial y^2)_n$ may be written and the Laplacian of temperature at n takes the form

$$(\nabla^2 t)_n = \left(\frac{\partial^2 t}{\partial x^2} + \frac{\partial^2 t}{\partial y^2}\right)_n$$

$$= \frac{2}{(1+f_x)(1+f_y)f_x f_y(\Delta x)^2}[f_x(1+f_x)t_b$$

$$+ f_y(1+f_y)t_a + f_x f_y(1+f_x)t_{n-\Delta y} + f_x f_y(1+f_y)t_{n-\Delta x}$$

$$- (f_x+f_y)(1+f_x)(1+f_y)t_n] \tag{7-25}$$

Combining with Eq. 7-14 and rearranging give

$$t_n = \frac{1}{(f_x + f_y)}\left[\frac{f_x}{(1 + f_y)} t_b + \frac{f_y}{(1 + f_x)} t_a + \frac{f_x f_y}{(1 + f_y)} t_{n-\Delta y} + \frac{f_x f_y}{(1 + f_x)} t_{n-\Delta x} + f_x f_y \Delta t_g\right] \tag{7-26}$$

where Δt_g is given by Eqs. 7-13. In the case where $f_y = 1.0$ and $t_b = t_{n+\Delta y}$, Fig. 7-8, Eq. 7-26 becomes

$$t_n = \frac{f_x}{2(1 + f_x)}(t_{n+\Delta y} + t_{n-\Delta y}) + \frac{1}{(1 + f_x)^2} t_a + \frac{f_x}{(1 + f_x)^2} t_{n-\Delta x} + \frac{f_x}{(1 + f_x)} \Delta t_g \tag{7-27}$$

Equation 7-27 reverts back to Eq. 7-16c for the case of $f_x = 1.0$ and $t_a = t_{n+\Delta x}$.

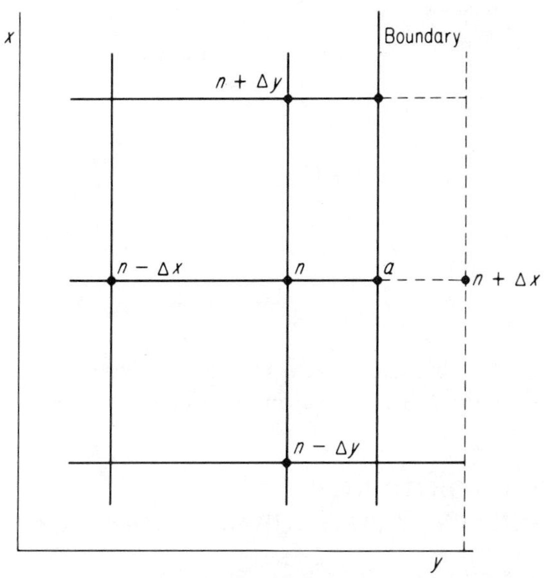

FIG. 7-8. Straight boundary, parallel to mesh and between nodal points.

In the above examples, nodal points *on* boundaries were pressumed to have known temperatures. The following three sections cover situations where this is not the case.

7-10. NODAL POINTS ON INSULATED BOUNDARIES

If a nodal point n falls on an insulated boundary such as in Fig. 7-9, a heat balance at n will be given in the manner of Eq. 7-22, where $q_{\dot{x}o} = 0$,

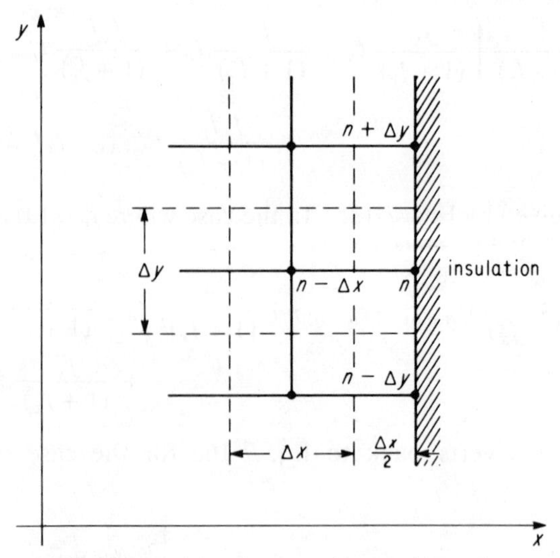

FIG. 7-9. Nodal point on an insulated boundary.

by

$$\frac{1}{2}\,q'''(\Delta x)^2 + k\left[(t_{n-\Delta x} - t_n) + \frac{1}{2}\,(t_{n+\Delta y} - t_n) + \frac{1}{2}\,(t_{n-\Delta y} - t_n)\right] = 0 \tag{7-28a}$$

or

$$t_n = \frac{t_{n+\Delta y} + t_{n-\Delta y} + 2t_{n-\Delta x}}{4} + \frac{1}{2}\,\Delta t_g \tag{7-28b}$$

This case is therefore identical to that of a nodal point along a symmetry line in that the temperature on one side of that line is counted twice.

7-11. NODAL POINTS ON A BOUNDARY WITH A SURFACE CONDUCTANCE *h*

In many cases the boundary of an element exchanges heat by convection with an adjacent fluid, or by radiation to another body, or both. The heat exchange is expressed in terms of a surface conductance h, Btu/hr ft² °F. A heat balance on a nodal point n on such a boundary, Fig. 7-10, is obtained in the same manner as for the insulated boundary, except that q_{xo} will be given by

$$q_{xo} = h(\Delta x)\,1\,(t_f - t_n) \tag{7-29}$$

where t_f is the temperature of the fluid (or body) with which the element exchanges heat. Thus

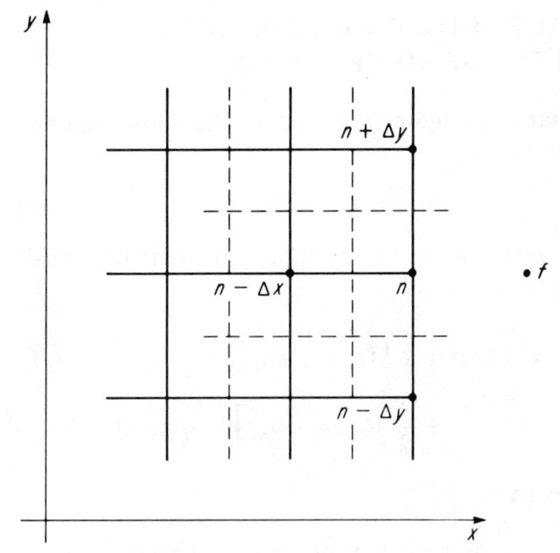

FIG. 7-10. Nodal point on a boundary with surface conductance h and fluid temperature t_f.

$$\frac{1}{2}\, q'''(\Delta x)^2 + k\left[(t_{n-\Delta x} - t_n) + \frac{1}{2}\,(t_{n+\Delta y} - t_n) + \frac{1}{2}\,(t_{n-\Delta y} - t_n)\right] + h(\Delta x)(t_f - t_n) = 0 \qquad \text{(7-30a)}$$

Dividing by k and noting that

$$\frac{h(\Delta x)}{k} = \text{Bi}_{\Delta x} \qquad \text{(7-31)}$$

where $\text{Bi}_{\Delta x}$ is a *Biot number** with Δx as a characteristic length, Eq. 7-30a is rearranged to

$$t_n = \frac{t_{n+\Delta y} + 2\text{Bi}_{\Delta x} t_f + t_{n-\Delta y} + 2t_{n-\Delta x}}{4 + 2\text{Bi}_{\Delta x}} + \frac{\Delta t_g}{2 + \text{Bi}_{\Delta x}} \qquad \text{(7-30b)}$$

Note that in the case of an insulated boundary, $\text{Bi}_{\Delta x} = 0$, and Eq. 7-30b simplifies to Eq. 7-28b.

* The Biot number has the same form as the Nusselt number Sec. 9-3. The difference is that in the Biot number h and k are for the fluid and solid respectively, while in the Nusselt number both are for the fluid.

7-12. NODAL POINTS ON A BOUNDARY WITH SPECIFIED HEAT FLUX

A heat balance is again obtained in the same manner as the above two cases, except that

$$q_{xo} = -\,q''(\Delta x)\,1 \tag{7-32}$$

where q'' is the heat flux at the boundary, Btu/hr ft², considered positive in the $+x$ direction. Thus

$$\frac{1}{2}\,q'''(\Delta x)^2 + k\left[(t_{n-\Delta x} - t_n) + \frac{1}{2}(t_{n+\Delta y} - t_n) + \frac{1}{2}(t_{n-\Delta y} - t_n)\right] - q''(\Delta x) = 0 \tag{7-33a}$$

which simplifies to

$$t_n = \frac{t_{n+\Delta y} + t_{n-\Delta y} + 2t_{n-\Delta x}}{4} - \frac{q''(\Delta x)}{2k} + \frac{1}{2}\,\Delta t_g \tag{7-33b}$$

Again this equation reduces to Eq. 7-28b for the insulated boundary when $q'' = 0$.

7-13. AN APPROXIMATE ANALYTICAL METHOD

It was indicated at the beginning of this chapter that some analytical solutions are available for complex geometries. There are, however, many problems that do not lend themselves to exact analytical solutions, or that have analytical solutions that are tedious and time consuming. Such problems may be solved by finite-difference methods, already outlined, but there also exist various approximate analytical techniques, the most important of which is the *integral* method. (Integral methods are also extensively used in convection heat transfer.) This method can best be illustrated by an example.

An unclad, long rectangular fuel element, Fig. 7-11, operating in the steady state is considered. All surfaces are kept at zero temperature. The following Poisson equation applies:

$$\frac{\partial^2 t}{\partial x^2} + \frac{\partial^2 t}{\partial y^2} + \frac{q'''}{k} = 0 \tag{7-34}$$

and the boundary conditions are

$$\left.\begin{aligned} t(a,y) &= 0 \\ t(x,b) &= 0 \end{aligned}\right\} \tag{7-35}$$

and, because of symmetry

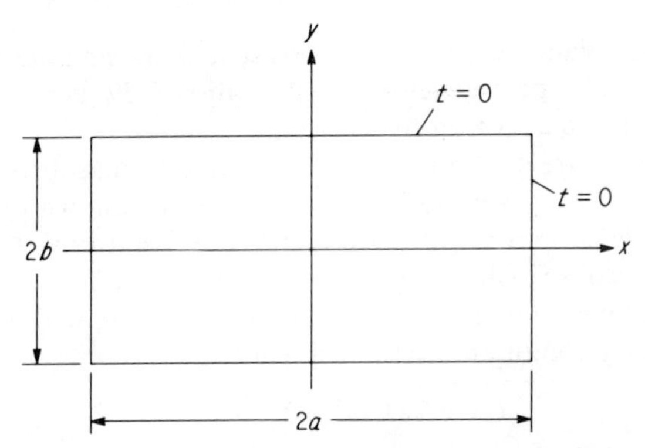

FIG. 7-11. Cross section of a long rectangular fuel element with heat generation and surfaces held at zero temperature.

$$\left.\begin{aligned} \frac{\partial t}{\partial x} &= 0 \qquad \text{at } x = 0 \\ \frac{\partial t}{\partial y} &= 0 \qquad \text{at } y = 0 \end{aligned}\right\} \tag{7-36}$$

The basic step in obtaining an approximate analytical solution of the above equation is to *assume* a temperature distribution. Since the exact solutions of many differential equations are in the form of infinite series, it is common to assume, for simplicity, a solution in the form of a finite-series polymonial. The coefficients of the polynomial are then evaluated using the original equation, boundary conditions and other conditions. The integral formulation of the problem is obtained by integrating Eq. 7-34 over the entire region of interest, for constant q''' and k, as

$$\int_{x=0}^{a}\int_{y=0}^{b} \frac{\partial^2 t}{\partial x^2}\, dydx + \int_{x=0}^{a}\int_{y=0}^{b} \frac{\partial^2 t}{\partial y^2}\, dydx + \frac{q'''}{k}\int_{x=0}^{a}\int_{y=0}^{b} dydx = 0 \tag{7-37}$$

The first two terms may be integrated by parts and the last integrated twice, so that

$$\int_{y=0}^{b} \left[\frac{\partial t}{\partial x}\right]_{x=0}^{a} dy + \int_{x=0}^{a} \left[\frac{\partial t}{\partial y}\right]_{y=0}^{b} dx + \frac{q'''ab}{k} = 0 \tag{7-38}$$

Using the boundary conditions of Eqs. 7-36, simplifies Eq. 7-37 to

$$\int_{y=0}^{b} \left.\frac{\partial t}{\partial x}\right|_{x=a} dy + \int_{x=0}^{a} \left.\frac{\partial t}{\partial y}\right|_{y=b} dx + \frac{q'''ab}{k} = 0 \tag{7-39}$$

It can be seen that the first two terms in Eq. 7-39 represent heat conducted into planes $x = a$ and $y = b$ respectively, and that the last term represents the total heat generation. Equation 7-39 therefore represents an overall energy balance on the element.

The temperature distribution is now assumed, and substituted into the integral Eqs. 7-38 or 7-39 above. There are two known procedures for making this assumption: the Ritz and the Kantorovich procedures. The Ritz procedure will be used here. In this, an $x - y$ temperature distribution that is symmetric about $x = 0$ and $y = 0$, and which satisfies all the boundary condition is assumed, such as

$$t(x,y) = C(a^2 - x^2)(b^2 - y^2) \tag{7-40}$$

The coefficient C is evaluated by substituting Eq. 7-40 into Eq. 7-39, giving

$$\int_{y=0}^{b} - 2Ca(b^2 - y^2)dy + \int_{x=0}^{a} - 2Cb(a^2 - x^2)dx + \frac{q'''ab}{k} = 0 \tag{7-41}$$

which integrates to

$$-2C\left\{a\left[b^2y - \frac{y^3}{3}\right]_0^b + b\left[a^2x - \frac{x^3}{3}\right]_0^a\right\} + \frac{q'''ab}{k} = 0 \tag{7-42}$$

Substituting the limits gives

$$-2Ca\,\frac{2b^3}{3} - 2Cb\,\frac{2a^3}{3} + \frac{q'''ab}{k} = 0$$

which is solved for C to give

$$C = \frac{3}{4}\,\frac{q'''}{k(a^2 + b^2)} \tag{7-43}$$

so that the temperature distribution is

$$t(x,y) = \frac{3}{4}\,\frac{q'''}{k(a^2 + b^2)}\,(a^2 - x^2)(b^2 - y^2) \tag{7-44}$$

An *exact* solution of this problem obtained by the method of *separation of variables* is given by Arpaci [40] as

$$t = \frac{q'''a^2}{k}\left\{\frac{1}{2}\left[1 - \left(\frac{x}{a}\right)^2\right] - 2\sum_{n=0}^{\infty}\frac{(-1)^n}{(\lambda_n a)^3}\left(\frac{\cosh \lambda_n y}{\cosh \lambda_n b}\right)\cos \lambda_n x\right\} \tag{7-45}$$

where $\lambda_n = (2n + 1)\pi/2a$ and $n = 0, 1, 2, \cdots$.

Note the difference in complexity between the approximate solution, Eq. 7-44 and the exact solution, Eq. 7-45. The method of separation of variables will be used in Sec. 8-11, in conjunction with a solution for the temperature distribution in a long cylindrical fuel element, cooled by a fluid and operating in the unsteady state. The result is given by Eq. 8-50. The steady-state solution for that element is easily obtained from Eq. 8-50 by putting $\theta^* = \infty$, reducing the term between the parentheses in that equation to unity, and resulting in Eq. 8-51.

7-14. ANALOGICAL TECHNIQUES

Thus far in the chapter we have covered solutions for the two-dimensional Poisson equation by numerical and by an approximate analytical technique. In this section we shall survey some of the other methods used in the solution. These are generally grouped under the title *analogical* methods because they rely on the analogy between the heat conduction equation and various other techniques and physical phenomena. Although the now almost universal availability of the computer has facilitated the use of numerical methods and made these methods nearly obsolete, they are still interesting and occasionally useful, and hence this survey.

Potential Field Plotting

This is a trial and error free-hand plotting technique for two-dimensional cases which is particularly applicable to boundaries at constant temperature. The field is divided into a number of *flow channels* between flow lines which begin perpendicularly at one isothermal boundary and end perpendicularly at another. The flow lines are intersected by a number of constant *potential lines* or *isotherms,* such that they together form a network of *curvilinear squares*. Several helpful guidelines are usually followed in constructing the network [36]. The accuracy of the method depends upon the number of channels and, of course, upon the accuracy of the individual doing the plotting. The method does not lend itself readily to internal heat generation. Figure 7-12 shows an example of such a plot.

Analog Plotting

This technique utilizes the similarities in form between the heat conduction, the hydrodynamic (inviscid fluid), the elasticity of membranes, and the electrostatic equations. There are two broad categories of analogs, *direct* and *indirect*. *Direct* analogs simulate the physical system in both the spatial and time domains. Fluid, membrane, and electrostatic analogs belong to this category.

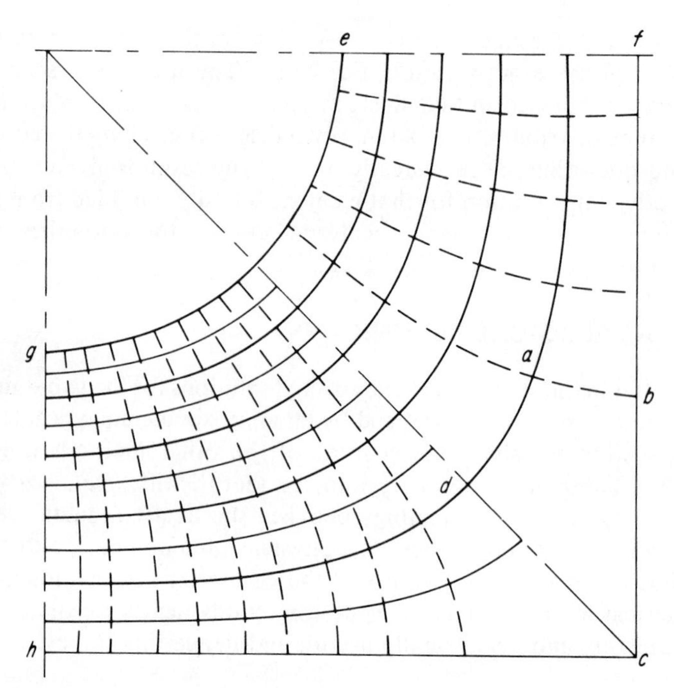

FIG. 7-12. Potential field plotting for a corner of a long square element with a circular hole. Coarse and fine grids shown. Surfaces *eg* and *fch* are at constant temperature. Surfaces *ef* and *gh* are surfaces of symmetry and therefore adiabatic. Solid lines are isotherms. Dashed lines are heat flow lines. Curvilinear squares drawn such that $ab + cd = ad + bc$.

In *fluid-flow analogy* [41], fluid analogs are based on the fact that the irrotational flow of an ideal fluid in a channel is governed by Laplace's equation, and the Poisson equation is simulated by the use of special fluids. The streamlines are analogous to heat flow lines and lines of constant velocity correspond to isotherms. Water or glycerol are used as an approximation to an ideal fluid. The streamlines are made visible by a soluble dye such as potassium permanganate which is made to form streaks in the fluid which are then photographed.

Membrane analogy [42,43], utilizes the property of an elastic membrane, a film of infinitesimal mass suspended with uniform tension at closed boundaries, that when subjected to tension forces greater than its weight and attached to boundaries of varying shape and elevation, would have its height at any point governed by Laplace's two-dimensional equation. The Poisson equation is simulated by applying a vertical distributed force to the membrane. A soap film makes an ideal membrane but has the obvious problems of durability and difficulty in measuring its heights above a datum plane (special optical and photographic techniques have been used). Thin rubber sheets have also been used as membrane material.

Nonelectrical direct analogs, as those two above, have the advantage of making the phenomena directly visible, which is not the case with direct electrical analogs. The latter, however, have the advantages of ease of recording, especially important in transient systems, and the availability of accurate and economical measuring and recording instrumentation. Hence, they are more widely used.

Direct electrical analogs rely on the fact that Ohm's law in electrical systems is analogous to Fourier's law in thermal systems. The voltage distribution in the analog is at all times proportional to the temperature distribution in the physical system, and the current to heat flow. In the designing of a direct electrical analog, analogous pairs of electrical and thermal parameters (such as electrical potential in volts and temperature in °F; and current in coulomb/sec, or ampere, and heat flow in Btu/hr), are related by scaling factors. Direct electrical analogs are constructed of electrically *conducting sheets, electrolytic fluids* (such as water-copper sulphate solution) and *lumped networks* made of electrical resistors and capacitors. In the former two, recording of equipotential and flow lines is accomplished by a probing stylus connected to a null potentiometer. Variable resistances can be obtained by punching holes or slits in the conducting sheets, varying the depth of the electrolytic fluid, or by using different-size resistors in the network. The latter, lumped networks, are not, of course, continuous and share the advantages and disadvantages of finite difference equations.

Indirect analogs, usually electrical, are composed of many components, each undertaking a mathematical operation in the equations of the physical system. Summers, multipliers, and integrators are examples of such components. Unlike the direct analogs, there is no direct proportionality between analog and physical system parameters. The subject of analog computation is wide and is beyond the scope of this book [44].

PROBLEMS

7-1. Derive the finite difference equation of temperature for a nodal point at the outer square corner of a body with heat generation, having thermal conductivity k, and cooled by a fluid at t_f with a uniform heat transfer coefficient h.

7-2. Repeat Prob. 1 but for an inside square corner of the body.

7-3. A flat plate fuel element is 0.5 in. thick. It has a thermal conductivity $k = 1.085$ Btu/hr ft°F, generates 1.5×10^7 Btu/hr ft^3 and is cooled by a gas at 1000°F with a uniform heat transfer coefficient h. Considering the end section

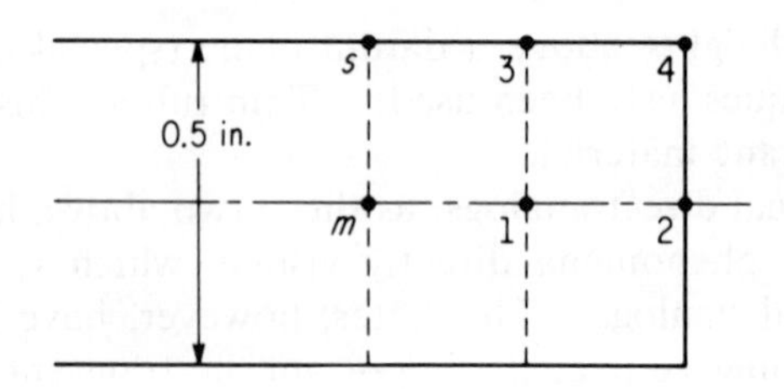

as shown, and assuming that the temperatures at *m-s* are related by one dimensional heat flow, and that the maximum temperature in the element should not exceed 4500°F, find the corresponding temperatures at points 1 through 4 by finite differences using (a) matrix, (b) iteration, and (c) relaxation solutions.

7-4. In the above problem find the heat transfer out of the end section, Btu/hr ft. Assume that at the section *m-s* heat flow was still one dimensional.

7-5. A long body has a cross section in the form of a right angle triangle with two equal sides, measuring 2 in. at the base. It generates 10^6 Btu/hr ft^3 and has a thermal conductivity 10.85 Btu/hr ft°F. All surfaces are held at 400°F. Choosing $\Delta x = 0.25$ in., find (a) the temperatures at the nodal points, and (b) the heat transferred from each side of the triangle in Btu/hr ft.

7-6. A very long fin is rectangular in cross section 0.48×0.24 in. It generates 2×10^6 Btu/hr ft^3. The fin base is at 1000°F. It is cooled by a gas at 600°F with a uniform heat transfer coefficient 100 Btu/hr ft^2 °F. Using a network with $\Delta x = 0.12$ in., write the necessary set of finite difference equations for the nodal points and solve by any one of the techniques at your command. k for the fin material = 10 Btu/hr ft °F.

7-7. Write the finite difference equation of temperature for a nodal point on an outside square corner of a body with heat generation for the cases where (a) both sides have the same specified heat flux, and (b) one side insulated and the other with a specified heat flux.

7-8. A long body has a rectangular cross section measuring 0.48×0.96 in. It has a thermal conductivity 2 Btu/hr ft °F and generates 10^6 Btu/hr ft^3. One of the longer sides of the rectangle is insulated. The remaining three have the same heat flux q''. Using $\Delta x = 0.24$ in. find the temperatures at the corresponding nodal points.

7-9. A long body has a cross section in the form of a right angle triangle with two equal sides. The body generates heat and is cooled by a fluid with a known heat transfer coefficient. A square grid is chosen so that it is parallel and perpendicular to the base of the triangle. Write the finite difference equation for the temperatures at nodal points (a) on the apex, and (b) at one of the other corners.

7-10. A long body with a cross section in the form of a right-angle triangle with two equal sides measures 2.4 in. at the base. Choosing $\Delta x = 0.6$ in. (use Prob. 7-9), find the temperatures at the nodal points if the body generates 2×10^6 Btu/hr ft^3, has a thermal conductivity 25 Btu/hr ft °F, and is cooled by a fluid at 500°F with a uniform heat transfer coefficient 2000 Btu/hr ft^2 °F.

7-11. A long body has a cross section measuring 0.4×0.6 in. The two ends are semi-circular. It generates 10^7 Btu/hr ft^3 and has a thermal conductivity

2.17 Btu/hr ft °F. All surfaces are held at 800°F. Choosing $\Delta x = 0.1$ in., find the temperatures at the corresponding nodal points.

7-12. A long body has a semi-circular cross section with a 0.6 in. diameter. It generates 5.75×10^6 Btu/hr ft³ and has a thermal conductivity 1.0 Btu/hr ft °F. All surfaces are held at 600°F. For $\Delta x = 0.1$ ft, find (a) the temperatures at the nodal points, and (b) the heat transferred from the base and from the circular side in Btu/hr ft.

7-13. A long fuel element has a rectangular cross section 1×2 in. It generates 10^7 Btu/hr ft³ and has a thermal conductivity of 1.085 Btu/hr ft °F. All surfaces are held at 1000°F. Find the heat flux, Btu/hr ft² at the center points of each side using (a) a finite difference solution with $\Delta x = 0.25$ in., and (b) the approximate analytical solution of Sec. 7-13.

7-14. Repeat any of the above finite difference problems using a finer mesh and an available digital computer.

7-15. Sketch a potential field plot for the cruciform of Fig. 7-4 if all outside surfaces are held at one temperature, and a small hole in the geometrical center is held at another. Assume no internal heat generation.

chapter **8**

Heat Conduction in Reactor Elements

IV. The Unsteady State

8-1. INTRODUCTION

In this chapter the behavior of the fuel-coolant combination in response to reactivity insertions, loss of coolant or other effects will be discussed. This behavior is important in evaluating such things as the time taken by the fuel to reach steady-state temperatures from a cold start if the neutron flux, and therefore the volumetric thermal source strength q''', are suddenly raised; the time taken by the fuel to reach meltdown temperatures after loss of coolant; the time taken by the coolant to reach equilibrium temperatures when the fuel temperature is suddenly raised, etc. Besides these operational and safety considerations, there exist nuclear considerations. One is the Doppler coefficient, Sec. 3-8. Also, if the coolant doubles as moderator, the moderator temperature coefficient of reactivity becomes involved. Problems would arise if the temperature of the moderator lags behind that of the fuel, since the normally negative moderator temperature coefficient may not come into play rapidly enough to help make the reactor self-controlling.

The effects of temperature transients, as well as lumped, numerical, graphical, and analytical techniques for evaluating these transients will be presented in this chapter.

8-2. THE LUMPED-PARAMETER TECHNIQUE

In evaluating temperature responses of bodies to environmental temperature changes, the *lumped-parameter* technique is often useful. The technique is illustrated by considering two bodies I and II which exchange heat with each other (they may be fuel and coolant) and which are assumed to be initially in temperature equilibrium at t_0. At arbitrary time $\theta = 0$, body II is suddenly raised (or lowered) to a constant uniform temperature t_2. If it is assumed that the internal conductive resistance of body I is negligibly small compared to the external resistance to heat transfer between the two bodies, all of body I, will always be at a uniform temperature $t_1(\theta)$ which will change gradually with time until it reaches t_2 at $\theta = \infty$, (Fig. 8-1). In other words, the thermal capacity of body I is treated as a single or lumped parameter. In this case the technique is

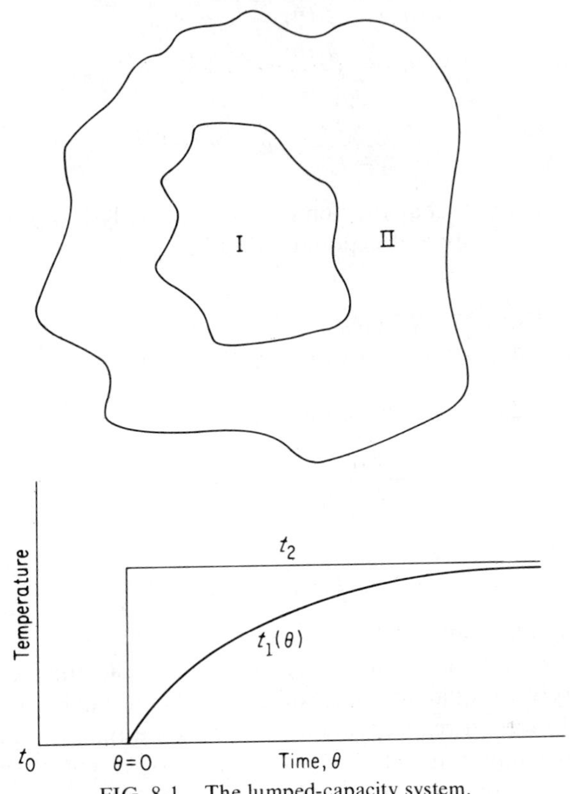

FIG. 8-1. The lumped-capacity system.

called the *lumped-capacity* technique. It is usually satisfactory for simple cases when the *Biot* number, Bi, representing the ratio of internal (conductive) to external (convective) resistance, is small, usually less than 0.1.*

An energy balance at time θ would be given by

$$\begin{pmatrix}\text{Sensible energy added to}\\ \text{body I during } d\theta\end{pmatrix} = \begin{pmatrix}\text{Heat transfer from}\\ \text{body II to body I}\end{pmatrix}$$

* The Biot number is given by hL/k, where h is the external heat transfer coefficient, k the thermal conductivity of the body and L is some significant length of the body (thickness of an infinite wall, radius of long cylinder or sphere, etc).

or

$$c_1\rho_1 V_1 dt_1(\theta) = hA_1[t_2 - t_1(\theta)]\,d\theta \tag{8-1}$$

where h is the heat transfer coefficient between the two bodies and c_1, ρ_1, V_1, and A_1 are the specific heat, density, volume and heat transfer area of body I respectively. Integrating,

$$\int_{t_1=t_0}^{t_1(\theta)} \frac{dt_1(\theta)}{t_2 - t_1(\theta)} = \frac{hA_1}{c_1\rho_1 V_1}\int_0^\theta d\theta$$

$$-\ln\frac{t_2 - t_1(\theta)}{t_2 - t_0} = \frac{hA_1}{c_1\rho_1 V_1}\theta$$

$$\frac{t_2 - t_1(\theta)}{t_2 - t_0} = e^{-(hA_1/c_1\rho_1 V_1)\theta} \tag{8-2a}$$

Equation 8-2a shows that the temperature of body I approaches that of body II asymptotically and exponentially, Fig. 8-1.

8-3. TIME-CONSTANT AND TEMPERATURE-COEFFICIENT EFFECTS

Equation 8-2a can be written in the form

$$\frac{t_1(\theta) - t_0}{t_2 - t_0} = 1 - e^{-\theta/\tau_1} \tag{8-2b}$$

where

$$\tau_1 = (c_1\rho_1 V_1/hA_1) \tag{8-3}$$

τ_1 is the *time constant* for body I. Likewise the time constant of body II, τ_2, would be given by $(c_2\rho_2 V_2/hA_2)$. The time constant of a body is therefore equal to the product of its thermal capacitance $c\rho V$ and external thermal resistance $1/hA$,* and is the time it takes that body to change in temperature by $1 - (1/e)$, or 63.2 percent of the maximum temperature change.

The time constant is a measure of the rapidity of the response of a body to environmental temperature changes. In reactors, the time constant of the moderator is usually longer than that of the fuel. If it is much longer, its temperature will lag considerably behind that of the fuel in case of a reactivity insertion.

A positive reactivity insertion increases the effective multiplication

* Note the analogy with an electrical system composed of a capacitor and resistor for which $(E_2 - E_1)/(E_2 - E_0) = e^{-\theta/CR}$ where E, C, and R are the potential, capacitance, and resistance of the system in electrical units. The product RC is the time constant of the system.

factor, k_{eff}, the neutron flux and hence the fission rate, power, and q'''. The fuel temperatures begin to rise. If the temperature coefficient of reactivity (Sec. 3-7) of the fuel is negative, k_{eff} will decrease. The rapidity with which the temperature coefficient comes into play is important. To avoid an unsafe condition (fuel meltdown, etc.), the time constants of fuel and moderator must be short so that core temperatures would follow power changes and the temperature coefficient takes effect rapidly. It also follows that a long time constant could cause power oscillations even though the temperature coefficient is negative.

If the temperature coefficient is positive, the temperatures will continue to rise. A reactor with an overall positive temperature coefficient, however, may not be unstable if the temperature coefficient due to the fuel alone is negative, and if the time constant of the fuel is shorter than that of the moderator. In this case a sudden reactivity (and power) increase causes the fuel temperature to rise immediately, while the moderator temperature and its positive coefficient lag behind. Such is usually the case in heterogeneous reactors with a large mass of moderator, or with separate moderator and coolant. An example is the sodium-cooled, graphite- or heavy-water moderated reactor. The reactor in the Sodium Reactor Experiment (SRE) [1] for example has a fuel with a negative temperature coefficient and 2-sec time constant, and a graphite moderator with a positive temperature coefficient and a 3-min time constant. A pressurized-water reactor using chemical shim may also have a positive overall temperature coefficient at the beginning of core life.

Positive temperature coefficients are, of course, partially counteracted by the Doppler coefficient (Sec. 3-8), and can be dealt with by proper reactor-control systems.

8-4. EVALUATION OF TEMPERATURE TRANSIENTS BY FINITE DIFFERENCES

The lumped-capacity technique, discussed in Sec. 8-2, neglects the conductive resistances within a body and treats its thermal capacity as a lumped parameter; it therefore results in body temperatures that are always uniform and vary only with time.

Where the conductive resistances are important, i.e., when the Biot number is large (greater than 0.1), the temperature of the body at any time is not uniform and is a function of both time and space coordinates. Exact analytical solutions are available in only a few cases of practical interest (see Sec. 8-10) and the results are often presented in the form of charts [45,46]. An analytical evaluation of the effects of containers (cladding) shows that these effects may or may not be ignored under certain conditions [47]. Numerical and graphical solutions, on the other hand, are available for a large number of practical problems. They are simple in nature, flexible, and are able to handle problems with complicated boundary conditions that cannot be tackled by analytical methods. They depend upon finite differences and give a continuing picture of temperature profiles at preselected time intervals. As in the steady state (Chap. 7), their accuracy depends upon the number of steps.

The numerical method is the more precise of the two because it can handle problems with varying physical properties more conveniently. With a digital computer, its speed and accuracy are greatly enhanced.

The applicable equation is the general heat-conduction equation

$$\nabla^2 t + \frac{q'''}{k} = \frac{1}{\alpha}\frac{\partial t}{\partial \theta} \tag{8-4}$$

The time variable is, as was the space variable (Chap. 7), broken into discrete intervals. Figure 8-2 shows a temperature-time relationship of one nodal point n, such as in Fig. 7-1, for which the time domain is broken into discrete intervals $\Delta\theta$.

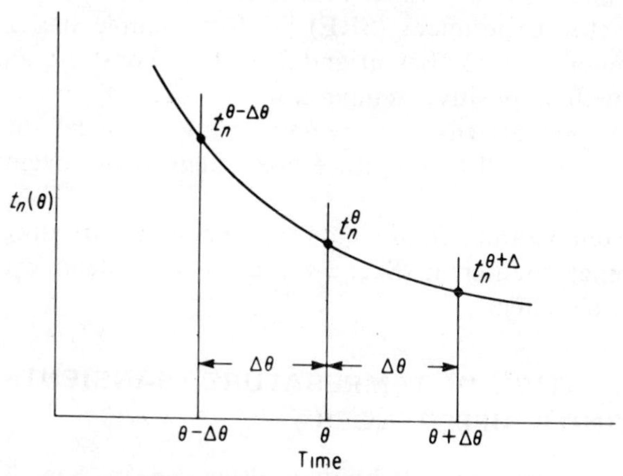

FIG. 8-2. Temperature-time history of one nodal point.

An equation analogous to 7-2 may now be written for the independent time variable in *forward** difference form as

$$\left.\frac{\partial t}{\partial \theta}\right|_{n,\theta} = \frac{t_n^{\theta+\Delta\theta} - t_n^{\theta}}{\Delta\theta} \tag{8-5}$$

* Forward differences result in errors that are proportional to $(\Delta x)^2$ and to $\Delta\theta$. Central and backward differences are also used. Discussion of these is beyond the scope of this book.

The One-Dimensional Case

The body is divided into a number of layers, bounded by vertical solid lines, of equal thickness Δx (Fig. 8-3). The temperature distribution at

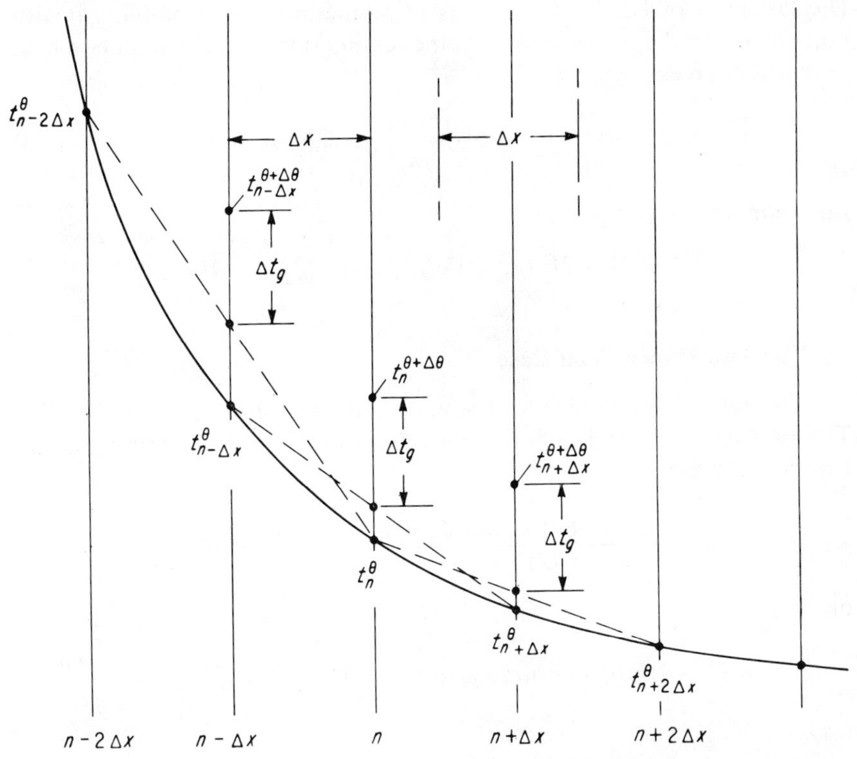

FIG. 8-3. Graphical determination of nodal temperatures after finite time increment $\Delta\theta$.

arbitrary time θ is shown by the solid curve. The nodal temperatures at θ, are called $t^{\theta}_{n-\Delta x}$, t^{θ}_{n}, $t^{\theta}_{n+\Delta x}$, etc. Combining Eq. 8-5 and Eq. 7-3 for the one-dimensional Laplacian in finite-difference form, here repeated,

$$\frac{\partial^2 t}{\partial x^2} = \frac{(t^{\theta}_{n+\Delta x} - t^{\theta}_{n}) - (t^{\theta}_{n} - t^{\theta}_{n-\Delta x})}{(\Delta x)^2} \quad [7\text{-}3]$$

with the general conduction Eq. 8-4 gives

$$\frac{(t^{\theta}_{n+\Delta x} - t^{\theta}_{n}) - (t^{\theta}_{n} - t^{\theta}_{n-\Delta x})}{(\Delta x)^2} + \frac{q'''}{k} = \frac{1}{\alpha}\,\frac{t^{\theta+\Delta\theta}_{n} - t^{\theta}_{n}}{\Delta\theta}$$

Rearranging and recalling that the thermal diffusivity $\alpha = k/\rho c$,

$$t^{\theta+\Delta\theta}_{n} = (1 - 2\text{Fo})t^{\theta}_{n} + \text{Fo}(t^{\theta}_{n-\Delta x} + t^{\theta}_{n+\Delta x}) + \frac{q'''\,\Delta\theta}{\rho c} \quad (8\text{-}6a)$$

where Fo is the *Fourier modulus,* dimensionless, given by

$$\text{Fo} = \frac{\alpha \Delta\theta}{(\Delta x)^2} \tag{8-7}$$

The last term of Eq. 8-6a has units of temperature. Combining it with Fo, Eq. 8-7, and Eq. 7-13b for Δt_g, the temperature increment due to heat generation, gives

$$\frac{q''' \Delta\theta}{\rho c} = \text{Fo}\, \frac{q'''(\Delta x)^2}{k} = 2\text{Fo}\, \Delta t_g \tag{8-8}$$

Equation 8-6a now becomes

$$t_n^{\theta+\Delta\theta} = (1 - 2\text{Fo})t_n^\theta + \text{Fo}(t_{n+\Delta x}^\theta + t_{n-\Delta x}^\theta) + 2\text{Fo}\, \Delta t_g \tag{8-6b}$$

The Two-Dimensional Case

The body is divided into a square grid with sides $\Delta x = \Delta y$, Fig. 7-2. The Laplacian of t in Eq. 8-4 was given by Eq. 7-15. Combining it with Eqs. 8-4 and 8-5 gives

$$\frac{t_{n+\Delta x}^\theta + t_{n-\Delta x}^\theta + t_{n+\Delta y}^\theta + t_{n-\Delta y}^\theta - 4t_n^\theta}{(\Delta x)^2} + \frac{q'''}{k} = \frac{t_n^{\theta+\Delta\theta} - t_n^\theta}{\alpha \Delta\theta}$$

or

$$t_n^{\theta+\Delta\theta} = (1 - 4\text{Fo})t_n^\theta + \text{Fo}(t_{n+\Delta x}^\theta + t_{n-\Delta x}^\theta + t_{n+\Delta y}^\theta + t_{n-\Delta y}^\theta) + \frac{q''' \Delta\theta}{\rho c} \tag{8-9a}$$

and using Δt_g from Eq. 8-8,

$$t_n^{\theta+\Delta\theta} = (1 - 4\text{Fo})t_n^\theta + \text{Fo}(t_{n+\Delta x}^\theta + t_{n-\Delta x}^\theta + t_{n+\Delta y}^\theta + t_{n-\Delta y}^\theta) + 2\text{Fo}\, \Delta t_g \tag{8-9b}$$

The Three-Dimensional Case

The body is divided into a cubical grid of sides $\Delta x = \Delta y = \Delta z$, Fig. 7-3. A similar treatment to the above two cases yields

$$t_n^{\theta+\Delta\theta} = (1 - 6\text{Fo})t_n^\theta + \text{Fo}(t_{n+\Delta x}^\theta + t_{n-\Delta x}^\theta + t_{n+\Delta y}^\theta + t_{n-\Delta y}^\theta + t_{n+\Delta z}^\theta + t_{n-\Delta z}^\theta) + 2\text{Fo}\, \Delta t_g \tag{8-10a}$$

The Fourier modulus in Eqs. 8-6, 8-9, and 8-10 can be arbitrarily chosen to conveniently relate the chosen space increment Δx to the time increment $\Delta\theta$, or vice versa. In Eq. 8-9b, for example, the coefficient of t_n^θ becomes negative for $\text{Fo} > 1/4$. This means that the higher the temperature at any position and time, t_n^θ, the lower will be the future temperature at the same position $t_n^{\theta+\Delta\theta}$. This would violate thermodynamic principles as well as lead to instabilities in which the numerical solution would not converge [48]. In Eqs. 8-6, 8-9 and 8-10, negative coefficients must be avoided and therefore the Fourier modulus must be chosen such that

$$\left.\begin{aligned} &\text{In one-dimensional cases, Fo} \leq \frac{1}{2} \\ &\text{In two-dimensional cases, Fo} \leq \frac{1}{4} \\ &\text{In three-dimensional cases, Fo} \leq \frac{1}{6} \end{aligned}\right\} \tag{8-11}$$

It can be shown that the lower the values of Fo chosen in each case, the more accurate are the results of the numerical process for the same Δx. Accuracy can also be improved, of course, by choosing as small Δx as possible. Both choices increase the number of steps necessary to reach a particular solution. It is evident, however, that choosing Fo at the limits of Eq. 8-11, such that

$$\left.\begin{aligned} &\text{In one-dimensional cases, Fo} = \frac{1}{2} \\ &\text{In two-dimensional cases, Fo} = \frac{1}{4} \\ &\text{In three-dimensional cases, Fo} = \frac{1}{6} \end{aligned}\right\} \tag{8-12}$$

greatly simplifies the equations, although it results in solutions that are accurate (especially for the first few time increments) only if Δx is very small. Equations 8-6, 8-9 and 8-10 now become:

One-dimensional

$$t_n^{\theta+\Delta\theta} = \frac{t_{n+\Delta x}^\theta + t_{n-\Delta x}^\theta}{2} + \Delta t_g \tag{8-6c}$$

Two-dimensional

$$t_n^{\theta+\Delta\theta} = \frac{t_{n+\Delta x}^\theta + t_{n-\Delta x}^\theta + t_{n+\Delta y}^\theta + t_{n-\Delta y}^\theta}{4} + \frac{1}{2}\Delta t_g \tag{8-9c}$$

Three-dimensional

$$t_n^{\theta+\Delta\theta} = \frac{t_{n+\Delta x}^\theta + t_{n-\Delta x}^\theta + t_{n+\Delta y}^\theta + t_{n-\Delta y}^\theta + t_{n+\Delta z}^\theta + t_{n-\Delta z}^\theta}{6} + \frac{1}{3}\Delta t_g \tag{8-10b}$$

so that the temperature at one nodal point at time $\theta + \Delta\theta$ is equal to the arithmetic average of the temperatures of the surrounding nodal points at time θ, plus a full, one-half or one-third of the temperature increment due to heat generation, respectively. The above equations can be written for a set of nodal points at each time interval and solved. The difference between this and the steady state (Chapter 7), is that here one usually begins with a known temperature profile and proceeds to determine usable temperature profiles at succeeding time intervals, after a change in heat generation, environmental conditions, or others. In the steady state, by comparison, one often begins with an assumed temperature profile, reduces one's errors in successive steps, and only a single solution (the last) is obtained.

Numerical methods can handle problems with complicated boundary conditions and with varying physical properties. (Graphical methods, suitable only for solving one-dimensional problems will be presented in Sec. 8-7.) The next section illustrates the numerical solution of a two-dimensional case.

8-5. NUMERICAL SOLUTION OF A TWO-DIMENSIONAL TRANSIENT CASE

The solution is based on Eq. 8-9c. An example will help illustrate the method.

Example 8-1. Consider the two-dimensional cruciform of Fig. 7-4, here repeated. The temperature in the steady state ($\theta = \infty$) with surfaces at 600°F and heat generation $q''' = 10^7$ Btu/hr ft³, was solved numerically by several methods in Chap. 7. Let us now assume that the cruciform was initially at a uniform temperature 600°F, and that q''' was suddenly imposed on it. The surfaces were made to remain at 600°F at all times. It is required to find the temperature distribution at successive intervals in that body. $k = 19.84$ Btu/hr ft°F, $\rho = 1984$ lb$_m$/ft³ and $c = 0.04$ Btu/lb$_m$ °F.

Solution. Choosing the same mesh of Fig. 7-4, so that $\Delta x = \Delta y = 0.2$ in., Δt_g remains the same at 70°F. Using Eq. 8-7 and

$$\text{Fo} = \frac{1}{4}$$

gives

$$\Delta\theta = \frac{(\Delta x)^2 \rho c}{4k} = \frac{(0.2/12)^2 \times 1984 \times 0.04}{4 \times 19.84}$$

$$= 2.778 \times 10^{-4} \text{hr} = 1 \text{sec}$$

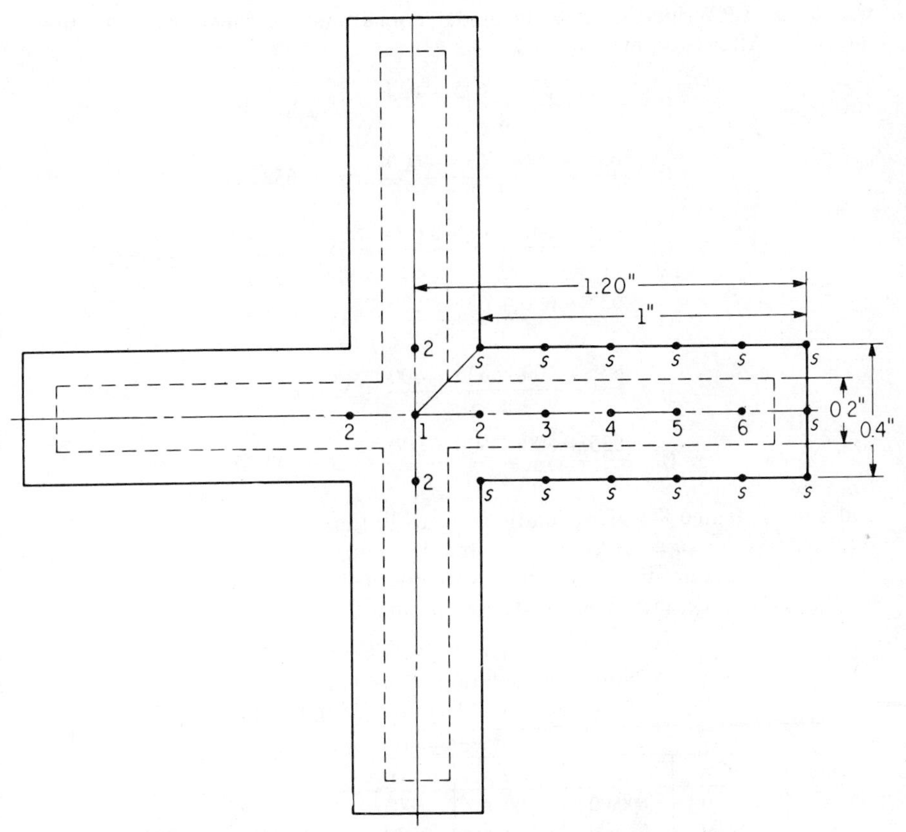

FIG. 7-4 (repeated). Two-dimensional (infinitely long) cruciform.

Sets of equations of the type of 8-9c are now written for the nodal points 1 through 6, one set for each time interval, and each set solved separately. Thus the temperatures after one time interval of 1 second would be given by

$$t_1^1 = \frac{600 + 600 + 600 + 600}{4} + 35 = 635°\text{F}$$

Likewise,

$$t_2^1 = 635°\text{F}$$
$$t_3^1 = 635°\text{F}$$
$$t_4^1 = 635°\text{F}$$
$$t_5^1 = 635°\text{F}$$
$$t_6^1 = 635°\text{F}$$

where the subscripts indicate the nodal points and the superscripts the time interval. After two time intervals,

$$t_1^2 = \frac{635 + 635 + 635 + 635}{4} + 35 = 670°\text{F}$$

$$t_2^2 = \frac{635 + 600 + 635 + 600}{4} + 35 = 652.5$$

$$t_3^2 = \frac{635 + 600 + 635 + 600}{4} + 35 = 652.5$$

$$t_4^2 = \frac{635 + 600 + 635 + 600}{4} + 35 = 652.5$$

$$t_5^2 = \frac{635 + 600 + 635 + 600}{4} + 35 = 652.5$$

$$t_6^2 = \frac{635 + 600 + 600 + 600}{4} + 35 = 643.75$$

and so on. Table 8-1 gives solutions up to 10 time intervals (10 sec) as well as the steady-state ($\theta = \infty$) solution found by matrix algebra (Sec. 7-4). More realistic temperature-time distributions would, of course, be obtained if the space, and therefore time, increments chosen were small.

TABLE 8-1
Numerical Solution of the Transient Two-Dimensional Case of Example 8-1

Time,	Node						
sec	1	2	3	4	5	6	Surface
0	600.00	600.00	600.00	600.00	600.00	600.00	600
1	635.00	635.00	635.00	635.00	635.00	635.00	600
2	670.00	652.50	652.50	652.50	652.50	643.75	600
3	687.50	665.63	661.25	661.25	659.06	648.13	600
4	700.63	672.19	666.72	665.08	662.34	649.77	600
5	707.19	676.84	669.32	667.27	663.71	650.59	600
6	711.84	679.13	671.03	668.26	664.46	650.93	600
7	714.13	680.72	671.85	668.87	664.80	651.12	600
8	715.72	681.49	672.40	669.16	665.00	651.20	600
9	716.49	682.03	672.66	669.35	665.09	651.25	600
10	717.03	682.29	672.84	669.44	665.15	651.27	600
–	–	–	–	–	–	–	–
∞	717.69	682.69	673.06	669.57	665.22	651.30	600

8-6. GRAPHICAL SOLUTION FOR A PLANE GEOMETRY AND NO BOUNDARY LAYER

Graphical methods are less convenient than numerical methods in that they cannot easily handle more than one dimension or variable properties. Like many analogical techniques (Sec. 7-14), their usefulness has been much diminished since the advent of the computer made numerical methods so much more convenient. They do, however, have the advantage of giving a continuing picture of temperature profiles at preselected time intervals.

This and succeeding sections will present a graphical method of solution of the one-dimensional case with different boundary conditions. The method extends the well-known *Schmidt* technique which has been widely used for solving unsteady state conduction problems with no heat generation, to the case where internal heat generation exists, which is of interest in nuclear reactor work.

The graphical method is based on $\mathrm{Fo} = 1/2$ and the one-dimensional equation 8-6c, here repeated.

$$t_n^{\theta + \Delta\theta} = \frac{t_{n+\Delta x}^{\theta} + t_{n-\Delta x}^{\theta}}{2} + \Delta t_g \qquad [8\text{-}6c]$$

Referring to Fig. 8-3, the temperature at n after one time interval $\Delta\theta$, $t_n^{\theta+\Delta\theta}$ is obtained by joining the temperature points of the preceding and succeeding nodal planes, $n - \Delta x$ and $n + \Delta x$, at θ by a straight line and marking the intersection of that straight line with nodal plane n (giving the first term at the right-hand side of Eq. 8-6c and would have been the desired temperature in the case of no heat generation), and marking a quantity of temperature equal to Δt_g above that point of intersection.

This process is repeated for all nodes within the wall. The new points similarly obtained as $t_n^{\theta+\Delta\theta}$ now form a new temperature profile that exists at time $\theta + \Delta\theta$. With this, a new start is made and a temperature profile at time $\theta + 2\Delta\theta$ is obtained, and so on.

Note that once Δx is fixed the temperature profiles at successive intervals of time are specified. However for high values of the thermal diffusivity α, a particular temperature level is reached in a shorter time, since the corresponding $\Delta\theta$ is inversely proportional to α, Eq. 8-7.

The above technique can be adapted to a wide variety of initial and boundary conditions. For example, the initial temperature distribution could be curved (Fig. 8-3), a sloping straight line, or uniform. The wall boundary temperatures at $x = 0$ and $x = L$, the thickness of the wall, may be fixed or could vary with time, so that it would acquire a new value at each new interval $\Delta\theta$. Also q''', and therefore Δt_g, could also be made to vary with x and θ.

In the case of constant heat generation and constant boundary temperatures t_0 and t_L (Fig. 8-4), the *steady state* is reached when the temperature no longer changes from one interval to another, or when $t_n^{\theta+\Delta\theta} = t_n^{\theta} = t_n$. This occurs when the difference between the value of the steady-state curve at n and that of a straight line connecting the two equidistant planes $n - \Delta x$ and $n + \Delta x$ is a constant value. In Fig. 8-4, for example,

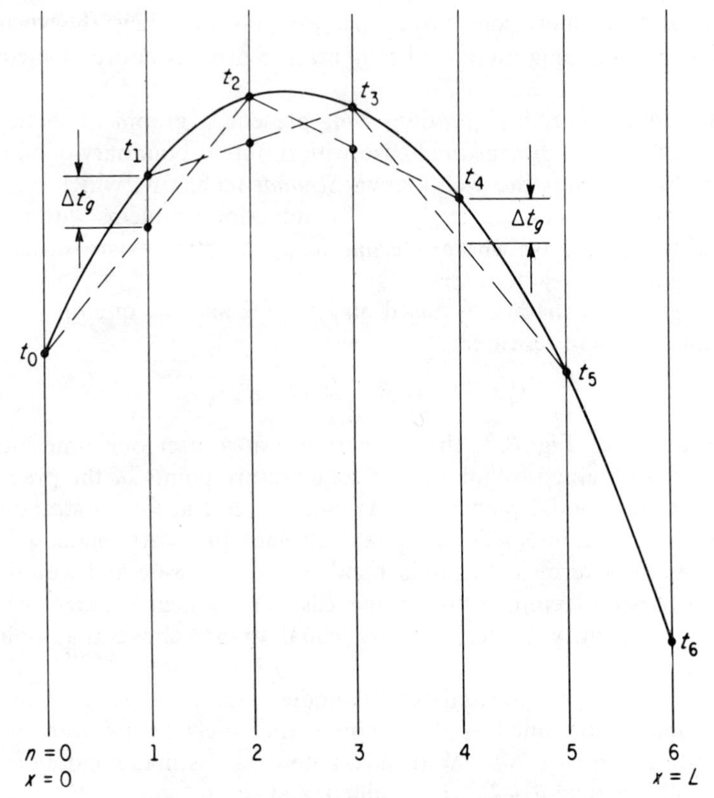

FIG. 8-4. Steady-state temperature distribution.

$$t_2 - \frac{t_1 + t_3}{2} = t_3 - \frac{t_2 + t_4}{2} = \cdots = \Delta t_g \tag{8-13}$$

This is a characteristic of a parabola. Remember that the steady-state temperature distribution of a fuel element with uniform heat generation is parabolic and for a plane wall, is given by $d^2t/dx^2 = -q'''/k$, Eq. 5-22.

The *heat conducted* at any plane at any time can, of course, be obtained from the slope of the temperature curve at that plane and time from

$$q''_{n,\theta} = -k \frac{dt}{dx}\bigg|_{n,\theta} \tag{8-14}$$

Example 8-2. A 1/2-in.-thick uranium metal fuel element is initially at a uniform 500°F and no heat generation. Suddenly the element was made to generate 2.28×10^7 Btu/hr ft³ but the element surface temperatures were held

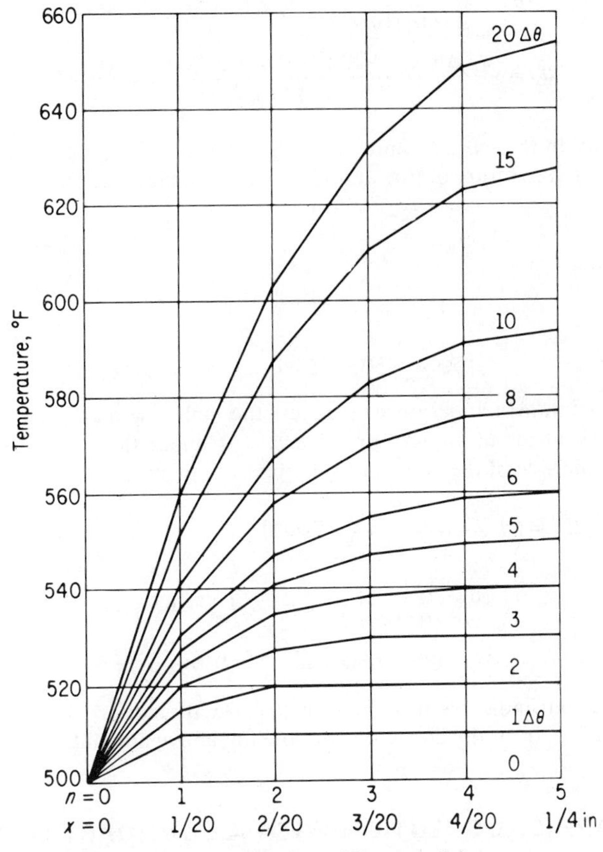

FIG. 8-5. Solution of Example 8-2.

at 500°F. When will the maximum temperature of the element reach the halfway mark, and what is the heat flux at the surface at that time? Assume the element to be an infinite plate. Take $k = 19.8$ Btu/hr ft°F, $\rho = 1{,}180$ lb_m/ft^3, $c = 0.04$ Btu/lb_m °F.

Solution. The slab is divided into 10 layers 1/20 in. thick each. Since the element is symmetrical, only half of it need be treated and an insulated boundary at $n = 5$ substitutes for the other half (Fig. 8-5):

$$\Delta x = \frac{1}{20} \text{ in.} = 0.004167 \text{ ft}$$

$$\alpha = \frac{k}{\rho c} = \frac{19.8}{1180 \times 0.04} = 0.419 \text{ ft}^2/\text{hr}$$

$$\text{Fo} = \frac{1}{2}$$

$$\Delta\theta = \frac{(0.004167)^2}{2 \times 0.419} = 2.08 \times 10^{-5} \text{ hr} = 75 \text{ msec}$$

$$\Delta t_g = \frac{q''' \Delta\theta}{\rho c} = \frac{2.28 \times 10^7 \times 2.08 \times 10^{-5}}{1180 \times 0.04} = 10°\text{F}$$

The solution for selected time intervals is shown in Fig. 8-5. The maximum (steady-state) temperature occurs at midplane, $n = 5$, and is given by Eq. 5-27:

$$t_5 = t_0 + \frac{q''' s^2}{2k}$$

$$= 500 + \frac{2.28 \times 10^7 \times (5 \times 0.004167)^2}{2 \times 19.8}$$

$$= 500 + 250 = 750°\text{F}$$

The maximum temperature reaches the halfway mark, i.e., $t_5^\theta = 500 + 250/2 = 625°\text{F}$ at about $14.6\,\Delta\theta$ or 1.1 sec. At about that time the heat transferred from one side of the element, q_s'' may be obtained from

$$q_s'' = -\left(k\,\frac{t_1^{15} - t_0^{15}}{\Delta x} + \frac{1}{2}\,\Delta x\, q'''\right)$$

$$= -\left(19.8\,\frac{551 - 500}{0.004167} + \frac{1}{2} \times 0.004167 \times 2.28 \times 10^7\right)$$

$$= -\,(242{,}350 + 47{,}500) = -\,289{,}850 \text{ Btu/hr ft}^2$$

The negative sign indicates heat flow in the $-x$ direction. The total absolute value of heat flux from the element is double the above amount.

8-7. GRAPHICAL SOLUTION WITH A BOUNDARY LAYER

In the above example the surface temperatures were known. The graphical method, however, can be easily extended to cover the more usual case where it is the fluid temperature, outside a boundary layer, or of a second radiative body, that is known. The expression for the surface temperature $t_0^{\theta+\Delta\theta}$ here depends on the fact that only half a solid layer, adjacent to the surface at $n = 0$, with half the usual heat capacity is involved, and on the assumption that the fluid has negligible heat capacity. A heat balance at the surface is given by

$$k\,\frac{t_1^\theta - t_0^\theta}{\Delta x} - h(t_0^\theta - t_f^\theta) + \frac{q''' \Delta x}{2} = \frac{c\rho\,\Delta x}{2\Delta\theta}(t_0^{\theta+\Delta\theta} - t_0^\theta) \qquad (8\text{-}15a)$$

where t_f^θ is the fluid bulk temperature (outside the boundary layer at time θ), and h is the heat-transfer coefficient by convection (or radiation). The above equation is rearranged to

$$\frac{1}{2\text{Fo}}(t_0^{\theta+\Delta\theta} - t_0^\theta) = t_1^\theta - t_0^\theta + \text{Bi}_{\Delta x}(t_f^\theta - t_0^\theta) + \Delta t_g \qquad \text{(8-15b)}$$

where Fo = the Fourier modulus, Eq. 8-7.

$\text{Bi}_{\Delta x} = h\Delta x/k$, a Biot number based on Δx as a characteristic length

Again taking Fo = 1/2 (it should be the same inside and outside the slab), Eq. 8-15b becomes

$$t_0^{\theta+\Delta\theta} = \left[t_1^\theta - \frac{\Delta x}{(k/h)}(t_0^\theta - t_f^\theta)\right] + \Delta t_g \qquad \text{(8-16)}$$

The graphical representation of Eq. 8-16 requires the addition of an extra layer, outside the wall, of thickness $k/h = \Delta x/\text{Bi}_{\Delta x}$, which has the units of length. t_0^θ and t_f^θ are now connected by a straight line, Fig. 8-6.

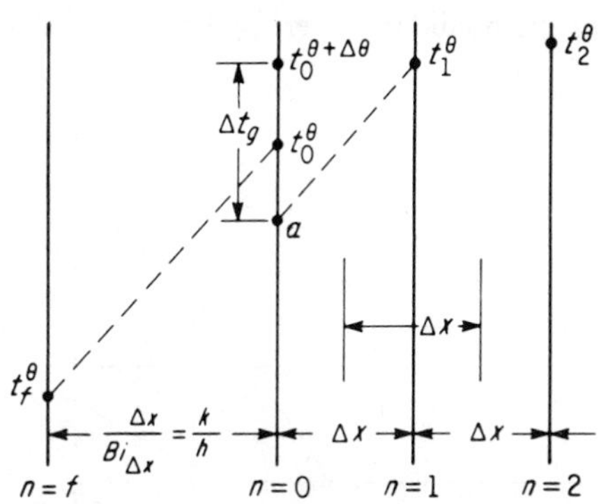

FIG. 8-6. Graphical determination of the surface temperature of a slab exchanging heat with a fluid across a boundary layer. Line $a - t_1^\theta$ is drawn parallel to line $t_f^\theta - t_0^\theta$.

Another straight line, parallel to it is then drawn from t_1^θ. The point of intersection of the latter with the $n = 0$ boundary (point a) represents the two terms between the brackets on the right-hand side of Eq. 8-16. The value of Δt_g is then added to obtain $t_0^{\theta+\Delta\theta}$.

One now proceeds to draw temperature gradients in the usual manner. A full Δt_g is added at plane 0 as at other planes, since no heat is generated in the fluid, and heat is generated in half a layer with half the normal heat capacity. As with wall surfaces, the fluid temperature can be held constant or made to acquire a new value at each time interval (to represent the heating effect from the preceding length of a fuel element, for example). The heat-transfer coefficient can also be made to vary with time so that the thickness of the extra layer, k/h, acquires a new value at each interval.

Example 8-3. A bare fuel element of the same dimensions and physical properties as that in Example 8-2 is cooled by a fluid. Initially the element and fluid were at 500°F. Suddenly heat is generated in the element at the same rate as in Example 8-2. The temperature of the fluid is held constant after heat generation. The heat transfer coefficient is 2376 Btu/hr ft °F. How much time does it take the mid-plane to reach the same temperature as in Example 8-2 (625°F) and what is the surface heat flux at that time?

Solution

$$(k/h) = \frac{19.8}{2376} = 0.00833\text{ft}$$

As in Example 8-2, $\Delta x = 0.004167$ ft, $\Delta\theta = 75$ milli-second and $\Delta t_g = 10$°F. The graphical solution is shown in Fig. 8-7 up to 20 $\Delta\theta$'s. The midplane reaches 625°F after about 13 $\Delta\theta$ or 0.975 sec, sooner than in Example 8-2, which is to be expected since the boundary layer represents an added resistance to heat transfer.

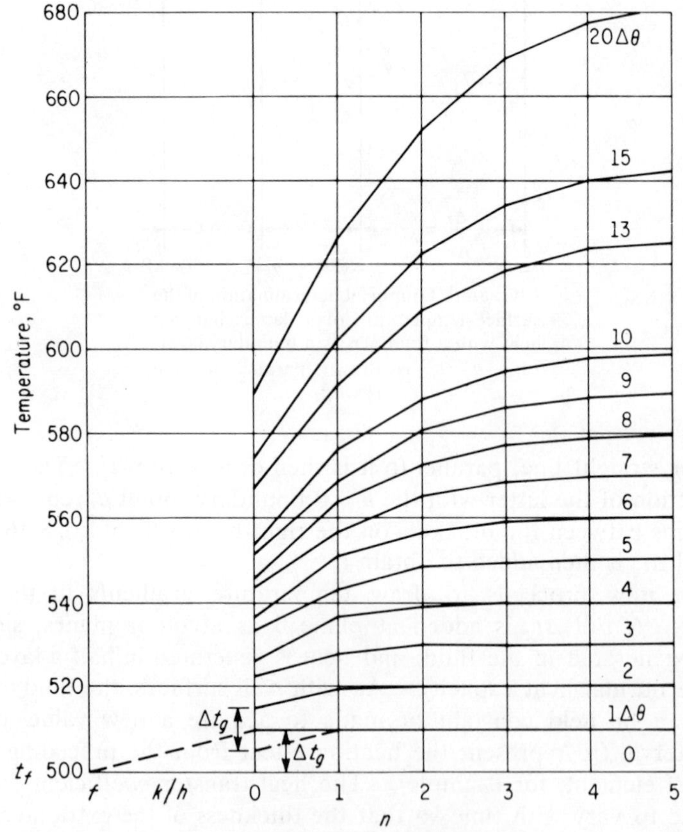

FIG. 8-7. Solution of Example 8-3.

The heat transferred from one side of the element q_s'' at 0.975 sec is

$$q_s'' = -\left(k\frac{t_1^{13} - t_0^{13}}{\Delta x} + \frac{1}{2}\Delta x\, q'''\right)$$

$$= -\left(19.8\,\frac{592 - 568}{0.004167} + \frac{1}{2} \times 0.004167 \times 2.28 \times 10^7\right)$$

$$= -(114{,}040 + 47{,}500) = -161{,}540 \text{ Btu/hr ft}^2$$

The same result could be obtained from

$$q_s'' = -h\,(t_0^{13} - t_f)$$

$$= -2376\,(568 - 500) = -161{,}570 \text{ Btu/hr ft}^2$$

as compared to $-289{,}850$ in Example 8-2.

8-8. THE CASE OF A COMPOSITE WALL

The graphical technique will now be extended to cover the case of a composite wall, such as a clad fuel element. Here the layer thickness Δx in the cladding and in the fuel would be different since Δx, for one material of constant k, really represents thermal resistance $\Delta x/k$. The temperature gradient at the interface between the two materials would be given by the condition that

$$-k_c\left(\frac{\partial t}{\partial x}\right)_c = -k_f\left(\frac{\partial t}{\partial x}\right)_f \tag{8-16a}$$

where the subscripts c and f refer to clad and fuel respectively, or

$$\left(\frac{\partial t}{\partial\, x/k}\right)_c = \left(\frac{\partial t}{\partial\, x/k}\right)_f \tag{8-16b}$$

a necessary condition to connect temperatures between one material and another in the graphical plot. The one-dimensional conduction equation is now written in terms of this thermal resistance for each material as

$$\frac{\partial t}{\partial \theta} = \frac{1}{k\rho c}\,\frac{\partial^2 t}{\partial (x/k)^2} + \frac{q'''}{\rho c} \tag{8-17a}$$

In finite-difference form this becomes

$$\frac{t_n^{\theta+\Delta\theta} - t_n^{\theta}}{\Delta\theta} = \frac{1}{k\rho c}\,\frac{t_{n-\Delta x}^{\theta} - 2t_n^{\theta} + t_{n+\Delta x}^{\theta}}{(\Delta x/k)^2} + \frac{q'''\,\alpha}{k} \tag{8-17b}$$

$\Delta\theta$ and $\Delta x/k$ are selected such that the Fourier modulus is 1/2, or

$$\Delta\theta = \frac{(\Delta x/k)^2\, k\rho c}{2} \tag{8-18}$$

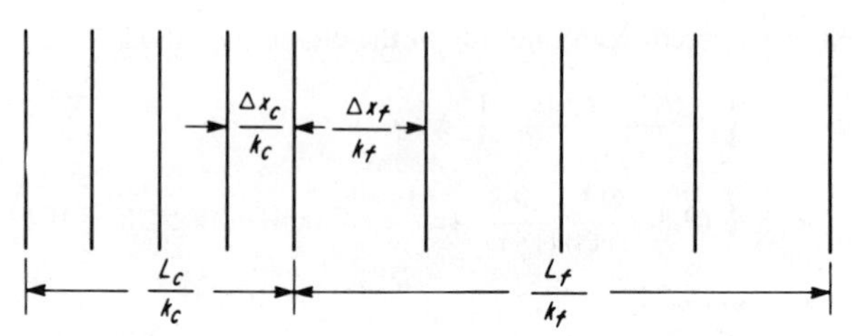

FIG. 8-8. Graphical representation of layers in a composite wall system.

Substituting this in Eq. 8-17, one obtains, for each material, an expression identical to that in Eqs. 8-6c. The relation between Δx for each material is obtained with the condition that $\Delta\theta$ must be the same in both, or

$$\frac{(\Delta x/k)_c^2 k_c \rho_c c_c}{(\Delta x/k)_f^2 k_f \rho_f c_f} = 1$$

Thus

$$\frac{(\Delta x)_c}{(\Delta x)_f} = \sqrt{\frac{\alpha_c}{\alpha_f}} \tag{8-19}$$

If the total thicknesses of clad and fuel, L_c and L_f, are known, the above relationship determines the relative number of layers each is to be divided into. The actual thickness of each layer, when plotted graphically, however, will be proportional to $(\Delta x)_c/k_c$ and $(\Delta x)_f/k_f$ to satisfy condition 8-16b at the interface. The total thickness of clad and fuel in the plot will be equal to the number of layers $L/\Delta x$, times the thickness of each layer $\Delta x/k$, or proportional to (L_c/k_c) for the cladding and (L_f/k_f) for the fuel, and the graphical plot will not represent the actual dimensions.

Where there is a fluid, the thickness of the extra layer would be proportional to $1/h$, [or $(\Delta x/k)/Bi_{\Delta x}$] instead of k/h (or $\Delta x/Bi_{\Delta x}$) as in the case of the single wall.

8-9. GRAPHICAL SOLUTION FOR CYLINDERS

The graphical solutions can be applied to hollow and solid cylinders such as fuel rods. For thin hollow cylinders, the plane wall solution is usually satisfactory. For solid and thick hollow cylinders the heat conduction equation in cylindrical form (Sec. 5-3), is used.

$$\frac{1}{\alpha}\frac{\partial t}{\partial \theta} = \frac{\partial^2 t}{\partial r^2} + \frac{1}{r}\frac{\partial t}{\partial r} + \frac{q'''}{k} \tag{8-20}$$

This equation is changed into a more suitable form for finite-difference solutions by introducing a new variable η such that

$$d\eta = \frac{dr}{r} \tag{8-21}$$

then

$$\frac{\partial t}{\partial r} = \frac{1}{r}\frac{\partial t}{\partial \eta}$$

and

$$\frac{\partial^2 t}{\partial r^2} = \frac{1}{r^2}\frac{\partial^2 t}{\partial \eta^2} - \frac{1}{r^2}\frac{\partial t}{\partial \eta} \tag{8-22}$$

The above equations are combined to give

$$\frac{\partial t}{\partial \theta} = \frac{\alpha}{r^2}\frac{\partial^2 t}{\partial \eta^2} + \frac{q'''}{\rho c} \tag{8-23a}$$

This equation can now be transformed into the finite-difference form

$$\frac{t_n^{\theta+\Delta\theta} - t_n^\theta}{\Delta\theta} = \frac{\alpha}{r^2}\frac{(t_{n+\Delta r}^\theta - t_n^\theta) - (t_n^\theta - t_{n-\Delta r}^\theta)}{(\Delta\eta)^2} + \frac{q'''}{\rho c}$$

or

$$t_n^{\theta+\Delta\theta} - t_n^\theta = \frac{2\alpha\Delta\theta}{(r\Delta\eta)^2}\left[\frac{(t_{n+\Delta r}^\theta + t_{n-\Delta r}^\theta)}{2} - t_n^\theta\right] + \frac{q'''\Delta\theta}{\rho c} \tag{8-23b}$$

The time interval is now chosen such that the Fourier modulus, which for a cylinder is given by

$$\text{Fo} = \frac{\alpha\Delta\theta}{(\Delta r)^2} \tag{8-24}$$

is equal to 1/2 or

$$\Delta\theta = \frac{(\Delta r)^2}{2\alpha} = \frac{(r\Delta\eta)^2}{2\alpha}$$

Equation 8-23b now becomes

$$t_n^{\theta+\Delta\theta} = \frac{t_{n+\Delta r}^\theta + t_{n-\Delta r}^\theta}{2} + \frac{q'''\Delta\theta}{\rho c} \tag{8-25a}$$

or

$$t_n^{\theta+\Delta\theta} = \frac{t_{n+\Delta r}^\theta + t_{n-\Delta r}^\theta}{2} + \Delta t_g \tag{8-25b}$$

which are identical in form to Eq. 8-6c.

The necessary condition for the graphical plot is that there would be no discontinuity in the temperature gradient between adjacent layers in the plot. This is obtained from the condition

$$\left[-k\,2\pi r\frac{\partial t}{\partial r}\right]_1 = \left[-k\,2\pi r\frac{\partial t}{\partial r}\right]_2 \tag{8-26a}$$

where the subscripts 1 and 2 refer to adjacent layers. For the same material, i.e., where $k_1 = k_2$, Eq. 8-26a simplifies to

$$\left(\frac{\partial t}{\partial r/r}\right)_1 = \left(\frac{\partial t}{\partial r/r}\right)_2 \tag{8-26b}$$

or

$$\left(\frac{\partial t}{\partial \eta}\right)_1 = \left(\frac{\partial t}{\partial \eta}\right)_2 \tag{8-26c}$$

The cylinder is now divided into a number of concentric layers of equal thickness Δr, which is analogous to dividing the plane wall into parallel layers of equal Δx. However, the scale of the plot is such that the widths of these layers is drawn proportional to $\Delta\eta$ or $\Delta r/r$ and the inner layers would be wider than the outer layers. If the layers are sufficiently thin, a reasonably accurate value of r for each layer would be its mean radius. Figure 8-9 is drawn for a hollow cylinder 0.5 in. inner and 1.5 in. outer radius, $k = 20$ Btu/hr ft°F, divided into 10 layers, $\Delta r = 0.1$ in. each, and cooled on the inside and outside with $h = 1200$ Btu/hr ft²°F.

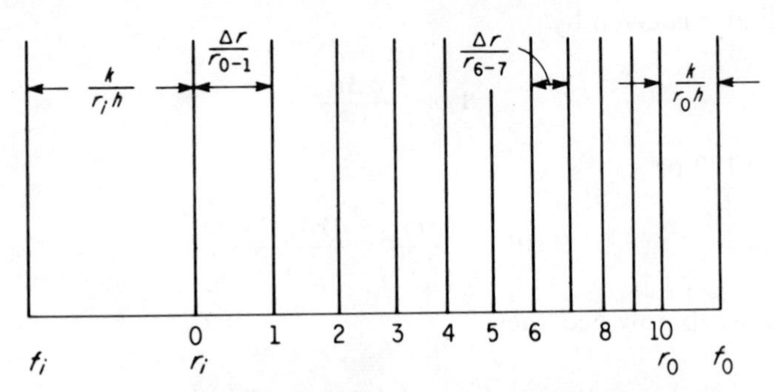

FIG. 8-9. Graphical representation of layers in a hollow cylinder with inner and outer surface conductances.

Cylinders with Boundary Layers

When a fluid exists in contact with the outer (or inner) radius of a cylinder, the boundary condition is similar to that of Eqs. 8-15, or for the case of the outer radius r_0,

$$-k\frac{\partial t}{\partial r}\bigg|_{r_0}^{\theta} = h(t_f^\theta - t_{r_0}^\theta) \tag{8-27a}$$

Combining with Eq. 8-21 gives

$$\frac{\partial t}{\partial \eta}\bigg|_{r_0}^{\theta} = \frac{t_{r_0}^\theta - t_f^\theta}{(k/r_0 h)} \tag{8-27b}$$

The graphical technique is now to add an extra layer, beyond the outer radius, with a width proportional to $(k/r_0 h)$. The rest of the solution is similar to that of a plane wall.

Composite Cylinders

In the case of a composite cylinder, such as a clad fuel, the thickness of layers would be based on the condition that $\Delta\theta$ is the same, or from Eq. 8-24:

$$\frac{(\Delta r)_c}{(\Delta r)_f} = \sqrt{\frac{\alpha_c}{\alpha_f}} \tag{8-28}$$

which is analogous to Eq. 8-19 for the plane composite wall. This determines the relative number of layers each is to be divided into. The widths of the layers in the plot, however, would be proportional to $(\Delta r/rk)_c$ for the cladding and $(\Delta r/rk)_f$ for the fuel.

Where there is a fluid adjacent to the outer radius of the cladding r_c the width of the extra layer would be proportional to $(1/hr_c)$.

8-10. AN EXACT ANALYTICAL METHOD

In this section an exact analytical solution by the method of *separation of variables* is presented. The method of separation of variables is used to solve both homogeneous and nonhomogeneous partial differential equations, but the nonhomogeneous equations must first be transformed into homogeneous equations. Also, essential to the method is the use of Fourier series. The following example illustrates the above method, as well as the technique of *normalization* of differential equations.

Consider the unsteady-state heat conduction in a long unclad cylindrical fuel element of radius r_0. The element is cooled by a fluid at a constant temperature t_f by a constant heat-transfer coefficient h. Initially the element was also at t_f. A sudden uniform heat generation q''' was then imposed on the element. The energy equation describing this system is

$$\frac{\partial^2 t}{\partial r^2} + \frac{1}{r}\frac{\partial t}{\partial r} + \frac{q'''}{k} = \frac{1}{\alpha}\frac{\partial t}{\partial \theta} \tag{8-29}$$

and the boundary conditions are

$$t(0,\theta) < \infty \tag{8-30}$$

$$-kA_c \frac{dt}{dr}\bigg|_{r=0} = hA_c(t - t_f) \tag{8-31a}$$

where $t(0,\theta)$ is the temperature at the center line of the element ($r = 0$) and time θ. A_c is the circumpherential area, at $r = r_0$.

It is often convenient to *normalize* an equation if the number of dimensional parameters is large. In normalization these parameters are transformed into nondimensional ones. In the process, nondimensional groups are often obtained, such as the well known Reynolds, Prandtl, and other numbers. Besides simplifying the solution, normalization, therefore, often provides an additional physical insight into the problem, even without completely solving it. The above set of equations is normalized by putting

$$\left.\begin{aligned} & t^* = \frac{t - t_f}{t_0}, \quad r^* = \frac{r}{r_0}, \quad \text{and} \quad \theta^* = \frac{\theta}{\theta_0} \\ \text{where} \quad & \\ & \theta_0 = \frac{r_0^2}{\alpha} \quad \text{and} \quad t_0 = \frac{q''' r_0^2}{k} \end{aligned}\right\} \tag{8-32}$$

Equations 8-29 and 8-31a after some rearranging now become

$$\frac{\partial^2 t^*}{\partial r^{*2}} + \frac{1}{r^*}\frac{\partial t^*}{\partial r^*} + 1 = \frac{\partial t^*}{\partial \theta^*} \tag{8-33}$$

and

$$\frac{\partial t^*(1,\theta^*)}{\partial r^*} = -\,\text{Bi}\, t^*(1,\theta^*) \tag{8-31b}$$

where Bi is a Biot number given by

$$\text{Bi} = \frac{hr_0}{k} \tag{8-34}$$

The problem requires a solution for the temperature distribution in space and time, i.e., for $t^*(r^*,\theta^*)$. The separation of variables technique assumes a product solution of the form

$$t^*(r^*,\theta^*) = R(r^*)\Theta(\theta^*) \tag{8-35}$$

where R is a function of r only, and Θ is a function of θ only.

Thus

$$\frac{\partial t^*}{\partial r^*} = \Theta \frac{dR}{dr^*}$$

$$\frac{\partial^2 t^*}{\partial r^{*2}} = \Theta \frac{d^2 R}{dr^{*2}}$$

and

$$\frac{\partial t^*}{\partial \theta^*} = R \frac{d\Theta}{d\theta^*}$$

Substituting in Eqs. 8-33 and 8-31b and rearranging,

$$\frac{1}{R}\left(\frac{d^2R}{dr^{*2}} + \frac{1}{r^*}\frac{dR}{dr^*}\right) = \frac{1}{\Theta}\frac{d\Theta}{d\theta^*} \tag{8-36}$$

and

$$\frac{dR(1)}{dr^*} = -\text{ Bi } R(1) \tag{8-37}$$

The left-hand side of Eq. 8-36 is a function of r only and cannot change with θ, and the right-hand side is a function of θ only and cannot change with r. Since both sides are equal, neither can change with r or θ, and each must be a constant. This constant is assumed to be $-\lambda^2$. Thus Eq. 8-36 becomes

$$\frac{d^2R}{dr^{*2}} + \frac{1}{r^*}\frac{dR}{dr^*} + \lambda^2 R = 0 \tag{8-38}$$

Equation 8-38 is a Bessel equation. It has as a solution

$$R(r^*) = C_1 J_0(\lambda r^*) + C_2 Y_0(\lambda r^*) \tag{8-39a}$$

where J_0 and Y_0 are the Bessel functions of the zero order, and first and second kinds respectively (Appendix C). C_1 and C_2 are constants to be found from the boundary conditions. Applying the first boundary condition, Eq. 8-30, at $r = 0$, gives

$$R(0) = C_1 J_0(0) + C_2 Y_0(0) < \infty \tag{8-39b}$$

from which $C_2 = 0$, since $Y_0(0) = -\infty$, Table C-3, Appendix. Equation 8-39a now becomes

$$R(r^*) = C_1 J_0(\lambda r^*) \tag{8-39c}$$

Applying the second boundary condition, Eq. 8-37, at $r = r_0$, and noting that $J_0' = -J_1$, Table C-1, Appendix, gives for Eq. 8-39c:

$$-C_1\lambda J_1(\lambda) = -\text{Bi}\, C_1 J_0(\lambda)$$

or

$$\text{Bi} = \lambda \frac{J_1(\lambda)}{J_0(\lambda)} \tag{8-40}$$

For a given value of Bi $(= hr_0/k)$, Eq. 8-40 has many solutions for λ. These are the eigenvalues of λ, given as $\lambda_1, \lambda_2, \cdots, \lambda_n, \cdots \infty$. Equation 8-40 is then written in the form

$$\text{Bi} = \lambda_n \frac{J_1(\lambda_n)}{J_0(\lambda_n)} \tag{8-41}$$

where n is an integer, 1, 2, 3, ..., ∞.
Some values of λ_n up to $n = 5$ are found in Table 8-2.

TABLE 8-2
First five roots of λ_n for $\text{Bi} = \lambda_n J_1(\lambda_n)/J_0(\lambda_n)$

Bi	λ_1	λ_2	λ_3	λ_4	λ_5
0.00	0.0000	3.8317	7.0156	10.1735	13.3237
0.02	0.1995	3.8369	7.0184	10.1754	13.3252
0.04	0.2814	3.8421	7.0213	10.1774	13.3267
0.06	0.3438	3.8473	7.0241	10.1794	13.3282
0.08	0.3960	3.8525	7.0270	10.1813	13.3297
0.10	0.4417	3.8577	7.0298	10.1833	13.3312
0.20	0.6170	3.8835	7.0440	10.1931	13.3387
0.30	0.7465	3.9091	7.0582	10.2029	13.3462
0.40	0.8516	3.9344	7.0723	10.2127	13.3537
0.50	0.9408	3.9594	7.0864	10.2225	13.3611
0.60	1.0184	3.9841	7.1004	10.2322	13.3686
0.70	1.0873	4.0085	7.1143	10.2419	13.3761
0.80	1.1490	4.0325	7.1282	10.2516	13.3835
0.90	1.2048	4.0562	7.1421	10.2613	13.3910
1.00	1.2558	4.0795	7.1558	10.2710	13.3984
2.00	1.5994	4.2910	7.2884	10.3658	13.4719
3.00	1.7887	4.4634	7.4103	10.4566	13.5434
4.00	1.9081	4.6018	7.5201	10.5423	13.6125
5.00	1.9898	4.7131	7.6177	10.6223	13.6786
6.00	2.0490	4.8033	7.7039	10.6964	13.7414
7.00	2.0937	4.8772	7.7797	10.7646	13.8008
8.00	2.1286	4.9384	7.8464	10.8271	13.8566
9.00	2.1566	4.9897	7.9051	10.8842	13.9090
10.00	2.1795	5.0332	7.9569	10.9363	13.9580
15.00	2.2509	5.1773	8.1422	11.1367	14.1576
20.00	2.2880	5.2568	8.2534	11.2677	14.2983
30.00	2.3261	5.3410	8.3771	11.4221	14.4748
40.00	2.3455	5.3846	8.4432	11.5081	14.5774
50.00	2.3572	5.4112	8.4840	11.5621	14.6433
100.00	2.3809	5.4652	8.5678	11.6747	14.7834

The solution of Eq. 8-39c is now constructed, using as many of these eigenvalues as necessary for accuracy, in the form

$$t^*(r^*,\theta^*) = \sum_{n=1}^{\infty} C_{1,n}(\theta) J_0(\lambda_n r^*) \tag{8-42}$$

where the first constant in Eq. 8-39c is now $C_{1,n}(\theta)$ and varies with n and θ. Equation 8-42 is solved by multiplying both sides by $r^* J_0(\lambda_m r^*)$ and integrating with respect to r^*. Thus

$$\int_0^1 r^* t^*(r^*, \theta^*) J_0(\lambda_m r^*) dr^* = \sum_{n=1}^{\infty} C_{1,n}(\theta) \int_0^1 r^* J_0(\lambda_m r^*) J_0(\lambda_n r^*) dr^* \tag{8-43}$$

The case where $m \neq n$ is the so-called *orthogonality relation* where the integral on the right-hand side of the above equation is zero. For the case where $m = n$,

$$\begin{aligned} \int_0^1 r^* t^*(r^*, \theta^*) J_0(\lambda_m r^*) dr^* &= C_{1,m}(\theta) \int_0^1 r^* J_0^2(\lambda_m r^*) dr^* \\ &= C_{1,m}(\theta) \left[\frac{J_0^2(\lambda_m) + J_1^2(\lambda_m)}{2} \right] \end{aligned} \tag{8-44}$$

Combining the above equation with Eq. 8-33, and differentiating the right-hand side with respect to θ^*, result in

$$\int_0^1 r^* \left(\frac{d^2 t^*}{dr^{*2}} + \frac{1}{r^*} \frac{dt^*}{dr^*} + 1 \right) J_0(\lambda_m r^*) dr^* = \frac{dC_{1,m}(\theta)}{d\theta^*} \left[\frac{J_0^2(\lambda_m) + J_1^2(\lambda_m)}{2} \right] \tag{8-45}$$

The left-hand side is integrated by parts resulting in an ordinary linear differential equation whose solution is

$$C_{1,m}(\theta) = \frac{2J_1(\lambda_m)}{(\lambda_m)^{1.5} [J_0^2(\lambda_m) + J_1^2(\lambda_m)]} + C_3 e^{-\lambda\theta^*} \tag{8-46}$$

where C_3 is a constant of integration, to be evaluated from the initial condition of the problem. In this example, this condition has been set such that the cylindrical element, at $\theta = 0$, was at the same temperature as the fluid. Thus the initial condition is

$$t^*(r^*,0) = 0 \tag{8-47}$$

At this initial condition $C_{1,m}(\theta) = 0$, so that

$$C_3 = \frac{-2J_1(\lambda_m)}{(\lambda_m)^3 [J_0^2(\lambda_m) + J_1^2(\lambda_m)]} \tag{8-48}$$

and $C_{1,m}(\theta)$ is now completely known. Equation 8-42 now becomes

$$t^*(r^*,\theta^*) = \sum_{n=1}^{\infty} J_0(\lambda_n r^*) \frac{2J_1(\lambda_n)}{(\lambda_n)^3[J_0^2(\lambda_n) + J_1^2(\lambda_n)]} (1 - e^{-\lambda_n^2\theta^*}) \tag{8-49}$$

which, with the help of the eigencondition 8-41, is simplified to

$$t^*(r^*,\theta^*) = \sum_{n=1}^{\infty} \frac{2\mathrm{Bi}}{\lambda_n^2(\lambda_n^2 + \mathrm{Bi}^2)} \cdot \frac{J_0(\lambda_n r^*)}{J_0(\lambda_n)} (1 - e^{-\lambda_n^2\theta^*}) \tag{8-50}$$

The steady-state solution is obtained by putting $\theta^* = \infty$ in Eq. 8-50, resulting in

$$t^*(r^*,\infty) = \sum_{n=1}^{\infty} \frac{2\mathrm{Bi}}{\lambda_n^2(\lambda_n^2 + \mathrm{Bi}^2)} \cdot \frac{J_0(\lambda_n r^*)}{J_0(\lambda_n)} \tag{8-51}$$

Example 8-4. A long cylindrical fuel element of radius 0.01 ft. and thermal conductivity 10 Btu/hr ft°F, initially at 500°F, is cooled by a fluid at 500°F, with a heat transfer coefficient $h = 2000$ Btu/hr ft²°F. Heat at the rate of 10^8 Btu/hr ft³ was suddenly generated by the element. Calculate (a) the maximum temperature the element would reach, (b) the maximum temperature of the surface, and (c) the maximum temperature after 0.36 sec of heat generation if the thermal diffusivity of the element is 1.0 ft²/hr.

Solution

(a) The maximum temperature in the fuel element is reached at

$$r^* = 0 \text{ and } \theta^* = \infty$$

$$\mathrm{Bi} = hr_0/k = \frac{2000 \times 0.01}{10} = 2.0$$

$$J(0) = 1.0 \text{ (Table C-3, Appendix)}$$

Thus Eq. 8-51 reduces to

$$t^*(0,\infty) = \sum_{n=1}^{\infty} \frac{2\mathrm{Bi}}{\lambda_n^2(\lambda_n^2 + \mathrm{Bi}^2)} \cdot \frac{1}{J_0(\lambda_n)}$$

The values of λ_n are obtained from Table 8-2. Using the first four eigenvalues gives

$$t^*(0,\infty) = \frac{2 \times 2}{1.5994^2(1.5994^2 + 2^2)J_0(1.5944)} + \frac{2 \times 2}{4.2910^2(4.2910^2 + 2^2)J_0(4.2910)}$$

$$+\frac{2 \times 2}{7.2884^2(7.2884^2 + 2^2)J_0(7.2884)}$$

$$+\frac{2 \times 2}{10.3658^2(10.3658^2 + 2^2)J_0(10.3658)}$$

$$= 0.5235 - 0.0267 + 0.0045 - 0.0013 = 0.500$$

Note that higher terms result in little change in the final result. Using Eqs. 8-32.

$$t^*(0,\infty) = \frac{t(0,\infty) - t_f}{(q'''r_0^2/k)}$$

or

$$t(0,\infty) = t^*(0,\infty)\,\frac{q'''r_0^2}{k} + t_f$$

$$= 0.500\,\frac{10^8(0.01)^2}{10} + 500$$

$$= 0.500 \times 1000 + 500$$

$$= 500 + 500 = 1000°\text{F}$$

This calculation is now repeated using the simple steady-state, one-dimensional solution, Eq. 5-51. Since $t(0,\infty)$ is the steady-state center-line temperature of the element, Eq. 5-51 is written, for no cladding, in the form

$$t(0,\infty) = t_m = \frac{q'''r_0^2}{2}\left(\frac{1}{2k} + \frac{1}{hr_0}\right) + t_f$$

$$= \frac{10^8(0.01)^2}{2}\left(\frac{1}{2 \times 10} + \frac{1}{2000 \times 0.01}\right) + 500$$

$$= 500 + 500 = 1000°\text{F}$$

which checks the above calculation.

(b) The maximum temperature of the surface is given by $t^*(1,\infty)$. Equation 8-51 reduces to

$$t^*(1,\infty) = \sum_{n=1}^{\infty} \frac{2\text{Bi}}{\lambda_n^2(\lambda_n^2 + \text{Bi}^2)}$$

Again, using the first four eigenvalues of λ_n, gives

$$t^*(1,\infty) = \frac{2 \times 2}{1.5994^2(1.5994^2 + 2^2)} + \frac{2 \times 2}{4.2910^2(4.2910^2 + 2^2)} +$$

$$\frac{2 \times 2}{7.2884^2(7.2884^2 + 2^2)} + \frac{2 \times 2}{10.3658(10.3658^2 + 2^2)}$$

$$= 0.2384 + 0.0097 + 0.0013 + 0.0003 = 0.2497$$

Thus

$$t^*(1,\infty) = t^*(1,\infty)\frac{q'''r_0^2}{k} + t_f$$

$$= 0.2497 \times 1000 + 500 = 749.7^\circ\text{F}$$

Again the same temperature may be obtained using the simple steady-state, one-dimensional solution, Eq. 5-45:

$$t(1,\infty) = t_s = t_m - \frac{q'''r_0^2}{4k}$$

$$= 1000 - \frac{10^8(0.01)^2}{4 \times 10}$$

$$= 1000 - 250 = 750^\circ\text{F}$$

and the two values compare closely.

(c)

$$\theta^* = \frac{\theta}{r_0^2/\alpha}, \text{ Eqs. 8-32}$$

$$= \frac{0.36/3600}{(0.01)^2/1.0} = 1, \text{ dimensionless}$$

Equation 8-50 is now rewritten for $r = 0$ as

$$t^*(0,\theta^*) = \sum_{n=1}^{\infty} \frac{2\text{Bi}}{\lambda_n^2(\lambda_n^2 + \text{Bi}^2)J_0(\lambda_n)} (1 - e^{-\lambda^2{}_n\theta^*})$$

Using computations from part (a),

$$t^*(0,\theta^*) = 0.5235(1 - e^{-2.558}) - 0.0267(1 - e^{-18.413}) + \cdots$$
$$= 0.5235 \times 0.9225 = 0.4829$$

Note that terms higher than the first are negligible. Thus

$$t(0,\theta) = 0.4829 \times 1000 \times 500 = 982.9^\circ\text{F}$$

PROBLEMS

8-1. A 2 in. diam. steel ball initially at a uniform 850°F, is suddenly subjected to an enviroment at 200°F. The natural convection heat transfer coefficient is 2 Btu/hr ft² °F. Find the time necessary for the ball to cool down to 300°F.

8-2. A 1 in.-diameter fuel element is made of a material that has average values of specific heat and density of 0.04 Btu/lb$_m$ and 700 lb$_m$ ft³. Prior to reactor startup the fuel and coolant were at 100°F. In preparation for startup coolant at 500°F was suddenly circulated through the core. The heat-transfer coefficient between coolant and fuel was 200 Btu/hr ft^2 °F. Assuming a lumped-capacity system, find (a) the time constant of the fuel, and (b) the time it takes the fuel to reach 499°F.

8-3. Derive the one-dimensional, finite-difference equation for the temperature $t_n^{\theta + \Delta\theta}$ at nodal plane n from the temperatures at nodal planes surrounding n at θ by using a heat balance. The body in question generates q''' and has density ρ, specific heat c, and thermal conductivity k.

8-4. A flat-plate fuel element 1.25 × 0.25 in. in cross section is initially at a uniform 1000°F. Suddenly heat was generated at the rate of 1.5×10^7 Btu/hr ft^3. All surfaces were maintained at 1000°F. Find by a numerical technique the time it takes the element to reach a maximum temperature 99.5 percent of the way to maximum steady-state temperature. $k = 1.085$ Btu/hr ft °F. $c = 0.06$ Btu/lb$_m$ °F. $\rho = 740$ lb$_m$/ft^3.

8-5. Derive the unsteady-state finite difference equation for the temperature of a nodal point on a surface cooled by a fluid at temperature t_f with a heat transfer coefficient h.

8-6. Derive the unsteady-state finite difference equation for the temperature of a nodal point on an insulated surface.

8-7. Derive the unsteady-state finite difference equation for the temperature of a nodal point on a surface with a specified heat flux q''.

8-8. Derive the unsteady-state finite difference equation for the temperature of a nodal point at the outside square corner of a body cooled by a fluid at temperature t_f with a uniform heat transfer coefficient h.

8-9. Repeat Prob. 8-7 but for a nodal point on an inside square corner of the body.

8-10. A cruciform similar to that in Fig. 7-4 is 2.4 in. on the flats and 0.8 in. thick. It is cooled by a fluid. The solid and fluid were initially at 600°F when 10^7 Btu/hr ft^3 where suddenly generated in it. The coolant remained at 600°F. Find the variations with time of maximum and minimum temperatures in the cruciform up to 5 seconds. $\rho = 496$ lb$_m$/ft^3. $c = 0.04$ Btu/lb$_m$ °F. Use $\Delta x =$ 0.4 in.

8-11. A plate type fuel element is composed of 0.4 in. thick fuel and 0.03 in thick cladding on each side. The fuel and cladding, respectively have $k = 10$ and 7.5 Btu/hr ft°F, $\rho = 625$ and 512.8 lb$_m$/ft^3 and $c = 0.04$ and 0.26 Btu/lb$_m$ °F. The fuel generates 10^7 Btu/hr ft^3 at a steady state. The cladding outer surface is always maintained at 600°F. The reactor is suddenly shut down. Ignoring decay heat in the fuel, find by the method of finite differences the time it takes the fuel midplane to cool down one-quarter of the way to the base temperature of 600°F.

8-12. Repeat the above problem taking into account decay heat at 3 percent of maximum volumetric thermal source strength.

8-13. A 1 in. diam. UO_2 fuel element has $q''' = 8.333 \times 10^6$ Btu/hr ft^3. It is cooled by water at 500°F with $h = 200$ Btu/hr ft^2 °F. A sudden loss of coolant accident occurred. Find by a graphical technique the time it takes the element to reach meltdown temperature (4980°F) if the reactor failed to scram immediately and no emergency cooling was provided. Take $k =$ 1 Btu/hr ft °F, $c = 0.06$ Btu/lb$_m$ °F and $\rho = 525$ lb$_m$/ft^3. Ignore the effect of cladding.

8-14. A 0.1 ft thick slab is originally at 500°F. Suddenly heat was generated uniformly at the rate $q''' = 10^7$ Btu/hr ft³. A pump was immediately started to cool both sides of slab. The fluid velocity however takes 10^{-5} hr to reach maximum according to

$$\frac{V}{V_{max}} = \left(\frac{\theta}{10^{-3}}\right)^{1.25} \quad \text{where } \theta \text{ is in hours.}$$

The maximum h (reached at 10^{-3} hr) is 2000 Btu/hr ft² °F. The slab material has $k = 10$ Btu/hr ft °F, $\rho = 1000$ lb$_m$/ft³ and $c = 0.02$ Btu/lb$_m$ °F. The coolant is nonmetallic (so that h is proportional to $V^{0.8}$) and remains always at 500°F. Using $\Delta x = 0.01$ ft, draw the first three temperature profiles in the slab.

8-15. A long cylindrical fuel element has a radius of 0.6 in. and a thermal conductivity of 1.25 Btu/hr ft °F. It is cooled by a gas. Both element and coolant were originally at 1000°F when heat was suddenly generated in it at therate of 5×10^6 Btu/hr ft³. One of the coolant pumps, however, failed so that a reduced heat-transfer coefficient of 250 Btu/hr ft² °F was obtained. Using the analytical solution of Sec. 8-10, find the time it takes the element to reach the melting temperature of 4500°F if no extra emergency cooling and no scram were initiated. $\rho = 676$ lb$_m$/ft³. $c = 0.060$ Btu/lb$_m$ °F. Ignore cladding.

chapter **9**

Heat Transfer and Fluid Flow, Nonmetallic Coolants

Single-Phase

9-1. INTRODUCTION

Nuclear reactors use a wide variety of coolants. A given reactor type can use any one of a number of coolants, but not all coolants are suitable for all types of reactors. For example, fast reactors cannot use coolants that have a high concentration of hydrogen or carbon (such as light water, heavy water, and organic liquids) because of their moderating power. Thus fast reactors are limited to molten metals (such as liquid sodium, potassium, etc.) and gases (such as helium or steam). Reactors using natural-uranium fuel cannot use coolants that have high neutron absorption cross sections (such as light water and other hydrogenous coolants). Thus they are limited to heavy water or to gases in conjunction with a solid moderator. In circulating-fuel systems the fuel, in fluid form, acts as its own coolant, i.e., carries its own heat produced in the core (which is essentially a bulge of critical dimensions in the fuel piping system) to the heat exchangers. Boiling-type reactors can use many types of liquid coolants, such as light water, heavy water, liquid metals, and organic coolants, but have so far mostly used light water because of its availability and the advanced state of knowledge concerning it.

It can be seen that the most suitable coolant for a given reactor depends upon the reactor type and the application for which it is put to use. In selecting coolants for power reactors, several characteristics are desirable:

1. Economic
   - (a) Low initial cost
   - (b) Availability (helium, heavy water, and liquid metals are less available than light water and other gases)
   - (c) Low pumping losses (these depend upon frictional losses and are highest for gases)

   (d) High heat-transfer coefficients
2. Physical
   (a) Low melting points (so that the coolant remains fluidized, if possible, during shutdown, and certainly at reactor inlet temperatures during operation)
   (b) Low vapor pressures (high boiling points) for liquid coolants (to avoid high pressurization)
   (c) Compatibility with fuel, cladding, heat exchangers, pumps, valves, and other structural materials (to minimize the use of expensive materials such as titanium, and molybdenum)
   (d) Good thermal stability (organic liquids, for example, are subject to decomposition at high temperatures)
3. Nuclear
   (a) Low neutron absorption cross sections
   (b) Moderating power to suit reactor type
   (c) Low induced radioactivity (or activity of short half-lives)
   (d) Good radiation stability (organic coolants, in particular, suffer from nuclear-radiation damage)

Needless to say, no single coolant possesses all the desirable characteristics. The selection of coolants is, like most engineering problems, a matter of compromise between usually conflicting requirements.

This chapter deals mostly with heat-transfer and fluid-flow characteristics of nonmetallic reactor coolants. Metallic coolants will be covered in the next chapter. It is, however, often desirable to make comparisons or take a unified approach to both nonmetallic and metallic coolants. When this is the case, such discussions will be found in this chapter.

9-2. HEAT REMOVAL AND PUMPING POWER

Coolants remove heat from fuel elements in the core and blanket (if any) and from thermal shields, pressure vessels, etc., and may transfer this heat to an intermediate or secondary coolant or to the working fluid in a heat exchanger. The heat transferred, q (Btu/hr), between the coolant and a surface (of a fuel element, heat exchanger tube, etc.) is given by Newton's law of cooling:

$$q = hA(t_w - t_f) \tag{9-1}$$

where h = coefficient of heat transfer by convection, Btu/hr ft² °F; also called surface conductance

A = area, ft², across which heat flows

t_w, t_f = wall-surface and fluid-bulk temperatures, °F

In most power-reactor systems the coolant is force-circulated (by a liquid pump or a gas blower), and forced-convection heat-transfer coefficients apply. Turbulent-flow conditions also predominate in such cases. Thus h is a function of coolant physical characteristics (thermal conductivity, specific heat, viscosity, etc.) as well as operating conditions (speed, pressure) and flow-channel geometry. Consequently a wide range of values for h is to be expected. Because of size limitations as well as the large thermal output of power reactors, values of h, much higher than those encountered in conventional engineering practice, are required. Table 9-1 gives representative ranges of h for four coolant categories under full load conditions normally encountered in power reactors.

TABLE 9-1
Coolant Performance in Power Reactors

Coolant	h, Btu/hr ft² °F	W'/q (relative)
Light and heavy water	5000-8000	1.0
Organic liquids (polyphenyls, etc.)	2000-3000	4-10
Liquid metals (sodium, sodium-potassium alloys, etc.)	4000-10,000	3-7
Gases (He, CO_2, N_2, air, etc.)	10-100	~ 100

The pumping work, W ft-lb$_f$/hr, required by the circulating coolant to overcome pressure losses through a complete loop (reactor, piping, heat exchangers, etc.) is given by*

$$W = \Delta p A_c V \tag{9-2}$$

where Δp = pressure drop through loop, lb$_f$/ft²
A_c = cross-sectional area of coolant passage, ft²
V = coolant speed, ft/hr

Δp is the algebraic sum of frictional drops, entrance and exit pressure changes, and losses through the entire loop. For the sake of simplicity, it will be assumed that the coolant loop is made entirely of a channel of fixed dimensions and that the coolant speed is constant. This simplification is justified because only comparisons between coolants are sought here.

Δp in turbulent flow can also be expressed by the Darcy formula

$$\Delta p = f \frac{L}{D_e} \frac{\rho V^2}{2g_c} \tag{9-3}$$

* It will be assumed here that the coolant remains in a single phase throughout the loop. The case of boiling and two-phase flow, as occurs in boiling-water reactors, is discussed in Chaps. 11 and 12.

where f = friction factor, dimensionless
L = channel length, ft
D_e = equivalent diameter of channel, ft, $= 4A_c/P$, where P is wetted perimeter, ft (See p. 245)
ρ = density of coolant, lb_m/ft^3
g_c = conversion factor $= 4.17 \times 10^8$ lb_m ft/lb_f hr^2

f, above, is the Darcy-Weisbach friction factor. It should not be confused with the often used Fanning friction factor $f' = f/4$. The well-known Moody chart, Appendix F, gives f for isothermal (little or no temperature difference between wall and fluid) circular-pipe flow.

Although the equivalent diameter is used in Eq. 9-3 for noncircular channels, f has been found to deviate from pipe values at the same Reynolds numbers [49]. For channels with smooth surfaces, the values of f are found to be lower than those given by the Moody smooth-surface line by about 3 percent in the case of triangular and by about 10 percent in the case of square cross sections, in the Reynolds number range of 10^4 to 2×10^5.

For smooth rectangular channels, as between plate-type fuel elements, the values of f where found to be the same as for pipe flow. For rough plates, however, the values are about 20 percent lower than the corresponding Moody values at the same relative roughnesses.

For flow past smooth-tube bundles, as in the case of cylindrical fuel elements, the values of f depend upon the lattice used. Data taken with clearance-to-tube-diameter ratios of 0.12 for triangular lattices and 0.12 and 0.20 for square lattices indicated that the values of f are some 3 percent below the Moody smooth line in the case of triangular lattices and as much as 10 percent below the same line for square lattices. Other tube-bundle arrangements, such as in annuli, and those including spacers or grids were also investigated [49].

In the case of nonisothermal flow of water in smooth and rough rectangular channels, f may be obtained by multiplying the values of the isothermal friction factor by $(1 - 0.0025\, \Delta t_{fm})$, where Δt_{fm} is the temperature drop in the film, i.e., the difference between the wall and fluid bulk temperatures (see below). The properties in the Reynolds number at which f is found are to be evaluated at the fluid bulk temperature.

For nonisothermal flow in the case of gases, no one simple correlation has found general acceptance, and the reader is referred to the references in the original article [49].

For highly turbulent flow in a circular channel with smooth surfaces (smooth surfaces are encountered in most reactor work), f may be given by [50]

$$f = \frac{0.184}{\mathrm{Re}^{0.2}} = \frac{0.184}{(D_e V \rho/\mu)^{0.2}} \tag{9-4}$$

where Re = Reynolds number, dimensionless

μ = absolute viscosity of coolant, lb_m/hr ft

Combining Eqs. 9-2 to 9-4 and rearranging give

$$W = 2.2 \times 10^{-10} \left(\frac{A_c L}{D_e^{1.2}}\right) (V^{2.8}) (\rho^{0.8} \mu^{0.2}) \tag{9-5}$$

where the first term in parentheses on the right-hand side of Eq. 9-5 pertains to channel geometry, the second to the operating conditions, and the third to the physical properties of the coolant.

It can be seen that, from the point of view of pumping power, a desirable coolant is one with a low value of $\rho^{0.8}\mu^{0.2}$. This criterion by itself, however, may not afford a good comparison between coolants since, to achieve desirable heat transfer, different coolants are pumped at different speeds. A better criterion, therefore, is the ratio of pumping power to heat removal. This will now be attempted.

The heat removed by the coolant, q Btu/hr, can be given by

$$q = \rho A_c V c_p \Delta t_f \tag{9-6}$$

where c_p = coolant specific heat at constant pressure, Btu/lb_m °F

Δt_f = coolant temperature rise in length L, °F

The dimensionless ratio $W'/q = W/Jq$, where J is an energy-conversion factor (778.16 lb_f-ft/Btu), is obtained by combining Eqs. 9-5 and 9-6 to give

$$\frac{W'}{q} = 2.83 \times 10^{-13} \left(\frac{L}{D_e^{1.2}}\right) \left(\frac{V^{1.8}}{\Delta t_f}\right) \left(\frac{\mu^{0.2}}{\rho^{0.2} c_p}\right) \tag{9-7}$$

where the quantities between the successive parentheses have the same meaning as for Eq. 9-5.

A second correlation involving fuel temperatures (limited by fuel material) and coolant temperature (affecting plant thermal efficiency) can be obtained by writing q in the form

$$q = hA\,\Delta t_m \tag{9-8}$$

where A = area across which heat is transferred between fuel and coolant in fuel channel or circumferential area of coolant channel = $PL = 4A_c L/D_e$, ft²

Δt_m = mean temperature difference between channel wall and coolant, °F

Combining Eqs. 9-5 and 9-8, substituting for A as above, and rearranging give

$$\frac{W'}{q} = 7.07 \times 10^{-14} \left(\frac{1}{D_e^{0.2}}\right) \left(\frac{V^{2.8}}{h \Delta t_m}\right) (\rho^{0.8} \mu^{0.2}) \tag{9-9}$$

Table 9-1 gives relative values of W'/q for tne four coolant categories, normalized to 1.0 for light and heavy water. It can be seen that, from the point of view of heat removal and pumping power, both kinds of water are superior to all other coolants, followed by liquid metals and the organics. Gases have high values of W'/q and differ widely among themselves [1,2].

9-3. HEAT-TRANSFER COEFFICIENTS: GENERAL

The heat-transfer coefficient h is defined by Newton's law of cooling, Eq. 9-1, here repeated in the form

$$q'' = \frac{q}{A} = h(t_w - t_f) \qquad [9\text{-}1]$$

where q'' is the heat flux, Btu/hr ft^2, at a particular cross section of a flow channel, at which the temperature of the wall is t_w and the temperature of the coolant fluid is t_f (see Fig. 9-1). Heat may flow from the wall to the fluid or vice versa. In this model t_w may be well defined. The fluid temperature is not constant over the cross section, however, and t_f is defined as that temperature which, when multiplied by the coolant mass-flow rate m and specific heat c_p, gives the heat transported by the coolant. Thus, between cross sections 1 and 2 of Fig. 9-1, the heat transferred is $\dot{m}c_p(t_{f2} - t_{f1})$.

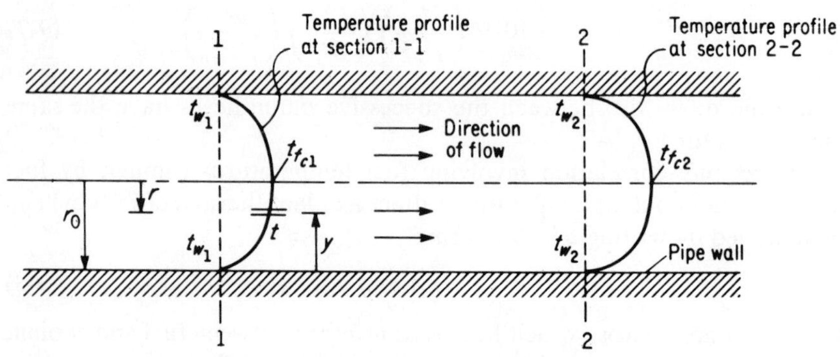

FIG. 9-1. Temperature profiles in pipe flow.

t_f is given various names, among which are the *bulk, mixed mean,* and *mixing-cup** temperature. In highly turbulent flow, the temperature profile is fairly flat over much of the cross section, and the bulk temperature is taken as equal to the temperature at the center of the channel t_{fc}.

* This name implies that t_f is that uniform temperature produced if the entire volume of the fluid passing the cross section during a given time is mixed together with negligible mixing work in an adiabatic cup.

If the bulk temperature varies along the axis of a flow channel, as it may well do in heat exchange, an average value of h would be based on the bulk temperature halfway between the cross sections considered.

In gas-cooled reactors, the gas speed may be high enough to create aerodynamic heating. This is discussed in Sec. 9-8.

The value of h is governed by many factors:

1. Operating factors
   (a) The geometrical shape of the channel
   (b) The flow rate of the coolant
   (c) The heat flux
   (d) The system temperature
2. Physical properties of the coolant fluid

In most reactor work, the flow of the fluid is forced and turbulent. The value of h in forced convection is governed by the thermal conductivity of the fluid as well as by those factors representing turbulence and operating conditions. In the correlations for h, it is given as part of the *Nusselt* number, a dimensionless group which includes the thermal conductivity of the fluid and the equivalent diameter of the channel. The Nusselt number is a function of the *Reynolds* and *Prandtl* numbers, a fact that can be proved by theoretical analysis, dimensional analysis, or by experiment. Thus

$$\text{Nu} = f(\text{Re},\text{Pr}) \tag{9-10}$$

where

$$\text{Nu} = \text{Nusselt number} = \frac{hD_e}{k} \quad \text{dimensionless} \tag{9-11}$$

$$\text{Re} = \text{Reynolds number} = \frac{D_e V \rho}{\mu} \quad \text{dimensionless} \tag{9-12}$$

$$\text{Pr} = \text{Prandtl number} = \frac{\nu}{\alpha} = \frac{c_p \mu}{k} \quad \text{dimensionless} \tag{9-13}$$

$$\nu = \text{kinematic viscosity of fluid} = \frac{\mu}{\rho}$$

$$\alpha = \text{thermal diffusivity of fluid} = \frac{k}{\rho c_p}$$

The Nusselt number is numerically equal to the ratio of the temperature gradient *at the wall* to a reference temperature gradient.

The Reynolds number is a measure of the ratio of inertial to viscous forces. A low Re means that any disturbances arising in the fluid are dampened and that the flow is laminar. At a *critical* value of Re, such disturbances are no longer dampened and transition from laminar to turbulent flow takes place. In turbulent flow the energy-transport (heat-transfer) mechanism is aided by eddies which in effect cause lumps of fluid, acting as energy carriers, to mix with the rest of the fluid. A high value of Re therefore means a high degree of turbulence, a high rate of mixing, and a large heat-transfer coefficient by convection.

The Prandtl number is a ratio of two molecular transport properties—the kinematic viscosity, which is a measure of the rate of momentum transfer between molecules and affects the velocity gradient, and the thermal diffusivity, which is a measure of the ratio of heat transfer to energy storage by molecules and affects the temperature gradient. Pr therefore relates the temperature gradient to the velocity gradient. Values of Pr much lower than unity mean that near the wall the temperature gradient is less steep than the velocity gradient.

Of the above, only the Prandtl number is made up entirely of physical properties (it is sometimes tabulated as a physical property). It is a function of the temperature of the fluid and, to a lesser extent, of its pressure.

Thus for one fluid operating at a given temperature, the Nusselt number is primarily governed by the Reynolds number. However, different coolants operating at the same Reynolds number exhibit heat-transfer characteristics strongly dependent on the Prandtl number. This dependence governs the type of correlation for h. This is discussed in the following section.

9-4. THE EFFECT OF THE PRANDTL NUMBER ON CONVECTIVE HEAT TRANSFER

Coolants used in nuclear reactors generally fall into two categories having widely differing Prandtl numbers (see Table 9-2). Gases, water, and the organics have Prandtl numbers in the neighborhood of unity (at power-reactor temperatures). Liquid metals have Prandtl numbers that are much lower (heavy oils have Pr much greater than 1 but will not be considered here).

TABLE 9-2
Prandtl Numbers of Some Reactor Coolants

Category	Coolant	Prandtl number
I	Water	4.25 at 100°F 0.87 at 500°F
	Organic coolants (Santowax O-M)	8.55 at 500°F 4.57 at 800°F
	Gases	0.68 to 0.77
II	Liquid metals	0.004 to 0.03

The correlations for h are of two different types for the two categories. Before introducing these correlations, a physical picture of the reasons for this difference will be introduced. This is a function of the effect of the parameters in the Prandtl number on the heat-transfer mechanism.

The first of these parameters is the *thermal diffusivity* α, in square feet per hour. It is equal to the ratio of the thermal conductivity k, in Btu/hr ft °F, to the *volumetric* specific heat ρc_p, in Btu/ft³ °F. The latter quantity is a measure of the capacity of a given volume of fluid to absorb and transport heat. These two parameters affect convection as follows. Unlike heat transfer in a solid (conduction), where only k is determining, convection takes place by two independent processes. One is due to k of the moving fluid, with conduction occuring normal to the area across which heat is being transferred. The other is due to turbulent diffusion of heat when transported by the fast randomly moving molecules of the fluid. The combined effect of these two processes may be thought of as a combined or *effective* thermal conductivity k_e given by

$$k_e = k + \rho c_p \epsilon_h \tag{9-14}$$

ϵ_h is called the *turbulent* or *eddy diffusivity of heat,* having the dimension of square feet per hour here. In the case where no heat is generated in the fluid, the heat flow per unit area, q'', in the fluid at y, in the y direction away from the wall (Fig. 9-1) is

$$q'' = -k_e \frac{dt}{dy} = -(k + \rho c_p \epsilon_h) \frac{dt}{dy}$$

$$= -\rho c_p \left(\frac{k}{\rho c_p} + \epsilon_h \right) \frac{dt}{dy}$$

Thus

$$\frac{q''}{\rho c_p} = -(\alpha + \epsilon_h) \frac{dt}{dy} \tag{9-15}$$

in which $\alpha (= k/\rho c_p)$ is due to heat conduction from molecule to adjacent molecule, normal to the area in question. α is sometimes called the *normal* thermal diffusivity.

The second of the parameters in the Prandtl number is the kinematic viscosity ν. An equation containing ν, analogous to Eq. 9-15, can be derived for the diffusion of momentum between the wall and the fluid by writing

$$\tau = \frac{1}{g_c} \mu_e \frac{dV}{dy} \tag{9-16}$$

where τ = shear stress (drag force per unit area), lb_f/ft^2
μ_e = *effective* viscosity, allowing for added effect of turbulence, lb_m/hr ft
V = speed of fluid at y (Fig. 9-1), ft/hr

μ_e, as k_e, is written in terms of two components as

$$\mu_e = \mu + \rho\epsilon_m \tag{9-17}$$

where μ = absolute viscosity of fluid
ϵ_m = *turbulent*, or *eddy, diffusivity of momentum,* having same dimensions as kinematic viscosity, ft²/hr; also called the *turbulent kinematic viscosity*

Equations 9-16 and 9-17 are combined to give

$$\frac{\tau g_c}{\rho} = (\nu + \epsilon_m)\frac{dV}{dy} \tag{9-18}$$

The similarity between Eqs. 9-15 and 9-18 is evident. Also note that ϵ_h and ϵ_m have similar dimensions (length²/time). There is good reason to believe that they are numerically equal. Experimental data indicate that their ratio ϵ_m/ϵ_h has maximum variance between 0.7 and 1.6. In pure laminar flow both are equal to zero. In highly turbulent flow their values are much greater than either α or ν, except in the case of liquid metals where α and ϵ_h may be of the same order of magnitude.

For most cases, other than liquid metals (and viscous oils), the ratio ϵ_m/ϵ_h may be taken as unity with little error. Equations 9-15 and 9-18 may now be combined to give

$$\frac{dt}{dV} = -\frac{\nu + \epsilon_m}{\alpha + \epsilon_h}\frac{q''}{\tau g_c c_p} \tag{9-19a}$$

For a Prandtl number of unity, that is, $\alpha = \nu$, and in case $\epsilon_m = \epsilon_h$, this equation reduces to

$$\frac{dt}{dV} = -\frac{q''}{\tau g_c c_p} \tag{9-19b}$$

This equation was originally obtained by Reynolds in what is called the *Reynolds analogy* of heat and momentum transfer. It is sometimes used to evaluate heat transfer in turbulent systems where the Prandtl number is equal to unity. Equation 9-19a indicates the degree to which the velocity-temperature relationships are dependent upon the Prandtl number.

Martinelli [51] has shown that for small temperature drops and constant fluid properties, and with some reasonable approximations, Eqs. 9-15 and 9-18 can be written for fully developed turbulent flow in a tube as*

$$\frac{q''_w}{\rho c_p}\left(1 - \frac{y}{r_0}\right) = -(\alpha + \epsilon_h)\frac{dt}{dy} \tag{9-20}$$

and

$$\frac{\tau_w g_c}{\rho}\left(1 - \frac{y}{r_0}\right) = (\nu + \epsilon_m)\frac{dV}{dy} \tag{9-21}$$

where q''_w and τ_w are the heat flow per unit area and the shear stress at the wall, r_0 is the radius of the tube, and y is the radial distance from the tube wall to the point in question at radius r, that is, $y = r_0 - r$ (Fig. 9-1). Martinelli obtained the value of ϵ_m from Eq. 9-21 and three velocity distributions in the tube representing three regions, a laminar sublayer, a buffer layer, and a turbulent core (Fig. 9-2). These velocity distributions hand been found from theoretical and experimental correlations. ϵ_h was assumed equal to ϵ_m. Equation 9-20 was then integrated to give three equations for the temperature distribution in the three above-mentioned layers. In the laminar layer, ϵ_h and ϵ_m were considered to be zero. In the turbulent core, α and ν were first considered to be negligible, as for nonmetallic coolants. Next, α was considered not negligible with respect to ϵ_h for the case of Pr less than unity, as for liquid metals.

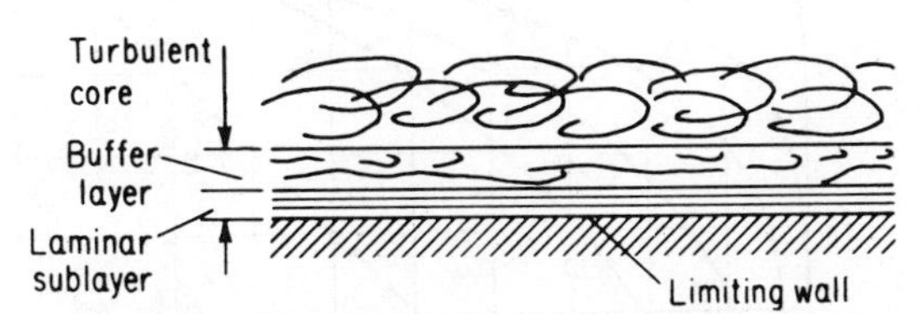

FIG. 9-2. Nature of turbulent flow parallel to a wall.

Figure 9-3 shows the result of this analysis. The ordinate is a dimensionless temperature function $(t_w - t)/(t_w - t_{fc})$, where t_w is the wall temperature, t is the fluid temperature at y, and t_{fc} is the fluid temperature at the center of the tube. The abscissa is the ratio of y/r_0. The different curves represent lines of constant Prandtl numbers, all for Reynolds numbers of 10,000 and 1,000,000.

The bulk temperature t_f of a fluid at any one cross section of the tube can be evaluated from the curves of Fig. 9-3. Such evaluation would show that the temperature of the fluid at any one point between the wall and the center line is closer to that of the wall, t_w, and that the change from t_w to t_{fc} is more linear, the lower the Prandtl number. This indicates

* Fully developed flow is defined in Sec. 9-6.

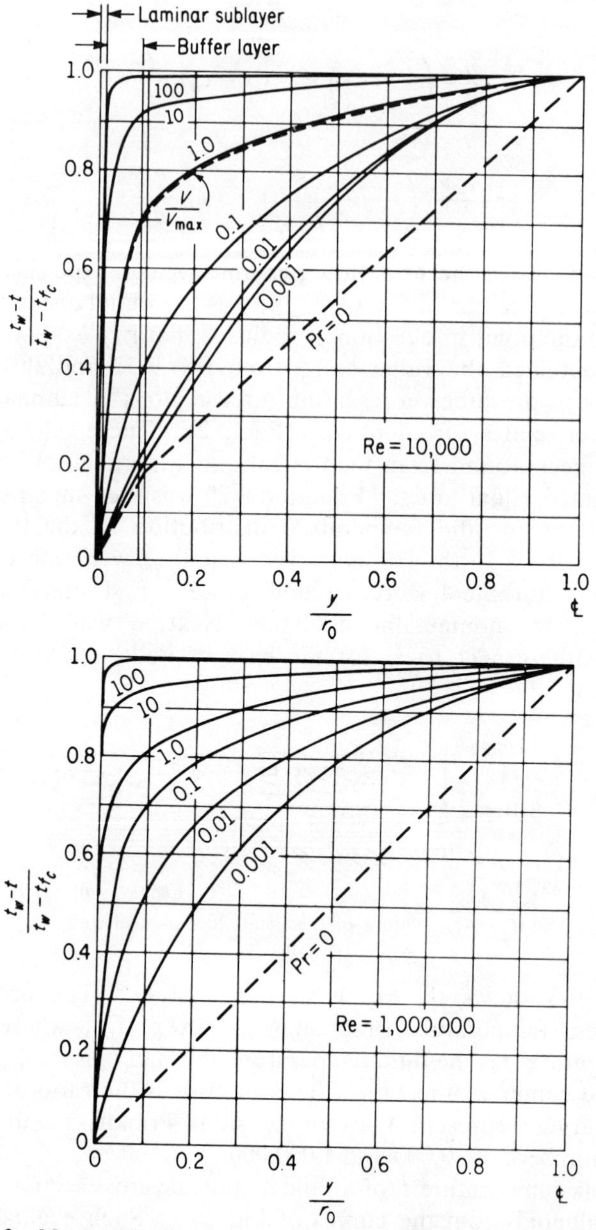

FIG. 9-3. Martinelli's solutions for the temperature gradients in pipe flow.

that the resistance to flow of heat in liquid metals is not found in the laminar sublayer or in the buffer layers, as is the case with other liquids, but is more evenly distributed throughout the cross section.

9-5. CONVECTIVE HEAT-TRANSFER COEFFICIENTS FOR NONMETALLIC COOLANTS

Nonmetallic fluids, such as light and heavy water and organic liquids and gases, have Prandtl numbers in the neighborhood of unity (Table 9-2). The correlations to follow are recommended for fully developed turbulent flow. These correlations are in the form of Eq. 9-10 and have been obtained theoretically, experimentally, and by dimensional analysis.

For nonmetallic coolants the Nusselt numbers are insensitive to wall-surface conditions. There is very little difference in Nusselt number between the cases of constant heat flux and constant wall temperature. This is not the case with metallic coolants (See Sec. 10-6).

Flow in Circular Tubes

1. The Dittus-Boelter equation [52]

$$\mathrm{Nu} = 0.023\,\mathrm{Re}^{0.8}\,\mathrm{Pr}^{0.4} \tag{9-22}$$

This is probably the best known and most widely used correlation. Here the physical properties in the dimensionless numbers are evaluated at the bulk temperature of the fluid. The correlation does not take into account the effect of the variation in temperature from the tube wall to fluid bulk on the physical properties. In other words, it is tailored for isothermal or nearly isothermal flow.

For special fluids this equation may be further simplified. For example, the quantity $k(\rho/\mu)^{0.8}(c_p\mu/k)^{0.4}$, containing only physical properties, may be combined into one temperature-dependent expression. For the case of water, Eq. 9-22 thus reduces [53] to

$$h = 0.00134(t + 100)\,\frac{V^{0.8}}{D^{0.2}} \tag{9-23}$$

If the temperature drop across the film does not exceed 10°F, t in Eq. 9-23 is the bulk temperature. Otherwise the *film temperature* is used. This is the arithmetic mean of the wall and fluid bulk temperatures, that is, $(t_w + t_f)/2$.

For most gases Pr varies between the narrow limits of 0.65 and 0.8. When raised to the power of 0.4 in Eq. 9-22, the difference between gases is further minimized, justifying the use of 0.86 as an average value of $\mathrm{Pr}^{0.4}$. Equation 9-22 thus reduces to

$$h = 0.020\,\frac{k}{D}\,\mathrm{Re}^{0.8} \tag{9-24}$$

an expression good for gas temperatures up to 2000°F.

2. The Sieder-Tate equation

For large temperature drops across the film, the physical property most affected is the viscosity. For such a case, the Sieder-Tate correlation is used.

$$\mathrm{Nu} = 0.023\ \mathrm{Re}^{0.8}\ \mathrm{Pr}^{0.4} \left(\frac{\mu_w}{\mu}\right)^{0.14} \tag{9-25}$$

Here the physical properties are evaluated at the fluid bulk temperature, except μ_w, which is evaluated at the temperature of the wall.

3. The Colburn equation

$$\mathrm{St}\ \mathrm{Pr}^{\frac{2}{3}} = 0.023\ \mathrm{Re}^{-0.2} \tag{9-26}$$

where St = Stanton number = Nu/(Re Pr) = $h/\rho V c_p$, dimensionless.

This is a more recent correlation, based on experimental data. It is good for Prandtl numbers between 0.5 and 100 and for cases where the temperature drop across the film is not very large. The physical properties are all evaluated at the film temperature, except for c_p in the Stanton number, which is to be evaluated at the fluid bulk temperature.

The difficulty in using the above two correlations stems from the fact that, in many practical problems, t_w and t_f are not both known in advance, necessitating a trial-and-error solution.

4. For the specific case of *organic coolants* the following equation is recommended [54]:

$$\mathrm{Nu} = 0.015\ \mathrm{Re}^{0.85}\ \mathrm{Pr}^{0.3} \tag{9-27}$$

5. For the specific case of heat transfer to *superheated steam* at high pressures (the case of steam-cooled reactors), McAdams, Kennel, and Addoms [55] have obtained the following correlation:

$$\mathrm{Nu} = 0.0214\ \mathrm{Re}^{0.8}\ \mathrm{Pr}^{\frac{1}{3}} \left(1 + \frac{2.3}{L/D_e}\right) \tag{9-28}$$

L is the length of the channel and D_e is its equivalent diameter (see below). The physical properties are all evaluated at the film temperature. Thermal conductivities of steam at high pressures and temperatures, on which the above correlation is based, are shown in Fig. 9-4.

Appendix E contains physical properties, needed for the evaluation of these and other correlations, of some liquid and gaseous coolants. Not all the properties of heavy water are known over as wide a range as those for light water. They do, however, vary little from those of light water, so that the use of Table E-1 for heavy water is justified.

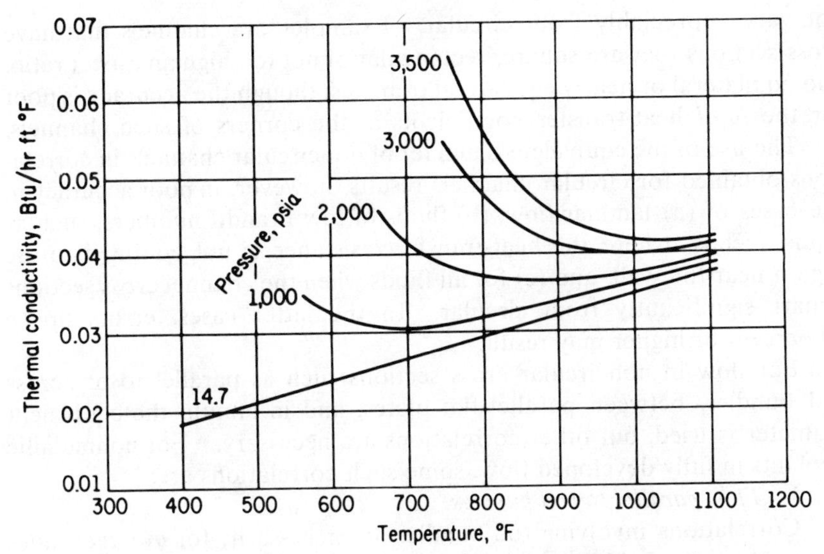

FIG. 9-4. Thermal conductivity of steam at high temperatures and pressures (Ref. 55).

Flow in Noncircular Passages

When flow occurs in passages other than circular, the concept of *equivalent diameter* D_e applies. Also called the effective diameter, it is used in place of the diameter D in the various correlations. It is defined as

$$D_e = 4 \text{ (hydraulic radius)} = 4 \frac{A_c}{P} \tag{9-29}$$

where A_c is the cross-sectional area of the flow channel. P is the wetted perimeter of the channel, including all surfaces wetted by the coolant such as those of fuel rods, channel walls, etc. It has the dimension of length. For a tube of Diameter D, completely filled with fluid.

$$A_c = \frac{\pi D^2}{4}, \; P = \pi D \text{ and } D_e = D$$

The equivalent diameter is that which would give the same ratio of channel volume to heat-transfer area in circular and noncircular channels. In a tube of diameter D, this ratio is $(\pi D^2 L/4)/\pi DL = D/4$. If the right-hand side of Eq. 9-29 is multiplied and divided by L, the ratio can be seen to also equal $A_c L/PL = D_e/4$.

Heat-transfer coefficients for flow in noncircular channels may be obtained from correlations for circular channels by using the equivalent diameter in these correlations, provided that the noncircular channels do not vary appreciably from circular. Examples are channels that have cross sections that are square, rectangular of not too high an aspect ratio, and equilateral or nearly equilateral triangles, though the accuracy is poor for the *local* heat-transfer coefficients at the corners of such channels.

The use of the equivalent diameter of noncircular channels in correlations obtained for circular channels results, however, in poor accuracy in the cases of (a) laminar flow, (b) fluids of low Prandtl numbers, such as liquid metals, where the heat-transfer resistance is not confined to the region near the wall, and (c) for all fluids when the channel cross sections depart significantly from circular. In the latter cases, errors up to 30 percent or higher may result.

For flow in noncircular cross sections such as parallel to or across rod bundles, between parallel flat plates, and in annuli, the equivalent diameter is used, but other correlations are necessary. For nonmetallic coolants in fully developed flow, some such correlations are:

1. *Flow parallel to rod bundles.*

Correlations involving rod bundles are necessarily for *average* values of the Nusselt number since local heat transfer as well as circumferential temperatures vary for each rod and from one rod to another. A correlation based on an analytical and graphical procedure by Deissler and Taylor [56] for fully developed turbulent flow parallel to rod bundles and for constant heat rate per unit length and constant peripheral surface temperature, is presented in Fig. 9-5. The figure gives the ratio of the

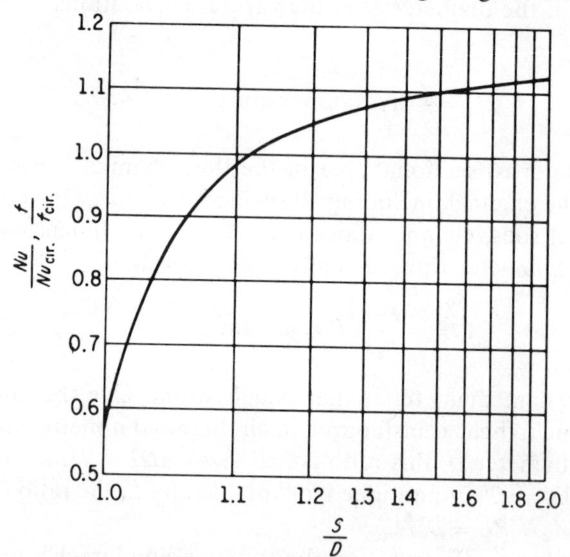

FIG. 9-5. Ratio of Nusselt number and friction factor for flow parallel to rod bundles to flow in circular tubes (Ref. 57).

desired Nusselt number to that obtained by using the equivalent diameter of the bundle in a correlation for circular tubes; as a function of S/D where S is the distance between rod centers, and D is the diameter of the individual rods. It was found that the influence of the Reynolds number on the curve is small, and that the friction factor behavior is virtually the same as that of the Nusselt number.

2. *Water flowing parallel to rod bundles.*

This is the case in water-cooled reactors using fuel rods. For this specific case the following correlation is recommended by Weisman [58]:

$$\text{Nu} = C\text{Re}^{0.8}\,\text{Pr}^{\frac{1}{3}} \tag{9-30}$$

where C is a constant which depends upon the lattice arrangement, given by

$$C = 0.042\,\frac{S}{D} - 0.024 \text{ for square lattices, } 1.1 \leqslant \frac{S}{D} \leqslant 1.3 \tag{9-31a}$$

$$C = 0.026\,\frac{S}{D} - 0.006 \text{ for triangular lattices, } 1.1 \leqslant \frac{S}{D} \leqslant 1.5 \tag{9-31b}$$

It was found that more-open lattices yielded higher heat-transfer coefficients but that triangular and square lattices gave essentially the same coefficients, provided that the ratio of water to total cross-sectional areas, in an infinite lattice, was the same.

3. *Flow across rod bundles.*

For this case, commonly found in heat exchangers, a correlation for turbulent flow, where the Reynolds number is equal to or greater than 6000, is given for 10 rows of tubes or more, by

$$\text{Nu}_D = 0.33\,C\,F\left(\frac{G_m D}{\mu}\right)^{0.6}\text{Pr}^{0.3} \tag{9-32}$$

where Nu_D is the Nusselt number based on a single tube diameter D, F is a factor equal to unity for 10 rows of tubes or more, G_m is the mass velocity, $\text{lb}_m/\text{hr ft}^2$ of minimum free area, and C is an empirical coefficient dependent upon the tube bundle arrangement and the Reynolds number. C has been found [36], however, not to deviate by more the 10 percent from unity for the pitch-to-diameter ratio, S/D, range 1.25–1.50. The physical properties in Eq. 9-32 should be evaluated at the film temperature. For less than 10 rows the factor F in Eq. 9-32 is given in Table 9-3.

G, the mass velocity of the fluid, $\text{lb}_m/\text{hr ft}^2$, is equal to the product of the fluid density and fluid speed. If these are materially affected by the existence of a film, G may be easily calculated from

$$G = \frac{\text{Fluid mass-flow rate, lb}_m/\text{hr}}{\text{Cross-sectional area of channel, ft}^2}$$

TABLE 9-3.
Factor F in Eq. 9-32*

Number of Rows	1	2	3	4	5	6	7	8	9	10
Triangular lattice	.68	.75	.83	.89	.92	.95	.97	.98	.99	1
Square lattice	.64	.80	.87	.90	.92	.94	.96	.98	.99	1

* From Ref. 36.

4. *Flow between parallel plates.*

This is the case of flow between platetype fuel elements, Fig. 9-6.

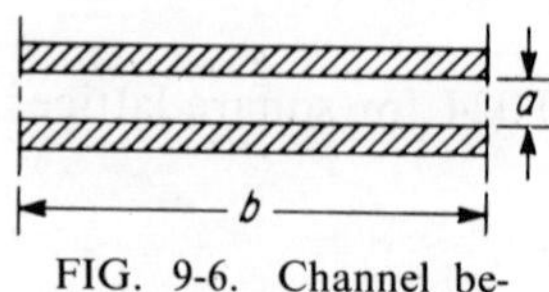

FIG. 9-6. Channel between two parallel plates.

Equation 9-22, here repeated, may be used.

$$\text{Nu} = 0.023 \text{ Re}^{0.8} \text{ Pr}^{0.4} \qquad [9\text{-}22]$$

Here D_e, to be used in Nu and Re, is given by

$$D_e = 4 \frac{ab}{2a + 2b} \qquad (9\text{-}33a)$$

If a is small compared with b, as is usually the case with parallel fuel plates, D_e reduces to

$$D_e = 2a \qquad (9\text{-}33b)$$

5. *Flow in annuli.*

For heat transfer at either inner or outer walls, Fig. 9-7, McAdams [50] recommends

$$\frac{h}{c_p G} \text{Pr}_m^{\frac{2}{3}} \left(\frac{\mu_w}{\mu}\right)^{0.14} = \frac{0.021(1 + 2.3D_e/L)}{(D_e G/\mu_m)^{0.2}} \qquad (9\text{-}34)$$

where

$$D_e = 4 \frac{(\pi/4)(D_2^2 - D_1^2)}{\pi(D_2 + D_1)} = D_2 - D_1 \qquad (9\text{-}35)$$

The subscripts w and m mean that the property in question should be evaluated at either the wall or film temperature, respectively. Others are evaluated at the fluid bulk temperature.

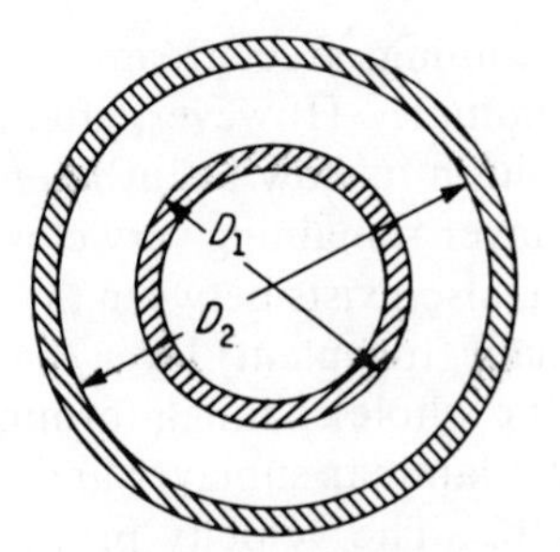

FIG. 9-7. Cross section of an annular channel.

9-6. THE EFFECT OF LENGTH AND SHAPE OF COOLANT CHANNEL

The correlations presented in the previous two sections give values of the heat-transfer coefficient h for fully developed turbulent flow. Fully developed turbulence occurs in a channel only after the fluid has traveled a sufficient length of that channel, measured from its entrance. This length is called the *entrance length,* usually given for tubes in terms of multiples of their diameters.

The attainment of fully developed turbulent flow is illustrated in Fig. 9-8 for a square-edged entrance channel, approached by a fluid at velocity V. In any type of flow (laminar or turbulent), the particles of the fluid immediately adjacent to the wall are always at zero velocity. Upon entering the channel, therefore, those particles that come into contact with the wall are slowed down. As the fluid moves upstream, more and more particles are retarded because of the shearing forces between slow and fast particles. A boundary layer is thus formed in which large velocity gradients exist. (The boundary layer may be loosely defined as that distance perpendicular to the wall at which the velocity of the fluid

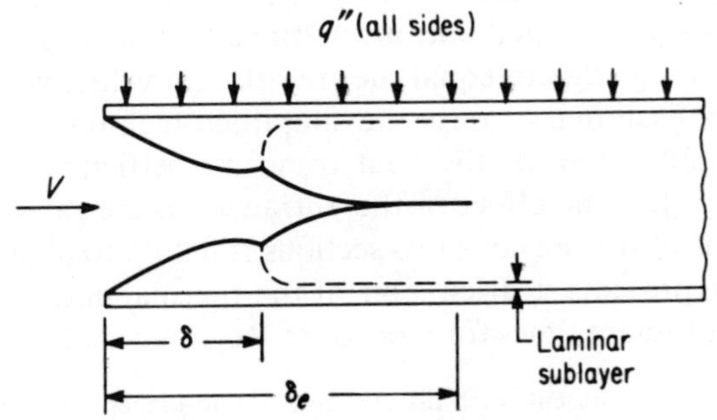

FIG. 9-8. Effect of square-edged channel on entrance conditions in turbulent flow.

is 99 percent of the incoming, free-stream velocity.) The flow within this boundary layer is laminar. However, after a certain distance δ from the leading edge, a transition in flow to turbulent conditions takes place, with only a laminar sublayer remaining very close to the wall*. A buffer zone or layer, not shown, also exists between the laminar sublayer and the turbulent layer. The main (turbulent) layer then continues to grow until at distance δ_e it engulfs the whole channel, forming a continuous turbulent core and leaving only the laminar sublayer and the buffer zone (Fig. 9-2). δ_e is the entrance length. The velocity profile is now established and remains unchanged from there on, except for heating and frictional effects.

A thermal boundary layer develops at the channel entrance, in a manner similar to the development of the hydrodynamic boundary layer discussed above. At the leading edge, the temperature is uniform across the channel. However, as the fluid travels, it becomes heated (or cooled) by heat transfer through the channel walls. A thermal boundary layer then forms, in which a large temperature gradient develops and increases in thickness until a fully developed temperature gradient exists across the channel (meaning that heat is transferred to, or from, the center of the channel).

The phenomena just described have a marked effect on the heat-transfer coefficient near the channel entrance. There a finite heat flux exists but only near zero temperature difference between the fluid and the wall. The convective heat-transfer coefficient would theoretically be nearly infinite. However, because of axial heat conduction through the wall (or through a nuclear fluel element) toward the entrance, the wall temperature actually rises above the temperature of the adjacent fluid, and the heat-transfer coefficient, though large, is finite.

As the temperature profile develops, the local heat-transfer coefficient decreases, until at a distance from the entrance equivalent to about 20 diameters (in a tube), the change in the local coefficient becomes negligible. It may now be added here that rounded entrances have almost the same effect as square-edged entrances, since they help the velocity of the fluid at the leading edge to equal the free-stream velocity.

Since it is customary to use the simplified technique of assuming an average constant value for the heat transfer coefficient along the entire channel (Chap. 13), the effect of the entrance on the values derived from the correlations of the previous two sections is felt up to about 40 diameters.

Kays [57] presents a discussion of the thermal entry lengths in both laminar and turbulent flows for a range of Reynolds and Prandtl numbers.

* At δ, the always-present small disturbances are no longer dampened by the viscous forces. The ratio of inertia to viscous forces (represented by the Reynolds number $DV\rho/\mu$) is sufficiently large at δ to cause the disturbances to grow.

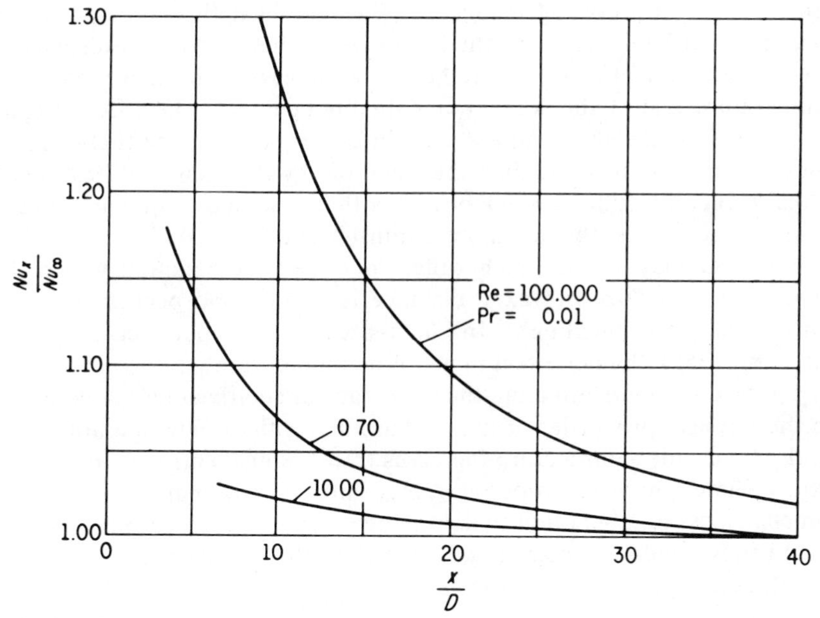

FIG. 9-9. Local-to-fully developed Nusselt number ratio as a function of length-to-diameter ratio and Prandtl number, at Re = 10^5 (Ref. 57).

Figure 9-9 shows the ratio of the *local* to fully developed Nusselt numbers at Re = 10^5 and for three typical Prandtl numbers for flow in a circular tube. The effect is more pronounced for low Prandlt number fluid fuels (liquid metals) and increases with the Reynolds number. For most other reactor coolants (Pr = 0.7) the effect is small and insensitive to Reynolds number. The *mean* Nusselt number over a length/diameter ratio of 100 for Pr = 0.7 shows a difference of about 2 percent from the fully developed value.

Another effect on the heat-transfer coefficient results from the added turbulence due to coiling. If the channel is in the form of a coil (as in a heat exchanger), the value of h for liquids may be increased, depending upon the tightness of the coil, by up to 20 percent.

9-7. THE EFFECTS OF AXIALLY NONUNIFORM HEAT-FLUX AND TEMPERATURE DISTRIBUTIONS

It has been mentioned that, for the sake of simplicity, a constant value of h is used in reactor fuel heat generation and removal. This assumption will now be examined.

The axial heat-flux distribution in reactor fuel channels can be approximated by a sinusoidal curve. This results in an axial change that affects h in the following manner. When the heat flux increases in an axial differential distance dz, the fluid passing that same axial distance is not immediately affected by the heat-flux increase. In other words, the temperature rise of the coolant lags slightly behind the heat flux. Thus at the end of dz, the temperature difference between the fuel-element surface and fluid is lower than that required by the increased heat flux.* Thus the actual value of h will be higher than assumed. Similarly, as the heat flux decreases, the actual heat-transfer coefficient will be low. The two effects may cancel each other in large-sized reactors where the change of heat flux with axial distance is smooth and occurs at a slow rate (small core buckling). In small-sized and compact reactors (with large buckling), the net effect may not be unimportant.

A second factor affecting the heat-transfer coefficient is the variation in the physical properties of the coolant fluid with axial temperature, i.e., as the fluid bulk temperature increases from channel entrance to channel exit. These physical properties enter into the evaluations of all the dimensionless numbers that make the heat-transfer correlations. Reference is now made to one of these equations, the Dittus-Boelter equation (9-22), here repeated:

$$\text{Nu} = 0.023\ \text{Re}^{0.8}\ \text{Pr}^{0.4} \qquad [9\text{-}22]$$

or

$$h = 0.023\ \frac{k}{D}\left(\frac{DV\rho}{\mu}\right)^{0.8}\left(\frac{c_p\mu}{k}\right)^{0.4}$$

In this equation, the value of $\text{Pr}^{0.4}$ for gases does not vary materially with temperature. For water, only a modest temperature rise is allowed through the reactor in order to keep uniform moderating power. (This is accomplished by a large water-circulation rate.) Thus no appreciable change in Prandtl number is encountered in this case either.† Also, since channel diameter D (or equivalent diameter) is usually constant, the quantity ρV, the mass-flow rate per unit cross-sectional area of channel, is constant. The value of h can therefore be thought of as varying directly with k and inversely with $\mu^{0.8}$.

In cases of large variations of these two quantities, the recommended procedure is to evaluate the physical properties of the fluid at a temperature halfway between the entrance and exit of the coolant channel. The arithmetic average temperature is a good approximation. Where the bulk temperature is used, this is given by

* Under ideal conditions, the equilibrium temperature difference between fuel surface and coolant fluid bulk is a cosine function of z or is directly proportional to the heat flux (Sec. 13-3).

† The Prandtl number of water changes from 0.85 to 0.83 with a temperature change from 450 to 500°F; $\text{Pr}^{0.4}$ then changes only by less than 1 percent.

$$t_f = \frac{t_{f_1} + t_{f_2}}{2}$$

where the subscripts 1 and 2 refer to entrance and exit conditions, respectively.

It should be noted that sometimes a trial-and-error procedure may have to be used in solving problems of heat transfer where all the temperatures of the system are not known prior to the solution. For example, if only the entrance temperature is known, the exit temperature is dependent upon h, which in turn can be evaluated only if both temperatures are known. The exit temperature may then be assumed, the heat flux h calculated, and the exit temperature recalculated. One or more iterations may be necessary until the assumed and calculated temperatures agree.

It must be added here that the results of the different correlations for heat transfer are no more accurate than the physical-property values used in them. Also, many of the cases encountered in practice may not be adequately covered by any one of the correlations presented. Indeed, special and new designs may have to await prototype construction and testing in order to evaluate accurately the heat-transfer coefficient encountered in them. The correlations presented should therefore be used with the understanding that they yield reasonably accurate results (within $\pm$ 10 percent) only in cases specifically suited to them. If used otherwise, the results that they yield should be regarded as rough approximations only.

9-8. THE EFFECT OF HIGH GAS VELOCITY

In order to gain as high a heat-transfer coefficient as is necessary for good reactor heat removal, gas coolants are usually pushed through fuel channels at very high velocities. These velocities may be an appreciable fraction of the velocity of sound at the pressure and temperature of the gas. In other words, the *Mach number* (the ratio of the gas velocity to the velocity of sound in it) may be of the order of 0.2 or more. Sonic velocity and Mach number are given by the following equations for a perfect gas:

$$\alpha = \sqrt{\gamma R g_c T_f} \tag{9-36}$$

and

$$M = \frac{V}{\alpha} = \frac{V}{\sqrt{\gamma R g_c T_f}} \tag{9-37}$$

where α = velocity of sound in gas, ft/hr
M = March number, dimensionless

γ = ratio of specific heats of gas, dimensionless
R = specific gas constant, ft-lb_f/lb_m °R
= universal gas constant divided by the molecular mass of gas
g_c = conversion factor, 4.17×10^8 ft lb_m/lb_f hr²
T_f = static bulk temperature of fluid on absolute scale °R
V = bulk speed of flowing gas stream, ft/hr

Now it is known that, if the gas is moving at velocity V with a temperature T_f, a temperature-sensing element (such as a thermometer or a thermocouple), moving with the gas at the same velocity, will register the temperature T_f. However, if this element is at a standstill, it acts as an obstruction at which the gas molecules in the stream come to rest. These molecules then impart their kinetic energy to the element which will then register a temperature higher than T_f.

Similar reasoning may be applied to the temperature of a gas moving axially at high speed but actually at rest near the walls of the channel. It has been shown that for fluids having Prandtl numbers near unity, such as gases, the heat-transfer coefficient depends upon the conditions within the boundary layer. When the bulk of a gas in a channel is flowing at a high axial speed, it has a steep velocity gradient in the radial direction. (Slip may occur but will be neglected here.) This velocity gradient results in the dissipation into the gas of the kinetic energy of the molecules near the wall, because of viscous shear forces. This shear work input causes the temperature to rise (aside from heat-transfer effects). The rise in the gas temperature by this mechanism naturally affects the heat-transfer coefficient.

In order to differentiate between this effect and the temperature gradient due to heat transfer across the boundary layer, temperature and velocity profiles are shown in Fig. 9-10 for an adiabatic wall, i.e., no heat transferred across the wall. T_f is the static temperature. The maximum theoretical temperature that can be attained at the wall is the *stagnation*

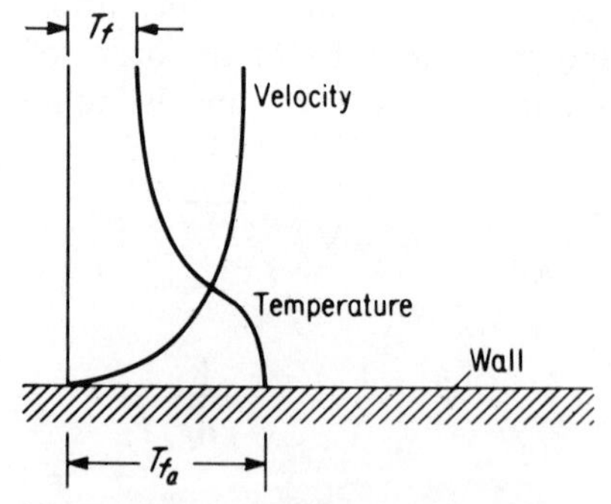

FIG. 9-10. Velocity and temperature profiles in adiabatic high-speed flow.

or *total temperature* T_{f_s} and may be calculated by the general energy equation. Thus for no change in potential energy, no work done, and no heat transferred,

$$h_{f_s} = h_f + (KE)_f$$

where h is the enthalpy of the gas, KE is the kinetic energy, and the subscripts f_s and f refer to stagnation and free stream, respectively. For a perfect gas, this equation becomes

$$c_p T_{f_s} = c_p T_f + \frac{V^2}{2 g_c J}$$

or

$$T_{f_s} = T_f + \frac{V^2}{2 g_c J c_p} \tag{9-38}$$

where c_p = specific heat at constant pressure of gas, Btu/lb$_m$ °R
J = energy conversion factor = 778-ft lb$_f$/Btu

The last term of Eq. 9-38 is called the *dynamic* or *velocity temperature.* Using the perfect gas relation

$$c_p = \frac{\gamma}{\gamma - 1} \frac{R}{J} \tag{9-39}$$

and combining with Eqs. 9-37 and 9-38 and rearranging give

$$T_{f_s} = T_f \left(1 + \frac{\gamma - 1}{2} M^2\right) \tag{9-40}$$

Equation 9-40 indicates that, for air flowing at a Mach number of 0.2 and at a relatively low temperature ($\gamma = 1.4$), the absolute stagnation temperature is some 0.9 percent higher than the absolute static temperature. Thus if T_f is 540°F (1000°R) the maximum theoretical temperature rise is about 8°F. This rise is important enough to warrant taking it into consideration in problems involving gas coolants flowing at high speeds.

The actual maximum temperature attained at the wall surface in the case of adiabatic flow parallel to the wall is lower than the value of T_{f_s} given by Eq. 9-40. This actual temperature is called the *adiabatic wall temperature* T_{f_a}. Theoretical analysis shows that $T_{f_a} = T_{f_s}$ only for Pr = 1.0. Gases, however, have Prandtl numbers less than unity, and T_{f_a} is less than T_{f_s}. The two are related by a *recovery factor* F_R, defined as

$$F_R = \frac{T_{f_a} - T_f}{T_{f_s} - T_f} \tag{9-41}$$

In turbulent flow, F_R usually falls between 0.88 and 0.90. (For turbulent flow the recovery factor for air has been shown to be approximately equal to $\text{Pr}^{\frac{1}{3}}$.)

Figure 9-11 shows temperature profiles for high-speed flow with and without heat transfer across the wall. Three cases are given: coolant fluid heated, i.e., heat flow from the wall to the fluid, adiabatic, and coolant fluid cooled.

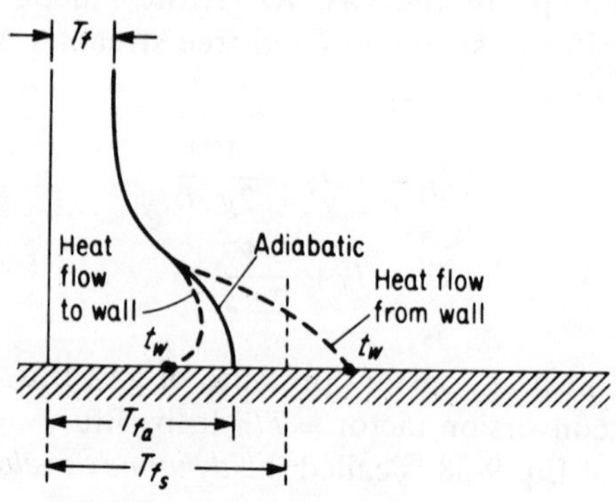

FIG. 9-11. Temperature profiles in adiabatic and nonadiabatic high-speed flow, Pr < 1.

It has been shown by Bialokoz and Saunders [59] that, to compute the heat-transfer coefficient in nonadiabatic high-speed gas flow, the Dittus-Boelter correlation (Eq. 9-22) can be used to obtain h, provided that, in calculating the heat flux, the adiabatic wall temperature T_{fa} computed as above is substituted for the free-stream temperature T_f. The heat flux is then given by

$$q'' = h(T_w - T_{fa}) \tag{9-42}$$

where T_w is the wall temperature in contact with the fluid.

Examination of the temperature gradient near the wall in the fluid-cooling curve of Fig. 9-11 reveals that, even though the wall temperature T_w may be higher than the fluid free-stream temperature T_f, heat may flow from the fluid to the wall. This is the so-called problem of aerodynamic heating within the boundary layer, a problem that is serious in high-speed projectiles, such as rockets and missiles reentering the atmosphere. In other words, heat transfer is no longer dependent upon the difference between the wall and free-stream temperatures $(T_w - T_f)$, as in low-speed flow, but rather on $T_w - T_{fa}$.

It should be noted that, as a gas coolant leaves a fuel channel and enters a compartment above the reactor core, it is substantially slowed down to zero velocity. Its temperature in that compartment is equal to the stagnation temperature.

PROBLEMS

9-1. A research reactor core is cubical, 20 ft on the side. The fuel elements are made of 1-in.-diameter solid natural-uranium metal rods. They are placed horizontally in the center of 3-in.-diameter graphite holes. Air is used as the coolant. It is forced at atmospheric pressure and an initial velocity of 15 fps. For the centermost fuel element, air enters at 80°F and leaves at 190°F. Calculate (*a*) the heat-transfer coefficient, and (*b*) the maximum neutron flux in the core, neglecting cladding and extrapolation lengths.

9-2. Compare the heat-transfer coefficients and the pumping power (hp) per 1,000-ft length of 1-in.-ID smooth-drawn tubing of the following coolants: (*a*) air at 10 atm, 100 fps, and 400°F; (*b*) CO_2 at the same conditions; (*c*) He at the same conditions; (*d*) water at 20 fps and 400°F; and (*e*) sodium at the same conditions. (Sodium for comparison-use Eq. 10-2.)

9-3. A nitrogen-cooled reactor uses UO_2 fuel. The fuel elements are 0.5 in. in diameter and are encased in a graphite can 0.1 in. thick. At a particular cross section the coolant pressure and bulk temperature are 6 atm and 1000°F. The stream velocity is 375 fps. At the same section the volumetric thermal source strength is 2×10^6 Btu/hr ft^3 and the maximum fuel temperature is 5000°F. Determine (*a*) the heat-transfer coefficient and (*b*) the necessary equivalent diameter of the coolant channel. Neglect contact resistance between the fuel and the graphite. Use $k_{\text{graphite}} = 35$ Btu/hr ft °F (aged).

9-4. In Prob. 6-1, if the iron shield is cooled with pressurized light water having a bulk temperature of 400°F and if the temperature within the shield is not to exceed 425°F, find the minimum water velocity necessary on the side facing the γ radiation. The coolant channel is narrow, 2 in. wide, but has large dimensions otherwise.

9-5. Helium coolant enters and leaves a reactor core at 400 and 1049°F, respectively. The mass rate of flow of helium is 100,000 lb$_m$/hr. The reactor core contains 400 fuel elements made of 3 percent enriched UO_2 in the form of vertical plates 4 in. wide and 0.135 in. thick (including 0.005-in.-thick cladding on each side). A solid moderator is used in the form of vertical plates interposed between the fuel plates so as to form coolant channels $\frac{1}{4}$ in. wide. The core is cylindrical, 8 ft in diameter and 8 ft high. Calculate the maximum neutron flux in the core. Use density of $UO_2 = 10.5$ g$_m$/cm^3. Neglect extrapolation lengths.

9-6. Light liquid water is used as a reactor coolant. In a particular channel, the average bulk temperature is 300°F. Determine the percent change in W'/q if it were to be used in the same channel, with the same mass-flow rate, the same mean temperature between ladding and coolant, but at average bulk temperatures of 200 and 400°F. Assume saturated conditions in all cases.

9-7. A pressurized-water reactor uses UO_2 cylindrical fuel pellets, 0.5 in. diam surrounded by a helium gap 0.003 in. wide and by Zirconium cladding 0.03 in. thick. The fuel rods are arranged in square lattice, 0.7 in. between centers. At a particular section, the bulk water temperature and velocity are 520°F and 15 fps respectively. $q''' = 5 \times 10^7$ Btu/hr ft^2. Find (*a*) the convective heat transfer coefficient, and (*b*) the minimum system pressure so that no boiling occurs in the film.

9-8. A fuel rod is composed of 0.45 in. diameter fuel ($k_f = 25.4$) with 0.025-in. thick cladding ($k_c = 45$). For metallurgical reasons the maximum temperatures in the fuel and cladding must not exceed 750°F and 600°F respectively. The rod is water-cooled. At a particular cross section the water temperature is 500°F. Find (*a*) the heat-transfer convective coefficient, and (*b*) the minimum water pressure to avoid boiling at that cross section.

9-9. A reactor core operating at 2000 psia contains parallel fuel rods having 0.55 in. fuel material diameter and 0.60 in. overall diameter. The rods are arranged in a square lattice with a spacing of 0.66 in. between centers. Water coolant flows parallel to the tubes. At a particular position in the core, the volumetric thermal source strength is 4×10^7 Btu/hr ft^3 and the coolant temperature is 540°F. Find the minimum water velocity at that position if the temperature difference across the film is not to exceed 50°F.

9-10. Helium at 300°F and 10 atm pressure flows in a 1 in.-diameter tube at 350 fps. The tube walls are at 280°F. Find the error in heat flux, Btu/hr ft^2, if aerodynamic heating is not taken into account.

chapter **10**

Liquid Metal Coolants

10-1. INTRODUCTION

Liquid (or *molten*) *metals* have characteristics that make them particularly suitable for use as primary and secondary coolants, and even as working fluids for certain types of reactors and power plants. Fast reactors, in particular, require coolants of no or low moderating capabilities and, because of their high power densities, coolants with high heat-transfer characteristics. The use of ordinary and heavy water and organic coolants in fast reactors is excluded because of their moderating capabilities. Gases have generally poor heat-transfer characteristics but low moderating capabilities, and therefore helium (the best from a heat transfer standpoint) and, to a lesser extent, steam are being considered as fast reactor coolants.

Liquid metals, on the other hand, generally have low moderating capabilities and excellent heat-transfer characteristics. They are therefore particularly suitable for use in fast reactors. They can also be used in thermal reactors in conjunction with a solid moderator (such as graphite) or a liquid moderator that is separated from the coolant (such as heavy water), when their use is necessitated by the need for exceptionally high heat-transfer rates.

Because of the importance of liquid metals in nuclear energy transport processes, this chapter is devoted to them.

10-2. SOME GENERAL CONSIDERATIONS

Many liquid metals have low absorption cross sections for neutrons in the fast- and intermediate-energy ranges. Some, however, have exceptionally high-neutron cross sections in the thermal-energy range and therefore may not be suitable for use in thermal reactors. Figure 10-1 shows total neutron cross sections of some liquid metals as a function of neutron energy.

In general liquid metals are not readily available in sufficient quantities in all parts of the world. They are relatively costly. Liquid-

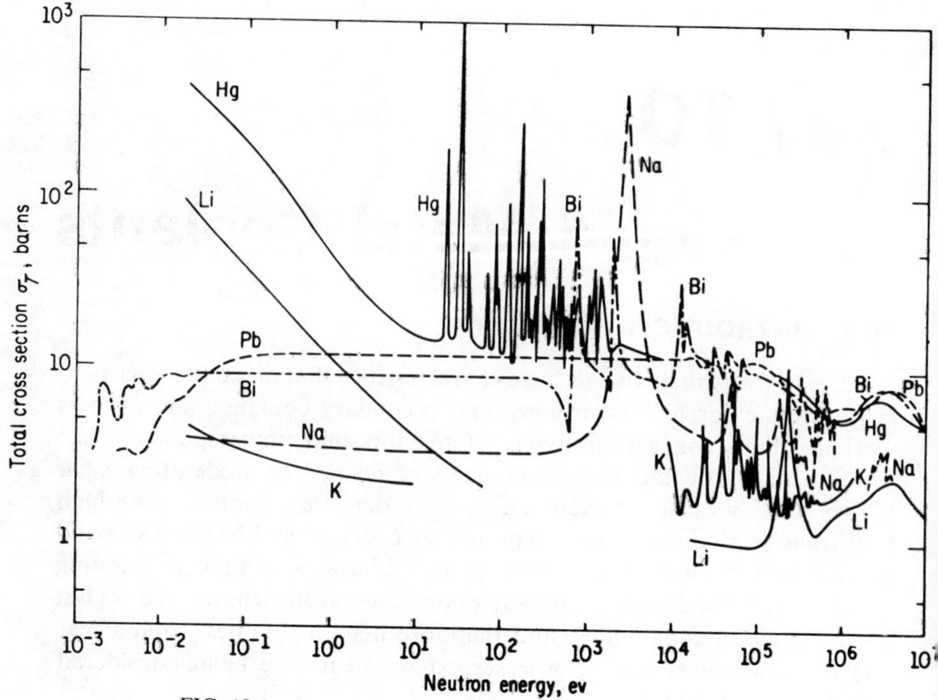

FIG. 10-1. Total neutron cross sections of some liquid metals.

metal-cooled reactors, however, are usually small in volume, and so are their primary-coolant systems. The liquid metals themselves are particularly resistant to radiation damage so that no appreciable replenishment costs are incurred (as in the case of organic coolants), although some degree of purification to remove corrosive oxides may be necessary. These factors indicate that the actual costs of the coolant in a liquid-metal-cooled power plant may not be too high. Liquid metals also have relatively low melting points and high boiling points (or low vapor pressures) and therefore remain in liquid form over a wide range ot temperatures. High reactor exit temperatures are therefore possible with low system pressures. This results in both high power cycle efficiencies and low pressures in the reactor vessel. This, in turn, makes possible reactor vessels that are low in cost or reactors of large sizes, contributing also to lowering power costs. Liquid metals also have excellent heat-transfer and fluid-flow characteristics. Their high thermal conductivities (Table 10-1) result in a reduced hot-spot factors (Sec. 13-6) and in lowering the temperature gradients in the core and consequently in a reduction of the probability of structural warpage, though a problem of thermal shock exists.

TABLE 10-1
Some Properties of Air, Water, and Selected Liquid Metals

Coolant	Temperature, °F	k, Btu/hr ft °F	Boiling point, °F	Melting point, °F
Air	600	0.027		
Water	600	0.356	212	32
Liquid sodium	600	43.3	1621	208
NaK (50% Na by mass)	600	15.69	1518	66
Liquid bismuth	600	9.77	2691	520
Mercury	300	9.31	675	-40

On the debit side, liquid metals suffer from high induced radioactivities (Sec. 10-3) and are generally chemically active (Sec. 10-4). It is therefore necessary in liquid-metal-cooled power plants to use special materials for surfaces in contact with the coolant and to use more than one coolant loop in the power plant and extensive shielding of the primary loop. These requirements tend to lower somewhat the good thermal efficiencies attainable with this type of power plant and to increase power costs.

Of the many available liquid metals, the alkalies, which include sodium, potassium, sodium-potassium mixtures (NaK), rubidium, cesium, and lithium, have received the greatest attention. Sodium, which is commercially available in solid brick form, is the least expensive. It has a good temperature range: 208°F melting point and 1621°F normal boiling point. The somewhat high melting point causes it to solidify at room temperatures, necessitating the use of external heating of the sodium system during extended shutdowns. Sodium has excellent heat-transfer characteristics and can be used with thermal as well as epithermal reactors. It has well-known corrosive characteristics and, by contrast with many other liquid metals, is compatible with a large number of materials.

10-3. INDUCED RADIOACTIVITY IN LIQUID-METAL COOLANTS

In general, liquid metals are troublesome because of the ease with which they become radioactive and the strong radiations and relatively long half-lives of their radioactive products. In large power plants, this usually necessitates the use of an intermediate coolant loop, also of liquid metal for good heat-transfer purposes, between the reactor or primary loop, where induced radioactivity exists, and the working fluid (steam). The reactions and activities of some liquid-metal coolants will now be discussed.

All naturally occurring sodium is made up of the isotope ${}_{11}Na^{23}$. It has a thermal-neutron absorption cross section of 0.53 barn (Appendix B) and a fast neutron cross section of about 1 millibarn at 0.25 Mev. When subjected to neutrons in a reactor core, it undergoes the reaction $Na^{23}(n,\gamma)Na^{24}$. The product Na^{24} is a radioactive isotope of about 15 hr half-life which emits γ radiation mainly of 2.76 and 1.38 Mev energy. It β-decays into stable $_{12}Mg^{24}$, which has a low neutron cross section. The level of activity of sodium, of course, depends upon the relative magnitude of the time spent within and outside the core.

Naturally occurring potassium is made of two stable isotopes $_{19}K^{38}$ ($\sim$93.10 percent) and K^{41} ($\sim$6.88 percent). The remainder (0.0118 percent) is radioactive K^{40} which is a β and γ emitter with a long half-life (1.32×10^9 yr) and thus has a very low level of activity. Of the three potassium isotopes only K^{41} is converted to a radioactive isotope upon neutron absorption. Potassium 41 has a thermal-neutron absorption cross section of 1.24 barn, converting to K^{42}, a β and γ emitter of 12.4-hr half-life, converting to stable $_{20}Ca^{42}$. While K^{40} has a large cross section (70 barns) for thermal-neutron absorption, converting to K^{41} which is susceptible as above, its abundance in nature is so low that its contribution to activity is unimportant.

The isotope $_3Li^7$ (92.58 percent of all lithium) has a 0.037-barn thermal cross section, converting to Li^8 of 0.845-sec half-life.

The thermal cross sections of $_{80}Hg^{196}$, Hg^{198}, and Hg^{202}, with abundances of 0.146, 10.02, and 29.8 percent, are 3,100, 20, and 4.5 barns, respectively. The resultant radioisotopes have half-lives of varying length, from 44 min to 47.9 days, and γ energies varying between 0.13 and 0.37 Mev.

Other reactions in heavy liquid metals occur in lead, $_{82}Pb^{208}(n,\gamma)Pb^{209}$ ($<$ 0.03 barn, 3.3-hr half-life), and in bismuth, $_{83}Bi^{209}(n,\gamma)Bi^{210}$ (0.034 barn). Bismuth 210 is a β emitter with a 5-day half-life, converting to Po^{210}, an α emitter of 138-day half-life converting to stable Pb^{206}. These reactions are serious enough to be taken into account if these liquid metals where used as reactor coolants.

Liquid metal neutron-absorption-cross-section data at fast-reactor energies (about 0.25 Mev average) are scant but are, in general, known to be 100 or 1,000 times lower than for thermal neutrons. This, however, is countered by the fact that neutron fluxes are about that much higher, or more, in fast reactors. Fast reactors, however, are smaller in size and require high coolant velocities, so that the residence time of the coolant within the reactor core is, on a percentage basis, much less than in a thermal reactor. The net effect depends, of course, upon the particular situation encountered. In general, however, fast reactors tend to induce lower coolant activities than thermal reactors.

10-4. LIQUID-METAL COMPATIBILITY WITH MATERIALS

Most materials suffer from attack or corrosion when subjected to liquid metals. The results of a large number of laboratory experiments on the effects of liquid metals on many materials are available in the literature [60-62]. Some of the materials that are particularly soluble in hot sodium, potassium, and NaK and are therefore unsuitable as high-temperature reactor materials are cadmium, antimony, bismuth, copper, lead, silicon, tin, and magnesium. On the other hand, nickel, Inconel, Nichrome, the Hastelloys, and nickel- and chromium-bearing steels (such as the Series 300 stainless steels and the columbium-bearing Type 347 stainless steel) are well suited for high-temperature use with the above liquid metals.

Probably the chief reason for attack or corrosion is the ability of sodium (and the sodium in NaK) to dissolve oxygen. This solubility increases rapidly with temperature, approaching a saturation of about 0.1 percent by mass at 900 to 1000°F. Oxygen reacts with sodium to form Na_2O which is highly corrosive. It is also relatively insoluble in sodium or NaK, especially at low temperatures, causing deposition in cooler passages. A little such deposition usually acts as a nucleus around which crystal growth of a mixture of approximately 20 percent Na_2O and 80 percent Na takes place, causing plugging in narrow cool passages. Because of this, a minimum of narrow passages should be designed into a sodium system, and when necessary they should be streamlined, restricted to the hotter regions of the system, or placed vertically to avoid precipitation.

The most troublesome effects of corrosion are the so-called *self-welding* and *thermal-gradient transfer* [60]. Self-welding results in the malfunction of such system components as pumps and valves. It is believed to be caused by the action of the oxygen-absorbing sodium reducing the oxides of surfaces in contact, when this sodium diffuses between these surface. Self-welding increases with sodium temperature and with contact pressure between the surfaces. In thermal-gradient transfer, materials are dissolved by the liquid metal in a high-temperature region, where the solubility is high, and precipitated in a cooler region, where the solubility is low. This, then, in effect results in the mass transport of materials by the liquid metal. While the rates of such mass transport are low, extended operation and coolant circulation may result in noticeable corrosion of the hotter regions and plugging in the cooler regions of the system.

It is obvious that absorption of oxygen by the coolant should be avoided. An inert-gas blanket should be provided over all free-sodium surfaces. The inert gas may be helium, which has the advantage of no induced radioactivity, or argon, which has the advantage of being heavier than air, thus facilitating the blanketing problem. Either of these two gases is expensive, and much attention is being given to nitrogen. Nitrogen is, however, soluble in sodium. This solubility, though low (less than 1 ppm), results in the mass transport of N_2 through the system (thought to be aided by the presence of calcium or carbon). This in turn results, at high temperatures, in nitriding and damage to thin-walled components such as cladding, valve bellows, etc.

Oxygen may enter the system via impurities in the gaseous blanket and during refueling or repair periods. Thus it is usually necessary to sample and analyze the coolant. Purification is done by bypassing a portion of the hot coolant (when oxide solubility is highest) and depositing the oxide in a cold trap. An oxide concentration below 30 ppm appears to give satisfactory results, although lower concentrations are easily attainable. Also purification of the gaseous blanket to remove air and water vapor may be necessary.

Sodium also reacts vigorously with most gases (except the inerts) and liquids. In the solid state, it is soft and silvery white but tarnishes readily when exposed to air, because of the formation of an oxide film on its surface. In the liquid state, also silvery white, when exposed to air it burns with a thick smoke of Na_2O. If such exposure is due to a fissure or hole in the container, the heat of combustion causes the sodium to reach its boiling point and to form a highly corrosive spongy mixture with its oxides. This mixture enlarges the original fissure, aggravating the situation.

Sodium also reacts vigorously with water according to

$$\left.\begin{aligned} Na + H_2O &\longrightarrow NaOH + \tfrac{1}{2}H_2 + 3600\ \text{Btu/lb}_m\ Na \\ Na + \tfrac{1}{2}H_2O &\longrightarrow \tfrac{1}{2}Na_2O + \tfrac{1}{2}H_2 + 1610\ \text{Btu/lb}_m\ Na \end{aligned}\right\} \\ Na + \tfrac{1}{2}H_2 \longrightarrow NaH \tag{10-1}$$

These reactions become explosive if they take place in a confined volume. If air is also present, the hydrogen liberated will burn, resulting in an additional 2300 Btu/lb_m Na. In sodium-water heat exchangers, care must be taken to choose materials of construction that will not be attacked by either fluid. One design, that of the Sodium Reactor Experiment [1] uses a double-walled tube construction with an intermediate high-thermal-conductivity fluid (mercury) in the annulus.

If H_2 and H_2O enter a sodium system in nonexplosive quantities (via impurities or leakage) they form NaH and NaOH. This appreciably increases the moderating ratio of the coolant, an effect that may not be desirable, especially in epithermal reactors.

Intermediate coolant loops in sodium-cooled reactor power plants guard against reactions between water and radioactive sodium. Such reactions result, among other things, in the radiolytic decomposition of steam-generator water by the strong γ radiation emitted by Na^{24}. The intermediate loops also ensure against water or hydrogen entering the reactor.

Fire-extinguishing materials for alkali metals include soda ash ($NaCO_3$) amd rock salt ($CaCO_3$). Alarm systems, such as that shown in Fig. 10-2 may be installed to warn against leakage.

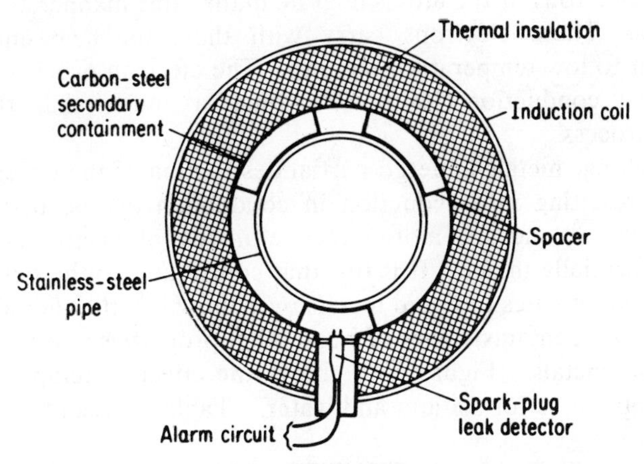

FIG. 10-2. Typical liquid-metal pipe section and alarm system.

Alkalies, other than sodium, have chemical reactions similar to those given above. Potassium, however, is more reactive than sodium, and NaK, with many of its mixtures liquid at room temperature, is more reactive than either of its components.

10-5. HEAT TRANSPORT PROPERTIES OF LIQUID METALS

Compared with nonmetallic fluids, liquid metals have low specific heats and viscosities and high thermal conductivities, resulting in low Prandtl numbers (Table 9-2). It has been indicated in the previous chapter that the normal diffusivity of heat of liquid metals is so high that it overshadows the turbulent diffusivity and that heat transfer correlations of liquid metals are therefore different from those of higher Prandtl-number fluids.

The reason for the high thermal conductivity of liquid metals can be briefly explained as follows: When in the solid state, materials have relatively high values of k. These are suddenly decreased on melting because of the partial but sudden destruction of the more or less orderly molecular structure of the solid. There remains only short-range molecular order or liquid crystals [63]. On evaporation, the molecular order completely breaks down, resulting in further sudden reduction in thermal conductivity. Conduction in gases (assuming no convective currents) is due solely to energy exchange by collision of the molecules during their random motion, a process of thermal diffusion.

Heat conduction in solid metals is due to two processes: (1) orderly crystal lattice vibration (as with nonmetals) and (2) electron motion, where the electrons of the atoms migrate in the same manner as electrical conduction. These electrons carry with them the heat energy and transport it to low-temperature regions. The electron process results in greater heat conduction (10 to 1,000 times as much) than the lattice vibration process.

On melting, metals undergo partial destruction of the orderly lattice structure, resulting in a reduction in conduction by the first process. However, the second and more effective process of electron conduction remains essentially intact. It is this that contributes to the much higher thermal conductivities of liquid metals over those of other liquids.

Table 10-1 contains values of k and other properties of air, water, and some liquid metals. Figure 10-3 shows the effect of temperature and phase change on k for sodium and water. Table 10-2 contains physical

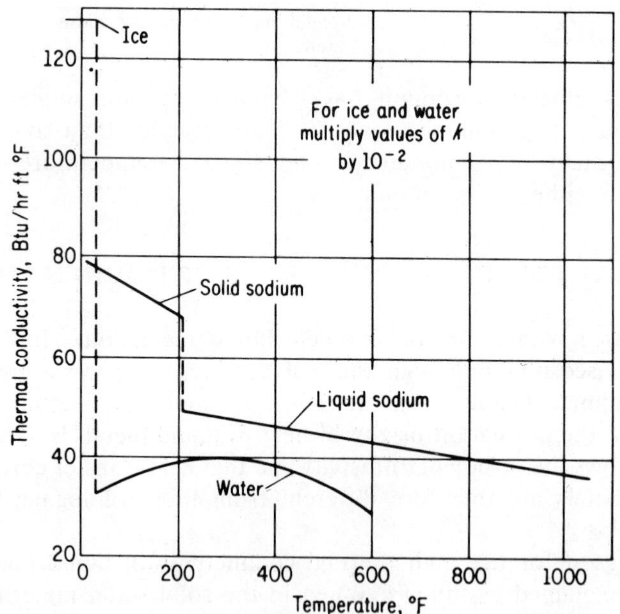

FIG. 10-3. Thermal conductivity of sodium and water in solid and liquid phases.

properties of liquid sodium, required in the determination of heat-transfer coefficients.

It should be mentioned here that water, if used at high enough speeds, is capable of heat-transfer coefficients almost as high as those of liquid metals. However, it requires pressurization and has nuclear characteristics (neutron absorption, moderating power) that may be undesirable, as in the cases of natural-uranium-fueled and fast reactors.

TABLE 10-2
Some Physical Properties of Liquid Sodium*

Temperature		c_p,	ρ,	μ,	k,	Pr
°C	°F	Btu/lb_m°F	lb_m/ft^3	lb_m/ft hr	Btu/hr ft °F	dimensionless
100	212	0.3305	57.87	1.706	.....	......
150	302	......	.....	1.309	.....	......
200	392	0.3200	56.44	1.089	47.11	0.0074
250	482	......	.....	0.949	.....	......
300	572	0.3116	55.06	0.835	43.75	0.0059
400	752	0.3055	53.63	0.687	42.15	0.0051
500	932	0.3015	52.07	0.588	38.61	0.0046
600	1112	0.2998	50.51	0.508	36.24	0.0042
700	1292	0.3003	48.88	0.450	34.10	0.0040
800	1472	0.3030	47.26	0.399	31.62	0.0038
900	1652	0.3079	.....	0.363	.....	......

* From Ref. 60.

10-6. HEAT-TRANSFER CORRELATIONS FOR LIQUID METALS

Before presenting heat-transfer coefficient correlations for liquid metals, it should be noted again that oxygen is soluble in sodium, with the solubility increasing rapidly with temperature, approaching a saturation of about 0.1 percent by mass at 900 to 1000°F. Oxygen reacts with sodium to form the oxide Na_2O, which, besides being highly corrosive, deposits on heat-transfer surfaces resulting in an added resistance to heat transfer. The heat-transfer coefficient for liquid sodium is therefore very sensitive to the oxygen content. Oxide contamination can decrease the Nusselt number by as much as 50 percent. In extreme cases, the presence of the oxide may result in plugging of the narrower heat transfer passages. Extreme care should therefore be taken, in obtaining correlations experimentally, to minimize or eliminate the oxide contamination. The correlations presented below are for clean systems.

Martinelli extended his analogy of heat and momentum transfer, Sec. 9-4, to cover the case of liquid metals in which normal heat conduction is not negligible when compared with eddy diffusivity. The results of this were presented by Lyon [60] in a simplified form, and with good accuracy when compared with experimental results. The resulting correlation, and others follow.

Flow in Circular Tubes

1. *Constant heat flux* along tube wall (Lyon-Martinelli):

$$Nu = 7 + 0.025\ Pe^{0.8} \tag{10-2}$$

2. *Uniform wall temperature* (Seban and Shimazaki [64]):

$$Nu = 5.0 + 0.025\ Pe^{0.8} \tag{10-3}$$

where Pe is the *Peclet* number, dimensionless.

$$Pe = Re\ Pr = \frac{DV\rho}{\mu}\frac{c_p\mu}{k} = \frac{DV\rho c_p}{k} \tag{10-4}$$

It can be seen that the above correlations include constants on the right-hand side (7.0 and 5.0), reflecting the fact that much heat transfer in liquid metals occurs even at low speeds (low Pe). The Peclet number is the product of the Reynolds and Prandtl numbers, so that the viscosity is eliminated. This reflects the fact that viscosity is not rate-determining in liquid-metal heat transfer.

Flow in Noncircular Channels

Use equivalent diameters (Sec. 9-5) where necessary.

3. *Flow parallel to rod bundles.*

Dwyer [65] presented the results of analytical work for fully developed turbulent flow and uniform heat flux. The work covered the effects of the Peclet number and the pitch-to-diameter ratio S/D where S is the distance between rod centers and D the diameter of the individual rods, for staggered rod bundles.

The results are given for $S/D = 1.2$ and 1.5 in Fig. 10-4. Comparisons between Dwyer's analytical work and experiments by Borishansky and Firsova [66] on sodium indicated fair agreement at relatively high values of the Peclet number, but poor agreement at low Pe, apparently because of oxide problems in the experiments.

4. *Flow across rod bundles.*

Experiments at Brookhaven National Laboratory [67, 68] on mercury flowing perpendicular to a rod bundle, 10 rows deep, of a staggered (equilateral triangle) lattice with $D = 0.5$ in. and $S/D = 1.375$, resulted in the following correlation for the average Nusselt number

$$Nu = 4.03 + 0.228\ (Re_m\ Pr)^{0.67} \tag{10-5}$$

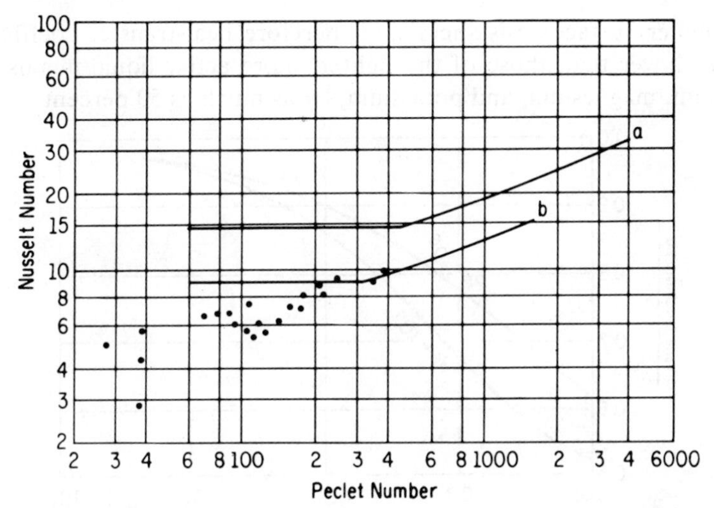

FIG. 10-4. Heat-transfer correlation for liquid metal flow parallel to rod bundles. S/D = (a) 1.5, (b) 1.2 (Ref. 65). Dots are experimental points (Ref. 66).

where Re_m is a Reynolds number based on the maximum speed through the bundle. The physical properties in Pr are taken at the average film temperature. Others are taken at the average bulk temperature.

5. *Flow between parallel plates.*

This is the case of flow between flat-plate fuel elements, Fig. 9-6. Seban (69) presented the following correlations:

(*a*) For constant heat flux through one wall only (the other is adiabatic),

$$\text{Nu} = 5.8 + 0.02\ \text{Pe}^{0.8} \tag{10-6}$$

(*b*) For constant heat flux through both walls, divide h obtained from Eq. 10-6 by the factor in Fig. 10-5.

6. *Flow in annuli* (Fig. 9-7):

(*a*) If D_2/D_1 is close to unity, use Eq. 10-6 for parallel plates.

(*b*) For $D_2/D_1 > 1.4$, Bailey [70] and Werner et al. [71] propose the following correlation:

$$\text{Nu} = 5.25 + 0.0188\ \text{Pe}^{0.8}\left(\frac{D_2}{D_1}\right)^{0.3} \tag{10-7}$$

Other analytical works on the heat transfer of liquid metals for flow between parallel plates and in concentric and eccentric annuli may be found in [72, 73] and related references.

It should be noted that, because of poor wetting characteristics, mercury and such other heavy liquid metals as lead and bismuth, usually have higher surface resistances and therefore heat-transfer coefficients that are lower than those of the lighter, more active liquid metals such as sodium, magnesium, and potassium, by as much as 50 percent.

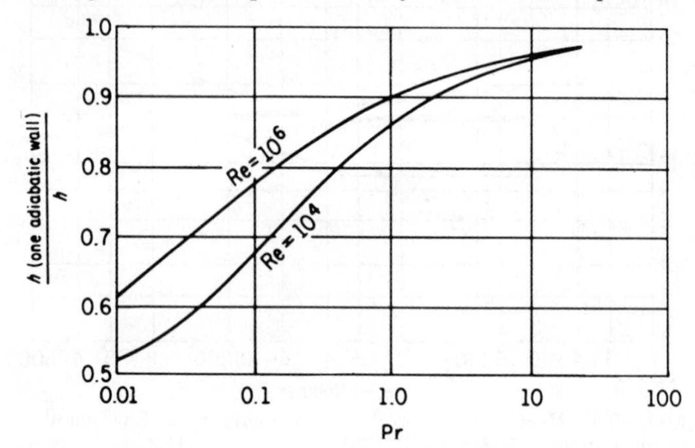

FIG. 10-5. Correction factors for Eq. 10-6 (Ref. 69).

It was also indicated, Sec. 9-5, that nonmetallic coolants yield Nusselt numbers that are not highly sensitive to the wall conditions. This is not the case with liquid metals [74]. Figure 10-6 shows the ratio

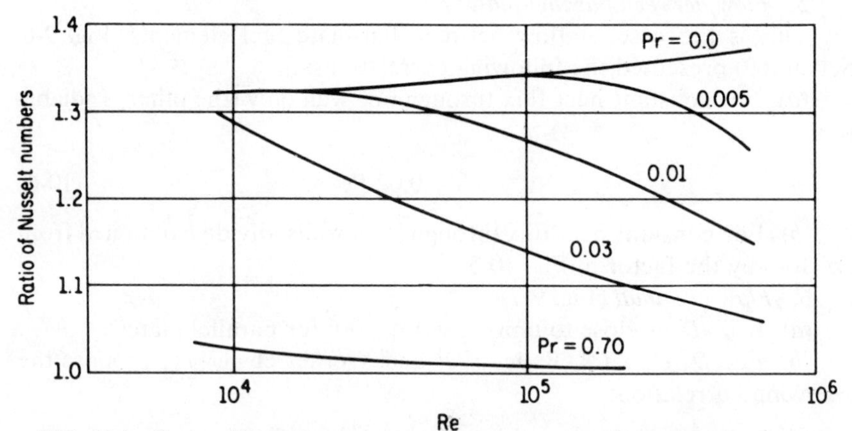

FIG. 10-6. Ratio of Nusselt number for constant heat flux to Nusselt number for constant wall temperature for fully developed flow in a circular tube (Ref. 74).

of Nusselt numbers for constant heat flux and for constant wall temperature as a function of the Reynolds number for various Prandtl numbers for fully developed flow in a circular tube. It is therefore necessary that care be taken in determining the Nusselt number that is commensurate with the wall conditions.

10-7. EFFECT OF AXIAL HEAT DIFFUSION

Correlations for the convective heat-transfer coefficient do not take into account the axial diffusion of heat within the fluid flowing through a channel due to its own thermal conductivity. This is a reasonable omission for nonmetallic fluids of low thermal conductivities. It is suggested, however, that in cases of Peclet numbers less than 35, which occur with the low Pr liquid metals moving at low velocities, axial diffusion of heat will have an effect on both the heat-transfer coefficients (Nusselt numbers) and the temperature distribution.

Burchill, Jones, and Stein [75] investigated this effect of axial heat diffusion in the fluid analytically, using as a model an annular duct of inner and outer wall radii, r_1 and r_2 (Fig. 10-7) with specified axial wall heat flux distributions $q_1''(z)$ and $q_2''(z)$, Btu/hr ft^2, at these walls, respectively. The model, which has long unheated inlet and outlet lengths, thus represents cylindrical ducts ($r_1/r_2 = 0$), annuli, and spaces between infinite parallel planes ($r_1/r_2 = 1.0$). In the analysis, the turbulent or eddy

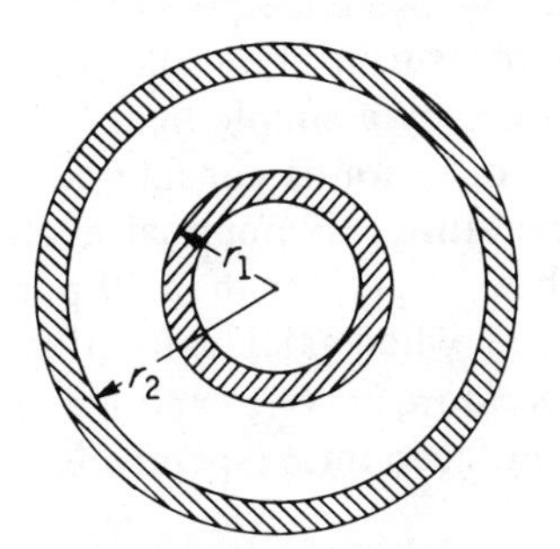

FIG. 10-7. Cross section of an annular channel.

diffusivity of heat was considered negligible compared to the molecular diffusivity of heat, a justifiable assumption for liquid metals (Pe < 100), and the velocity distribution was considered uniform (slug flow) a reasonable assumption for turbulent flow. The results of this work may be applied to predictions which neglect the effects of axial diffusion of heat, such as the correlations presented in the previous section, to estimate the extent of these effects.

The principal conclusions of this work were summarized as follows:

(a) Axial diffusion has no significant influence on the heat-transfer coefficients in turbulent liquid-metal flows for Pe > 35.

(b) At lower Peclet numbers the Nusselt number at either radius may be evaluated, for fully developed heat transfer, and *both* walls with uniform heat fluxes q_1'' and q_2'', from

$$\frac{1}{\mathrm{Nu}_i^*} = \frac{1}{\mathrm{Nu}_i} + \frac{4\left[\left(\frac{r_1}{r_2}\right) q_1'' + q_2''\right]}{\left[1+\left(\frac{r_1}{r_2}\right)\right] q_i'' \, \mathrm{Pe}^2} \tag{10-8}$$

where Nu_i^* and Nu_i are the Nusselt numbers at the inner *or* outer surfaces ($i = 1$ or 2), with and without axial heat diffusion taken into account, respectively. When only *one* wall, i, has heat flux, Nu_i^* can be obtained from

$$\frac{1}{\mathrm{Nu}_i^*} = \frac{1}{\mathrm{Nu}_i} + \frac{4m_i}{\mathrm{Pe}^2} \tag{10-9}$$

where

$$m_i = 2(r_1/r_2)/(1 + r_1/r_2) \qquad \text{for } i = 1 \tag{10-10a}$$

$$= 2/(1 + r_1/r_2) \qquad \text{for } i = 2 \tag{10-10b}$$

In either case, Nu* is the Nusselt number which is determined by most experimenters by measuring the wall temperatures and determining the fluid bulk temperature by a simple heat balance. It is therefore the one in which axial heat diffusion is considered for the wall temperature, but not the bulk temperature. Numerical evaluations showed errors in Nusselt numbers which were greater than 10 percent for $\mathrm{Pe} < 11$ and less than 1 percent for $\mathrm{Pe} > 35$, when axial heat diffusion was neglected.

(c) The fluid temperatures $t^*(r,z)$ and $t(r,z)$, with and without axial diffusion being taken into account, respectively, are related by

$$t^*(r,z) = t(r,z) + \frac{2D_e\left[\left(\frac{r_1}{r_2}\right) q_1''(z) + q_2''(z)\right]}{\left[1+\left(\frac{r_1}{r_2}\right)\right] k(\mathrm{Pe})^2} \tag{10-11}$$

where z = axial position
D_e = equivalent diameter of the duct, ft,
k = thermal conductivity of fluid, Btu/hr ft°F

The above expression has been found to be independent of the mode of heat addition, the Peclet number or the length of the heated section.

10-8. SOME HEAT-TRANSFER CORRELATIONS OF LIQUID METAL FLUID FUELS

Liquid-metal fluid fuels, such as uranium dissolved in molten bismuth, are among the fuels considered for use in fluid-fueled reactors. They pose interesting problems of fluid flow with simultaneous heat generation.

As a result of experimental investigation and analysis Muller [76] obtained a correlation for the forced convection heat-transfer coefficient h for liquid-metal fuels. This correlation gives h as part of a new Nusselt number that includes the effect of internal heat generation, as follows:

$$\frac{1}{\mathrm{Nu}_g} = \frac{1}{\mathrm{Nu}} + \frac{2q''_w}{q'''r_0} Y \tag{10-12}$$

where Nu_g = Nusselt number with internal heat generation, dimensionless
Nu = Nusselt number for liquid metals as given by Lyon's correlation (Eq. 10-2), dimensionless
q''_w = heat flux at wall, Btu/hr ft²
q''' = volumetric thermal source strength, Btu/hr ft²
r_0 = tube radius ft
Y is given by the empirical correlation:

$$Y = \frac{1}{53.0 + 0.152\ \mathrm{Pe}^{0.92}} \tag{10-13}$$

where Pe is the Peclet number, dimensionless.

It can be seen from Eq. 10-12 that internal heat generation decreases the effective Nusselt number, and consequently h, in the case of wall heating, i.e., heat transferred out of the system (q''_w positive), and increases it in the case of wall cooling (q''_w negative).

Some work on natural-convection heat transfer in fluids with internal heat generation can be found in the literature [77].

10-9. THE HEAT-TRANSFER MECHANISM IN FLUID-FUEL SYSTEMS

Fluid fuels present a unique problem in that they are moving heat-transfer agents that are simultaneously sources of heat. The general relationships for the case of a liquid-metal fuel flowing inside a pipe will now be obtained with the help of Fig. 10-8. The relationships also apply to all other fluid fuels, including molten-salt fuels, fuels in aqueous solutions, etc. [2]. Consider a differential element in the fluid in the form of a cylindrical shell of radius r, thickness Δr, and axial length Δz. A heat balance on this element follows:

1. Heat transported by the fluid within the element:

$$2\pi r\,\Delta r\, V\rho c_p \frac{\partial t}{\partial z}\,\Delta z \tag{10-14}$$

2. (Heat conducted radially across inner face, at r) – (heat conducted radially across outer face, at $r + \Delta r$)

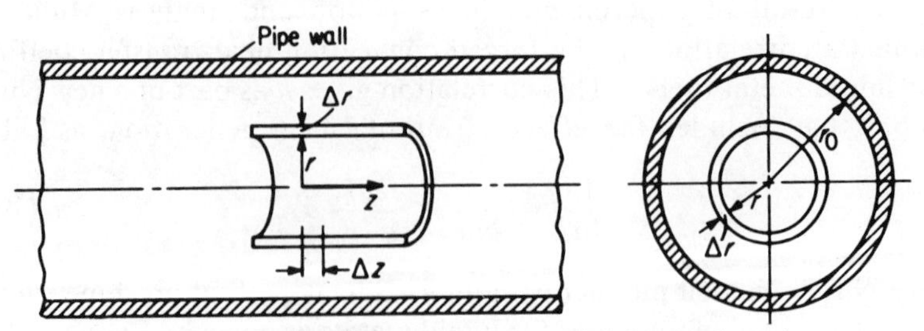

FIG. 10-8. Fluid with internal heat generation flowing in a pipe.

$$2\pi k\,\Delta z\left(r\,\frac{\partial^2 t}{\partial r^2}+\frac{\partial t}{\partial r}\right)\Delta r \tag{10-15}$$

3. Heat generated by the fluid within the element:

$$2\pi r\,\Delta r\,\Delta z\,q''' \tag{10-16}$$

where V = local fluid speed at r, ft/hr
ρ = density of fluid, lb_m/ft^3
c_p = specific heat of fluid, Btu/lb_m °F
t = fluid temperature at r, °F
z = axial distance along pipe, measured from arbitrary starting point, ft
k = thermal conductivity of fluid, Btu/hr ft °F (assumed constant)
q''' = volumetric thermal source strength of fluid (Sec. 4-2), Btu/hr ft³ (assumed constant)

Combining these items so that $1 = 2 + 3$ and rearranging give the general relationship

$$\frac{\partial^2 t}{\partial r^2}+\frac{1}{r}\frac{\partial t}{\partial r}+\frac{q'''}{k}=V\left(\frac{1}{\alpha}\frac{\partial t}{\partial z}\right) \tag{10-17}$$

where α = thermal diffusivity of fluid = $k/\rho c_p$, ft²/hr.

Solutions for Eq. 10-17 backed by some experimental work, in the two cases of laminar and turbulent flow have been given by Poppendiek [78]. The solutions apply to flow in long smooth pipes where the thermal and hydrodynamic flow patterns have been established, to the case of uniform wall heat flux so that $\partial t/\partial z$ = const, and to the case where the physical properties of the fluid are constant and independent of temperature. The two solutions will now be presented.

The Case of Laminar Flow

In laminar flow, the velocity at any radius r within a pipe cross section may be given by

$$V = 2\overline{V}\left[1 - \left(\frac{r}{r_0}\right)^2\right] \tag{10-18}$$

where $\overline{V}$ = mean fluid velocity in pipe, ft/hr
r_0 = pipe radius, ft

Combining with Eq. 10-17 gives, for laminar flow,

$$\frac{\partial^2 t}{\partial r^2} + \frac{1}{r}\frac{\partial t}{\partial r} + \frac{q'''}{k} = (r_0^2 - r^2)\left(\frac{2\overline{V}}{r_0^2 \alpha}\frac{\partial t}{\partial z}\right) \tag{10-19}$$

Introducing the two boundary conditions, (1) uniform wall heat flux, q''_w Btu/hr ft² at $r = r_0$, and (2) wall temperature t_w as a reference temperature within the fluid, Eq. 10-19 is then solved to give

$$\frac{t_w - t}{q''' r_0^2/2k} = \frac{1-2F}{2}\left[\left(\frac{r}{r_0}\right)^2 - 1\right] + \frac{F}{4}\left[\left(\frac{r}{r_0}\right)^4 - 1\right] \tag{10-20}$$

where F is [1 − (fraction of heat generated within moving fluid that is transferred at wall)], dimensionless. F is given by

$$F = 1 - \frac{2q''_w}{q''' r_0} \tag{10-21}$$

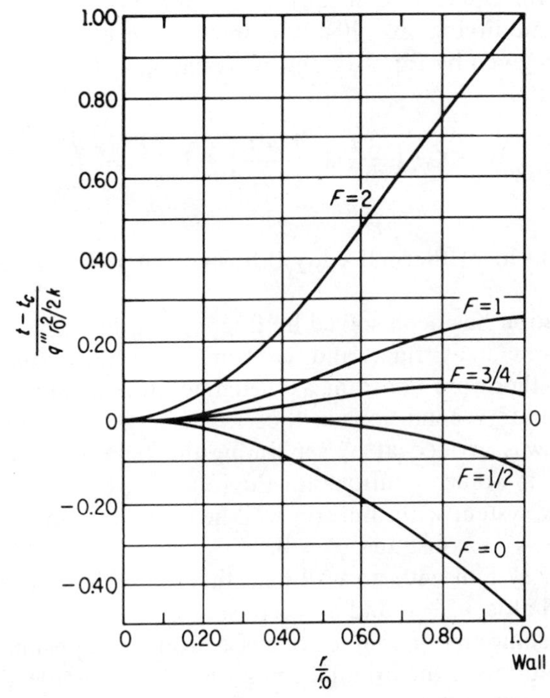

FIG. 10-9. Solution of laminar-flow case (Ref. 78).

Note that q''_w is positive if heat is transferred in the $+r$ direction, i.e., out of the pipe, and negative when transferred into the pipe. $F=1$ is the case of adiabatic wall. $F=0$ is the case where all the heat generated within the fluid is transferred out of the pipe.

The difference between the wall temperature t_w and the fluid bulk temperature t_f is given by

$$t_w - t_f = \frac{q''' r_0^2}{k} \frac{11F - 8}{48} \tag{10-22}$$

Figure 10-9 is a plot of the dimensionless temperature difference $(t - t_c)/(q''' r_0^2/2k)$, where t_c is the fluid temperature at the pipe center line, versus r/r_0 for different values of the dimensionless function F. It is interesting to note that for an adiabatic wall, and even for moderate fractions of heat generated that is transferred out (such as $F = 3/4$), the temperature rises toward the wall, reflecting the effect of slower fluid motion there.

The Case of Turbulent Flow

The general equation of heat transfer in a pipe for this case may be obtained by modifying Eq. 10-17 to include an effective thermal conductivity k_e as given by Eq. 9-14 and rearranging. Thus

$$(\alpha + \epsilon_h)\left(\frac{\partial^2 t}{\partial r^2} + \frac{1}{r}\frac{\partial t}{\partial r}\right) + V\frac{\partial t}{\partial z} = \frac{q'''}{\rho c_p} \tag{10-23}$$

where ϵ_h is the local thermal eddy diffusivity at radius r, in square feet per hour.

This equation has been solved by Poppendiek for an established flow region, uniform heat flux, and constant volumetric thermal source strength (and therefore constant axial temperature gradient) and for the same two boundary conditions as those used in the case of laminar flow. The solution was arrived at by separating the general equation into two simple boundary-value equations for the following cases:

1. A flow system with uniform wall heat flux but no volumetric heat source, that is, $q''_w =$ const and $q''' = 0$
2. A flow system with no wall heat flux but with uniform volumetric heat source, that is, $q''_w = 0$ and $q''' =$ const

The two solutions of the above cases are then superimposed to give the general solution of the original equation. For example, the difference between the temperature at some radius r and the temperature at a reference point is equal to the sum of the two temperature differences given by the two solutions for the same radius. The two solutions will now be presented.

Solution 1. This is the conventional solution of the usual case of heat transfer in turbulent-flow systems with no heat source and uniform wall heat flux. Temperature and velocity distributions for this case have already been discussed. The wall heat flux is expressed by the equation

$$q''_w = h(t_f - t_w) \tag{10-24}$$

where h is the convective-heat-transfer coefficient for the particular fluid and operating conditions used. h is given by the Nusselt number of the system as follows:

$$h = \mathrm{Nu}\,\frac{k}{2r_0} \tag{10-25}$$

Solution 2. The solution for the radial heat flux distribution in this case is given in Fig. 10-10 for different values of the Reynolds number.

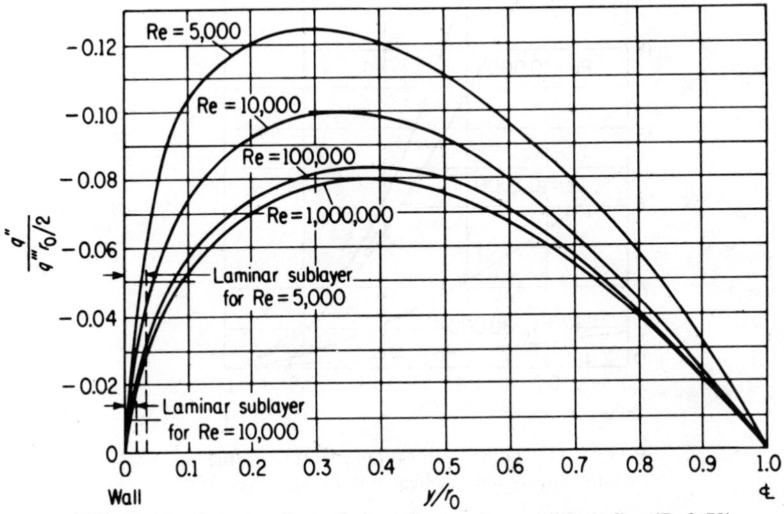

FIG. 10-10. Solution for turbulent flow and no wall heat flux (Ref. 78).

The ordinate is the dimensionless heat flux $q''/(q'''r_0/2)$, where q'' is the heat flux at any radial position r. Note that the ordinate values are always negative, meaning that heat flux is toward the pipe center line, and that it is equal to zero at the wall (as stipulated in this solution) and at the pipe center line (because of symmetry). The solution for the radial temperature distribution in this case is given by Fig. 10-11 in which a dimensionless temperature difference is plotted against radial position for various values of the Reynolds number, but only for a Prandtl number equal to 0.1. Figure 10-11 thus shows the general slope of the temperature profile within the pipe.

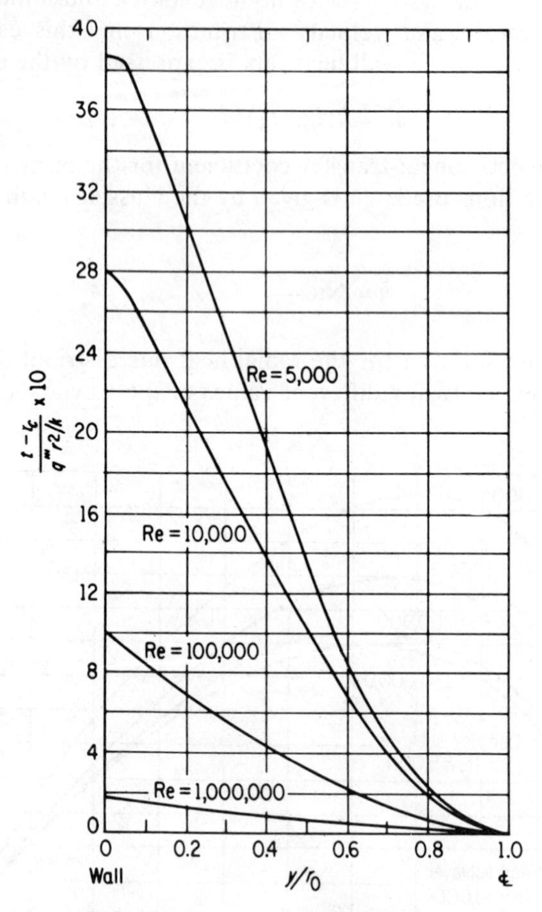

FIG. 10-11. Dimensionless temperature profile in turbulent flow and no heat flux at wall (Ref. 78).

The difference between the pipe wall and fluid bulk temperatures for case 2 only (compare with that given by Eq. 10-22 for the case of laminar flow) is given by Fig. 10-12 in which a dimensionless temperature difference between wall and fluid bulk temperatures is plotted versus Reynolds number for different values of the Prandtl number.

Again, the rise of the mixed mean temperature in the general case is obtained by adding the temperature rise in Eq. 10-24 to that obtained in Fig. 10-12.

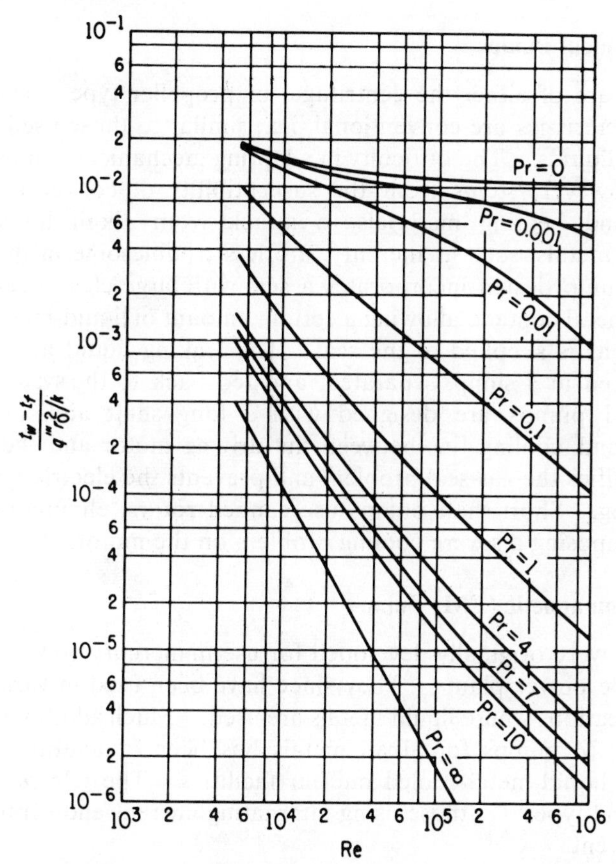

FIG. 10-12. Wall-bulk temperature difference for various Reynolds and Prandtl numbers in turbulent flow and no wall heat flux (Ref. 78)

10-10. LIQUID-METAL PUMPS

It has been shown (Sec. 9-2) that the pumping requirements of liquid metals are generally lower than those of other coolants (with the exception of water). Sodium and NaK have the lowest requirements among the liquid metals. Sodium also shows the highest convective-heat-transfer coefficient h, followed by Pb and PbBi and by NaK, in that order. The much lower pumping requirement of NaK, however, allows it to be circulated at a higher velocity, resulting in a higher value of h, while still having a lower pumping requirement than PbBi. (Because of their low specific heat, however, Na and NaK show the lowest amount of heat transported per unit of liquid temperature rise of all the liquid metals.)

Liquid-metal pumps are of two types: (1) mechanical and (2) electromagnetic.

Mechanical Pumps

These are of either the centrifugal or propeller type. Their design and characteristics are conventional, i.e., similar to those used for water and other liquids. The difficulty in adapting mechanical pumps to liquid metals, however, stems from the susceptibility of conventional shaft-packing and bearing materials to attack from alkali liquid metals. Nonalkali metals, such as mercury, are less troublesome in this respect.

One pump design incorporates a seal with large clearances to avoid metal-to-metal contact, allowing a certain amount of liquid-metal leakage. An inert gas is supplied to the seal. The leaking liquid metal and gas are collected in a sump, separated, and fed back to the system. Some mechanical pumps are designed with a long shaft and an inert-gas chamber and cooling fins between the driving motor and the impeller. This simplifies the gas-seal problem and prevents the electric motor from overheating. Short-shaft pumps use canned rotors, eliminating the gas seal but imposing a severe cooling problem on the motor.

Electromagnetic (EM) Pumps

These were originally developed for use in mercury boilers of binary-vapor-cycle power plants. They since have been used in various industrial applications where liquid metals are used. Much additional research work on EM pumps for alkali metals has been inaugurated since the advent of liquid-metal-cooled nuclear facilities. The EM pumps are oi two general types: (1) those using alternating current and (2) those using direct current.

Alternating-current pumps are used only where dc and mechanical pumps are undesirable. They have the advantage of utilizing common power sources (dc pumps usually require special large-current-low-voltage power) but are generally more complex and more expensive and require special cooling. Except in the large sizes, they are usually inferior to dc pumps in size, mass, and efficiency.

10-11. DIRECT-CURRENT PUMPS*

These are also known as dc Faraday pumps. They operate on the same principles as those of the dc motor. Their theory is derived from the basic principles of electromagnetism.

Figure 10-13 shows a wire conductor carrying current of I_e amp in a magnetic field of flux density of B gauss. It has an effective length, within the magnetic field, of a cm. In the direction shown, it is

* This section is based on material in Ref. 60.

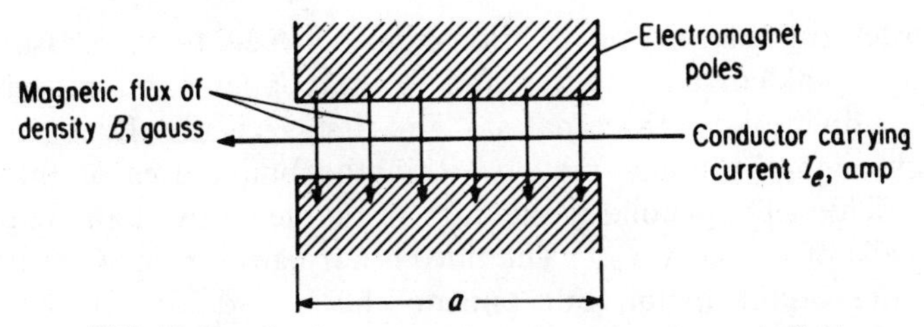

FIG. 10-13. Conductor carrying current in a magnetic field.

subjected to a force of F dynes, acting to move it perpendicular to and out of the plane of the paper (the right-hand rule). F is given by

$$F = \frac{BI_e a}{10} \tag{10-26}$$

Since liquid metals are good conductors of electricity, the above conductor can be replaced with liquid metal contained in a rectangular duct of length a and height b cm (Fig. 10-14). The current I_e is maintained within the liquid metal via copper-bar conductors brazed to the sides of the duct. The same force F is applied to the liquid metal, moving it through the duct perpendicular to and out of the plane of the paper. We thus have a liquid-metal pump with no moving parts.

The pressure rise in the pump, Δp dynes/cm², is given by

$$\Delta p = \frac{F}{\text{cross-sectional area of duct}} = \frac{F}{ab} = \frac{BI_e a}{10ab} = \frac{BI_e}{10b} \tag{10-27}$$

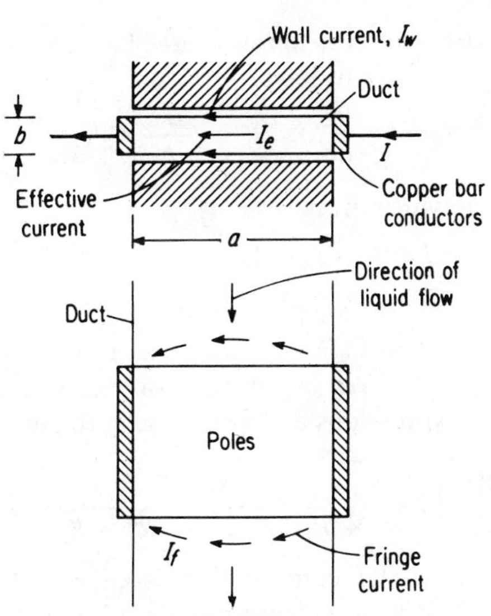

FIG. 10-14. Liquid metal carrying current in a magnetic field.

In order to develop the relationships between flow, pressure, and power, the total current input I, fed to the copper bars, must first be defined. I divides into three parts. The first, I_e, is the effective current passing through the liquid metal, within the boundaries of the strong magnetic field. The second is the current that leaks through the top and bottom walls of the duct, I_w. The third is a fringe current, I_f, that passes through the liquid metal, but outside the boundaries of the strong magnetic field, in the vicinity of the duct entrance and exit to the pump system.

The corresponding network of this system is shown in Fig. 10-15.

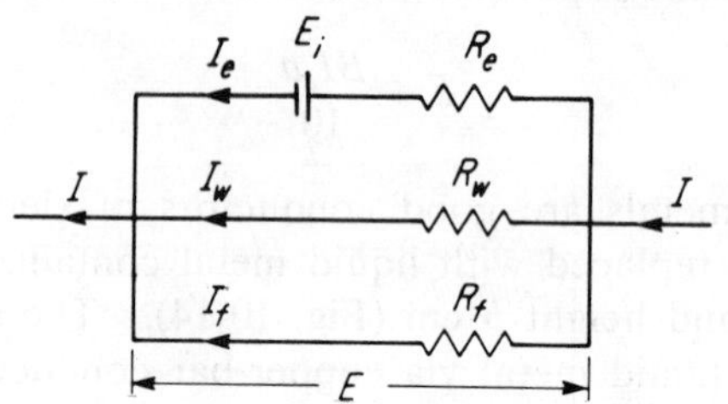

FIG. 10-15. Electrical network of EM pump.

E is the overall voltage drop between the bars. E_i is an induced voltage, developed whenever a conductor of electricity is moving in a magnetic field. If V is the velocity of the conductor, in centimeters per second, then

$$E_i = BaV \times 10^{-8} \text{ volt} \tag{10-28}$$

This expression can be written in terms of a volumetric flow rate (capacity) q cm³/sec, where $q = Vab$, as follows:

$$E_i = Ba \frac{q}{ab} \times 10^{-8} = \frac{Bq}{b} \times 10^{-8}$$

Solution of the network gives

$$I = I_e + I_w + I_f$$
$$E = I_e R_e + E_i = I_w R_w = I_f R_f$$

where R_e, R_w, and R_f are resistances, in ohms, in the effective, wall, and fringe paths, respectively. An expression for q can now be obtained by combining the above expressions and rearranging to give

$$q = \frac{10^8 b}{B}\left[I \frac{R_w R_f}{R_w + R_f} - \frac{10 \Delta p b}{B}\left(\frac{R_w R_f}{R_w + R_f} + R_e\right)\right] \tag{10-29}$$

where $R_w R_f/(R_w + R_f)$ is the combined resistance in the walls and the fringes. This equation is now rewritten by adding the necessary conversion factors to transform it into engineering units. Thus

$$q' = \frac{4 \times 10^6 b'}{B}\left[I\frac{R_w R_f}{R_w + R_f} - \frac{1.75 \times 10^6 \Delta p' b}{B}\left(\frac{R_w R_f}{R_w + R_f} + R_e\right)\right] \tag{10-30}$$

where now q' is in gallons per minute, b' is in inches, and $\Delta p'$ is in pounds force per square inch. This equation can be rearranged to give an explicit expression for the pressure rise:

$$\Delta p' = \frac{5.7 \times 10^{-7} BI}{b'} \frac{R_w R_f}{R_w R_f + R_e R_w + R_f R_e} - \frac{B^2 q'}{7 \times 10^{12} b'^2 [R_w R_f/(R_w + R_f) + R_e)]} \tag{10-31}$$

These relationships show that the pressure-capacity characteristics of a dc EM pump are linear with a negative slope for constant current input. (This contrasts with the curved, constant-rpm characteristics of a centrifugal pump.) They also show that the capacity is a direct function of the current at any one pressure or head. Note that, as given, the expressions do not include the duct width as a parameter, although it affects the various resistances in the system. The actual pressure developed in the pump is less than that given above by the hydraulic losses in the duct. These losses should be evaluated separately. Figures 10-16 and 10-17 show the characteristics of a 300-gpm, 40-psi (design point) EM pump used in the Experimental Breeder Reactor, EBR-I [1].

The wall and fringe-path resistances should be as high as possible. The duct walls must therefore be thin, of a material of high electrical resistivity as well as high resistance to attack by liquid metals. Ducts are usually made of Series 300 stainless steel, Nichrome, or Inconel X, and are formed by pressing thin-walled tubing $\frac{1}{40}$ to $\frac{1}{16}$ in. thick into the

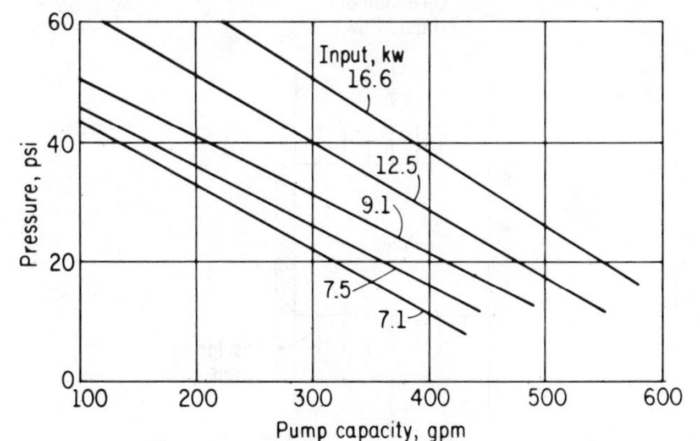

FIG. 10-16. Head-capacity characteristics of dc pump used in EBR-I (Ref. 60).

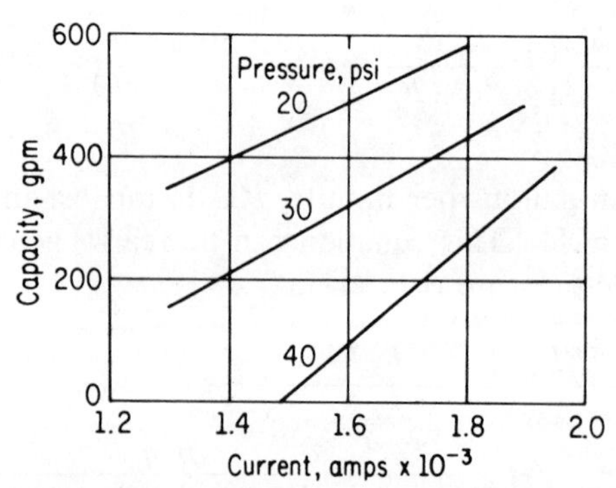

FIG. 10-17. Capacity-current characteristics of dc pump used in EBR-I (Ref. 60).

required shape. R_f may be increased by installing insulating vanes in the fringe-path area, parallel to flow (Fig. 10-18).

R_e and R_w are calculated from the characteristics and dimensions of the liquid metal and walls. R_f is more difficult to evaluate and is usually measured experimentally. The resistance between the copper bars and the liquid metal in the duct should be kept to a minimum by good bonding and by keeping the walls thin. The bars are usually made of silver or nickel alloy and are brazed to the side walls. A design using closed boxes welded to the side walls and filled with liquid metal has proven successful at high temperatures.

The current requirements of dc pumps are quite large, ranging between 1,000 amp for small (5 to 10 gpm) pumps to many thousands of amperes for higher capacities. Voltage drops, on the other hand, are extremely low, being of the order of 1 or 2 volts. The large-current-low-

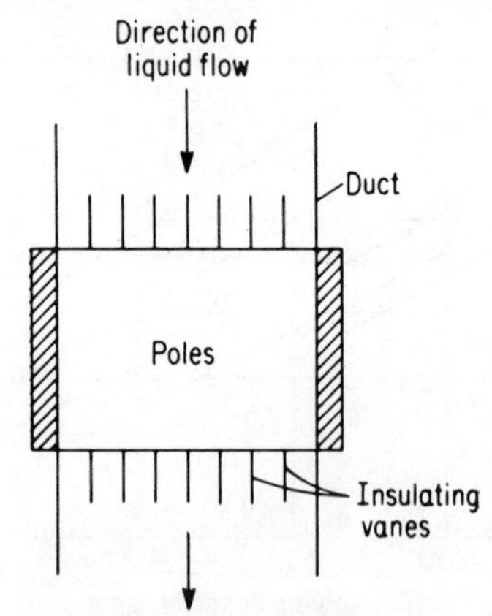

FIG. 10-18. Insulating vanes to reduce fringe current.

voltage requirement is not easily obtained. Transformer-rectifier systems operating under such conditions are very inefficient. However, *homopolar* or *unipolar* generators [79, 80] utilizing NaK brushes have been designed with efficiencies greater than 80 percent and up to 100,000-amp capacities.

The power requirement, W watts, of a dc pump, excluding that required for the magnetic field windings, is

$$\begin{aligned} W &= EI \\ &= 2.5 \times 10^{-7} \frac{Bq'I}{b'} + 1.75 \times 10^{6} \frac{\Delta p' b' I R_e}{B} \end{aligned} \tag{10-32}$$

The hydraulic power developed, W_h, is given by

$$W_h = 0.435 \, \Delta p' q' \tag{10-33}$$

The efficiency of the pump, η, excluding the power necessary to produce the magnetic field, is therefore given by

$$\eta = \frac{0.435 \, \Delta p' q'}{2.5 \times 10^{-7}(Bq'I/b') + 1.75 \times 10^{6}(\Delta p' b' I R_e/B)} \tag{10-34}$$

This expression can also be written in a form excluding q' (by substituting for it from Eq. 10-30) as follows:

$$\eta = \frac{1.74 \times 10^{6} \Delta p' b' \{R_w R_f/(R_w + R_f) - 1.75 \times 10^{6}(\Delta p' b'/BI)[R_w R_f/(R_w + R_f) + R_e]\}}{R_w R_f/(R_w + R_f)(BI - 1.75 \times 10^{6} \Delta p' b')} \tag{10-35}$$

Actual efficiencies of dc pumps are 15 to 20 percent for small pumps and 40 to 50 percent for large sizes. Including losses in the generator, bus, etc., overall efficiencies of the pumping system are of the order of 10 to 40 percent, increasing with pump size.

10-12. SOME DESIGN AND OPERATIONAL PROBLEMS OF DC PUMPS

Flux Distortion

This is a problem similar to that encountered in electromagnetic machinery. Figure 10-19 shows the directions of magnetic flux, liquid-metal flow, and current. The current shown is going *into* the plane of the paper, concentrated in one point for simplicity. In this situation, a magnetic field is induced around the wire and distorts the main magnetic flux. The latter is thus strengthened at the duct entrance to the pump and weakened at the duct exit. This effect is not serious in small pumps,

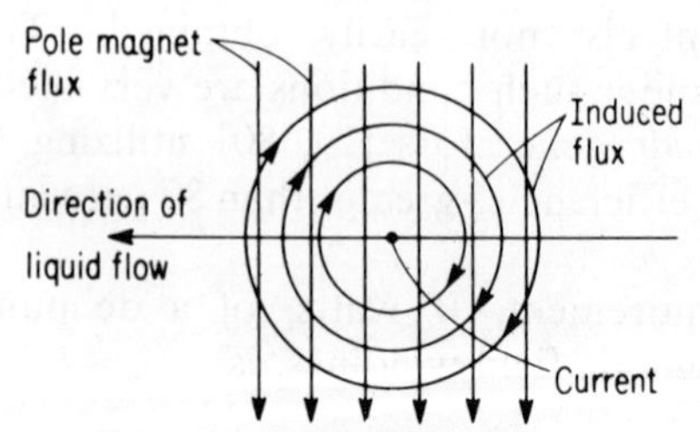

FIG. 10-19. Magnetic-flux distortion.

operating with large magnetic fluxes and small currents. It may, however, seriously affect the operation of large pumps operating with large currents. In such a case, magnetic-field compensation is necessary. This may be accomplished by either one of two modifications:

1. Bringing the current back in a direction opposite to its direction in the liquid metal, with the help of a U-shaped conductor placed within the magnetic poles (Fig. 10-20). This has the effect of creating an induced magnetic field equal and opposite to that causing the distortion.

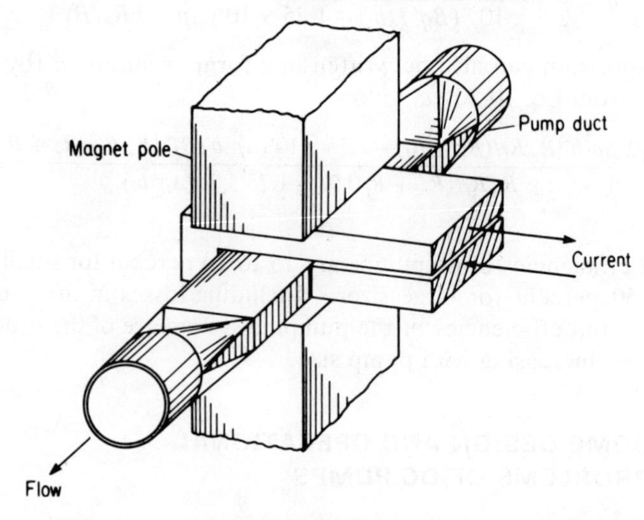

FIG. 10-20. Dc pump, showing U-current flow path for field compensation (Ref. 60).

2. Tapering the magnetic poles and duct, so that the gap between the poles is wider at the duct entrance (where the flux tends to be strengthened) and narrower at the exit (Fig. 10-21). This has the effect of increasing the liquid-metal velocity near the duct exit and causing the induced electromagnetic force (function of velocity) to remain constant along the duct length.

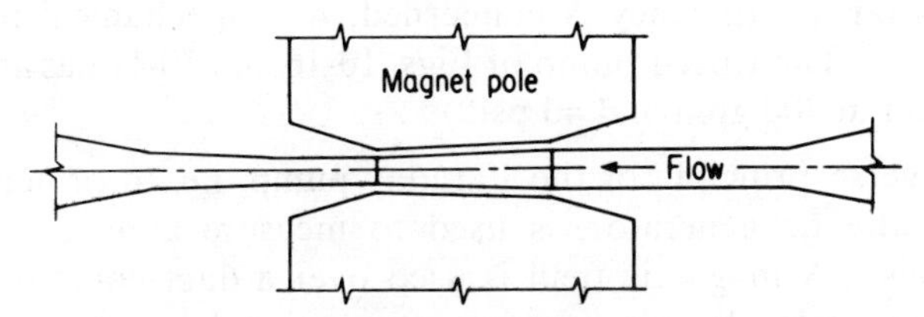

FIG. 10-21. Tapered magnetic poles, duct, and duct overhang for field compensation and fringe current utilization (Ref. 60).

Fringe Current

To make use of the fringe current, a pole overhang is sometimes incorporated in the design of the pump (see Fig. 10-21). The poles are extended beyond the duct length to furnish magnetic flux over the fringe current. An extra taper in the overhang is provided to allow a tapering off of the field intensity with the tapering off of the fringe current. The fringe current, on the other hand, may be reduced by insulating vanes (Fig. 10-18).

Gas Entrainment

Performance of EM pumps is severely impaired by the presence of gas bubbles in the liquid metal, since gas entrainment reduces the electrical conductances of the liquid. Adequate removal of the gas from the system, as well as precaution against gas entrainment, is necessary.

Field Winding

A separately excited field winding of many turns may use available high-voltage current. This presents electrical insulation problems of high-temperature operation and radiation damage. A series-wound field uses the same large current passing through the liquid metal in the pump. This requires the use of only a few turns (down to one or two) of fairly large copper conductors (several inches square).

In general, dc EM pumps are simple and reliable, present a minimum of insulation problems, are suitable for high-temperature work, and require no cooling beyond natural convection and thermal radiation. These pumps have no shaft seal problems, no moving parts, and no electric motor to be shielded against radiation, and they can be located anywhere in the system. They can be controlled fairly easily over a wide range of operating conditions by varying the current input. They do, however, require special current-generating equipment and are competitive, as far as efficiency is concerned, with mechanical pumps only in large sizes. The EBR-I pump of Figs. 10-16 and 10-17 has an efficiency of 42 percent at 300 gpm and 40 psi.

The reverse principle of the Faraday pump, i.e., a principle similar to that of the dc generator, is used to measure and control flow of liquid metals. A magnetic field is used over a duct containing flowing liquid metal. An electric current, proportional to the flow rate, is induced perpendicular to the magnetic flux lines and the direction of flow. This current may then be used to indicate, record, and/or control flow.

10-13. ALTERNATING CURRENT PUMPS

Alternating current electromagnetic pumps are similar to dc pumps except that the magnetic field and the current in the fluid reverse direction in synchronism. The two must be properly phased for maximum pumping work. This can be accomplished by connecting the field winding in series with the liquid current as in the dc pump.

Ac pumps have the advantage of easier supply of the large-current, low-voltage requirement than dc pumps. This requirement is simply met by a step-down transformer. The liquid, current and field interactions of ac pumps are complex and no rigorous analysis is available. Ac pumps have the disadvantages of large power losses and, hence, lower efficiencies than dc pumps. Efficiencies vary from 2 percent for small-size pumps to a little over 20 percent for large pumps. Small pumps, while low in efficiency, have the advantage that they may be operated directly from a single-phase 60-cycle main.

Induction-type electromagnetic pumps develop their large current requirement directly in the liquid by the process of electromagnetic induction. The duct is placed in a region of changing field intensity such that induced currents in it would interact with the main field and pump the fluid. Induction-type pumps may be designed in many geometrical forms. The simplest and most widely used is the *linear induction* pump, Fig. 10-22. Here the flattened duct is placed between two cores which contain three-phase ac winding, similar to that found in the stator of an ac induction motor except that, in the pump, the winding is flat, not circular, and a sliding rather than rotating magnetic field is obtained.

The principles of the ac induction pump are similar to those of the ac induction motor except that the current paths and field distribution are not as well defined, and that the magnetic gap is much larger so that the field losses are higher. Power losses due to the open-ended cores also exist and the efficiency is therefore much lower than for a motor.

The pressures developed in an ac induction pump are roughly propor-

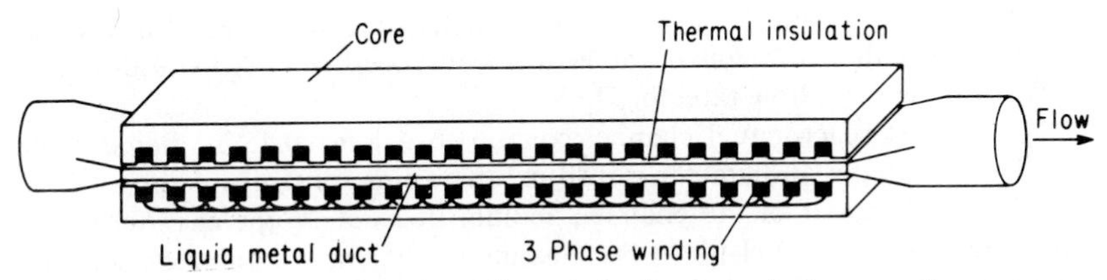

FIG. 10-22. Alternating-current linear induction-type electromagnetic pump.

tional to the relative speed of the liquid and the field. High efficiencies of about 40 percent, comparable to those in dc pumps especially in the larger sizes, are attainable. The greatest advantage of this type is that they do not require special current generating and busing equipment, since the large currents required are developed directly in the liquid. One disadvantage is the necessity of a special cooling system for the cores for operation at high liquid temperatures. The power supply is simply available 60-cycle ac.

Another geometrical arrangement of ac induction is found in the *Einstein-Szilard* pump. In this, the fluid flows through the annulus between cylindrical inner and outer cores containing toroidal windings which produce a moving field similar to the linear induction pump. The induced current in the liquid, however, has a circular path, so that end losses are avoided. This type of pump is more costly and has many construction problems.

Other ac pump types such as the *helical* pump, which consists of a helical duct in which the liquid is forced to rotate and therefore move ahead, and the *rotating-field* pump in which the liquid is driven in a spiral path by a mechanically driven magnet have suffered from fabrication difficulties and low efficiencies and have not been developed fully.

Only the dc pump and the ac linear induction pump have received wide acceptance to date. They have comparable efficiencies. The former operates at low voltages and does not require heavy insulation which is susceptible to thermal and radiation damage. The latter, on the other hand, has the convenience of not requiring a special power supply.

PROBLEMS

10-1. A hypothetical liquid metal has $Pr = 0$ and $k = 50$ Btu/hr ft °F. It flows through a 1-in.-diameter tube with a Reynolds number of 1,000,000. The tube-wall temperature is 1,000°F. The temperature of the fluid halfway between the wall and the centerline is 900°F. Find (a) the temperature at the centerline of the tube, (b) the bulk temperature of the fluid, and (c) the heat flux in Btu/hr ft^2.

10-2. Liquid sodium flows in a 1 in. diameter tube at 932°F. For the case of constant heat flux, the convective heat-transfer coefficient is 5,000 Btu/hr ft^2 °F. What is the mass flow rate, lb$_m$/hr?

10-3. A fast reactor fuel channel contains 6 ft long, 0.45 in. diameter fuel pins. The fuel material diameter is 0.40 in. The pins are staggered on a 0.54 in. pitch. If at the center of the channel, sodium flows at 20 fps and 700°C, find the heat generated per fuel pin if the maximum temperature difference between cladding and coolant is 25°F. Ignore the extrapolation lengths.

10-4. A sodium-cooled reactor core contains plate-type fuel elements, separated by plate-type graphite moderator with 0.30 in. gaps. The fuel material thickness is 0.25 in. If at a particular position in the core, sodium flows at 20 fps and 1110°F, find the maximum volumetric thermal source strength in the fuel if the cladding temperature is not to exceed 1160°F at the same position.

10-5. Liquid sodium flows in a 1 in. diameter tube at 3.76 fps. At a particular cross section, the sodium was at 500°C and the tube walls were at 400°C. Find the variations in sodium temperature, wall temperature, and heat flux, Btu/hr ft^2, for the cases of (a) uniform heat flux, and (b) uniform wall temperature.

10-6. A 0.60 in. diameter fuel rod is axially situated in a 1.2 in. diameter channel, the surfaces of which may be considered insulated. Sodium flows through the channel at 0.25 fps and 500°C. Find the percentage error in the heat transfer coefficient if axial heat diffusion in the coolant is not taken into account.

10-7. A fast fluid-fueled reactor uses uranium metal dissolved in liquid bismuth with a U^{235} density of 10^{20} nuclei/cm^3. The fuel thermal conductivity is 9 Btu/hr ft °F. The fission cross section is 5 b. The core can be approximated by a cylinder 3 ft in diameter. At a particular plane in the core, the neutron flux was flattened to 10^{13} and the fuel bulk temperature is 782.5°F. If the fuel flow is assumed to be laminar, find for that plane (a) the wall temperature in case the walls were adiabatic, (b) the centerline temperature for the preceding case, and (c) the percent heat generated that must be removed if, for structural reasons, the wall temperature should not exceed 700°F.

10-8. A dc electromagnetic pump is to produce a 100 psi pressure rise for sodium flowing in a 5 × 25 cm duct. The poles are 25 × 25 cm. A homopolar generator capable of generating 2000 amps at 2 volts is available. Ignoring all but the effective current, and ignoring all the mechanical and electrical losses (i.e. 100 percent efficiency) find the necessary magnetic flux density (gauss), and the flow (gpm). 1 dyne/cm^2 = 1.45×10^{-5} psi. Electrical resistivity of liquid sodium (at 350°C) is 18.44×10^{-6} ohm-cm.

10-9. A liquid-metal dc electromagnetic pump consumes 4,000 amps at 2 volts. The magnetic flux density is 50,000 gauss and the duct is 50 × 10 cm in cross section. The resistance in the liquid metal path is 10^{-5} ohm. The pressure rise in the pump is 25 psi. Find (a) the total fringe and wall currents, and (b) the liquid metal flow in cm^3/sec.

chapter 11

Heat Transfer with Change in Phase

11-1. INTRODUCTION

The importance of the phenomena of heat transfer and fluid flow with change in phase (from liquid to vapor) of a coolant in nuclear work arises from the fact that many reactors use a liquid coolant that is subjected to high heat fluxes. In boiling-type reactors, the change in phase is deliberately designed into the system and should be controlled. In other liquid-cooled reactors, such as the pressurized-water reactor and organic-cooled reactors, local boiling at the fuel surface may be allowed at high loads to increase heat transfer. The change in phase of the coolant bulk in these reactors is, however, avoided in normal operation but may be allowed only under emergency and transient conditions. When the coolant acts also as moderator, bubble formation accompanying phase change reduces the moderating power of the coolant. Also, liquid-cooled reactors are heat flux-limited by the nature of the boiling mechanism (burnout).

In any case the behavior of a coolant undergoing a change in phase must be understood. This chapter discusses the mechanism of boiling and its reverse, condensation. The processes of boiling in a flow system are discussed in the next chapter.

11-2. PROCESSES OF PHASE CHANGE

Some of the processes associated with change in phase from liquid to vapor are as follows:

Evaporation

This is simply the process of conversion of liquid into vapor.

Boiling

This is the process in which vapor forms within a continuous liquid phase. Boiling takes many forms. It can be classified as *pool* boiling and *volume,* or *bulk*, boiling. Pool boiling is the process in which the formation of vapor is due to heat added to the liquid by a surface in contact with or submerged within the liquid. Volume boiling occurs within the bulk of the liquid because of heat generation within the liquid by chemical or nuclear reaction (such as in homogeneous fluid-fueled *water-boiler* reactors).

Boiling is further classified as *nucleate* and *film* boiling. These are shown in Fig. 11-1. In nucleate boiling, bubbles of vapor are formed within the liquid around a small nucleus of vapor or gas. (Vapor-bubble formation and growth will be discussed in more detail later.) Thus we may have *pool nucleate* boiling or *volume nucleate* boiling. Film boiling, on the other hand, is the formation of a continuous film of vapor that blankets the heating surface. Film boiling is, therefore, usually associated only with pool rather than volume boiling. Under certain conditions, nucleate and film boiling coexist. This state has been given many names; among them are *partial film* boiling, *partial nucleate* boiling, *transition film* boiling, *unstable nucleate film* boiling, and others.

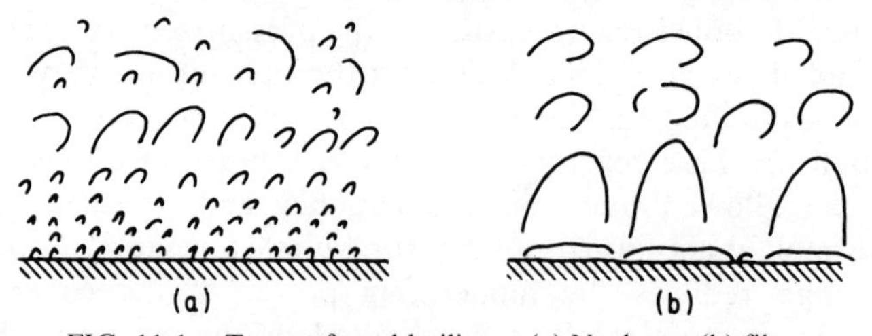

FIG. 11-1. Types of pool boiling. (a) Nucleate; (b) film.

Boiling may be further classified as *saturated* and *subcooled,* or *surface,* boiling. These two classifications refer to the temperature of the bulk of the liquid. In saturated boiling, the bulk of the liquid is at the saturation temperature corresponding to the pressure of the system (e.g., for water it is 212°F at atmospheric pressure and 544.6°F at 1,000 psia). In subcooled boiling, the bulk of the liquid is at lower temperatures than those given above. It should be noted that, to generate a bubble in either case, the heating surface must be at a temperature higher than saturation, and consequently some of the liquid, which is immediately adjacent to that surface, is also at higher than saturation temperatures (Sec. 11-3). In other words, we have some superheated liquid, although the bulk of the liquid may be saturated or subcooled. In saturated boiling, the bubbles formed rise to the liquid surface where they are detached. In subcooled boiling, the bubbles begin to rise but may collapse before they reach the liquid surface. In nonflow systems the effect is continuously to heat the liquid toward saturation, where the first form, saturation boiling, takes place. Both saturated and subcooled boiling may be of either the nucleate or film type. However, volume boiling is normally only of the saturated type. If a homogeneous heat-generating fluid is in a subcooled state, it must reach the saturation state first before volume boiling takes place.

Two-phase Flow

This is the case in which both the vapor and liquid move together in a channel. Two-phase flow may be classified as *diabatic* or *adiabatic,* that is, boiling may or may not take place. Examples of diabatic flow are to be found in the riser tubes in water tube steam generators and in coolant channels between nuclear fuel elements in a boiling-water reactor. In these cases, vapor is continuously formed as the mixture of liquid and vapor rises. Two-phase flow may be further classified as *single component* flow as in the above cases with water and steam, or two-*component* flow, as with water and air. In all cases the velocities of the two phases are usually not equal, but the vapor rises faster than the liquid. It is said that *slip* exists between the vapor and the liquid. The ratio of the velocities of the vapor and the liquid is called the *slip ratio.* No slip means that the slip ratio is 1. Two-phase flow will be discussed in more detail in the next chapter.

Condensation

This is the reverse of evaporation, or simply the process of conversion of vapor back to a liquid. Condensation may be *filmwise* when there is a continuous flow of liquid over the cooling surface, or *dropwise* when the vapor condenses in drops and the cooling surface is not completely covered by liquid.

11-3. BUBBLE FORMATION, GROWTH AND DETACHMENT

The processes of bubble formation and growth (also called nucleation), and detachment in pool boiling will now be discussed.

Nucleation Aids

Bubbles are formed due to liquid superheat adjacent to the heating surface. The degree of liquid superheat, and therefore the formation mechanism is aided by (1) gas or vapor present in the liquid, (2) surface scratches or cavities in the heating surface, and (3) the wetting characteristics of the surface. These are called *nucleation aids* or *centers.*

Gaseous nucleation centers may be in the form of dissolved air coming out of solution in the heated liquid. In a nuclear reactor, the number of nucleation aids is increased, over those existing in a conventional boiler, by the presence of ionizing radiations and by oxygen and hydrogen resulting from the radiolysis of water in water-cooled reactors. The ions acting as nucleation centers further help bubble motion because of the electrical repulsive forces of the now-charged bubbles. Experiments in boiling-water reactors have shown that they have higher heat-transfer coefficients than those encountered in conventional boiling apparatus.

Scratches and *cavities* act as nucleation centers because they entrain vapor and can seldom be completely filled with liquid because of surface tension effects. The crevices are also centers of relatively high temperatures, a characteristic that aids nucleation. The growth and detachment of a bubble from a crevice within a wetted surface are illustrated in Fig. 11-2.

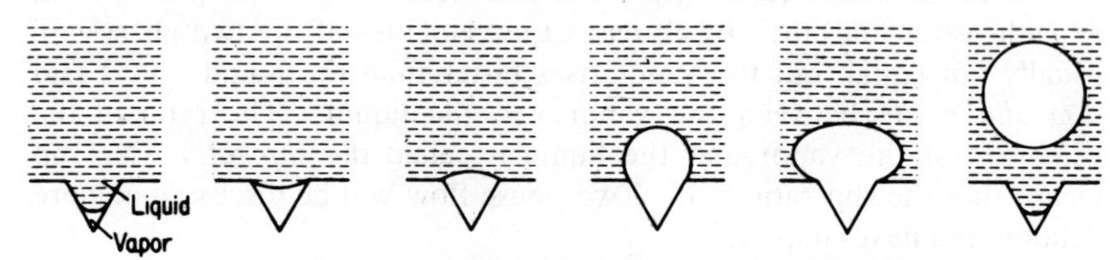

FIG. 11-2. Bubble growth and detachment from a crevice.

The third nucleation aid depends upon the *wetting characteristics* of the heating surface. The degree of wetting of a surface may be illustrated with reference to Fig. 11-3 in which a vapor bubble at the heating surface is shown. Three surface-tension forces have to be considered in studying the equilibrium of this bubble. These are:

σ_{fg} = surface tension acting in the liquid-vapor interface for a free bubble. The magnitude of this surface tension depends upon the properties of the liquid and vapor.

σ_{fs} = surface tension acting in the liquid-surface interface, along the heating surface. It depends on the properties of the liquid and the surface.

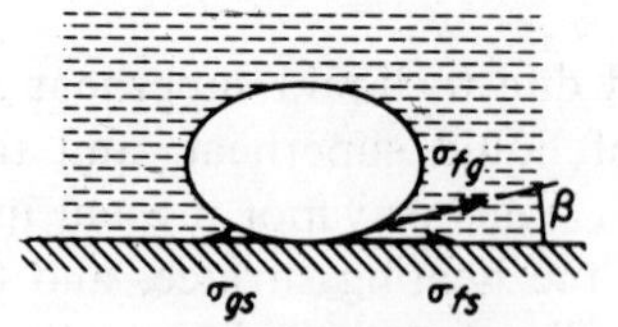

FIG. 11-3. Bubble attached to heating surface.

σ_{gs} = surface tension acting in the vapor-surface interface, along the surface. It depends on the properties of the vapor and the surface.

Considering the direction along the surface, the following relationship holds:

$$\sigma_{gs} = \sigma_{fs} + \sigma_{fg} \cos \beta \tag{11-1}$$

where β is the angle that the vapor bubble makes with the surface. The magnitude of this angle depends upon the relative magnitudes of the three surface tensions. If $\sigma_{gs} = \sigma_{fs}$, then $\beta = 90°$. This is the borderline case between wetted and unwetted surfaces: If β is less than 90°, the surface is said to be wetted, and if β is greater than 90°, the surface is considered unwetted, Fig. 11-4. The degrees of wetting and unwetting depend upon how far β deviates from 90°. Note that the terms wetting and unwetting refer to the liquid phase.

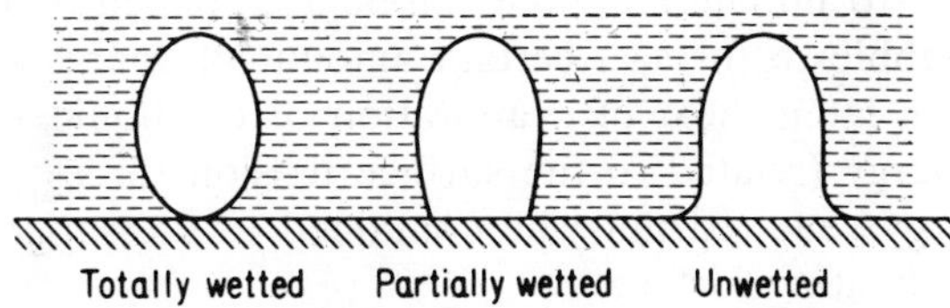

FIG. 11-4. Effect of heating-surface condition.

Bubble Growth

The dynamic *growth* of a vapor bubble strongly depends upon the contact angle β. Maximum rate of growth occurs when the surface is totally unwetted. (In this case the degree of superheat required to initiate a bubble is least.) From a heat-transfer standpoint, however, a surface that is poorly wetted over its entire area is not suitable because the vapor bubbles are difficult to detach from the surface, grow rapidly, coalesce, and form a continuous film barrier to heat flow from the heating surface to the bulk of the liquid.

A desirable boiling surface from a wetting point of view is one that is partially wetted (β less than 90°) over most of its area, with the rest of the area composed of tiny unwetted patches which act as nucleation centers and where bubbles begin to form. As they grow, they overlap to a wetted surface where they can detach. Unwetted patches on a generally wetted surface are usually caused by impurities in the surface material or by gas adsorption at a point where the material crystal lattice is faulty.

Another desirable boiling surface would be one that is also partially wetted over most of its area but contains scratches or tiny cavities which act as nucleation centers.

Naturally, not all the nucleation centers are equally effective. For example, a surface with a normal finish contains a wide spectrum of cavity sizes. It is a common experience to see bubbles come out of only the same few nucleation centers at low heat fluxes and an increasing number of nucleation centers come into play as the heat flux is raised. The degree of effectiveness of each, dependent on its physical characteristics, determines the number of degrees of superheat of the adjacent liquid necessary for bubble formation. A superheated liquid is one whose temperature is higher than the saturation temperature corresponding to its pressure.

Liquid Superheat

That the liquid adjacent to the heating surface must become superheated in order to initiate a bubble can be ascertained by noting that in order for a bubble to form, a certain amount of energy is required to supply both the latent heat of evaporation of the liquid vaporized and the surface energy (related to the surface tension forces of the liquid). The total energy required must be contained in the superheated liquid adjacent to the heating surface.

The degree of liquid superheat necessary to form a bubble may be very high in a clean system. A clean system is defined as one in which no vapor or gases exist and in which the heating surface is perfectly smooth (such as clean, unscratched glass) and thus contains little or no nucleation aids. Such a system does not exist in practice, and the degree of superheat required is much less than that required by a clean system.

The degree of superheat associated with an idealized conical cavity is believed [81] to be given by

$$T - T_{sat} = \frac{2\sigma T_{sat}}{\lambda \rho_g r_c} \tag{11-2}$$

where T and T_{sat} are the superheat and saturation liquid temperatures, σ is the surface tension, λ the latent heat of vaporization, ρ_g the density of the vapor and r_c the radius of mouth of the cavity. Liquid superheat results in a steep temperature gradient near the heating surface. Figure 11-5 shows a temperature profile encountered in water boiling above a horizontal heating surface at atmospheric pressure. Under these conditions the degree of superheat is seen to be about 16°F.

In the case of *liquid metals,* the situation is complicated by the fact that, due to the low Prandtl numbers, temperature profiles are much less steep at the heating surface than in the case of nonmetallic liquids (Sec.

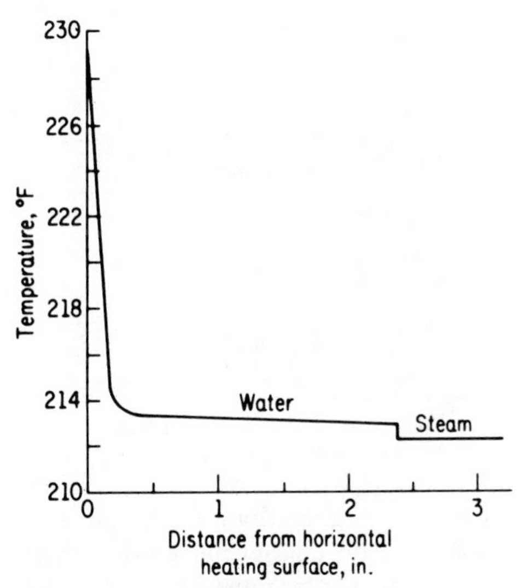

FIG. 11-5. Water temperature above heating surface at atmospheric pressure (Ref. 35).

9-4). Also, liquid metals (the alkali ones) generally have good wetting characteristics. This in turn makes cavities, especially the larger one, more likely to be flooded, so that there is a dearth of nucleation centers. All these combine to make the degree of superheat required to initiate boiling much greater in liquid metals than in other liquids. Experimental data has been obtained for the case of liquid potassium in forced-convection flow [82]. The degrees of superheat were found to decrease with pressure and to increase with the degree of subcooling. Measured degrees of superheat ranged from 17 to 117°F for the boiling pressure range 8–24 psia (corresponding to approximately 1300–1600°F). A correlation, also based on an idealized conical cavity, was proposed.

A small degree of superheat is also associated with liquid adjacent to a free vapor bubble. Figure 11-6 shows a vapor bubble of diameter D, free-floating within a continuous liquid phase. The forces acting on that bubble are (1) the force due to the pressure of the vapor (vapor pressure) p_g (lb_f/ft^2) within the bubble, (2) the force due to the pressure of liquid p_f (lb_f/ft^2) acting on the bubble from the outside, and (3) the surface tension force σ_{fg} (lb_f/ft) on the liquid-vapor interface of the bubble. Considering one-half of the bubble, a balance of the forces acting on it is given by

$$\frac{\pi D^2}{4}(p_g - p_f) = \pi D \sigma_{fg}$$

or

$$p_g - p_f = \frac{4\sigma_{fg}}{D} \tag{11-3}$$

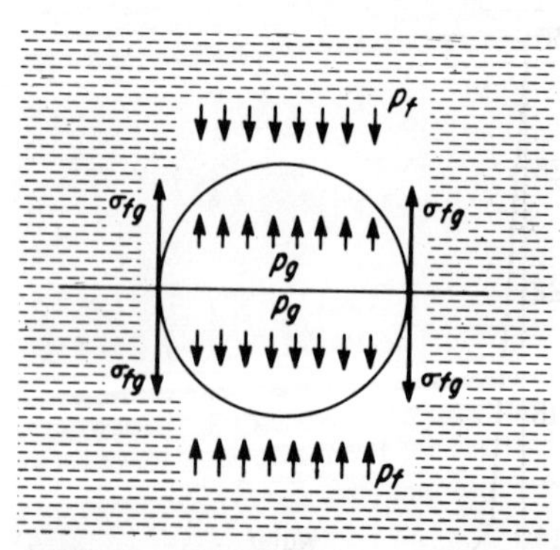

FIG. 11-6. Forces on a free bubble within a liquid.

Thus the vapor pressure is higher than the liquid pressure. Since both the vapor and liquid must be in thermal equilibrium (i.e., at the same temperature), and since the vapor must at least be saturated, the liquid must be at a temperature greater than the saturation temperature corresponding to pressure p_f, and is therefore superheated.

Bubble Detachment

The process of bubble *detachment* from the heating surface causes liquid agitation and therefore has a great effect on the heat-transfer coefficient in boiling-heat transfer. A bubble detaches when it grows large enough so that its buoyancy just overcomes the capillary force holding it to the heating surface. The greater the ratio between the bubble volume and the contact area, the more easily can the bubble detach from the surface. Thus a wetted surface makes it easier for bubbles to detach themselves than a nonwetted surface. It is an experimental fact that a larger number of smaller bubbles are generated and detached from a generally wetted than nonwetted surface.

Figure 11-7 shows the agitation of the liquid phase caused by bubble formation, growth, and detachment (or collapse in subcooled boiling). This agitation causes the hot liquid in the vicinity of the heating surface to be pushed away from that surface and, as the bubbles detach, colder liquid to rush back and fill the resulting void. Thus there is a continuous mixing which increases as more bubbles are formed (i.e., as the heat flux is increased) and which is the reason for the large heat-transfer coefficients possible with nucleate boiling (below). As the bubbles rise in the liquid they may remain separate (at low heat fluxes) or they may interfere with others (as their number increases, at high heat fluxes). In the latter case, an oscillating vapor column is formed, causing the heat-transfer coefficient to increase with heat flux.

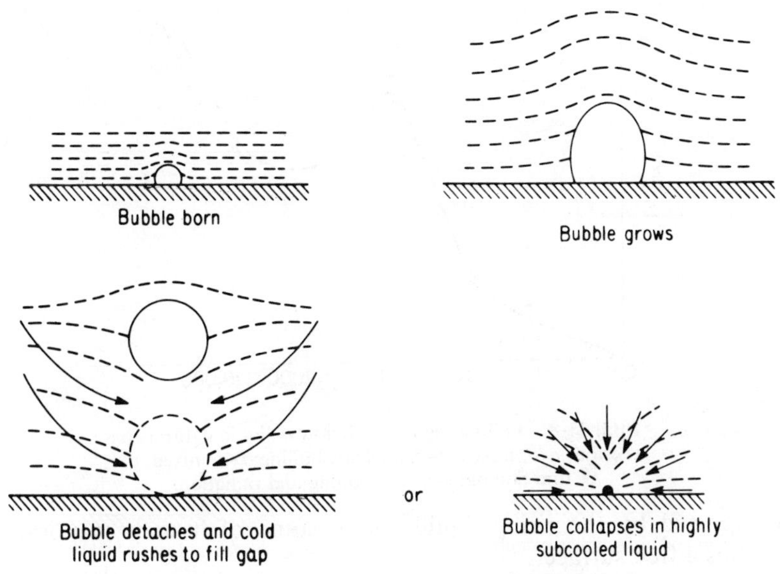

FIG. 11-7. Liquid agitation in boiling.

11-4. THE BOILING REGIMES

Figure 11-8 shows the conventional representation of the heat flux q'' (rate of heat transfer per unit surface area) versus Δt, the difference between the surface and the liquid bulk temperatures. The curve is for nonflow pool saturated boiling, so that the liquid bulk is at the saturation temperature corresponding to its pressure and Δt is the wall superheat. Because of the large changes in the magnitudes of both q'' and Δt, the curve is usually plotted on log-log coordinates. The heat-transfer coefficient h can be plotted on the same graph by dividing q'' by the corresponding Δt ($h = q''/\Delta t$). The resulting curve would have the same general shape as the q''-Δt curve but would be displaced from it. There are several regimes whose existence was originally suggested by Nukiyama [83] and later verified experimentally by McAdams et al. [84] and Farber and Scorah [85]. The curve may be divided into six regimes. The discussion in this section, unless otherwise noted, assumes that Δt is the independent variable and can be closely controlled.

In the regime preceding *a*, the heating-surface temperature is nearly at, or only a few degrees above, the liquid temperature. Thus there is insufficient liquid superheat and no bubble formation. Heat transfer in this regime is therefore by liquid natural convection only. In natural convection, q'' is approximately proportional to Δt to the 1.25 power and consequently increases slowly with it. The heat thus transferred is

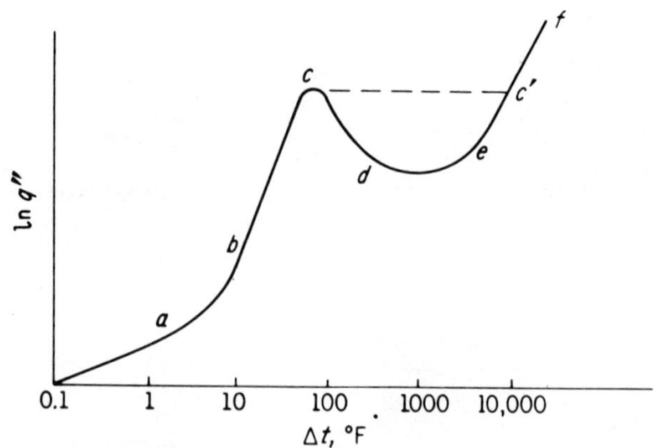

FIG. 11-8. Boiling regimes: Below *a*, liquid natural convection; *a–b*, mixed; *b–c*, nucleate boiling; *c–d*, mixed; *d–e*, film boiling; *e–f*, film boiling and radiation.

transported by the rising liquid where some of it may evaporate if it reaches a free surface.

As the heating-surface temperature and thus Δt are increased, the heat flux increases and regime *a–b* is reached. In this the heat flux is still small, only a small number of nucleation centers come into play, and only a few bubbles are formed. These bubbles rise quickly toward the surface and may collapse on the way. The slight agitation thus encountered causes q'' (and h) to increase more rapidly with Δt. The regime *a–b* is mixed, and its heat transfer is due to a combination of liquid natural convection and bubble agitation.

As Δt is further increased, regime *b–c* is reached. In this, the number of nucleation centers, and consequently the number of bubbles, increase rapidly with Δt. The increasing number of bubbles causes considerable agitation and turbulence of the liquid in the boundary layer and consequently an increasing heat flux with Δt. Regime *b–c* is called the *nucleate-boiling* regime and gives large rates of heat flux with moderate values of Δt, that is, moderate heating-surface temperatures.

The heat flux at point *c* is given various names; among them are the *maximum nucleate, critical, peak* heat flux, and *departure from nucleate boiling, DNB*. The temperature difference corresponding to point *c* is likewise called the *maximum nucleate* or the *critical* temperature difference. Reference to point *c* will be made a little later.

As the temperature difference is increased beyond the critical value, regime *c–d* is reached. Here the bubbles become so numerous that they begin to coalesce and clump near the heating surface. In this case a portion of the heating surface gets blanketed with vapor. The vapor blankets act as heat insulators (vapor being a poor heat-transfer agent).

Regime *c–d* can be experimentally obtained only if the surface temperature can be closely controlled (such as by heating it on the opposite side with a condensing vapor at fixed pressures, with little temperature drop through the solid). However, if the heat flux is the controlled parameter instead operation in regime *c–d* becomes unstable. A small increase in Δt, for example, causes a momentary increase in heat flux, causing more blanketing and a reduction in heat flux. The vapor blankets thus periodically collapse and reform. Regime *c–d* is given various names, among which are the *partial nucleate, partial film, mixed nucleate-film,* and *transition film* boiling.

In regime *d–e*, a continuous blanket of vapor forms over the heating surface, and the heat-transfer coefficient reaches a low value. Regime *d–e* is called the *stable film* or simply the *film*-boiling regime.

In regime *e–f*, the temperature of the heating surface is so high that thermal radiation from the surface comes into play. Thermal radiation is a strong function of temperature, and the heat flux rises rapidly with Δt. The exact location of point f depends, of course, on the thermal radiative emission and absorption characteristics of the surface and liquid. Regime *e–f* may be called the *film and radiation* regime.

It can be seen that the same heat flux could be obtained at as many as three heating surface temperatures. It is probably obvious, however, that it is most desirable to operate a boiling system in the nucleate boiling regime *b–c* only, since large heat fluxes can be obtained in a stable manner with moderate heating surface superheats.

It should be noted that, although the above discussion pertains to nonflow saturated pool boiling at atmospheric pressure, a similar mechanism occurs at other pressures, and in subcooled boiling.

A correlation for the heat flux q'' in the nucleate regime for pool-type boiling on clean surfaces, good for many liquids and where the effect of pressure is taken into account via its effect on the physical properties (such as density) that it contains, is given by Rohsenow [86] in the form

$$\frac{c_{pf}\Delta t}{h_{fg}\mathrm{Pr}^{1.7}} = C_{sg}\left[\frac{q''}{\mu_f h_{fg}}\sqrt{\frac{g_c\sigma_{fg}}{g(\rho_f-\rho_g)}}\right]^{0.33} \tag{11-4}$$

where q'' = required heat flux, Btu/ft^2

c_{pf} = specific heat of saturated liquid, Btu/lb$_m$ °F

h_{fg} = latent heat of vaporization at system pressure, Btu/lb$_m$

Pr = Prandtl number of saturated liquid, dimensionless

Δt = difference between heating-surface temperature and saturation temperature, °F

μ_f = viscosity of liquid, lb$_m$/hr ft

g_c = conversion factor, 4.17×10^8 lb_m ft/lb_f hr^2

g = acceleration of gravity, ft/hr^2

σ_{fg} = surface tension at the liquid-vapor interface, lb_f/ft

ρ_f, ρ_g = densities of saturated liquid and vapor phases, lb_m/ft^3

C_{sg} = dimensionless constant determined empirically for different surfaces and fluids

The value of C_{sg} depends on a number of factors, the most important of which is the degree of wettability between the heating surface and the fluid, but is fairly independent of system pressure. Its value for various fluid surface–heat flux–pressure combinations can be obtained by noting q'' and its corresponding Δt in only one experimental test. A value of 0.014 is suggested for the water-stainless steel system [87].

Figure 11-9 shows the agreement of Eq. 11-4 with pool boiling data

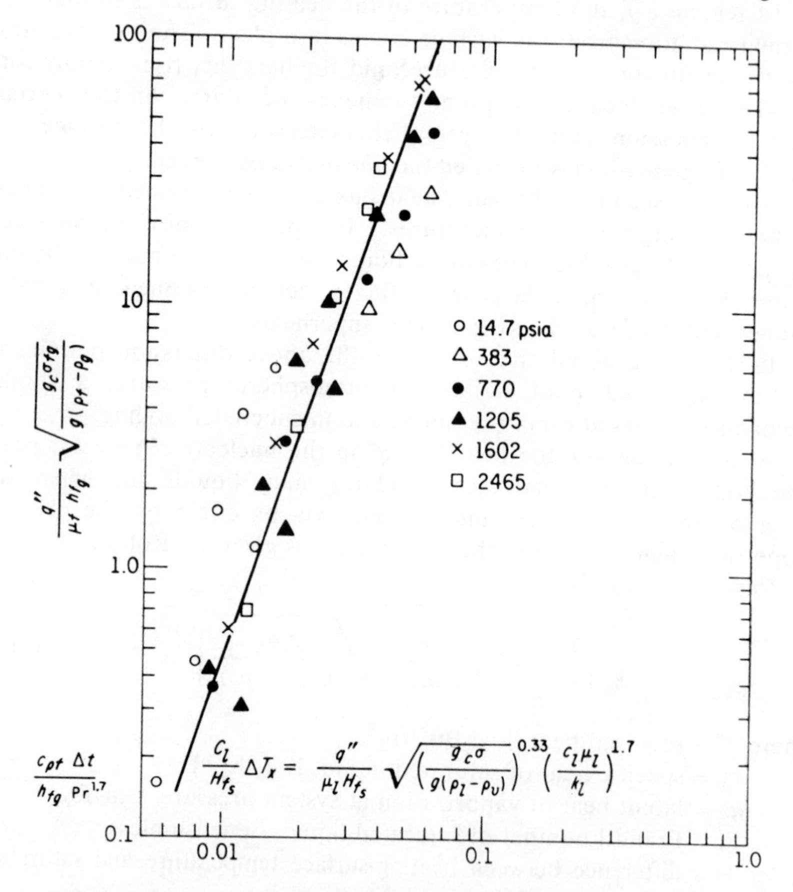

FIG. 11-9. A correlation of experimental pool boiling data for water with Eq. 11-4 (Ref. 86).

of water at various pressures. It is to be noted that Rohsenow's correlation predicts the shape of the nucleate boiling regime with good accuracy but not its beginning or end. The effect of contaminated surfaces has been found to vary the exponent of Pr in a range between 0.8 and 2.0.

11-5. THE BOILING CRISIS AND BURNOUT

It was stated earlier that the whole of the curve of Fig. 11-8 could be obtained if the temperature of the heating surface were closely controlled. In most practice, however, such as if the heating surface were that of a nuclear fuel element, it is the heat flux that is controlled.

Thus heat flux (the ordinate in Fig. 11-8) really becomes the independent variable and temperature difference (the abscissa) becomes the dependent variable. In this case, when the critical heat flux at c and the corresponding temperature are reached, a further increase in heat flux results in a sudden jump from c to c' into the film-radiation regime e–f. This causes the temperature difference to change abruptly from that at c to c'. The heating-surface temperature at c' is so high that it most likely exceeds safe limits. The surface in such a case is said to burn out. *Burnout** in a nuclear fuel element may result in fuel-cladding rupture, accompanied by the release into the coolant of large quantities of radioactive gases and solids. In any case, the heat flux corresponding to the surface temperature at burnout is called the *burnout heat flux*. This may be the flux corresponding to points c and c' but not necessarily so; that is, burnout may occur at any point on the curve, depending upon the material of the heating surface and the operating conditions. In most cases, however, burnout occurs when the heat flux at c, called the *critical heat flux* q''_c, is exceeded. Critical and burnout heat fluxes are often used to describe the same heat flux at c. The two terms should not be so confused in the mind of the reader, however.

Besides critical and burnout heat flux, the conditions at c have also been called *departure from nucleate boiling*, DNB, and, more recently, the *hydrodynamic crisis* and the *boiling crisis*.

The *flow boiling crisis* occurs in the general manner shown by Fig. 11-10 [88]. Curve a shows the fluid bulk temperature versus channel length, from subcooled inlet conditions, through the two-phase region (where it remains constant), and rising again in the superheat region. Curve b shows the channel wall temperature at low heat fluxes, where burnout is not likely to occur. This curve begins moderately above that of the fluid because of the high heat-transfer coefficients associated with nucleate boiling, but rises at higher vapor qualities where film boiling begins to take place and in the superheat region, where gas heat transfer

* Not to be confused with *burnup*, Sec. 1-9.

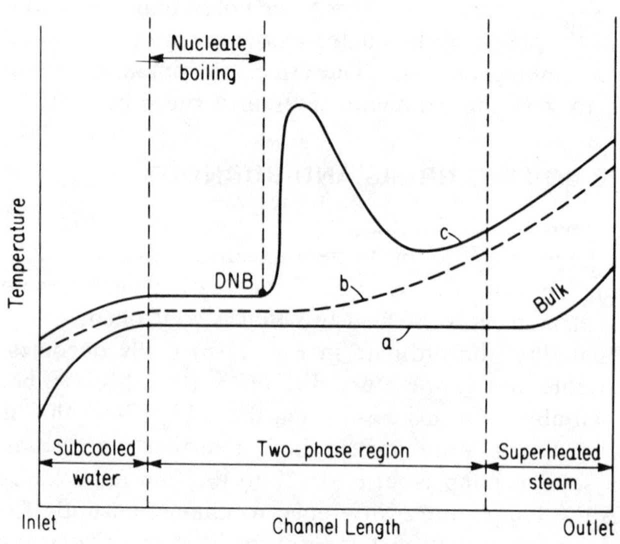

FIG 11-10. Flow boiling crisis (Ref. 88).

coefficients are low. Curve *c* shows the state of affairs at high heat fluxes. The point of departure from nucleate boiling (DNB) occurs suddenly rather than gradually, resulting in a jump in channel wall temperatures that may cause burnout. Downstream of this point, the wall temperature comes down to near values associated with nucleate boiling. The height and width of the jump depend on many parameters.

11-6. PARAMETRIC EFFECTS ON BOILING AND THE BOILING CRISIS

While the heat flux in the nucleate boiling regime is primarily dependent upon the mechanism of bubble agitation and consequently upon surface conditions, i.e., the number of active nucleation centers, and bubble dynamics, it is also affected by other parameters. The critical heat flux q_c'' seems to be also determined by hydrodynamic conditions [89].

A primary effect in *nonflow boiling* is the system pressure. Pressure affects both the nucleate boiling regime and q_c'' because it affects the vapor density, the latent heat of vaporization, and, because it changes the boiling temperature, the surface tension. Experiments by Cichelli and Bonilla [90] on pool boiling of water and a variety of organic liquids and by Kazakova [91] and by Lukomskii [92], both on water, showed a shift in the nucleate boiling regime and showed that q_c'' increased with pressure up to a maximum, at some optimum pressure, and then decreased. This optimum pressure was in all cases approximately equal to one-third to 40 percent of the critical pressure* of the liquid in question. Figures 11-11 and 11-12 show the results of some of the above experiments. (Figure 11-9 is a replot of the data in Fig. 11-11.) In all cases, the critical heat flux at the optimum pressure is four to ten times that at very low or very high pressures. The solid line in Fig. 11-12 represents the average for clean surfaces. For contaminated surfaces q_c'' is to be increased by 15 percent.

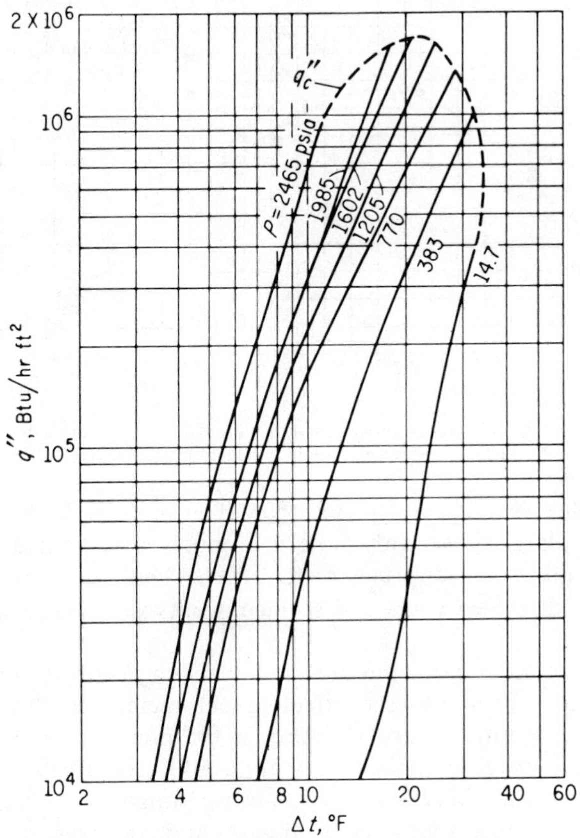

FIG. 11-11. Nucleate-boiling lines at various pressures for water in pool boiling, showing critical-heat-flux locus (Ref. 86).

It is advantageous, then, to operate a pool boiling system at or near the optimum pressure. This is the condition of maximum critical heat flux, and thus a correspondingly high actual operating heat flux can be tolerated with safety.

* *Critical pressure* is used here in the thermodynamic sense. It is the pressure at which the vapor and liquid phases of the substance become identical. For water it is 3206.2 psia.

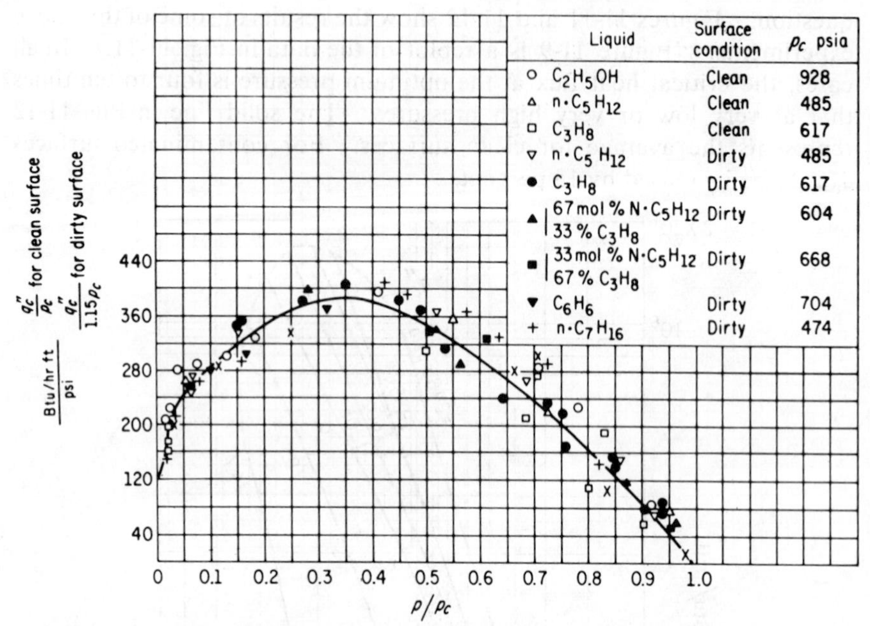

FIG. 11-12. Critical heat flux versus pressure ratio for various liquids in pool saturated boiling. p_c = thermodynamic critical pressure (Ref. 90).

The value of q_c'' has also been found to increase with certain additives, particularly those with molecules much heavier than the boiling fluid, with ultrasonic and electrostatic fields, but to decrease with the presence of dissolved gases and surface agents which reduce the surface tension.

There are many more parametric effects on *flow boiling* and the flow boiling crisis. Primarily the affecting parameters are heat flux, mass velocity, inlet enthalpy, and enthalpy at the point of crisis, the size and shape of the flow channel, and pressure. Of lesser effects are the type of heating, (uniform, sinusoidal, etc.), surface roughnesses, unheated walls near the critical-flux point, dissolved gases and additives, etc.

Unlike nonflow boiling, it is difficult to ascertain, independently, the effects of all of these parameters. For example, in flow boiling with known uniform heat flux, the exit enthalpy at which crisis is likely to occur, depends on both the mass velocity and the inlet enthalpy (inlet subcooling or quality) and, to a lesser extent, the pressure. It is therefore difficult to obtain, for example, the effect of pressure on q_c'' as was done for saturated pool boiling (Fig. 11-12), and there is much disagreement in the literature on the magnitude, and sometimes even on the trend, of the effects of these parameters on the flow-boiling crisis [88, 93].

Figure 11-13 shows the effects of subcooling and velocity in flow (forced convection) boiling. The data is for water flowing in annuli heated electrically by the inner tube, at pressures between 30 and 90 psia. The dotted lines here represent forced convection heat transfer (as contrasted to natural convection in Fig. 11-8), while the solid lines represent the nucleate boiling regime. It can be seen that the nucleate boiling heat fluxes are only dependent on the degree of subcooling and independent of the velocity. Both the onset of boiling and the critical heat flux q_c'', however, are dependent upon both the degree of subcooling and velocity, increasing with both.

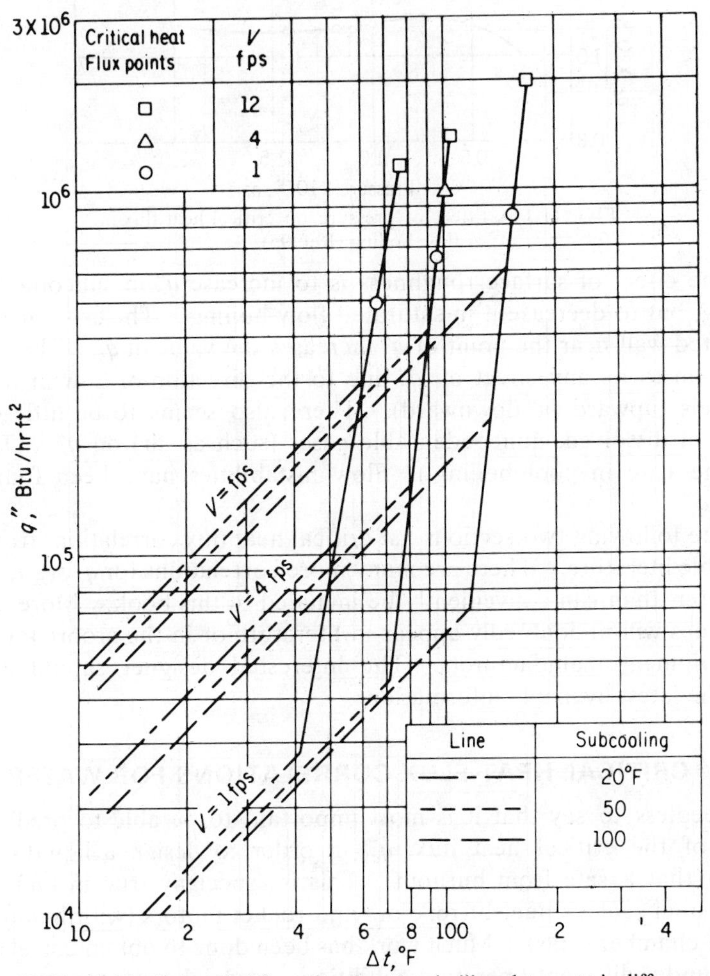

FIG. 11-13. Subcooled forced convection boiling data. Δt is difference between surface and water bulk temperature (Ref. 94).

Correlation of various data on flow in vertical round tubes [88, 93] showed that q_c'' increases with mass velocity, decreases with inlet temperature, inlet quality and tube diameter, and decreases somewhat with pressure. The effect of pressure on q_c'' in upward flow in a round tube is shown in Fig. 11-14 [95]. As can be seen in this case, the effect is rather slight at reactor operating pressures.

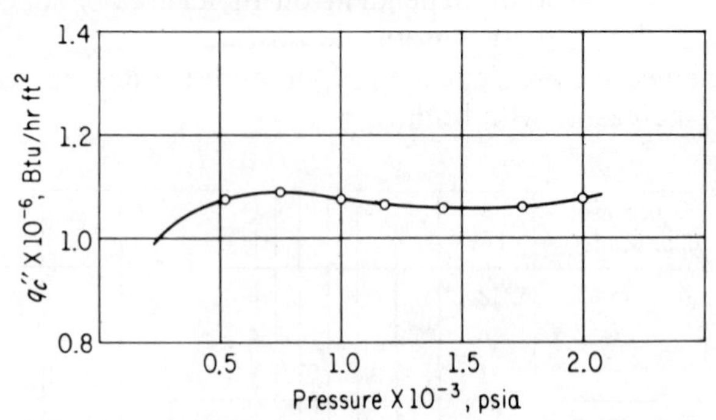

FIG. 11-14. Effect of pressure on critical heat flux in flow boiling (Ref. 95).

The effect of surface roughness is to increase q_c'' in subcooled flow boiling, but to decrease it in saturated flow boiling. The existence of an unheated wall near the point of q_c'' increases the value of q_c''. There does not seem to be any effect on q_c'' due to the direction of flow in vertical channels (upward or downward). There also seems to be little or no effect of dissolved, noncondensable gases (such as air) on q_c''. (This is not the case in pool boiling.) Flow instabilities have been found to lower q_c''.

The following two sections list critical heat flux correlations from the available literature. There are many more correlations for q_c'' particularly for water, than can conveniently be included in this book. More up-to-date ones will undoubtedly appear in literature or in the proprietary files of equipment manufacturers. The interested designer should always seek the latest available information.

11-7. CRITICAL HEAT-FLUX CORRELATIONS FOR WATER

Needless to say that it is most important to be able to predict the value of the critical heat flux, q_c'', in order to design a liquid-cooled system that is safe from burnout. This is especially true in high heat-flux system such as nuclear reactors and rocket motors (with liquid fuel-cooled chamber walls). Much work has been done to obtain correlations for q_c'' under different operating conditions, especially after the advent of nuclear reactors.

The correlations may be divided into categories according to whether the liquid is saturated or subcooled and whether it is in nonflow (pool) or flow boiling.

1. Saturated Nonflow Pool Boiling

(*a*) Rohsenow and Griffith [96] proposed the following dimensional correlation for pool saturated boiling, based in part on theoretical analysis and on experimental data, such as those in Fig. 11-12. It is good for water as well as a variety of organic liquids:

$$q''_c = 143\, h_{fg}\rho_g \left(\frac{\rho_f - \rho_g}{\rho_g}\right)^{0.6} \left(\frac{g}{g_c}\right)^{0.25} \tag{11-5}$$

where h_{fg} = latent heat of vaporization at system pressure, Btu/lb_m

ρ_f, ρ_g = densities of the saturated liquid and saturated vapor, respectively, at system pressure, lb_m/ft^3

g = gravitational acceleration, ft/hr^2

g_c = conversion factor = $4.17 \times 10^8\ \text{lb}_m\,\text{ft/lb}_f\,\text{hr}^2$

Equation 11-5 shows that, since an increase in pressure decreases h_{fg} and ρ_f and increases ρ_g, there exists an optimum value of pressure at which q''_c is maximum. This is in agreement with the data of Fig. 11-12. For water the optimum pressure is around 1200 psia and the maximum q''_c, in standard gravity, is about 1.2×10^6 Btu/hr ft^2. Equation 11-5 also shows that q''_c increases with the gravitational field. This is thought to be due to increased separation between the two phases because of the increased difference between the gravitational forces on them due to their different densities. This effect is in agreement with experimental data [97].

(*b*) Kutateladze [98] formulated the following correlation for pool saturated boiling by the use of a combination of dimensional analysis and experimental data. The correlation agrees well with the data of Fig. 11-12, showing a maximum critical heat flux at a critical pressure ratio about 0.3.

$$q'' = C h_{fg}\rho_g^{0.5}\left[\sigma(\rho_f - \rho_g) g g_c\right]^{0.25} \tag{11-6}$$

C has been obtained for various horizontal wires and disks. It varies between 0.13 and 0.19, depending on surface conditions, and has an average value of 0.14.

2. Subcooled Nonflow Pool Boiling

(*a*) If the liquid is subcooled, higher critical heat fluxes are attained. Zuber et al. [89] extended Kutateladze's correlation to the case of pool subcooled boiling by adding heat conduction and subcooling terms. The resulting correlation which agrees well with subcooled data on water and alcohol [99] is

$$\frac{q_c'', \text{ subcooled}}{q_c'', \text{ saturated}} = 1 + \frac{5.3}{h_{fg}\rho_g}(k_f \rho_f c_{pf})^{0.5}\left[\frac{(\rho_f - \rho_g)g}{\sigma g_c}\right]^{0.25} \times \left[\frac{\sigma(\rho_f - \rho_g)gg_c}{(\rho_g)^2}\right]^{-1/8}(t_{sat} - t_b) \tag{11-7}$$

where k_f = thermal conductivity of the liquid Btu/hr ft°F
t_b = bulk temperature, °F

(*b*) A simpler correlation for the ratio q_c'' in subcooled to q_c'' in saturated pool boiling is given by Ivey and Morris [100] as

$$\frac{q_c'', \text{ subcooled}}{q_c'', \text{ saturated}} = 1 + 0.1\left(\frac{\rho_g}{\rho_f}\right)^{0.25}\left[\frac{c_{pf}\rho_f(t_w - t_b)}{h_{fg}\rho_g}\right] \tag{11-8}$$

where t_w = wall temperature, °F.

3. Subcooled Flow Boiling

Subcooled boiling is likely to occur in liquid-cooled reactors when the heat flux is high enough so that the outer temperature of the cladding is above the saturation temperature corresponding to the system pressure, while the coolant bulk temperature remains below the saturation temperature. This is the case of pressurized-water reactors operating at high loads at which time such subcooled boiling is encouraged and results in high heat-transfer coefficients. Knowledge of q_c'' therefore, is as necessary as in boiling-water reactors, and both types are q_c'' – limited. The values for q_c'' are, however, affected by the degree of subcooling, and the correlations are different for subcooled and saturated boiling in flow systems.

(*a*) The ANL correlation: Jens and Lottes [101] represented a large number of experimental data of subcooled water flowing upward through a vertical tube under various operating conditions, including high pressures, by the following correlation:

$$q_c'' = C\left(\frac{G}{10^6}\right)^m (t_{sat} - t_b)^{0.22} \tag{11-11}$$

where t_{sat}, t_b are the saturation and bulk temperatures at the point of burnout, °F respectively
C, m are constants, depending upon the pressure, Table 11-1

TABLE 11-1
Constants in the Jens and Lottes Correlation

Water Pressure, psia	$C \times 10^{-6}$	m
500	0.817	0.160
1,000	0.626	0.275
2,000	0.445	0.500
3,000	0.250	0.730

The mass velocity range covered by the above correlation was 0.96×10^6 to 7.8×10^6 $lb_m/hr\ ft^2$. The range of subcooling was 5.5 to 163°F. It can be seen that the correlation is unsuitable for saturated boiling, since it results in zero q''_c at zero subcooling.

(*b*) *The Westinghouse Correlation*

As a result of work done at the Westinghouse Atomic Power Division (WAPD) the following correlation was developed from available uniform heat flux data with water [102]:

$$\begin{aligned}\frac{q''_c}{10^6} &= [(2.022 - 0.0004302p) + (0.1722 - 0.0000984p)e^{(18.177 - 0.004129p)x}] \\ &\times [(0.1484 - 1.596x + 0.1729|x|)G \times 10^{-6} + 1.037] \\ &\times [1.157 - 0.869x] \times [0.2664 + 0.8357e^{-3.151D_e}] \\ &\times [0.8258 + 0.000794(h_f - h_i)] \qquad (11\text{-}12)\end{aligned}$$

This correlation covers the ranges: $G = 0.5 \times 10^6$ to 5.0×10^6 $lb_m/hr\ ft^2$, $p = 800$ to 2,000 psia, channel length $L = 10$ to 79 in., $D_e = 0.2$ to 0.7 in., heated-to-wetted-perimeter ratio $= 0.88$ to 1.0, inlet enthalpy $h_i \geqslant 400$ Btu/lb_m, local quality $x_{loc} = -0.15$ to $+0.15$ (negative quality means subcooled liquid).

For nonuniform heat flux, the above q''_c is divided by a factor F given by

$$F = \frac{C}{q''_{l_c}(1 - e^{-Cl_c})} \int_0^{l_c} q''(z) e^{-C(l_c - z)}\, dz \qquad (11\text{-}13)$$

where $C = 0.44(1 - x_c)^{7.9}/(G \times 10^{-6})^{1.72}$, in.$^{-1}$.

l_c = channel length at which DNB takes place, in.

q''_{l_c} = heat flux at l_c, Btu/hr ft^2

z = variable distance from channel entrance, in.

x_c = quality at DNB

The above correlation correlates most existing heat flux data to within ± 20 percent. Other available subcooled flow boiling correlations are the Bernath [103] and Zenkevich [104] correlations.

4. Saturated Flow Boiling

This is sometimes referred to in the literature as bulk boiling (which is at variance with the definition of that term given in Sec. 11-2). Saturated boiling occurs when the bulk temperature of the fluid is equal to the saturation temperature corresponding to the system pressure. The bubbles formed at the heating surface therefore do not collapse when they enter the mainstream of flow as they do in subcooled flow boiling. Saturated flow boiling occurs in the boiling heights of boiling-water-reactor channels (Sec. 12-4).

(*a*) *General Electric Correlations*

The following correlations are design (lower envelope of data) equations [105]. For 1,000 psia,

$$
\begin{aligned}
q''_c &= 0.705 \times 10^6 + 0.237G, && \text{for } x < x_1 \\
q''_c &= 1.634 \times 10^6 - 0.270G - 4.71 \times 10^6 x && \text{for } x_1 < x < x_2 \\
q''_c &= 0.605 \times 10^6 - 0.164G - 0.653 \times 10^6 x && \text{for } x > x_2
\end{aligned}
\tag{11-14}
$$

At other pressures p, psia,

$$q''_c = q''_c(\text{at 1,000 psia}) + 440(1000 - p) \tag{11-15}$$

where $x_1 = 0.197 - 0.108 \times 10^{-6} G$
$x_2 = 0.254 - 0.026 \times 10^{-6} G$

The above correlations apply over a pressure range of 600 to 1450 psia, $G = 0.4 \times 10^6$ to 6.0×10^6 $\text{lb}_m/\text{hr ft}^2$, quality x = negative (subcooled water) to 0.45, $D_e = 0.245$ to 1.25 in., and channel length $L = 29$ to 108 in.

(*b*) *The Westinghouse Correlation*

This correlation, instead of q''_c as such, predicts the steam-water enthalpy, h_c, at which burnout is likely to occur in terms of the inlet enthalpy, h_i, and other parameters [102].

$$
\begin{aligned}
h_c = h_i &+ 0.529(h_f - h_i) + (0.825 + 2.3e^{-17D_e})h_{fg}e^{-1.5 \times 10^{-6} G} \\
&- 0.41 h_{fg} e^{-0.0048 L/D_e} - 1.12 h_{fg} \frac{\rho_g}{\rho_f} + 0.548 h_{fg}
\end{aligned}
\tag{11-16}
$$

where D_e is in inches. This correlation applies for flow in circular and rectangular channels and along rod bundles, for uniform and monuniform heat flux over a pressure range of 800 to 2,750 psia, $h_i = 400$ Btu/lb_m to saturation, $G = 0.4 \times 10^6$ to 2.5×10^6 $\text{lb}_m/\text{hr ft}^2$, local heat flux $= 0.1 \times 10^6$ to 1.8×10^6 Btu/hr ft², $D_e = 0.1$ to 0.54 in., $L = 9$ to 76 in., heated-to-wetted-perimeter ratio $= 0.88$ to 1.0, and exit quality $= 0$ to 0.9. This equation correlates most data to within ± 25 percent.

11-8. CRITICAL HEAT-FLUX CORRELATIONS FOR LIQUIDS OTHER THAN WATER

Few correlations have been suggested which are suitable for coolants other than water.

(*a*) For liquid metals, Noyes [106] obtained q_c'' data for *sodium in nonflow pool boiling* at pressures between 0.5 and 1.5 psia (corresponding to saturation temperatures of 1130 and 1250°F respectively). He recommends the correlation

$$q_c'' = 0.144 h_{fg} \rho_g \left(\frac{\rho_f - \rho_g}{\rho_g} \right)^{0.5} \left(\frac{g g_c \sigma}{\rho_f} \right)^{0.25} \mathrm{Pr}^{-0.245} \tag{11-17}$$

where Pr is the Prandtl number.

(*b*) Colver [107] reported data on q_c'' for *potassium in saturated nonflow pool boiling* at pressures between 0.15 and 22 psia (corresponding to saturation temperatures between 800 and 1480°F respectively). His data appears to be well correlated by the expression

$$q_c'' = (4 \times 10^5) p^{0.16} \tag{11-18}$$

where q_c'' is in Btu/hr ft^2 and p is the pressure in psia.

(*c*) For *mercury*, Kutateladze [108] obtained data on magnesium amalgams (mixtures with mercury). Magnesium acts as a wetting agent for mercury and increases its q_c''. Concentrations of 0.01 percent and 0.03 percent by mass of Mg in Hg show q_c'' values of 0.05×10^6 and 0.12×10^6 Btu/hr ft^2 respectively. It is difficult to obtain nucleate boiling of pure mercury (without a wetting agent).

(*d*) For *liquid metals in flow boiling*, data by Hoffman [109] on boiling potassium (with high exit qualities) were found to be well represented by correlations originally developed by Lowdermilk, Lanzo, and Siegel [110] representing low-pressure (under 100 psia) forced-flow of water in tubes. Their correlations were obtained for the ranges (water) of inlet velocity = 0.1 to 98 ft/sec, inlet subcooling = 0 to 140°F, tube diameter = 0.051 to 0.188 in., and length/diameter ratio $L/D = 25$ to 250. (Because of low pressures, these correlations were not suitable for power-reactor work with water, but were suitable in safeguards investigations in loss of pressure accidents, in rocket-motor cooling, etc.). These correlations are

$$q_c'' = 270\, G^{0.85} D^{-0.2} \left(\frac{L}{D} \right)^{-0.85} \qquad \text{for } \frac{G}{(L/D)^2} < 150 \tag{11-19}$$

and $$q_c'' = 140\, G^{0.5} D^{-0.2} \left(\frac{L}{D}\right)^{-0.15} \quad \text{for } \frac{G}{(L/D)} > 150 \tag{11-20}$$

One peculiarity of boiling-liquid-metal systems is the relatively large change in saturation temperature with small changes in pressure (as compared to water). Thus a small pressure decrease due to flow friction or rise in elevation results in a drop in saturation temperature (i.e., increases liquid superheat) which could be severe enough to cause an "explosion type" of flow and oscillations.

(*e*) For *flow boiling* of *organic liquids* used in organic cooled reactors, the polyphenyls, Core and Sato [111] suggested the following two correlations from experimental data on the unirradiated coolants (irradiation causes molecular damage to organic coolants).

For diphenyl:

$$q_c'' = 454\, \Delta t_{\text{sub}}\, V^{0.63} + 116{,}000 \tag{11-21}$$

For Santowax R:

$$q_c'' = 552\, \Delta t_{\text{sub}}\, V^{2/3} + 152{,}000 \tag{11-22}$$

These two correlations have limits of uncertainly of $\pm$ 106,000 and $\pm$ 160,000 Btu/hr ft^2 respectively. They cover the ranges: (a) for diphenyl: subcooling $\Delta t_{\text{sub}} = 0$ to 328°F (also low quality tests), $V = 0.5$ to 17 ft/sec, $p = 23$ to 406 psia, and coolant bulk temperature = 510 to 831°F; (b) for Santowax R: $V = 5$ to 15 ft/sec, $p = 100$ psia, and coolant bulk temperature = 595 to 771°F. All tests were made on coolant flowing in an annular passage formed by a 0.46-in. ID electrically-heated tube containing and unheated 0.188-in.-diameter coaxial rod.

Critical heat fluxes of polyphenyls are in the neighborhood of 10^5 Btu/hr ft^2, Fig. 11-15, several times lower than for water. They are appreciably affected by impurities, especially those that affect the saturation temperature and consequently the degree of subcooling in Eqs. 11-21 and 11-22.

(*e*) Nucleate-pool-boiling heat-transfer studies of systems consisting of heated $\frac{1}{16}$-and $\frac{1}{8}$-in. platinum tubes submerged in aqueous ThO_2 *slurries* (fuel material in suspension in a liquid carrier) resulted in the following correlation [113]:

$$q'' = a(\Delta t)^n \tag{11-23}$$

where q'' = boiling heat flux, Btu/hr ft^2

a, n = const

Δt = temperature difference between the tube surface and bulk of slurry

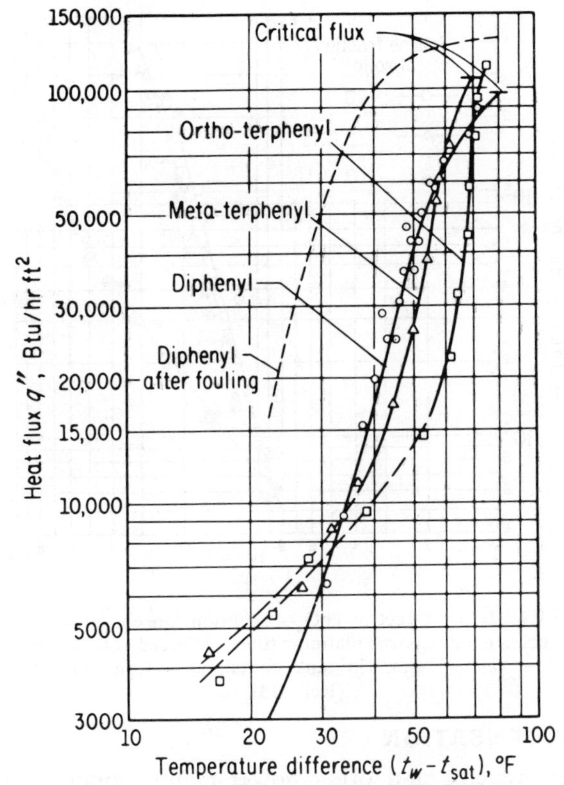

FIG. 11-15. Nucleate-boiling and critical heat fluxes of polyphenyl organic coolants at atmospheric pressure (Ref. 112).

The results of the investigation are shown in Fig. 11-16. Note that n, the slope of the lines, decreases with the volume fraction of the solid particles. At a temperature difference of 10°F, the heat flux was found to be 10^4 Btu/hr ft², fairly independent of concentration.

In the slurry system investigated, it was noticed that, as the boiling process progressed, a soft film, less that $\frac{1}{32}$ in. thick, surrounded the heating surface. This film resulted in a steady increase in Δt by 5 to 6°F/hr. The film was not in the nature of a hard cake and did not adhere to the surface of the heating tube after it was removed from the slurry.

The heat flux q_c'' for low solid-particle concentrations (around 200 g_m Th/kg$_m$ H_2O) was found to be roughly the same as that for water at comparable operating conditions. At higher concentrations, however, it was much less. A representative case, where the concentration was 1,000 g_m Th/kg$_m$ H_2O, showed that q_c'' was only 210,000 Btu/hr ft², compared with 490,000 Btu/hr ft² for water at the same conditions.

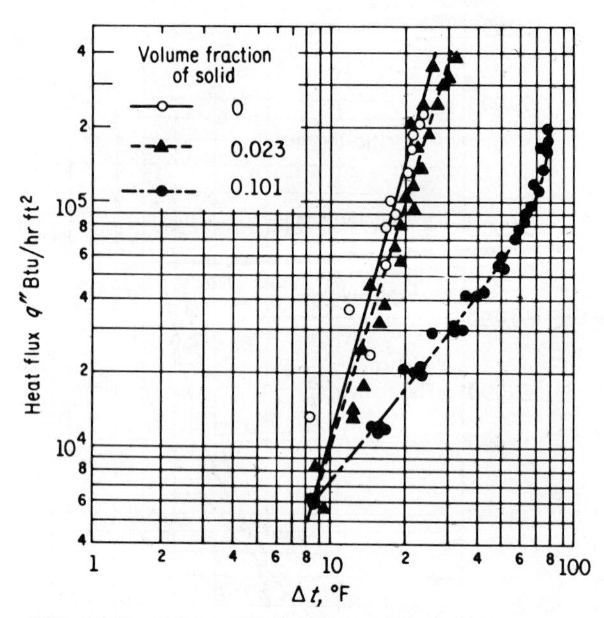

FIG. 11-16. Effect of ThO_2-suspension concentration on nucleate boiling from platinum tube. (Forced-convection flow rate ~ 1 gpm in tank containing ~ 3 gal slurry.) (Ref. 113).

11-9. CONDENSATION

In nuclear-reactor and other power plants, condensation occurs in turbine condensers and other heat exchangers. It takes place when a wet, saturated, or slightly superheated vapor (such as steam) comes in contact with a surface at a temperature lower than the saturation temperature corresponding to the pressure of the vapor. The resulting liquid is called the *condensate*. Condensation is classified as (a) *filmwise*, and (b) *dropwise*.

Filmwise Condensation

In filmwise condensation, the condensate flows in a continuous liquid film, simply called the *film*, over the cooling surface and usually flows downward by gravity. If the surface is vertical or inclined the film thickness increases downward. If the film is not too thick or the vapor velocity not too high, the condensate velocity profile $V(x,y)$ within the film, varying between zero at the wall and maximum at the vapor-liquid interface, Fig. 11-17, is laminar. Heat transfer between the vapor and the cooling surface occurs through the film by conduction only and the temperature gradient $t(y)$ through the film is linear between that of the wall t_w and the saturated vapor t_s.

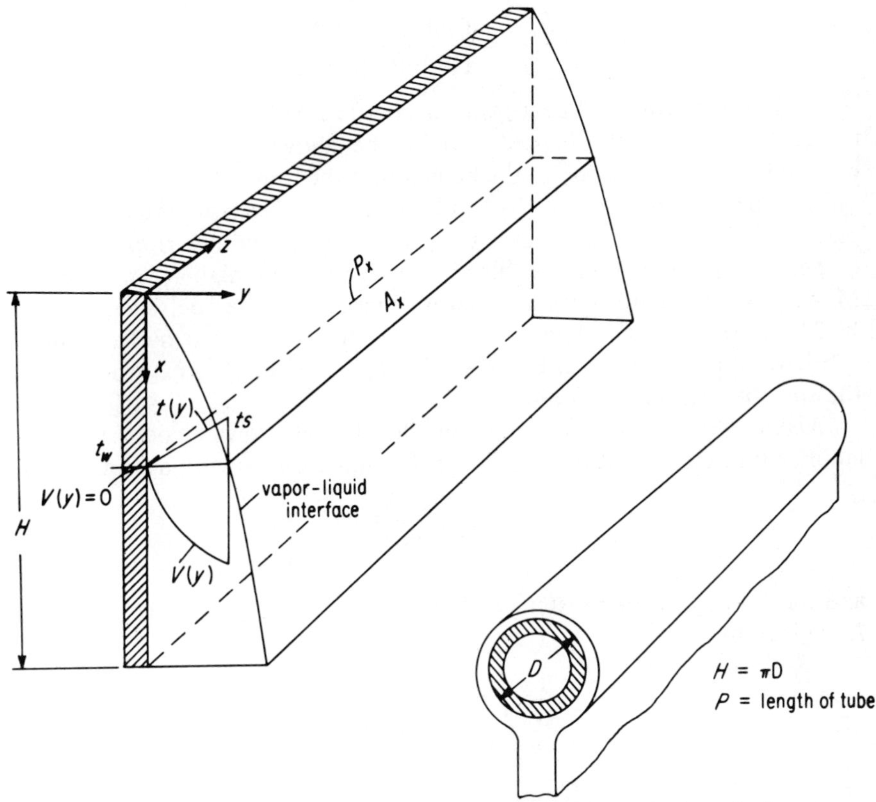

FIG. 11-17. Laminar condensation over vertical surface or horizontal tube.

Theoretical analysis of filmwise condensation was originally presented in 1916 by Nusselt [114]. In his classic work on condensation, Nusselt analyzed the case of condensation on a vertical surface with no vapor motion and extended this analysis to the case of laminar film-condensation inside vertical tubes, but with vapor motion. In this he assumed that the shear stress at the vapor-liquid interface was proportional to the pressure drop for tube flow containing vapor only. Experimental work, however, yielded heat-transfer coefficients that were higher than those calculated from the Nusselt model. Many researchers modified several of Nusselt's assumptions, such as those pertaining to the mass balance inside the tube, film turbulence, shear stress at the vapor-liquid interface, etc. One of the latter analyses is that of Rohsenow, Webber, and Ling [115].

Nusselt's analysis of laminar-film condensation, which may be found in most heat-transfer texts, yields the following correlation for the local Nusselt number Nu_x and the local heat-transfer coefficient h_x

$$\mathrm{Nu}_x = \frac{h_x x}{k_l} = \left[\frac{\rho_l(\rho_l - \rho_g) g h_{fg} x^3}{4\mu_l k_l (t_s - t_w)}\right]^{\frac{1}{4}} \tag{11-24}$$

Actual experimental heat-transfer coefficients have often been found to be higher than those predicted by the above correlation. This is believed to be mainly due to the behavior of the film. Pure film condensation, for example, is a rarity, and some dropwise condensation takes place. Also, it is often found that the films contained ripples which helped liquid mixing in the film. These result in higher values of h. McAdams [50] therefore suggested an increase in the coefficient in Eq. 11-24 by 20 percent. Another correction, recommended by Rohsenow [116], would replace h_{fg} in Eq. 11-24 by $h_{fg} + 0.68\, c_{p_l}\,(t_s - t_w)$ where c_{p_l} is the specific heat of the liquid.

Using McAdams' correction, the local heat-transfer coefficient for laminar film-condensation over a vertical surface thus would be given by

$$h_x = 1.2 \left[\frac{\rho_l(\rho_l - \rho_g) g h_{fg} k_l^3}{4\mu_l x (t_s - t_w)}\right]^{\frac{1}{4}} \tag{11-25}$$

and the average heat-transfer coefficient $\bar{h}$ for a vertical surface of height H is given by

$$\bar{h} = \frac{1}{H}\int_0^H h_x dx = \frac{4}{3}\, h_{x=H} \tag{11-26}$$

or

$$\bar{h} = 1.60 \left[\frac{\rho_l(\rho_l - \rho_g) g h_{fg} k_l^3}{4\mu_l H (t_s - t_w)}\right]^{\frac{1}{4}} \tag{11-27}$$

where the properties with the subscript l refer to the liquid and are evaluated at the arithmetic average of the saturated vapor and wall temperatures, i.e., at $(t_s + t_w)/2$. If the surface is inclined by an angle φ with the horizontal, the value of H, above, should be replaced by $H/\sin\varphi$. The above relationships are valid for both vertical and inclined plates and the inside and outside of vertical tubes if the tube diameter is large compared to the film thickness. The above correlations show that the local heat-transfer coefficient h_x decreases with x (and $\bar{h}$ decreases if the total surface height H is increased). This is to be expected since the film thickens, and therefore the thermal resistance increases, with x. The correlations also show that h_x and $\bar{h}$ decrease if the degree of subcooling $(t_s - t_w)$ is increased, that is, if the wall temperature is lowered. This unexpected effect is again due to increased thickening of the film.

In condensation, as in convection, the Reynolds number is a useful parameter. The local Reynolds number, at x, is given by

$$\text{Re}_x = \frac{D_{e_x}\rho_l V_x}{\mu_l} = \frac{4A_x\rho_l V_x}{P_x\mu_l} = \frac{4\dot{m}_x}{P_x\mu_l} \tag{11-28}$$

where D_{e_x} is the local equivalent diameter, in feet, equal to 4 times the flow area A_x divided by the wetted perimeter at x, P_x; the symbol $\dot{m}_x$ is the total condensate mass-flow rate at x, lb_m/hr; and V_x is the average condensate velocity at x, ft/hr. Note that the wetted perimeter is equal to the width of the surface perpendicular to flow direction. A mass balance for the condensation of saturated vapor between $x = 0$ and $x = H$, gives

$$\bar{h}A_H(t_s - t_w) = \dot{m}_H h_{fg} \tag{11-29}$$

where A_H is the total heat-transfer surface area between $x = 0$ and $x = H$, equal to HP_x, and $\dot{m}_H$ is the total mass flow rate of the condensate at H. Combining Eqs. 11-28 and 11-29 results in a form of the Reynolds number at $x = H$, given by

$$\text{Re}_H = \frac{4H\bar{h}(t_s - t_w)}{\mu_l h_{fg}} \tag{11-30}$$

Equation 11-27 can now be rewritten for vertical surfaces, in terms of this Reynolds number as

$$h = 1.87\left[\frac{\rho_l(\rho_l - \rho_g)gk_l^3}{\mu_l^2}\right]^{\frac{1}{3}}(\text{Re}_H)^{-\frac{1}{3}} \tag{11-31}$$

For inclined surfaces the right-hand side of the above equation is multiplied by $(\sin\varphi)^{1/3}$

The average heat-transfer coefficient for filmwise condensation of saturated vapor on the outside of a horizontal tube of diameter D, Fig. 11-17, is given by

$$\bar{h} = 0.725\left[\frac{\rho_l(\rho_l - \rho_g)gh_{fg}k_l^3}{D\mu_l(t_s - t_w)}\right]^{\frac{1}{4}} \tag{11-32}$$

$\bar{h}$ for a horizontal tube is considerably larger than when the tube is placed in a vertical position because of the reduced flow path and much thinner film. If, as often is the case in condensers, condensation occurs over a stack of horizontal tubes such that the condensate from one tube drips over a lower one, $\bar{h}$ can conservatively be estimated by replacing D in the above correlation by DN, where N is the total number of tubes.

Equation 11-32 may be rewritten in terms of the condensate Reynolds number for a tube, Re_D, which is obtained by replacing H in Eq. 11-30 by πD, giving

$$\text{Re}_D = \frac{4\pi D\bar{h}(t_s - t_w)}{\mu_l h_{fg}} \tag{11-33}$$

Combining Eqs. 11-32 and 11-33 and rearranging thus give

$$\bar{h} = 1.52 \left[\frac{\rho_l(\rho_l - \rho_g) g k_l^3}{\mu_l^2} \right]^{\frac{1}{3}} (\mathrm{Re}_D)^{-\frac{1}{3}} \tag{11-34}$$

Effect of Film Turbulence

If the film becomes turbulent, the assumptions in the Nusselt model do not apply, and the Nusselt correlations yield overly conservative results.

Film turbulence usually occurs when the film thickness becomes sufficiently large, a situation that may occur near the lower part of vertical or inclined surfaces (and seldom occurs in the case of single small-diameter horizontal tubes). When turbulence occurs in the film, heat transfer will no longer solely occur by conduction in the film (resulting in a large thermal resistance to heat flow) but also by eddy diffusivity in the film, and the local heat-transfer coefficient increases. The heat-transfer

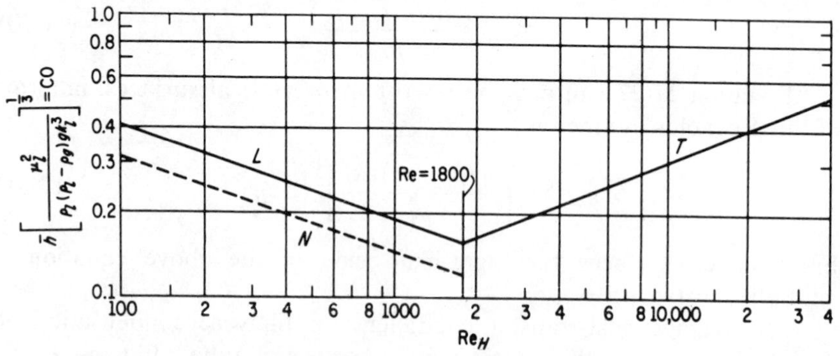

FIG. 11-18. Condensation vs. Reynolds numbers for a flat vertical surface. N = Nusselt, L = recommended laminar, T = recommended turbulent.

coefficient now increases with x (since the level of turbulence increases as the film thickens), a trend opposite to that in laminar flow, Fig. 11-18. The transition from laminar to turbulent flow occurs at a local film Reynolds number of about 1800*. A correlation for the average heat-transfer coefficient in turbulent film condensation is given by Kirkbride [117] as

$$\bar{h} = 0.0076 \left[\frac{\rho_l(\rho_l - \rho_g) g k_l^3}{\mu_l^2} \right]^{\frac{1}{3}} (\mathrm{Re}_H)^{0.4} \tag{11-35}$$

Notice the similarity between Eqs. 11-31, 11-34, and 11-35. The dimensionless quantity $\bar{h}/[\rho_l(\rho_l - \rho_g) g k_l^3/\mu_l^2]^{1/3}$ is often called the *condensation number* Co and the three correlations are often expressed in the form

* This is analogous to the case of the transition from laminar to turbulent flow of a fluid flowing over a surface in convection.

$$\text{Co} = a(\text{Re})^b \tag{11-36}$$

where a and b are constants.

Figure 11-18 shows a plot of the theoretical (Nusselt) and recommended correlations for laminar film condensation together with the correlation for turbulent condensation over a flat vertical surface vs. Reynolds number.

Effect of Superheated Vapor

All the above correlations were obtained for the case of condensation of saturated vapor. If the vapor is superheated, the degree of superheat must first be removed by a convective process between vapor and the condensate at their interface. The interface temperature still must correspond to saturation at the system pressure. If the degree of superheat is not too great, or if the convective transfer is high, the effect of superheat may be ignored and the above correlations all apply.

Effect of High Vapor Velocity

It was indicated earlier that Nusselt had extended his analysis of laminar flow inside vertical tubes to the case where the vapor velocity is appreciable. In this case the frictional drag between vapor and liquid is no longer negligible and a shear stress is introduced at the vapor-liquid interface. If vapor flows upward, it will tend to retard and therefore thicken the film and reduce the heat transfer coefficient. Downward flow, on the other hand, thins the film and results in higher heat-transfer coefficients. High vapor velocities also cause the transition from laminar to turbulent flow to occur at Reynolds numbers around 300 (instead of 1800).

A correlation by Carpenter and Colburn [118] for the condensation of steam and hydrocarbon vapors in downward flow in vertical tubes is

$$\bar{h} = 0.046\,\bar{G}\left(\frac{c_{p_l}^2 \rho_l f}{\rho_g \text{Pr}_l}\right)^{\frac{1}{2}} \tag{11-37}$$

where $\bar{G}$ = mass velocity given by $\left(\dfrac{G_1^2 + G_1G_2 + G_2^2}{3}\right)^{\frac{1}{2}}$ where G_1 and G_2 are the mass velocities of the vapor at top and bottom of tube respectively, lb_m/hr ft^2

f = fanning pipe friction factor evaluated at average vapor velocity, dimensionless (1/4 of the values in App. F)

Pr_l = Prandtl number of liquid, dimensionless

The above correlation was obtained for a 0.5 in. ID, 8-ft-long tube with inlet vapor velocities up to 0.14 ft/hr. All physical properties are to be evaluated at $(0.25\,t_s + 0.75\,t_w)$.

Heat-Transfer Coefficients Range

$\bar{h}$ in filmwise condensation of pure steam over horizontal tubes 1-3 in. diameter, is approximately 2000–4000 Btu/hr ft^2 °F. For organic compounds, the values are less by about an order of magnitude. The presence of appreciable amounts of noncondensable gases, such as atmospheric air seeping into a turbine condenser which is held at high vacuum (1 psia pressure or less), complicates the analysis [50]. In general, it reduces the heat-transfer coefficients because such gases blanket the cooling surfaces and considerably increase the heat- and mass-transfer resistances. Such noncondensables must therefore be vented or pumped out such as by the use of air ejectors with turbine condensers.

Liquid-Metal Film Condensation

Little information is available on the condensation of metallic vapors. The above correlations have not therefore been found adequate for the prediction of heat-transfer coefficients in liquid metal condensation. The discrepancy seems to be due to the existence of a thermal resistance at the vapor-liquid interface manifesting itself in fairly large temperature drops there [119, 120].

Dropwise Condensation

In dropwise condensation no continuous liquid film blankets the cooling surface and much of the latter remains bare. The vapor condenses in the form of drops which originate at surface nuclei, grow and then detach from the surface under the influence of gravity. The process has similarities to its opposite, nucleate boiling. Dropwise condensation occurs when the cooling surface is prevented from being wetted such as by a contaminant.

Because of the mechanism of dropwise condensation, heat-transfer coefficients about 4 to 20 times those of filmwise condensation have been obtained. The advantages of dropwise condensation are therefore obvious. It has, however, been reliably obtained with steam only, and under carefully controlled conditions. Additives, to promote dropwise condensation by preventing the condensate from wetting the surface, have been used with varying degrees of success. They are effective for limited periods of time. Such additives may not be desirable anyway, such as in boiling-reactor systems. Unlike filmwise condensation, Shea and Krase [121] reported that the heat-transfer coefficient in dropwise condensation increases with the temperature difference $(t_s - t_w)$ to a maximum, then decreases because of increased condensation rates which blanket the surface and cause an approach to filmwise condensation conditions. They also showed a large velocity effect. The presence of noncondensable gases affects dropwise condensation in a manner similar to that for filmwise condensation.

Because of the difficulty of maintaining reliable dropwise condensation in practice, filmwise heat-transfer coefficients are recommended for design purposes.

PROBLEMS

11-1. If the surface tension between the liquid and vapor for water at 212°F is 4.03 lb_f/ft, calculate the amount of liquid superheat necessary to generate a 4.68×10^{-3} in. diameter bubble at atmospheric pressure (average).

11-2. In an experiment on pool boiling of water, the heat flux and water temperature and pressure were simultaneously increased so that saturation boiling occured at all times. Burnout occured when the pressure reached 300 psia. Assuming for simplicity that burnout heat transfer occured solely by radiation, and that the radiation heat transfer coefficient is 20 Btu/hr ft² °F, estimate the temperature of the heating surface at burnout.

11-3. What should the maximum allowable volumetric thermal source strength be for a flat-plate fuel element operating in saturated pool boiling water on the lunar surface where the gravitational forces are one sixth those on earth? The system pressure is 100 psia. The element measures 4×0.25 in. Ignore cladding and use a 2:1 safety factor.

11-4. Water flows at 2,000 psia and 500°F through a hollow cylindrical fuel element having inner and outer diameters of 1 and 1.28 in. respectively, including 0.030 in. cladding on both sides. The outer surface is surrounded by graphite and may be considered insulated. For a water velocity of 30 fps, estimate the volumetric thermal source strength that would cause burnout.

11-5. A 5-ft high boiling-water reactor channel operates at 600 psia. Water enters the channel at 470°F and 4 fps. 2×10^6 Btu/hr are generated sinusoidally in the channel. The channel may be assumed circular, 2-in. in diameter, with fuel cross-sectional area of 0.0218 ft². Neglect the extrapolation lengths. Assuming that burnout will probably occur at the center plane, determine whether the channel is safe.

11-6. A boiling-water reactor channel operating at 1,000 psia is 7.68 ft high and contains 0.80 in. diameter fuel rods (including 0.025 in. cladding). Water enters the channel saturated at 15 fps. The flow area per fuel element is 0.0018 ft². It is assumed for simplicity that energy is generated sinusoidally in the channel with a maximum volumetric thermal source strength of 2×10^7 Btu/hr ft³, and that the extrapolation lengths may be ignored. Calculate the critical heat fluxes at the entrance, center, and exit of the channel. Sketch the variation of critical heat flux and actual heat flux versus height up the channel and calculate the minimum safety factor (ratio of critical-to-actual fluxes) in the channel.

11-7. Liquid sodium flows at 20 fps and 700°C inside a 4-ft long hollow cylindrical fuel element having diameters 1 and 0.5 in. respectively. The outside surface of the element may be considered insulated. Using a safety factor of 2, what should be the highest value of volumetric thermal source strength to avoid burnout?

11-8. A pressurized-water reactor channel operating at 2000 psia contains fuel with a total surface area of 2.7 ft^2 and has a flow area of 0.01 ft^2. 4×10^4 lb_m/hr of water enter the channel at 580°F. When 2.33×10^6 Btu/hr were generated in the channel, the heat flux at midplane was just sufficient to cause burnout. Find the ratio of maximum-to-average heat flux at these conditions.

11-9. Saturated steam at 1000 psia enters the top of a 1 in. diameter, 12 ft long vertical tube at 10 fps. The tube walls are held at 530°F. Estimate the heat transfer, Btu/hr, and the mass flow rates of steam and water at the tube exit, lb_m/hr.

11-10. A turbine condenser operating at 1 psia contains 400 2.4 in. diameter, 20 ft long horizontal tubes arranged in 40 rows, 10 tubes deep. Cooling water flows inside the tubes in a one-pass arrangement with a heat transfer coefficient of 312 Btu/hr ft^2 °F. The conductive resistance in the tube walls may be ignored. If the tube walls are at 98°F, calculate (a) the amount of steam condensed, lb_m/hr, and (b) the average cooling water temperature.

chapter **12**

Two-Phase Flow

12-1. INTRODUCTION

Two-phase flow phenomena are of utmost importance in liquid-cooled reactors. When a flowing coolant undergoes a partial change in phase, two-phase flow occurs, giving rise to interesting heat-transfer and fluid-flow problems. The phenomena occur regularly in boiling-water reactors, and in pressurized-water and other liquid-cooled reactors at high power densities. In general, two-phase flow implies the concurrent flow of two phases, such as liquid and its vapor (one component), liquid and gas (two component), gas and solid, etc. This chapter deals mainly with the first type, liquid and vapor flow. Flow in reactor channels, past restrictions, and critical flow will be discussed.

Two-phase flow takes several forms, shown in Fig. 12-1. These have been observed only in adiabatic two-phase, two-component flow, such as with air and water. However, there is good reason to believe that they exist in nonadiabatic two-phase flow. *Bubble flow* is the case in which individual dispersed bubbles move independently up the channel. *Plug, or slug, flow* is the case where patches of coalesced vapor fill most of the channel cross section as they move upward. Plug flow has been reported as being both a stable and an unstable transition flow between bubble flow and the next type, annular flow. In *annular flow* the vapor forms a

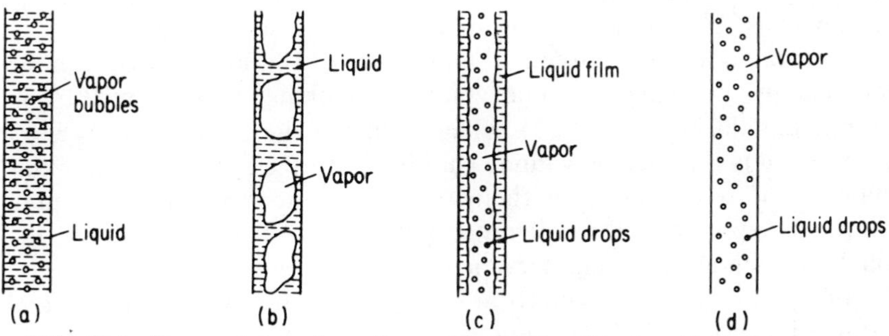

FIG. 12-1. Types of two-phase flow. (a) Bubble; (b) plug (slug); (c) annular (channelized); (d) fog.

continuous phase, carrying only dispersed liquid droplets, and travels up the channel core, leaving an annulus of superheated liquid adjacent to the walls. The fourth type shown in Fig. 12-1 is called *fog, dispersed,* or *homogeneous* flow. This is the opposite of the first type, bubble flow, in that in the latter case the vapor fills the entire channel and the liquid is dispersed throughout the vapor in the form of individual droplets.

Little is now known about the exact type of two-phase flow taking place in high-pressure reactor systems. It is conceivable, however, that it is determined by the *void fraction* (volumetric ratio of vapor to vapor and liquid). It is also believed that the type of flow is a function of the Froude number (a measure of the ratio of inertia to gravity forces).

12-2. QUALITY AND VOID FRACTION IN A NONFLOW SYSTEM

The importance of coolant-density and void-fraction studies in a boiling-reactor core will be discussed in Chapter 14. It suffices now to point out that, among several things, when a moderator boils, vapor voids displace this moderator and affect the reactivity of the system. Consequently only vapor of extremely low quality (a few percent) is to be produced in the core. The *quality* x of a vapor-liquid mixture in a nonflow system, or where no gross relative motion between the vapor and liquid phases exists, is defined as

$$x = \frac{\text{Mass of vapor in mixture}}{\text{Total mass of mixture}} \tag{12-1}$$

The *void fraction* α is defined as

$$\alpha = \frac{\text{Volume of vapor in mixture}}{\text{Total volume of liquid-vapor mixture}} \tag{12-2}$$

The term *void* here is somewhat misleading, since there is actually no void. However, the vapor present in "voids" has no moderating value because of its relatively low density.

The relationship between x and α in a nonflow system can be obtained by assuming a certain volume containing 1 lb_m of mixture in thermal equilibrium (Fig. 12-2). That volume will be equal to $(v_f + xv_{fg})$ ft^3, where v is the specific volume (ft^3/lb_m). In the notation used in this chapter, the subscripts f, g (below), and fg refer to saturated liquid, saturated vapor, and the difference between the two, respectively, the usual notation used in steam practice.

In an equilibrium mixture, the two phases are saturated liquid and saturated vapor and the volume of vapor present is equal to its mass, x lb_m, times its specific volume v_g. α is thus given by

$$\alpha = \frac{xv_g}{v_f + xv_{fg}} \tag{12-3a}$$

This equation can also be written in the form

$$\alpha = \frac{1}{1 + \left(\frac{1-x}{x}\right)\left(\frac{v_f}{v_g}\right)} \tag{12-3b}$$

where the specific volumes are all taken at the system pressure from appropriate thermodynamic-property tables, such as the Keenan and Keyes steam tables in Appendix D. Equation 12-3b serves to show the large values of α associated with small values of x, especially at low pressures.

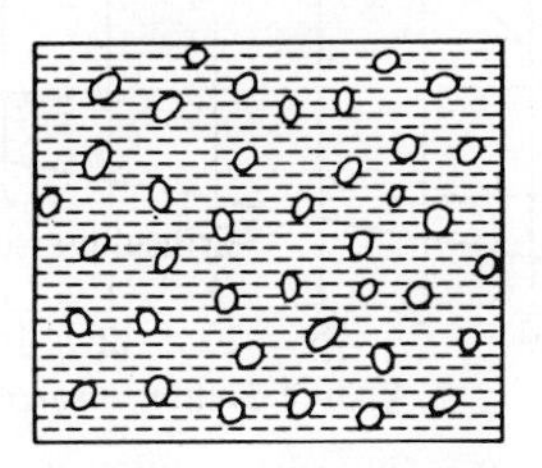

FIG. 12-2. A mixture of liquid and vapor in a nonflow system. Mass of mixture = 1 lb_m; composed of x lb_m saturated vapor plus $1 - x$ lb_m saturated liquid; volume of mixture = $v_f + xv_{fg}$.

Example 12-1. Calculate α corresponding to $x = 2$ percent for ordinary water at atmospheric pressure.

Solution. From the steam tables, at atmospheric pressure,

$$v_f = 0.01672 \text{ ft}^3/\text{lb}_m \qquad \text{and} \qquad v_g = 26.800 \text{ ft}^3/\text{lb}_m$$

$$\alpha = \frac{1}{1 + [(1 - 0.02)/0.02]0.01672/26.800} = 0.971 \quad \text{or} \quad 97.1 \text{ percent}$$

Thus a small steam fraction by mass corresponds to a very large fraction by volume. The difference decreases at high pressure, however. Figure 12-3 shows calculated values for α versus x for light water at various pressures. Examination of the curves reveals the following:

1. For constant x, α decreases with pressure. As the pressure approaches the critical pressure (3,206 psia for ordinary water, 3,212 psia for heavy water) α rapidly approaches x. At the critical pressure the two phases are indistinguishable and $\alpha = x$.
2. For any one pressure, $d\alpha/dx$ decreases with x.

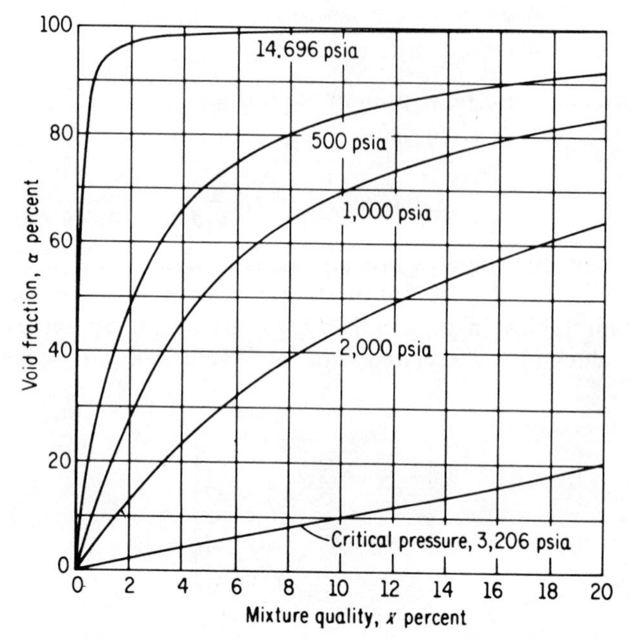

FIG. 12-3. α versus x for nonflow water system.

3. At low values of x (such as those used in boiling-type reactors), $d\alpha/dx$ increases as the pressure decreases and becomes very severely large at low pressure. This has a bearing on reactor stability [2].

12-3. THE FLOW SYSTEM

In the above calculations it was assumed that no relative motion existed between the two phases, i.e., between the vapor bubbles and the liquid. However, if a two-phase mixture is moving, say in a vertical direction as between fuel elements, the vapor, because of its buoyancy, has a tendency to slip past the liquid, i.e., move at a higher velocity than that of the liquid. While there are variations in velocity within each phase, no accurate method is now available to predict phase and velocity distributions in an actual two-phase flow system. The *lumped-system* solution in which each phase is assumed to move at one speed, however, has been found satisfactory in most cases. In the lumped system, a *slip ratio* S, equal to 1.0 in nonflow or homogeneous flow and greater than 1.0 in nonhomogeneous two-phase systems is used. It is defined as the ratio of the average velocity of the vapor V_g to that of the liquid V_f. Thus

$$S = \frac{V_g}{V_f} \tag{12-4}$$

The slip ratio modifies the relationship between void fraction and quality developed in the previous section. This will now be shown with the help of Fig. 12-4. Figure 12-4a shows a two-phase mixture flowing upward in a channel. A certain section, between the dotted lines, small enough so that x and α remain unchanged, is considered.

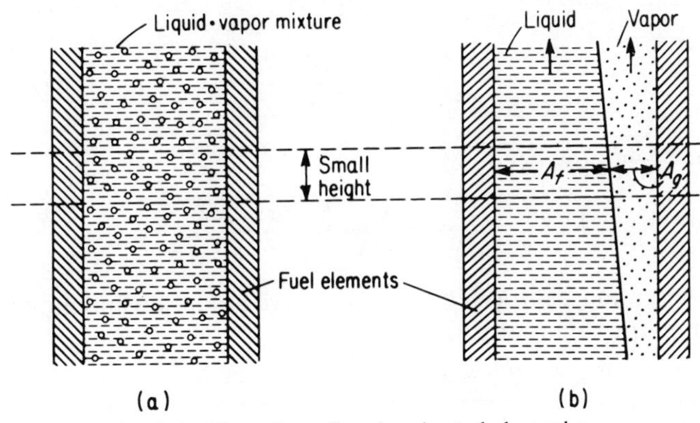

FIG. 12-4. Two-phase flow in a heated channel.

In a flow system, the quality at any one cross section is defined by

$$x = \frac{\text{Mass-flow rate of vapor}}{\text{Mass-flow rate of mixture}} \tag{12-5}$$

Thus if the total mass flow of the mixture is $\dot{m}_t$ (lb_m/hr), the vapor-flow rate is $x\dot{m}_t$ and the liquid-flow rate is $(1 - x)\dot{m}_t$, where x is the quality at the particular section in question. Applying the continuity equation, the velocities of vapor and liquid are given by

$$V_g = \frac{v_g \dot{m}_t}{A_g}$$

and

$$V_f = \frac{v_f(1 - x)\dot{m}_t}{A_f}$$

where A_g and A_f are the cross-sectional areas of the two phases, perpendicular to flow direction, if the two phases are imagined to be completely separated from each other (Fig. 12-4b). Combining the above equations gives

$$S = \frac{V_g}{V_f} = \frac{x}{1 - x} \frac{A_f}{A_g} \frac{v_g}{v_f} \tag{12-6}$$

The void fraction in the section considered is the ratio of the vapor-phase volume to the total volume within the section. In the small section of channel considered, this is the same as the ratio of cross-sectional area of vapor A_g to the total cross-sectional area of the channel. Thus

$$\alpha = \frac{A_g}{A_g + A_f}$$

or

$$\frac{A_f}{A_g} = \frac{1-\alpha}{\alpha} \tag{12-7}$$

Equation 12-4a now becomes

$$S = \frac{x}{1-x}\,\frac{1-\alpha}{\alpha}\,\frac{v_g}{v_f} \tag{12-8}$$

This equation can be rearranged to give a relationship between α and x, including the effect of slip, as

$$\alpha = \frac{1}{1 + \left(\frac{1-x}{x}\right)\left(\frac{v_f}{v_g} S\right)} = \frac{1}{1 + \left(\frac{1-x}{x}\right)\psi} \tag{12-9}$$

and

$$x = \frac{1}{1 + \left(\frac{1-x}{x}\right)\frac{1}{\psi}} \tag{12-10}$$

where

$$\psi = \frac{v_f}{v_g} S \tag{12-11}$$

Note that the relationship between α and x for no slip ($S = 1$) given by Eqs. 12-3 is a special case of the general relationship given by Eqs. 12-9.

The effect of slip is to decrease the value of α corresponding to a certain value of x below that which exists for no slip. This can be seen if Eq. 12-9 is examined. At constant pressure and quality, the factor $(1-\alpha)/\alpha$ is directly proportional to S. Thus, α decreases with S. A high S is thus an advantage from both the heat-transfer and moderating-effect standpoints. Figure 12-5 shows α versus x for light water at 1,000 psia and several slip ratios.

S has been experimentally found to decrease with both the system pressure and the volumetric flow rate and to increase with power density. It has also been found to increase with the quality at high

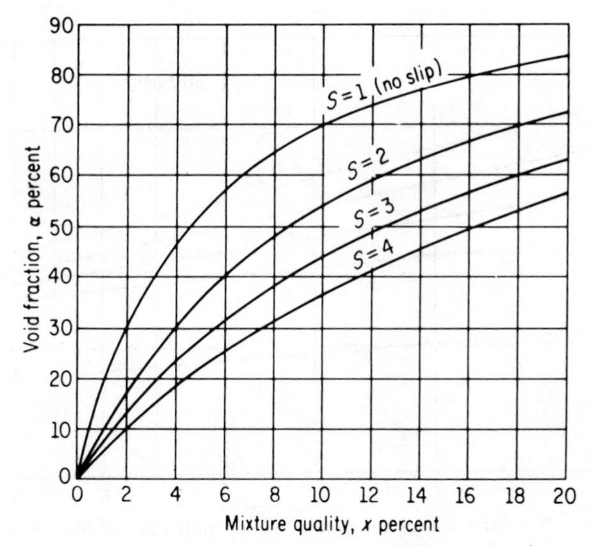

FIG. 12-5. α versus x for water at 1,000 psia and various slip ratios in a flow system.

pressures but to decrease with it at very low pressures [122]. Figures 12-6 to 12-8 show some of these effects. Figure 12-9 gives α versus $x\rho_f/\rho_g$ for various values of V_i^2/D, where V_i is the inlet velocity, at a pressure of 32 atm [123], from which S may be computed.

As a function of the channel length, S has been found to increase rapidly at the beginning and then more slowly as the channel exit is

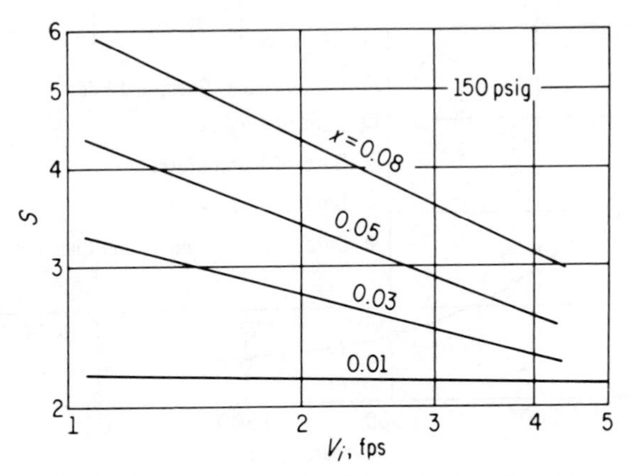

FIG. 12-6. Working curves for predicting slip ratios at 150 psig (Ref. 122).

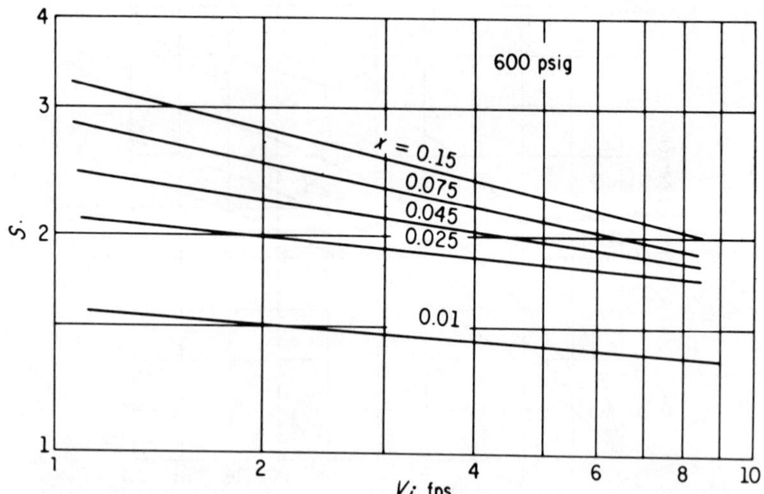

FIG. 12-7. Working curves for predicting slip ratios at 600 psig (Ref. 122).

approached. At the exit itself, turbulence seems to cause a sudden jump in the value of S. This effect is shown in Fig. 12-10 [124].

Von Glahn [125] proposed the following empirical relationship between x and α, based on much of the experimental data available at

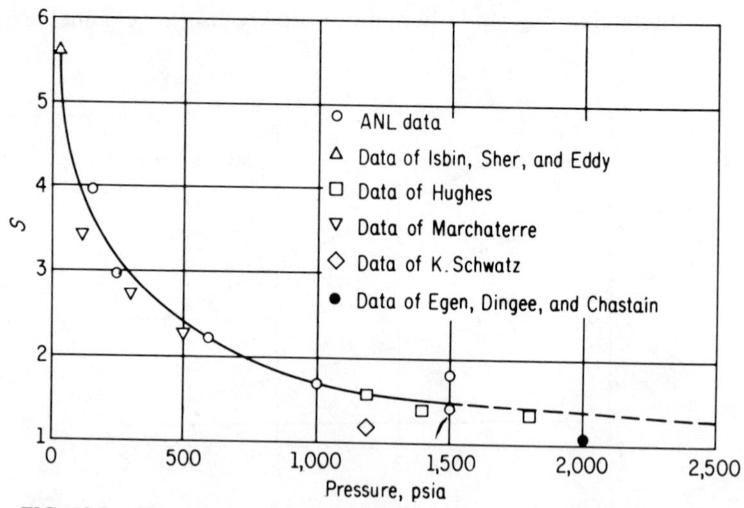

FIG. 12-8. Slip ratio as a function of pressure for $V_i \approx 2$ ft/sec and $x \approx 0.05$ (Ref. 122).

the time for water and covering a wide range of operating conditions and channel geometries:

$$\frac{1}{x} = 1 - \left(\frac{v_g}{v_f}\right)^{0.67} \left[1 - \left(\frac{1}{\alpha}\right)^{(v_g/v_f)^{0.1}}\right] \tag{12-12}$$

Experimental data or theoretical correlations for S covering all possible operating and design variables do not now exist. In boiling-reactor studies, values for S may be estimated from data that closely approach those of interest. In this a certain amount of individual judgment is necessary. Otherwise, experimental values of S under similar conditions of a particular design must be obtained. This procedure is usually expensive and time-consuming but may be necessary in some cases. The importance of obtaining accurate values of S may best be emphasized by the following: One step in the procedure of core channel design in to set a maximum value of α at the channel exit.

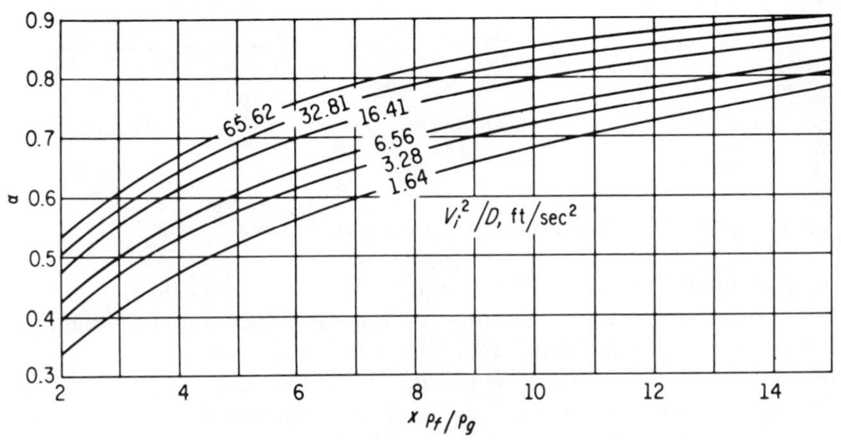

FIG. 12-9. Void fraction as a function of V_i^2/D and $x\ \rho_f/\rho_g$ for vertical tubes ($p = 32$ atm) (Ref. 123).

This is usually determined from nuclear (moderation) considerations. A corresponding value of x, at the selected S, is then found from the above equations. The latter determines the heat generated in the channel.

In design, the usual procedure is to assume a constant value of S along the length of the channel. This, of course, is a simplification, which may introduce further error into the results. In Fig. 12-10, however, S is seen to be fairly constant over most of the channel length, indicating this assumption to be a good one.

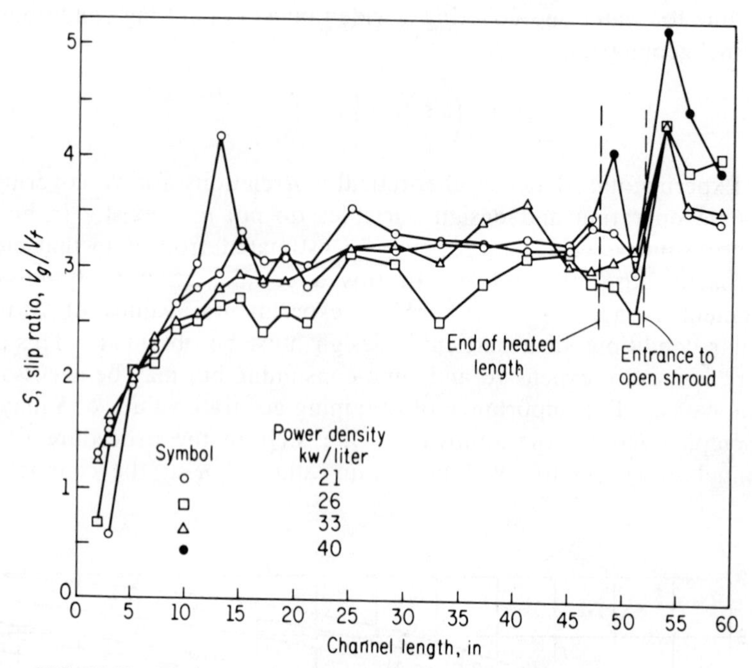

FIG. 12-10. Typical plots of slip ratio versus channel length at 114.9 psia (Ref. 124).

12-4. BOILING AND NONBOILING HEIGHTS

A boiling fluid in a flow system, such as a coolant flowing through the various channels of a boiling-reactor core, encounters several resistances to flow, manifesting themselves in pressure drops. Two of the largest are (1) the pressure drop due to friction (Sec. 12-5) and (2) the pressure drop caused by the acceleration of the coolant that undergoes an increase in volume as it receives heat in the channel (Sec. 12-6). Other pressure drops are due to the obstruction of flow by submerged bodies such as spacers, handles, tie plates, etc., and to abrupt changes in flow areas, such as at entrances and exits to the core, downcomers, etc. (Secs. 12-7 to 12-10).

In order to evaluate correctly the above-mentioned and other pressure drops as well as the average density in a boiling channel (necessary, among other things, for evaluating the driving head in a natural-circulation reactor), it is necessary to calculate the nonboiling height in the channel. The *nonboiling height* H_0 (Fig. 12-11) is that in which only the sensible heat is added to the incoming subcooled coolant at the channel bottom.

At $z = H_0$, the coolant becomes saturated. The remainder of the channel is that in which boiling takes place and is called the *boiling height* H_B.

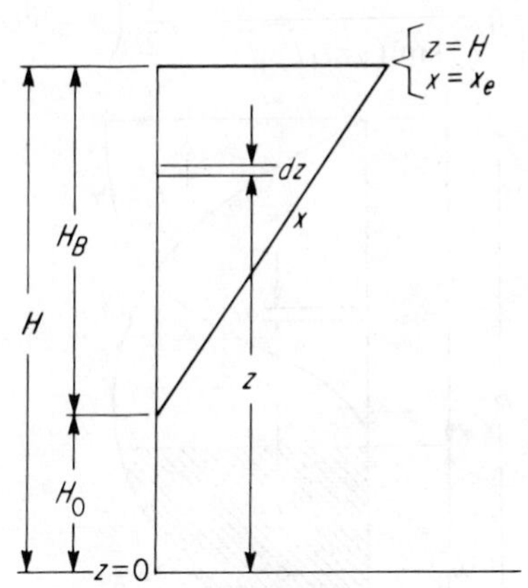

FIG. 12-11. Variation of quality with height in a uniformly heated channel.

Some subcooled boiling may, of course, occur in H_0 but will have little effect on pressure drops or density. The ratios H_0/H and H_B/H may be evaluated from the ratio of sensible heat added to total heat added in the channel:

$$\frac{q_s}{q_t} = \frac{h_f - h_i}{(h_f + x_e h_{fg}) - h_i} \tag{12-13}$$

where q_s = sensible heat added per pound mass of incoming coolant, Btu/lb_m

q_t = total heat added in channel per pound mass of incoming coolant, Btu/lb_m

h_i = enthalpy at inlet of channel, Btu/lb_m

The ratio q_s/q_t is related to the height ratio H_0/H according to the mode of heat addition in the channel. In the case of uniform heat addition (Fig. 12-11),

$$\frac{q_s}{q_t} = \frac{H_0}{H} \tag{12-14}$$

and $H_B/H = 1 - H_0/H$. In the case of sinusoidal heat addition, with extrapolation lengths neglected, Fig. 12-12, the area under the q''' or q'(heat added per unit length of channel) curve is proportional to q_t. The area bounded by $z = 0$ and $z = H_0$ is equivalent to q_s. Thus

$$\frac{q_s}{q_t} = \frac{\int_0^{H_0} q_c' \sin (\pi z/H)\, dz}{\int_0^{H} q_c' \sin (\pi z/H)\, dz} = \frac{1}{2}\left(1 - \cos \frac{\pi H_0}{H}\right) \tag{12-15}$$

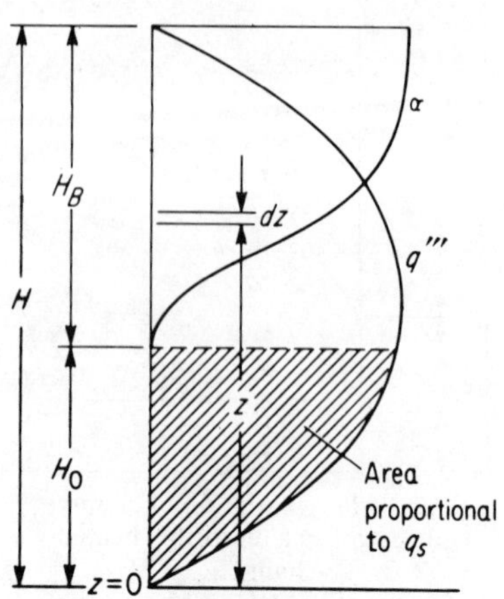

FIG. 12-12. Void fraction with sinusoidal heat addition in a boiling channel.

where q_c' is the heat addition per unit length of channel at the channel center, $z = H/2$, Btu/ft. H_0/H (and H_B/H) may then be evaluated from Eq. 12-15.

For other modes of heat addition that can be represented by a simple function, a similar procedure may be used. Otherwise, a stepwise or graphical solution becomes necessary.

12-5. THE FRICTION DROP IN A TWO-PHASE CHANNEL

The friction drop in a boiling channel of height H, $(\Delta p_f)_H$, is composed of friction due to single-phase (liquid) flow in H_0, $(\Delta p_{sp})_{H_0}$, and two-phase flow in H_B, $(\Delta p_{TP})_{H_B}$, or

$$(\Delta p_f)_H = (\Delta p_{SP})_{H_0} + (\Delta p_{TP})_{H_B} \tag{12-16}$$

Single-Phase Friction

The single-phase friction pressure drop in H_0 $(\Delta p_{SP})_{H_0}$ is evaluated by the standard methods used in single-phase flow, such as the Darcy formula. Thus

$$(\Delta p_{SP})_{H_0} = f_0 \frac{H_0}{D_e} \frac{\bar{\rho}_0 \bar{V}_0^2}{2g_c} \tag{12-17}$$

where f_0 = friction factor in H_0, dependent upon average Reynolds' number and wall roughness in H_0, dimensionless, (Appendix F)

D_e = equivalent diameter of channel, ft

$\bar{\rho}_0$ = average liquid density in H_0, lb_m/ft^3

$\bar{V}_0$ = average liquid velocity in H_0, ft/hr

g_c = conversion factor, 4.17×10^8 lb_m ft/lb_f hr^2

$\bar{\rho}_0$ may be evaluated, for linear change in density in H_0, or for small degrees of subcooling, quite closely from

$$\bar{\rho}_0 = \frac{1}{2}(\rho_i + \rho_f) = \frac{1}{2}\left(\frac{1}{v_i} + \frac{1}{v_f}\right) \quad (12\text{-}18)$$

where subscripts i and f again refer to channel inlet and liquid saturation. The specific volumes v_i and v_f are obtained at t_i and t_f respectively. Likewise $\bar{V}_0$ may be evaluated from

$$\bar{V}_0 = \frac{1}{2}(V_i + V_{f_0}) \quad (12\text{-}19a)$$

where V_{f_0} = the velocity of the saturated liquid at $z = H_0$. For a constant area channel and steady flow,

$$\bar{V}_0 = \frac{1}{2} V_i \left(1 + \frac{v_f}{v_i}\right) \quad (12\text{-}19b)$$

The average Reynolds number in H_0, at which f_0 is evaluated is also given by

$$\overline{\text{Re}}_0 = \frac{D_e \bar{V}_0 \bar{\rho}_0}{\bar{\mu}_0} = \frac{D_e V_i \rho_i}{\bar{\mu}_0} \quad (12\text{-}20)$$

where $\bar{\mu}_0$, the average liquid viscosity, is obtained quite closely from $\frac{1}{2}(\mu_i + \mu_f)$.

Two-Phase Friction

Two-phase flow friction is greater than single-phase friction for the same height and mass-flow rate. The difference appears to be a function of the type of flow (see Fig. 12-1), and results from increased flow speeds. It is experimentally determined by measuring the total pressure drop in a two-phase flow system and subtracting calculated drops due to single phase in H_0, acceleration (Sec. 12-6), and inlet, exit, contraction and expansion losses (Secs. 12-7 to 12-10).

The two-phase friction drop in H_B, $(\Delta p_{TP})_{H_B}$, is usually evaluated by first calculating a single-phase pressure drop in H_B assuming that only saturated liquid of the same total mass-flow rate exists in the channel. This is given by

$$(\Delta p_{SP})_{H_B} = f_B \frac{H_B}{D_e} \frac{\rho_f V_{f_0}^2}{2g_c} \tag{12-21}$$

where f_B is the friction factor in H_B, a function of a single (liquid)-phase Reynolds number in H_B, $D_e \rho_f V_f / \mu_f = D_e \rho_i V_i / \mu_f$.

A *two-phase friction multiplier, R,* greater than 1, is then multiplied by the above to give $(\Delta p_{TP})_{H_B}$. Thus

$$\bar{R} = \frac{(\Delta p_{TP})_{H_B}}{(\Delta p_{SP})_{H_B}} \tag{12-22}$$

and the total friction pressure drop in the channel, Eq. 12-16, is now given by

$$(\Delta p_f)_H = \left(f_0 \frac{H_0}{D_e} \frac{\bar{\rho}_0 \bar{V}_0^2}{2g_c}\right) + \bar{R}\left(f_B \frac{H_B}{D_e} \frac{\rho_f V_f^2}{2g_c}\right) \tag{12-23}$$

If the degree of subcooling is not too great, Eq. 12-23 could be written, with little error, in the form

$$(\Delta p_f)_H = \left(f_B \frac{\rho_f V_f^2}{2g_c D_e}\right)(H_0 + \bar{R} H_B) \tag{12-24}$$

Many attempts at finding a correlation for $\bar{R}$ have been made with varying degrees of success. The most widely accepted correlation is that of Martinelli and Nelson [126]. It is based upon previous correlations of two-phase turbulent flow by Martinelli and represents an extrapolation of pressure-drop data obtained for isothermal flow of air and various liquids at low pressures and temperatures. The Martinelli-Nelson correlation is presented in the form of a chart (Fig. 12-13) giving $\bar{R}$ as a function of pressure and exit quality.

As seen, the Martinelli-Nelson correlation shows that $\bar{R}$ is a function of quality and pressure but is independent of mass flow rate. While that correlation is widely used at reactor operating conditions, an experimental investigation by Huang and El-Wakil [127] on a two-phase flow loop with electrically heated rods (simulating fuel elements) at low pressures and consequently high void fractions (as would be expected in a loss of pressure accident) has shown that the two-phase friction multiplier is a function of quality, pressure and also mass velocity at these conditions. It has also shown that the Martinelli-Nelson correlation greatly underestimates that multiplier under these same conditions. Figure 12-14 shows the two-phase friction multiplier at the relatively low pressures of 25, 35, and 50 psia for various mass flow rates, as well as the

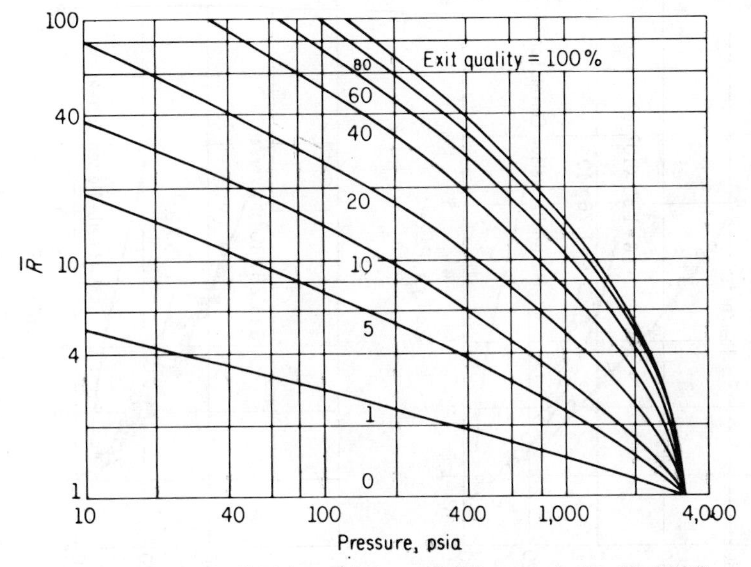

FIG. 12-13. Martinelli-Nelson friction-multiplier correlation (Ref. 126).

Martinelli-Nelson multiplier under the same conditions for comparison.

Another correlation for $\bar{R}$, given as a simple function of the exit void fraction, α_e, but one that lacks a pressure term in it, has been derived by Lottes and Flinn [128] on the basis of an annular-flow model (Fig. 12-1), uniform heating, and very low qualities. The model was conceived when plots of measured values of $\bar{R}$ versus various parameters for a wide range of operating conditions but with uniform heat flux indicated that, for low qualities, $\bar{R}$ is a function only of $1/(1 - \alpha_e)$.

It was theorized then that the increased friction drop was due to the increased liquid velocity caused by the reduced liquid flow area. Note that $1/(1 - \alpha_e)$ is the ratio of the total channel area to the liquid flow area at the exit. The friction in the annular-flow model considered would be mainly due to the friction between the liquid and the channel walls.

The above model is, of course, restrictive, and Lottes and Flinn reported later [129] that their correlation represents most data more qualitatively than quantitatively. Its derivation, however, is useful in understanding some two-phase flow concepts and will be presented here. Since the liquid velocity is variable along the channel, the Darcy formula is written in the differential form:

$$dp_{TP} = f_B \frac{\rho_f dz}{D_e} \frac{V_f^2(z)}{2g_c} \tag{12-25}$$

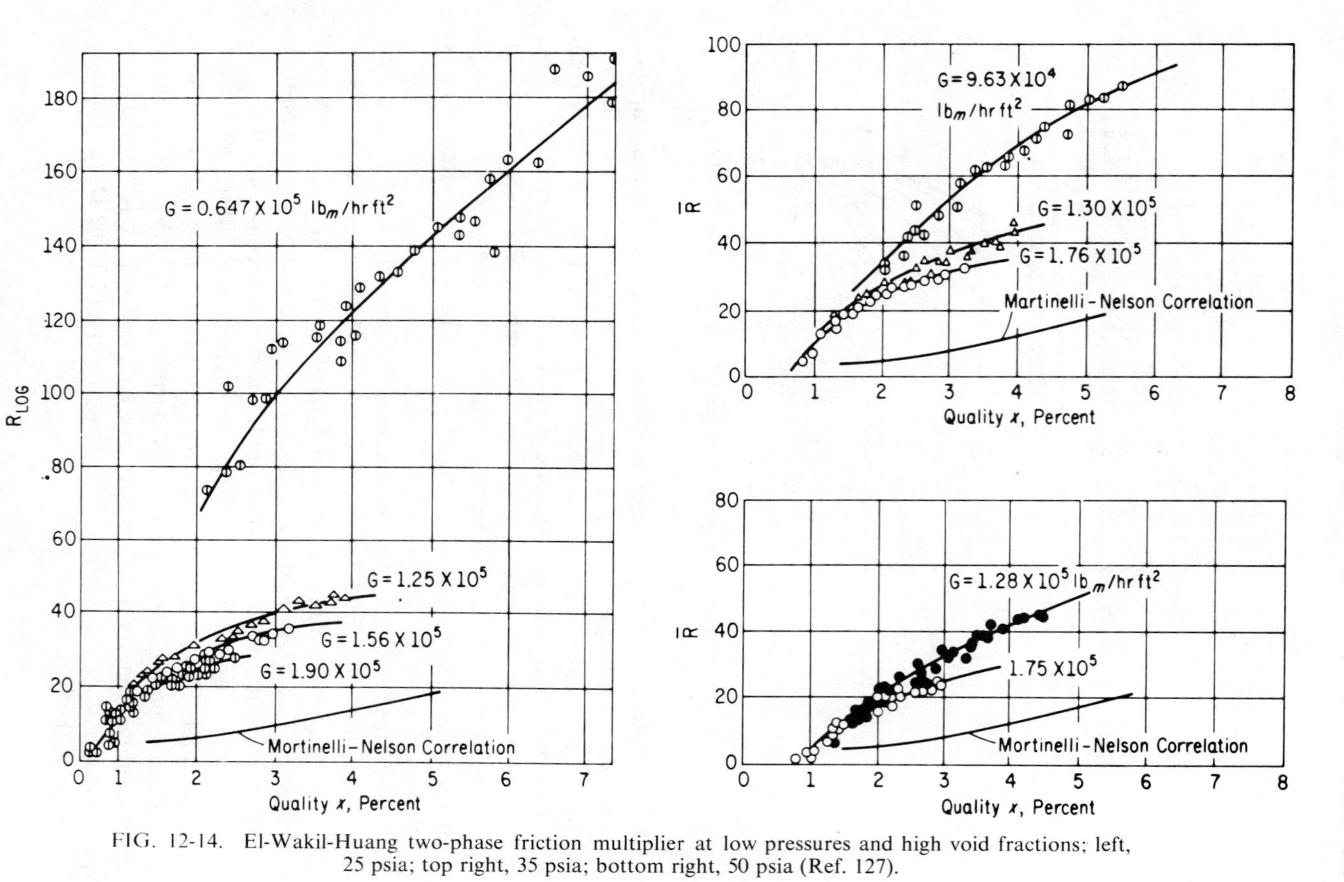

FIG. 12-14. El-Wakil-Huang two-phase friction multiplier at low pressures and high void fractions; left, 25 psia; top right, 35 psia; bottom right, 50 psia (Ref. 127).

Then

$$(\Delta p_{TP})_{H_B} = f_B \frac{\rho_f}{2g_c D_e} \int_{H_0}^{H} V_f^2(z)\, dz \tag{12-26a}$$

where $V_f(z)$ = speed of the saturated liquid at z, feet per hour and f_B = friction factor in H_B, assumed constant. To evaluate the integral, a relationship between $V_f(z)$, z and α_z, the void fraction at z, will be obtained. From the continuity equation, it can be shown that the mass-flow rates of the total mixture, $\dot{m}_t$, and of the saturated liquid at z, $\dot{m}_f(z)$, are given by

$$\dot{m}_t = V_{f_0} A \rho_f$$

and

$$\dot{m}_f(z) = (1 - x_z)\dot{m}_t = V_f(z) A (1 - \alpha_z)\rho_f$$

so that

$$\frac{V_f(z)}{V_{f_0}} = \frac{1 - x_z}{1 - \alpha_z} \tag{12-27}$$

where x_z is the quality at z.

For small values of x, $1 - x \cong 1$. Thus

$$(\Delta p_{TP})_{H_B} = f_B \frac{\rho_f V_{f_0}^2}{2g_c D_e} \int_{H_0}^{H} \frac{1}{(1 - \alpha_z)^2}\, dz \tag{12-26b}$$

For the case of uniform heat flux, the quality is a linear function of z, Fig. 12-11, or

$$x_z = x_e \frac{z - H_0}{H_B} \tag{12-28}$$

Equation 12-10 can be modified, for small values of x, to

$$x_z = \frac{\alpha_z}{1 - \alpha_z} \psi \tag{12-29}$$

and similarly

$$x_e = \frac{\alpha_e}{1 - \alpha_e} \psi \tag{12-30}$$

If S is assumed constant over H_B, Eqs. 12-28 to 12-30 can be combined to give

$$z - H_0 = \left(\frac{1 - \alpha_e}{\alpha_e}\right) \left(\frac{\alpha_z}{1 - \alpha_z}\right) H_B \tag{12-31}$$

from which

$$dz = \left(\frac{1-\alpha_e}{\alpha_e}\right)\left(\frac{1}{1-\alpha_z}\right)^2 H_B d\alpha_z \tag{12-32}$$

Substituting this equation in Eq. 12-26*b* gives

$$(\Delta p_{TP})_{H_B} = f_B \frac{H_B}{D_e} \frac{\rho_f V_{f_0}^2}{2g_c} \int_0^{\alpha_e} \left(\frac{1-\alpha_e}{\alpha_e}\right)\left(\frac{1}{1-\alpha_z}\right)^4 d\alpha_z \tag{12-26c}$$

The portion outside the integral sign on the right-hand side of the above equation is recognized as $(\Delta p_{SP})_{H_B}$, Eq. 12-21. The integral is therefore the two-phase multiplier $\bar{R}$. Thus

$$\bar{R} = \left(\frac{1-\alpha_e}{\alpha_e}\right)\int_0^{\alpha_e}\left(\frac{1}{1-\alpha_z}\right)^4 d\alpha_z$$

$$= \left(\frac{1-\alpha_e}{\alpha_e}\right)\left[\frac{1}{3}\left(\frac{1}{1-\alpha_z}\right)^3\right]_{\alpha_z=0}^{\alpha_z=\alpha_e}$$

$$= \left(\frac{1-\alpha_e}{\alpha_e}\right)\left[\frac{1}{3}\left(\frac{1}{1-\alpha_e}\right)^3 - \frac{1}{3}\right] \tag{12-33a}$$

This equation can be rearranged to the more convenient form

$$\bar{R} = \frac{1}{3}\left[1 + \left(\frac{1}{1-\alpha_e}\right) + \left(\frac{1}{1-\alpha_e}\right)^2\right] \tag{12-33b}$$

which is the Lottes-Flinn correlation for $\bar{R}$.

Example 12-2. A 6-ft-high boiling-water reactor channel has an equivalent diameter of 0.145 ft, and fuel cladding corresponding to smooth-drawn tubing. It receives heat sinusoidally and operates at a pressure of 1,000 psia, an exit quality of 8 percent, an inlet velocity of 3 fps, and inlet water temperature of 522°F. Compute the friction drop in the channel. Assume that the friction multipliers apply to sinusoidal heating.

Solution. Using Eq. 12-13 and the steam tables (Appendix D),

$$\frac{q_s}{q_t} = \frac{542.4 - 514.3}{(542.4 + 0.08 \times 649.4) - 514.3} = 0.351$$

Thus

$$0.351 = \frac{1}{2}\left(1 - \cos\frac{\pi H_0}{H}\right),$$

or

$$\cos\frac{\pi H_0}{H} = 0.298$$

from which $H_0/H = 0.404$, $H_0 = 2.424$ ft, and $H_B = 3.576$ ft.

We now evaluate f_0 at a Reynolds number corresponding to the average flow conditions in H_0. The product ρV is constant along the channel. The inlet liquid density ρ_i (at 522°F) is $1/v_i = 1/0.0209 = 47.847$ lb$_m$/ft³. Thus

$$\text{Re}_0 = \frac{D_e \rho_i V_i}{\frac{1}{2}(\mu_f + \mu_i)} = \frac{0.145(3 \times 3600)47.847}{\frac{1}{2}(0.233 + 0.244)} = 314{,}165$$

where μ_f is evaluated by interpolation from Table E-1, Appendix. The relative roughness corresponding to smooth-drawn tubing is $0.000005/0.145 = 0.0000345$. Thus, from Appendix F,

$$f_0 = 0.0143$$

$$\rho_f = \frac{1}{v_f} = \frac{1}{0.0216} = 46.296 \text{ lb}_m/\text{ft}^3$$

$$\bar{\rho}_0 = \frac{1}{2}(\rho_i + \rho_f) = \frac{1}{2}(47.847 + 46.296) = 47.072 \text{ lb}_m/\text{ft}^3$$

$$V_{f_0} = V_i \frac{v_f}{v_i} = 3 \times \frac{0.0216}{0.02094} = 3.095$$

$$\bar{V}_0 = \frac{1}{2}(V_i + V_{f_0}) = \frac{1}{2}(3 + 3.095) = 3.047 \text{ fps}$$

The single-phase friction drop in H_0 is

$$(\Delta p_{SP})_{H_0} = f_0 \frac{H_0}{D_e} \frac{\bar{\rho}_0 \bar{V}_0^2}{2g_c} = 0.0143 \frac{2.424}{0.145} \frac{47.072(3.047 \times 3600)^2}{2 \times 4.17 \times 10^8}$$
$$= 1.623 \text{ lb}_f/\text{ft}^2 = 0.0113 \text{ lb}_f/\text{in}^2$$

To obtain the two-phase friction drop in H_B:

$$\text{Re}_B = \frac{D_e \rho_i V_i}{\mu_f} = \frac{0.145 \times 47.847(3 \times 3600)}{0.233} = 321{,}580$$

For the same roughness as above, and at Re_B, f_B is obtained from Appendix F.

$$f_B = 0.0142$$

The two-phase friction multiplier may now be obtained from the Martinelli-Nelson Chart, Fig. 12-13. Thus

$$\bar{R} = 2.8$$

The two-phase friction drop in H_B is

$$(\Delta p_{TB})_{H_B} = \bar{R} f_B \frac{H_B}{D_e} \frac{\rho_f V_{f_0}^2}{2g_c} = 2.8 \times 0.0142 \frac{3.576}{0.145} \frac{46.296(3.095 \times 3600)^2}{2 \times 4.17 \times 10^8}$$
$$= 1.756 \text{ lb}_f/\text{ft}^2 = 0.0469 \text{ lb}_f/\text{in}^2$$

Thus the total friction drop in the channel is

$$(\Delta p_f)_H = 0.0113 + 0.0469 = 0.0582 \text{ lb}_f/\text{in}^2$$

In the above case, the degree of subcooling is not too great and the above computations may be simplified by assuming all densities, viscosities and velocities to be those for saturated liquid. Thus

$$
\begin{aligned}
(\Delta p_f)_H &= f\frac{\rho_f V_{f_0}^2}{D_e 2g_c}(H_0 + \overline{R}H_B) \qquad [12\text{-}24] \\
&= 0.0143\,\frac{46.296(3.095 \times 3600)^2}{0.145 \times 2 \times 4.17 \times 10^2}(2.424 + 2.8 \times 3.576) \\
&= 8.452\ \text{lb}_f/\text{ft}^2 = 0.0587\ \text{lb}_f/\text{in}^2
\end{aligned}
$$

If the Lottes-Flinn correlation were to be used under these conditions, a value of slip ratio should first be obtained, as from Fig. 12-7. (A method of correcting S values for pressures other than the 600 psig shown is outlined by Marchaterre and Petrick [122].) Say $S = 1.9$.

Thus, for $x_e = 0.08$, α_e is obtained with the help of Eq. 12-9 as 0.486. Thus

$$
\overline{R} = \frac{1}{3}\left[1 + \left(\frac{1}{1-0.486}\right) + \left(\frac{1}{1-0.486}\right)^2\right] = 2.243
$$

giving a total friction drop of 0.0493 psia.

12-6. THE ACCELERATION PRESSURE DROP

When a coolant expands (or contracts) because of heating, it has to accelerate as it travels through a channel. There will therefore be a force F equal to the change in momentum of the fluid. This force equals an *acceleration pressure drop* times the cross-sectional area of the channel. This pressure drop is usually small in single-phase flow, but can be quite large in two-phase flow. In reactor work where single-phase fluid enters the channel and two-phase fluid, with a high exit void fraction, leaves it, the change in momentum is particularly important. In such a case

$$
F = \Delta p_a A_c = \frac{\dot{m}_f V_{fe}}{g_c} + \frac{\dot{m}_g V_{ge}}{g_c} - \frac{\dot{m}_t V_i}{g_c} \qquad (12\text{-}34)
$$

where Δp_a = acceleration pressure drop, due to change in momentum, between inlet and exit of channel, lb_f/ft^2

A_c = cross sectional area of channel, in square feet, assumed constant

$\dot{m}_t, \dot{m}_f, \dot{m}_g$ = total mass flow rate and mass-flow rates of saturated liquid and vapor, at exit of channel, lb_m/hr

V_i, V_{fe}, V_{ge} = velocities of water at inlet, of saturated liquid and of saturated vapor at exit, ft/hr

g_c = conversion factor, $4.17 \times 10^8\ \text{lb}_m\text{ft}/\text{lb}_f\text{hr}^2$

Equation 12-34 can be written in the form

$$\Delta p_a = \frac{(1-x_e)\dot{m}_t V_{fe}}{g_c A_c} + \frac{x_e \dot{m}_t V_{ge}}{g_c A_c} - \frac{\dot{m}_t V_i}{g_c A_c}$$

$$= \frac{G}{g_c}\left[(1-x_e)V_{fe} + x_e V_{ge} - V_i\right] \tag{12-35}$$

where G = mass velocity = total mass-flow rate per unit cross-sectional area of channel = $\dot{m}_t/A_c$, (lb_m/hr ft²), a constant quantity.

From the continuity equation,

$$V_{fe} = \frac{\dot{m}_f v_f}{A_{fe}} = \frac{(1-x_e)\dot{m}_t v_f}{A_{fe}} = \frac{(1-x_e)\dot{m}_t v_f}{(1-\alpha_e)A_c} = \frac{(1-x_e)G v_f}{1-\alpha_e} \tag{12-36}$$

Similarly,

$$V_{ge} = \frac{x_e G v_g}{\alpha_e} \tag{12-37}$$

and

$$V_i = G v_i \tag{12-38}$$

where A_{fe} = cross-sectional area of liquid at channel exit

v_f, v_g, v_i = specific volumes of saturated liquid, saturated vapor, and subcooled liquid at channel inlet.

Thus

$$\Delta p_a = \frac{G^2}{g_c}\left[\frac{(1-x_e)^2}{1-\alpha_e} v_f + \frac{x_e^2}{\alpha_e} v_g - v_i\right] \tag{12-39}$$

where x_e and α_e are related by Eqs. 12-9 and 12-10. Equation 12-39 is usually written in the form

$$\Delta p_a = r\frac{G^2}{g_c} \tag{12-40}$$

where r is an *acceleration multiplier,* having the units ft³/lb_m and given by

$$r = \left[\frac{(1-x_e)^2}{1-\alpha_e} v_f + \frac{x_e^2}{\alpha_e} v_g - v_i\right] \tag{12-41}$$

The value of r increases rapidly with α and only slightly with S. It is independent of the mode of heat addition (uniform, sinusoidal, etc.) in the channel.

Example 12-3. Compute the acceleration pressure drop for the conditions of Example 12-2.

Solution

$$G = \rho_i V_i = 47.847(3 \times 3{,}600) = 0.5167 \times 10^6 \text{ lb}_m/\text{hr ft}^2$$

$$r = \left[\frac{(1-0.08)^2}{1-0.486} 0.0216 + \frac{0.08^2}{0.486} 0.4456 - 0.02094\right] = 0.0205 \text{ ft}^3/\text{lb}_m$$

$$\Delta p_a = 0.0205 \frac{(0.5167 \times 10^6)^2}{4.17 \times 10^8} = 13.12 \text{ lb}_f/\text{ft}^2 = 0.0911 \text{ lb}_f/\text{in.}^2$$

12-7. TWO-PHASE FLOW PRESSURE DROP AT RESTRICTIONS

Two-phase mixtures commonly encounter sudden area changes, such as when they enter and leave fuel channels, pass by spacers, and others. These can be grouped under the headings abrupt expansion, abrupt contraction, orifices and nozzles. The pressure changes associated with these are usually small. Their magnitude could, however, as in natural-circulation systems, be large relative to the driving pressures (Sec. 14-4), and therefore must be evaluated as accurately as possible since they can materially influence coolant flow rates and consequently power output.

As with friction, these pressure changes are higher in two-phase than in single-phase flow because of the increased speed of the liquid caused by the presence of the voids. It also follows that the larger the void fraction, the larger the pressure change.

In the analysis it is necessary to evaluate integrated momentum and kinetic energy before and after the area change. These, however, cannot be evaluated with complete accuracy because detailed velocity distributions and the extent of equilibrium between the phases are not well known. Experimental data and empirical or semiempirical approaches are often resorted to.

The following analysis assumes (a) one-dimensional flow, (b) adiabatic flow across the area change, so that x is constant, and (c) pressure changes that are small compared to the total pressure so that ρ_f and ρ_g do not change (incompressible flow).

The following continuity equations, derived with reference to Fig. 12-4b, will aid in the analysis.

In general,

$$\left.\begin{aligned} \dot{m}_t &= \rho A V \\ \dot{m}_f &= (1-x)\dot{m}_t = (1-\alpha)\rho_f A V_f \\ \text{and} \qquad \dot{m}_g &= x\dot{m}_t = \alpha \rho_g A V_g \end{aligned}\right\} \qquad (12\text{-}42)$$

so that

$$\left.\begin{aligned} V_{f_1} &= \frac{1-x}{1-\alpha_1}\frac{\dot{m}_t}{\rho_f A_1}, \qquad V_{f_2} = \frac{1-x}{1-\alpha_2}\frac{\dot{m}_t}{\rho_f A_2} \\ \text{and} \qquad V_{g_1} &= \frac{x}{\alpha_1}\frac{\dot{m}_t}{\rho_g A_1}, \qquad V_{g_2} = \frac{x}{\alpha_2}\frac{\dot{m}_t}{\rho_g A_2} \end{aligned}\right\} \qquad (12\text{-}43)$$

where the subscripts 1 and 2 refer to before and after the area change respectively.

It will be found instructive to begin the analysis with the case of single-phase incomprenssible flow, then follow up with the case of two-phase flow. This is the procedure that will be followed in each of the cases. The following three sections will deal with sudden expansions and contractions, and with orifices.

12-8. PRESSURE RISE DUE TO SUDDEN EXPANSION

Single-Phase Flow

For the case of *single-phase* flow, in a sudden expansion, Fig. 12-15,

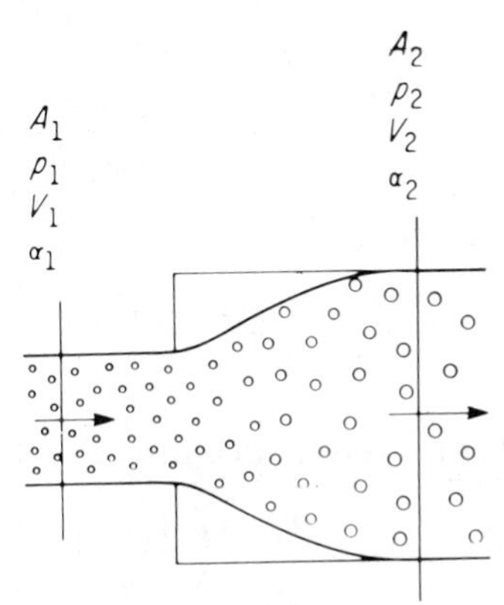

FIG. 12-15. Two-phase flow in sudden expansion.

the Bernoulli energy equation, written for incompressible (ρ = constant) one-dimensional* flow, is

$$p_1 + \frac{\rho V_1^2}{2g_c} = p_2 + \frac{\rho V_2^2}{2g_c} + k\,\frac{\rho V_1^2}{2g_c} \tag{12-44}$$

where the change in elevation between 1 and 2 is neglected and where all the terms have the dimensions of pressure. The last term represents a pressure loss because of viscous dissipation as the flow slows down. The symbol k is a dimensionless *loss coefficient,* also called the *Borda-Carnot coefficient*. Its value is obtained with the help of a momentum balance in which the upstream pressure p_1 is assumed to still act on the expanded

* This is an approximation, since the flow in this case is not truly one dimensional.

area A_2 immediately after expansion, since the flow lines there are essentially the same as at section 1. Thus

$$p_1 A_2 - p_2 A_2 = \frac{\dot{m}_t}{g_c}(V_2 - V_1) \tag{12-45a}$$

with the help of the continuity equation $\dot{m}_t = \rho A_2 V_2$, Eq. 12-45a may be written in the form

$$p_1 + \frac{\rho V_1 V_2}{g_c} = p_2 + \frac{\rho V_2^2}{g_c} \tag{12-45b}$$

Comparing Eqs. 12-44 and 12-45b gives

$$k \frac{\rho V_1^2}{2g_c} = \frac{\rho V_1^2}{2g_c} - \frac{\rho V_1 V_2}{g_c} + \frac{\rho V_2^2}{2g_c}$$

or

$$k = 1 - 2\left(\frac{V_2}{V_1}\right) + \left(\frac{V_2}{V_1}\right)^2 = \left(1 - \frac{V_2}{V_1}\right)^2 = \left(1 - \frac{A_1}{A_2}\right)^2 \tag{12-46}$$

giving the pressure-loss term as

$$k \frac{\rho V_1^2}{2g_c} = \frac{\rho (V_1 - V_2)^2}{2g_c} \tag{12-47}$$

The total pressure change is obtained directly from Eq. 12-45b as

$$p_2 - p_1 = 2\left[\frac{A_1}{A_2} - \left(\frac{A_1}{A_2}\right)^2\right]\frac{\rho V_1^2}{2g_c} \tag{12-48a}$$

$$= \frac{1}{g_c}\left(\frac{1}{A_1 A_2} - \frac{1}{A_2^2}\right)\frac{\dot{m}_t^2}{\rho} \tag{12-48b}$$

or by combining Eqs. 12-44 and 12-46 to give

$$p_2 - p_1 = \frac{\rho (V_1^2 - V_2^2)}{2g_c} - \frac{\rho (V_1 - V_2)^2}{2g_c} \tag{12-48c}$$

Since A_1 is less than A_2, the right-hand side of the above equation is positive, indicating a *net pressure rise* in the case of area expansion. This can be considered to be made of a pressure *rise* due to kinetic-energy change, obtained by putting $k = 0$ in Eq. 12-44 and given by the first term on the right-hand side of Eq. 12-48c, and a pressure *drop* due to viscous dissipation, given by the last term in Eq. 12-48c. This loss can be reduced in magnitude if the expansion were made gradual, to avoid flow separation, resulting in what is called a *subsonic diffuser*.

Two-Phase Flow

In *two-phase* flow, Fig. 12-15, several methods have been suggested for the calculation of the total pressure change across a sudden expansion. Lottes [130] compared four of these and recommended one by Romie. In this, a momentum balance is written assuming, as in single-phase above, that p_1 still acts on A_2 immediately after expansion, as follows

$$p_1A_2 + \frac{\dot{m}_{f_1}V_{f_1}}{g_c} + \frac{\dot{m}_{g_1}V_{g_1}}{g_c} = p_2A_2 + \frac{\dot{m}_{f_2}V_{f_2}}{g_c} + \frac{\dot{m}_{g_2}V_{g_2}}{g_c}$$

Using the relationships of 12-43 and rearranging give

$$p_2 - p_1 = \frac{\dot{m}_t^2}{g_c}\left\{\frac{(1-x)^2}{\rho_f}\left[\frac{1}{(1-\alpha_1)A_1A_2} - \frac{1}{(1-\alpha_2)A_2^2}\right] + \frac{x^2}{\rho_g}\left(\frac{1}{\alpha_1A_1A_2} - \frac{1}{\alpha_2A_2^2}\right)\right\} \tag{12-49}$$

To evaluate the above equation, a relationship between α_2 and α_1 is needed. Based upon experiments by Petrick with air and water in vertical columns near atmospheric pressure, the following relationship was suggested:

$$\alpha_2 = \frac{1}{(p_2/p_1)[(1-\alpha_1)/\alpha_1](A_2/A_1)^{0.2} + 1} \tag{12-50}$$

α decreases somewhat across an expansion, with the percent decrease becoming smaller the higher the value of α_1 and the higher the area ratio. Table 12-2 shows values of α_2 vs. α_1 computed for $p_2/p_1 \simeq 1$, the case of interest in reactor work. For area ratios 0.5 or larger, and high void fractions, the change may be ignored, i.e., $\alpha_1 = \alpha_2 = \alpha$, and Eq. 12-49 reduces to

$$p_2 - p_1 = \frac{1}{g_c}\left(\frac{1}{A_1A_2} - \frac{1}{A_2^2}\right)\dot{m}_t^2\left[\frac{(1-x)^2}{\rho_f(1-\alpha)} + \frac{x^2}{\rho_g\alpha}\right] \tag{12-51}$$

As in single-phase, there is a net pressure rise across an expansion. It can be seen that for the same mass-flow rate and areas, and since x is smaller than α, the pressure rise in two-phase flow, Eq. 12-51, is greater than in single-phase flow, Eqs. 12-48 and is larger the larger the void fraction. The two-phase equation 12-51 can be reduced to the single phase equation 12-48b by putting $x = 0$ and $\alpha = 0$.

TABLE 12-2
Effect of Area Expansion on Void Fraction in Vertical Air-Water Mixtures*

A_1/A_2	α_1, percent	α_2, percent	Percent Change
0.2	10	7.4	-26
	20	15.3	-23.5
	40	32.0	-20
	60	52.0	-13.3
0.5	10	9.0	-10
	20	18.0	-10
	40	36.6	-8.5
	60	57.0	-5

* From Ref. 130.

12-9. PRESSURE DROP DUE TO SUDDEN CONTRACTION

In both single- and two-phase flow; area contractions result in *vena contracta* of area A_0 following the contraction (Fig. 12-16). The fluid then expands to the reduced area A_2. The losses from the contraction to the vena contracta are negligibly small and the contraction losses are assumed to be totally due to the expansion from the vena contracta to where the fluid fills A_2.

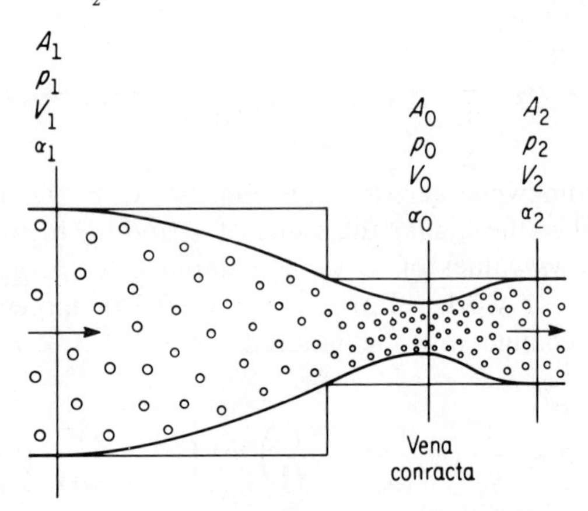

FIG. 12-16. Two-phase flow in sudden contraction.

Single-Phase Flow

Again, it is instructive to begin with *single-phase* flow. The pressure change in sudden contraction is given by a Bernoulli equation, similar to

Eq. 12-44 of sudden expansion, except that, since the losses are from the vena contracta to the reduce area, the loss term is customarily given in terms of the downstream velocity. Thus

$$p_1 + \frac{\rho V_1^2}{2g_c} = p_2 + \frac{\rho V_2^2}{2g_c} + k\,\frac{\rho V_2^2}{2g_c}$$

or

$$p_2 - p_1 = \frac{\rho(V_1^2 - V_2^2)}{2g_c} - k\,\frac{\rho V_2^2}{2g_c} \tag{12-52a}$$

$$= \left[1 - (k+1)\left(\frac{A_1}{A_2}\right)^2\right]\frac{\rho V_1^2}{2g_c} \tag{12-52b}$$

k, the dimensionless loss-coefficient, is here a function of A_0 and A_2 and is given by

$$k = \left(\frac{A_2}{A_0} - 1\right)^2 \tag{12-53}$$

Since A_0 is not known with certainty, k is usually given by

$$k = a\left[1 - \left(\frac{A_2}{A_1}\right)^2\right] \tag{12-54}$$

where a is a dimensionless number, less than unity and reported variously between 0.4 and 0.5. Using 0.4 and combining Eqs. 12-52a and 12-54 and rearranging give

$$p_2 - p_1 = \frac{\rho(V_1^2 - V_2^2)}{2g_c} - 0.4\left[1 - \left(\frac{A_2}{A_1}\right)^2\right]\frac{\rho V_2^2}{2g_c} \tag{12-55a}$$

and since $A_1V_1 = A_2V_2$,

$$p_2 - p_1 = \frac{\rho(V_1^2 - V_2^2)}{2g_c} + 0.4\,\frac{\rho(V_1^2 - V_2^2)}{2g_c} \tag{12-55b}$$

The above equation shows that the pressure change in sudden expansion is composed of a change of pressure due to change in kinetic energy and a loss term, which, in this case, is 40 percent of the kinetic energy term. Since $V_2 > V_1$ both terms are negative, so that sudden contractions result in a pressure drop. The above equations can be written in the convenient forms

$$p_2 - p_1 = 1.4\left[1 - \left(\frac{A_1}{A_2}\right)^2\right]\frac{\rho V_1^2}{2g_c} \tag{12-55c}$$

and

$$p_2 - p_1 = \frac{0.7}{g_c}\left(\frac{1}{A_1^2} - \frac{1}{A_2^2}\right)\frac{\dot{m}_t^2}{\rho} \tag{12-55d}$$

Two-Phase Flow

In *two-phase a* was found to be more nearly 0.2 [130]. Also the vapor is able to accelerate more readily than the liquid, resulting in a considerable decrease in the void fraction at the vena contracta. Beyond the vena contracta, however, shear forces between liquid and wall and between liquid and vapor steady the flow and cause α_2 to assume substantially the same value as α_1 so that $\alpha_1 = \alpha_2 = \alpha$. If ρ is replaced by the density of the two-phase mixture

$$\rho = (1 - \alpha)\rho_f + \alpha\rho_g \tag{12-56}$$

and the product $\alpha\rho_g$ is ignored, being relatively small at pressures far removed from critical, and if the pressure losses are attributed to the liquid alone, the pressure change in two-phase flow would be given by

$$p_2 - p_1 = 1.2\left[1 - \left(\frac{A_1}{A_2}\right)^2\right]\frac{\rho_f V_{f_1}^2}{2g_c}(1 - \alpha) \tag{12-57a}$$

or with the help of Eqs. 12-43, by

$$p_2 - p_1 = \frac{0.6}{g_c}\left(\frac{1}{A_1^2} - \frac{1}{A_2^2}\right)\frac{\dot{m}_t^2}{\rho_f}\frac{(1 - x)^2}{(1 - \alpha)} \tag{12-57b}$$

Since A_2 is less than A_1, contractions result in pressure drops in both single- and two-phase flow. A comparison of Eqs. 12-55 and 12-57 shows that for the same mass-flow rate and areas, and for most values of x and α encountered in reactor work, the two-phase contraction pressure losses are larger than in single-phase and are larger the larger the void fraction.

12-10. TWO-PHASE FLOW IN ORIFICES

In general, the main object of orifices is to measure flow rates, which are functions of the square roots of pressure drops across them. Orifices, however, approximate many flow restrictions, such as spacers in reactor cores and others. It is the intent here to evaluate the pressure drop across orifices.

Single-Phase Flow

Starting as usual with *single phase,* the mass-flow rates for liquid and vapor are obtained from the continuity and energy equations, and given by the well-known equations

$$\dot{m}_f = A'_0\sqrt{2g_c\rho_f(\Delta p_{SP})_f} \tag{12-58}$$

and
$$\dot{m}_g = A'_0 \sqrt{2 g_c \rho_g (\Delta p_{SP})_g} \tag{12-59}$$

where
$$A'_0 = C_d \frac{A_0}{\sqrt{1 - \left(\frac{A_0}{A_c}\right)^2}} \tag{12-60}$$

(Δp_{SP}) is the single-phase pressure drop, A_0 and A_c are the cross sectional areas of the orifice and channel, Fig. 12-17, and C_d is a coefficient of discharge, obtained from experiments or published data [131] and is a function of the type of orifice, pressure-tap locations, etc.

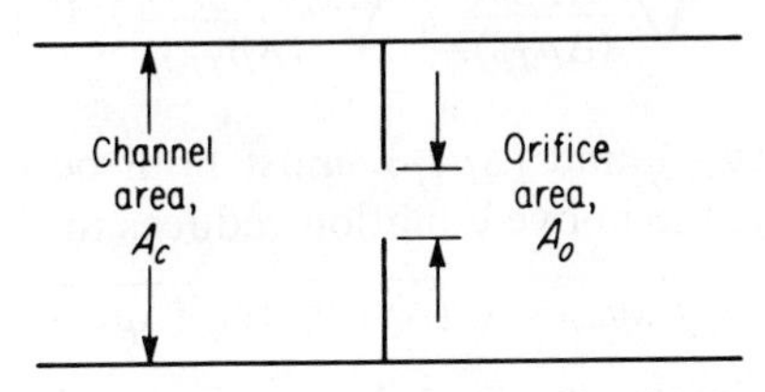

FIG. 12-17. Channel with thin plate orifice.

Two-Phase Flow

In *two-phase* flow, two similar expressions can be written for *coexistent* liquid and vapor phases of the above mass-flow rates as

$$\dot{m}_f = A'_{0_f} \sqrt{2 g_c \rho_f (\Delta p_{TP})_f} \tag{12-61}$$

and
$$\dot{m}_g = A'_{0_g} \sqrt{2 g_c \rho_g (\Delta p_{TP})_g} \tag{12-62}$$

$(\Delta p_{TP})_f$ and $(\Delta p_{TP})_g$ are the two-phase pressure drops due to liquid and vapor respectively. A'_{0_f} and A'_{0_g} are given by

$$A'_{0_f} = C_d \frac{A_{0f}}{\sqrt{1 - \left(\frac{A_0}{A_c}\right)^2}} \tag{12-63}$$

and
$$A'_{0_g} = C_d \frac{A_{0g}}{\sqrt{1 - \left(\frac{A_0}{A_c}\right)^2}} \tag{12-64}$$

where A_{0_f} and A_{0_g} are the flow areas within the orifice occupied by liquid and vapor respectively. Thus

$$A'_{0f} + A'_{0g} = C_d \frac{(A_{0f} + A_{0g})}{\sqrt{1 - \left(\frac{A_0}{A_c}\right)^2}}$$

and since $A_{0f} + A_{0g} = A_0$, comparison with Eq. 12-60 gives

$$A'_{0f} + A'_{0g} = A'_0 \tag{12-65}$$

Assuming C_d does not change materially, Eqs. 12-58 to 12-62 can be combined with 12-65 to give

$$\sqrt{\frac{(\Delta p_{SP})_f}{(\Delta p_{TP})_f}} + \sqrt{\frac{(\Delta p_{SP})_g}{(\Delta p_{TP})_g}} = 1$$

Noting that $(\Delta p_{TP})_f$ and $(\Delta p_{TP})_g$ must both be equal to a two-phase pressure drop Δp_{TP}, the above equation reduces to

$$\sqrt{\Delta p_{TP}} = \sqrt{(\Delta p_{SP})_f} + \sqrt{(\Delta p_{SP})_g} \tag{12-66}$$

Δp_{TP} is therefore simply obtained from calculated pressure drops of the liquid and vapor phases as if they were flowing alone through the orifice. Their respective flow rates are obtained from the total flow rate and quality.

Experimental work by Murdock [132] on a wide range of operating conditions, showed that the above equation should be modified to

$$\sqrt{\Delta p_{TP}} = 1.26\sqrt{(\Delta p_{SP})_f} + \sqrt{(\Delta p_{SP})_g} \tag{12-67}$$

12-11. CRITICAL FLOW

When the back pressure p_b is reduced below a constant upsteam pressure p_0 in a flow system, line 1 in Fig. 12-18, flow begins and a pressure gradient is established in the connecting channel between p_0 and a pressure p_e at the exit of the channel. The flow increases as p_b is reduced further, line 2. p_e remains equal to p_b to a point. This point, represented by line 3, is reached when p_b is reduced sufficiently to cause the flow velocity at the exit of the channel to equal that of the speed of sound at the temperature and pressure at the exit of the channel, V^* At that point the mass flow attains a maximum value.

Further reduction in p_b results in no further increase in mass flow rate or decrease in p_e, lines 4 and 5. The flow remains at the above maximum value and is said to be *critical.* The channel, and flow, are also said to be *restricted* or *choked.* As p_b is reduced and p_e remains constant, free expansion of the fluid between p_e and p_b occurs outside the channel and the flow takes on a paraboloid shape there.

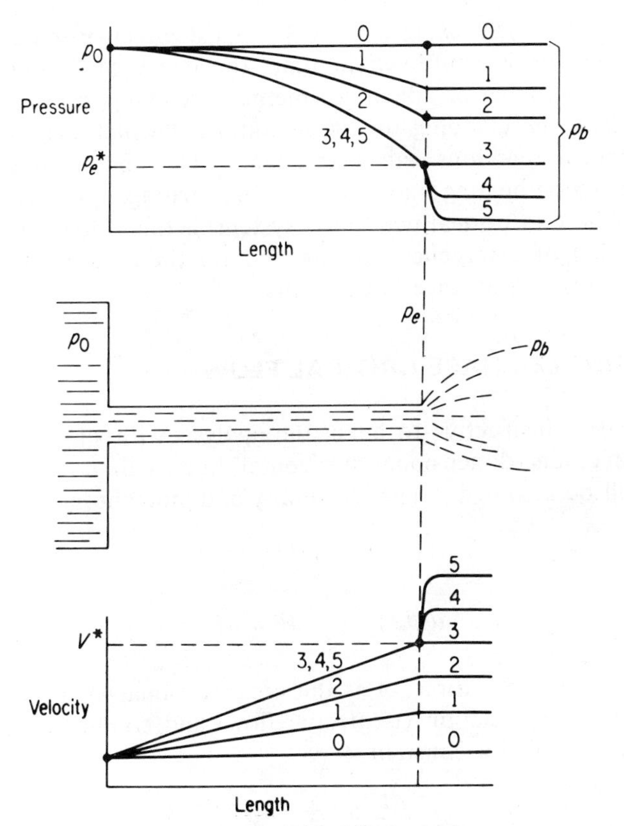

FIG. 12-18. Critical flow model.

This phenomenon occurs in both single- and two-phase flow. It has been well studied in single-phase, particularly gas flow, both theoretically and experimentally. It occurs, and is utilized, in many flow and measuring systems. In two-phase flow, serious theoretical and experimental studies of the phenomenon have been made only in recent years, although the phenomenon has long been observed in boiler and turbine systems, flow of refrigerants and rocket propellants, and many others.

In nuclear work, the phenomenon is of utmost importance in safety considerations of both boiling and pressurized systems. A break in a primary coolant pipe causes two-phase critical flow in either system since even in a pressurized water reactor, the reduction of pressure of the hot coolant from about 2,000 psia to near atmospheric causes flashing and two-phase flow. This kind of break results in a rapid loss of coolant and is considered to be the *maximum credible accident* in power reactors built to date.

The loss of coolant exposes the core to a steam environment. Even when the reactor is shutdown in time, the decay of fission products would, in the absence of adequate emergency cooling, release sufficient energy to heat the cladding to a temperature at which cladding metal-steam chemical reactions can take place. The combined decay and chemical reaction heating may result in core melting. An evaluation of the rate of flow in critical two-phase systems is therefore of importance for the design of emergency cooling and for the determination of the extent and causes of damage in accidents.

12-12. SINGLE-PHASE CRITICAL FLOW

Again it is instructive to begin the analysis with single-phase flow. Compressible, one-dimensional horizontal flow with no work or heat transfer will be assumed. The continuity and momentum equations are

$$\dot{m} = \rho A V \tag{12-68}$$

and

$$d(pA) + \frac{\dot{m}}{g_c}\, dV = dF \tag{12-69}$$

where F is a frictional force, A is the cross-sectional area and p is the pressure. Ignoring friction (isentropic flow) and combining the above two equations yields for constant A:

$$\frac{dV}{dp} + \frac{g_c}{\rho V} = 0 \tag{12-70}$$

The continuity equation is now differentiated with respect to p (with $dA/dp = 0$) to give

$$\frac{1}{\dot{m}}\frac{d\dot{m}}{dp} = \frac{1}{\rho}\frac{d\rho}{dp} + \frac{1}{V}\frac{dV}{dp} \tag{12-71}$$

When critical flow is reached the mass flow rate becomes a maximum, $\dot{m}_{\text{max}}$, and $d\dot{m}/dp = 0$ in isentropic flow. Thus

$$\frac{1}{V}\frac{dV}{dp} = -\frac{1}{\rho}\frac{d\rho}{dp}$$

or

$$\frac{dV}{dp} = -\frac{\dot{m}_{\text{max}}}{\rho^2 A}\frac{d\rho}{dp} \tag{12-72}$$

Combining Eqs. 12-68, 12-70, and 12-72 and rearranging give

$$\left(\frac{\dot{m}_{max}}{A}\right)^2 = g_c\rho^2 \frac{dp}{d\rho} \tag{12-73a}$$

The above equation is written in the more usual form

$$G_{max}^2 = -\,g_c \frac{dp}{dv} \tag{12-73b}$$

where G is the mass velocity m/A and v the specific volume ($v = \rho^{-1}$, $d\rho/dv = -\rho^2$). Equation 12-73b is identical to the equation for the speed of sound in isentropic flow, indicating that sonic velocity is reached in critical flow.

We shall now consider the energy equation (no heat transfer or work)

$$dh + \frac{VdV}{g_cJ} = 0 \tag{12-74}$$

which integrates to

$$h_0 = h + \frac{V^2}{2g_cJ} \tag{12-75}$$

where h and h_0 are the specific enthalpy and stagnation enthalpy of the fluid and J is Joule's equivalent, 778.16 ft lb_f/Btu. Thus

$$V = \sqrt{2g_cJ(h_0 - h)} \tag{12-76}$$

The above equation applies whether the process is reversible or not. For an ideal gas $dh = c_pdT$, so that it becomes

$$V = \sqrt{2g_cJc_p(T_0 - T)}$$

$$= \sqrt{2g_cJc_pT_0\left(1 - \frac{T}{T_0}\right)} \tag{12-77}$$

where T_0 is the stagnation temperature. Recalling that for an ideal gas $T/T_0 = (p/p_0)^{(\gamma-1)/\gamma}$ (reversible) and $\rho = p/RT = (p_0/RT_0)(p/p_0)^{1/\gamma}$, where γ is the ratio of specific heats, and applying the continuity equation, Eq. 12-77 becomes

$$\dot{m} = \frac{Ap_0}{RT_0}\sqrt{2g_cJc_pT_0\left[\left(\frac{p}{p_0}\right)^{2/\gamma} - \left(\frac{p}{p_0}\right)^{\frac{\gamma+1}{\gamma}}\right]} \tag{12-78}$$

where p_0 is the stagnation pressure and R the gas constant. This equation plots as *abd* in Fig. 12-19 if the pressure ratio p/p_0 were assumed to vary between 1.0 and 0. However, only the back-pressure ratio p_b/p_0 can vary between 1.0 and 0. The channel exit pressure p_e follows p_b as it is lowered until it reaches the critical value p_e^* indicated by the *critical pressure ratio* $r_p^* = p_e^*/p_0$, Fig. 12-19, and remains fixed for all values of

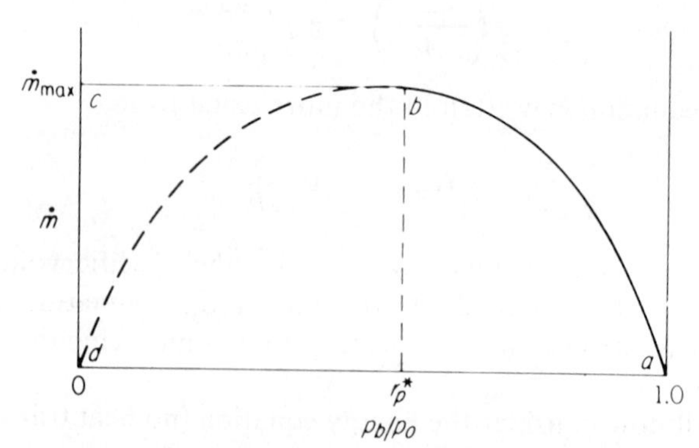

FIG. 12-19. Effect of pressure ratio on mass-flow rate.

p_b below this as indicated earlier. Also, the mass-flow rate increases as p_b/p_0 is lowered, reaches its maximum value $\dot{m}_{max}$ at r_p^* and remains fixed for all values of p_b/p_0 below r_p^*. The actual flow curve, therefore is *abc* in Fig. 12-19. The flow between *b* and *c*, again, is critical, restricted, or choked.

The value of r_p^* can be obtained by differentiating $\dot{m}$ in Eq. 12-78 with respect to p and equating to zero. This results in

$$r_p^* = \left(\frac{2}{\gamma + 1}\right)^{\gamma/(\gamma - 1)} \tag{12-79}$$

A representative value of r_p^* is 0.53 for air at low temperatures ($\gamma = 1.4$). Thus if the upstream pressure is 100 psia, for example, a back pressure of 53 psia or lower would cause critical or choked flow. While steam is not a perfect gas, r_p^* for single-phase steam flow is approximated from the perfect gas relationship Eq. 12-79 by replacing γ with 1.3 for superheated and supersaturated or metastable (a case where actual condensation lags behind theoretical condensation in rapid expansion) steam, resulting in an r_p^* value of 0.545.

12-13. TWO-PHASE CRITICAL FLOW

For frictionless liquid and vapor flow, the momentum equations, for the same pressure drop, are

$$d(pA_f) + \frac{1}{g_c}\,d(\dot{m}_f V_f) = 0$$

or

$$d(pA_f) + \frac{1}{g_c}\,d(\rho_f A_f V_f^2) = 0 \tag{12-80}$$

and

$$d(pA_g) + \frac{1}{g_c}\,d(\rho_g A_g V_g^2) = 0 \tag{12-81}$$

where now the flow areas, densities and mass-flow rates of liquid and vapor are variable. The above equations are added to give

$$dp = -\frac{1}{g_c A}\, d[(\rho_f A_f V_f^2) + (\rho_g A_g V_g^2)] \tag{12-82}$$

where $A_f + A_g = A$, the channel total area. Using the continuity equations for the two phases, Eqs. 12-43, and the relationships $A_g/A = \alpha$ and $A_f/A = (1 - \alpha)$, Eq. 12-82 becomes

$$dp = -\frac{G^2}{g_c}\, d\left[\frac{(1-x)^2}{\rho_f(1-\alpha)} + \frac{x^2}{\rho_g \alpha}\right] \tag{12-83a}$$

where G is the total mass velocity $\dot{m}_t/A$, and x and α are the quality and void fraction respectively. Using the lumped model of Fig. 12-4b, it can be shown that the specific volume of the mixture is related to v_f and v_g by

$$v = v_f \frac{1-x}{1-\alpha} = \frac{1-x}{\rho_f(1-\alpha)} \tag{12-84a}$$

and

$$v = v_g \frac{x}{\alpha} = \frac{x}{\rho_g \alpha} \tag{12-84b}$$

so that the quantity between the brackets in Eq. 12-83a is simply equal to v. Equation 12-83a reduces to

$$G^2 = -g_c \frac{dp}{dv} \tag{12-83b}$$

which is identical in form to Eq. 12-73b for single-phase flow.

12-14. TWO-PHASE CRITICAL FLOW IN LONG CHANNELS

It is assumed, as in single-phase flow, that critical flow occurs when the pressure gradient at channel exit has reached a maximum value. In long channels, residence time is sufficiently long and thermodynamic equilibrium between the phases is attained. The liquid partially flashes into vapor as the pressure drops along the channel, and the specific volume of the mixture v attains a maximum value at the exit. Since v is a function of both x and α, it must be a function of the slip ratio S. Different values of S therefore result in different values of G. Maximum pressure gradient (and maximum G) are therefore obtained at a slip ratio when $\partial v/\partial S = 0$. This model, called the *slip equilibrium model,* is suggested by Fauske [133]. It assumes thermodynamic equilibrium between the two phases, and therefore applies to long channels.

The slip ratio, Eq. 12-8, is now combined with v, the quantity in the brackets in Eq. 12-83a, to eliminate α, giving

$$v = \frac{1}{S}\,[v_f(1-x)S + v_g x]\,[1 + x(S-1)] \tag{12-85}$$

thus

$$\frac{\partial v}{\partial S} = (x - x^2)\left(v_f - \frac{v_g}{S^2}\right) \tag{12-86}$$

Equating the above to zero, the value of S giving maximum flow, S^*, is

$$S^* = \sqrt{\frac{v_g}{v_f}} \tag{12-87}$$

G_{max} can now be obtained by combining Eqs. 12-83b and 12-85 as

$$G^2_{max} = \frac{-g_c}{-\dfrac{d}{dp}\left\{\dfrac{1}{S}\,[v_f(1-x)S + v_g x]\,[1 + x(S-1)]\right\}}$$

giving finally [102]

$$G^2_{max} = -\,g_c S^* \Bigg/ \left\{[(1 - x + S^* x)x]\,\frac{dv_g}{dp} + [v_g(1 + 2S^* x - 2x) + v_f(2xS^* - 2S^* - 2xS^{*2} + S^{*2})]\,\frac{dx}{dp}\right\} \tag{12-88}$$

in which the term dv_f/dp has been neglected. The value of dv_g/dp can be approximated as $\Delta v_g/\Delta p$ for small Δp at p, or obtained from Fig. 12-20 for the water-steam system. The value of dx/dp at constant enthalpy is obtained from

$$h = h_f + x h_{fg}$$

Rearranging,

$$x = \frac{h - h_f}{h_{fg}} \tag{12-89}$$

Differentiating with respect to p,

$$\frac{dx}{dp} = d(h/h_{fg})/dp - d(h_f/h_{fg})/dp$$

$$= \frac{1}{(h_{fg})^2}\left[\left(h_{fg}\,\frac{dh}{dp} - h\,\frac{dh_{fg}}{dp}\right) - \left(h_{fg}\,\frac{dh_f}{dp} - h_f\,\frac{dh_{fg}}{dp}\right)\right]$$

The enthalpy of the water-steam mixture will not change with pressure, so that $dh/dp = 0$. Using $h_{fg} = h_g - h_f$, so that $dh_{fg} = dh_g - dh_f$, we can rewrite the above equation in the form

$$\frac{dx}{dp} = -\left(\frac{1-x}{h_{fg}}\frac{dh_f}{dp}\right) - \left(\frac{x}{h_{fg}}\frac{dh_g}{dp}\right) \tag{12-90}$$

As indicated above, the derivatives dh_f/dp and dh_g/dp are sole functions of the pressure. They are given in Fig. 12-20 for the ordinary-

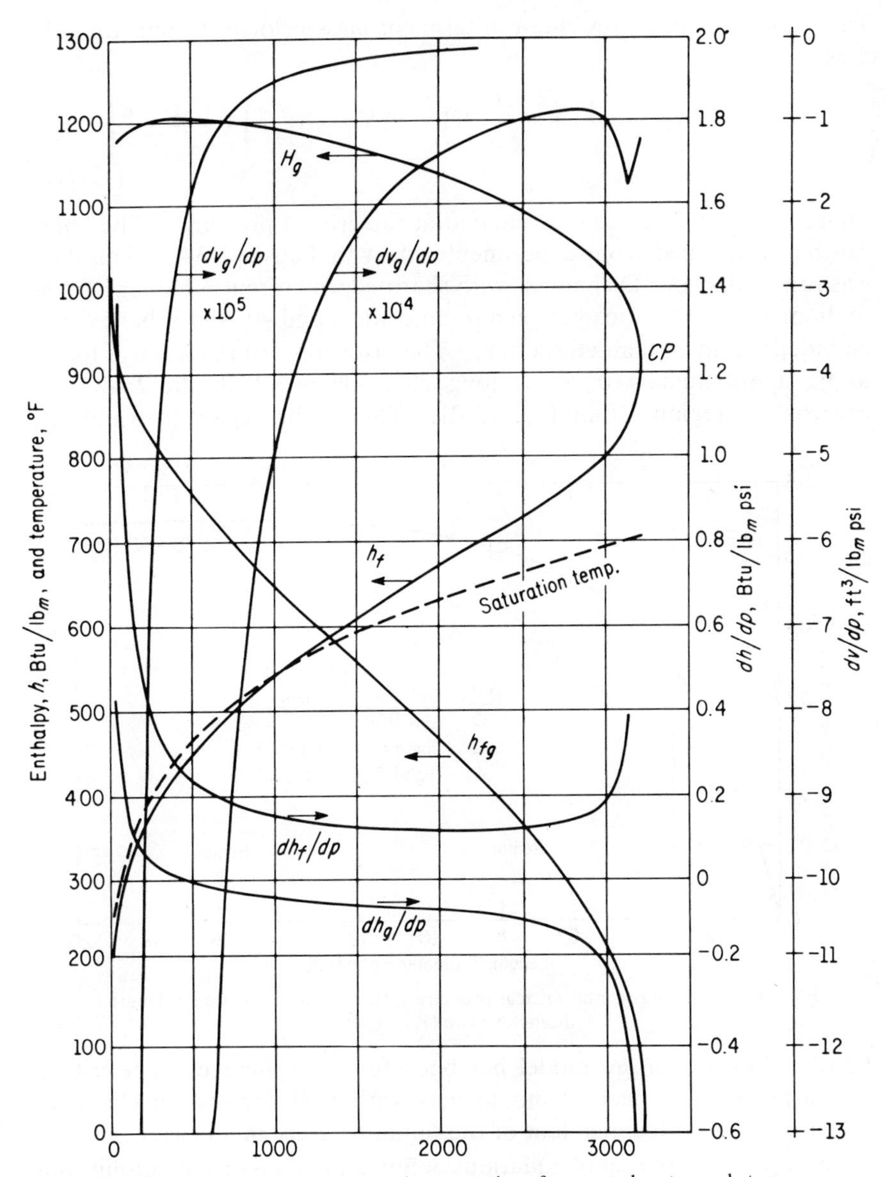

FIG. 12-20. Some thermodynamic properties of saturated water and steam.

water-steam system. The value of x is obtained from the energy equation 12-75 written for the two phases as follows

$$h_0 = (1-x)\left(h_f + \frac{V_f^2}{2g_cJ}\right) + x\left(h_g + \frac{V_g^2}{2g_cJ}\right) \tag{12-91a}$$

This equation can be rewritten in terms of mass velocity G and slip ratio S as

$$h_0 = (1-x)h_f + xh_g + \frac{G^2}{2g_cJ}\,[(1-x)Sv_f - xv_g]^2\left[x + \frac{1-x}{S^2}\right] \tag{12-91b}$$

where v_f, h_f, v_g and h_g are evaluated at the critical pressure. The latter can be determined from experimental data by Fauske [134]. This data was run on 0.25 in. ID channels with sharp-edged entrances having length-to-diameter, L/D, ratios between 0 (an orifice) and 40, and is believed to be independent of diameter alone. The critical pressure ratio was found to be approximately 0.55 for long channels in which the L/D ratio exceeds 12, region III in Fig. 12-21. This is the region in which the

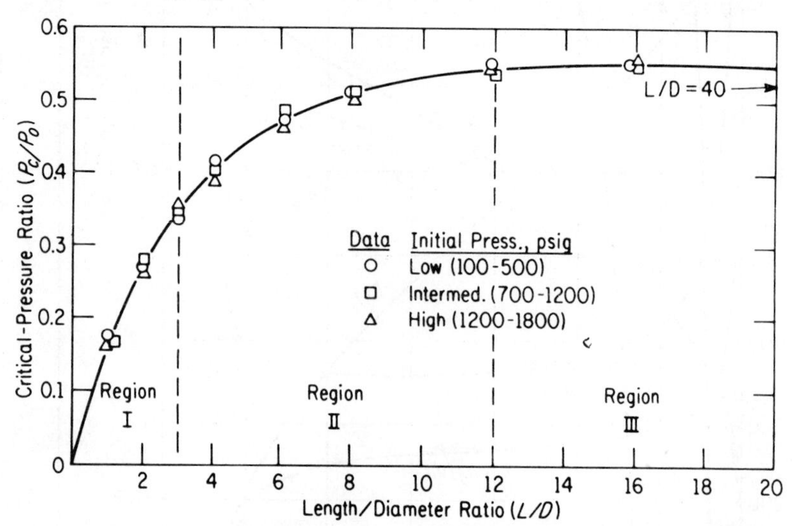

FIG. 12-21. Experimental critical pressure ratio data as a function of length / diameter ratio (Ref. 134).

Fauske slip-equilibrium model has been found applicable. The critical pressure ratio has been found to vary with L/D, for shorter channels, but appears to be independent of the initial pressure in all cases.

Solutions for the set of equations defining the Fauske slip-equilibrium model have been prepared by Fauske and are presented in Fig. 12-22.

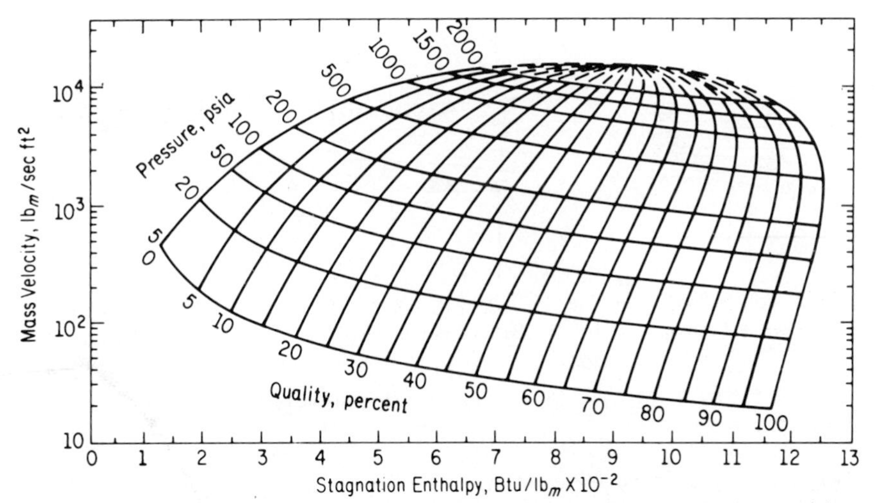

FIG. 12-22. Predictions of critical steam-water flow rates with slip equilibrium model (Ref. 134).

The critical flow is described by the local conditions at the channel exit. The flow is seen to increase with increasing pressure and with decreasing quality at the exit.

The Fauske model assumes thermodynamic equilibrium (no metastability, below), a case which due to the duration of flow, applies to long flow channels. Experimental data by many investigators showed the applicability of the Fauske model to L/D ratios above 12.

12-15. TWO-PHASE CRITICAL FLOW IN SHORT CHANNELS

Flashing of liquid into vapor, if thermal equilibrium is maintained, occurs as soon as the liquid moves into a region at a pressure lower than its saturation pressure. Flashing, however, could be delayed because of the lack of nuclei about which vapor bubbles may form, surface tension which retards their formation, due to heat-transfer problems, and other reasons. When this happens, a case of *metastibility* is said to occur. Metastability occurs in rapid expansions, particularly in short flow channels, nozzles, and orifices.

The case of *short channels* has not been completely investigated analytically. The experimental data obtained in [134] covered both long and short tubes, $0 < L/D < 40$. For L/D between 0 and 12 the critical pressure ratios depend upon L/D, unlike long channels, Fig. 12-21.

For *orifices* ($L/D = 0$) the experimental data showed that because residence time is short, flashing occurred outside the orifice (Fig. 12-23*a*)

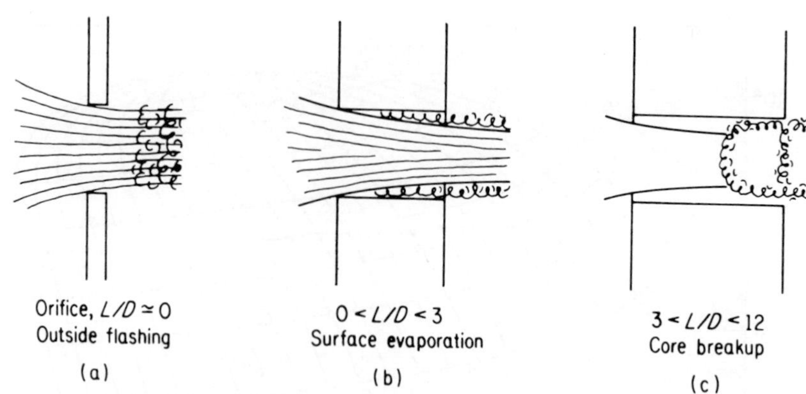

FIG. 12-23. Two-phase critical flow in orifices and short channels.

and no critical pressure existed. The flow is accurately determined from the incompressible flow orifice equation

$$G = 0.61\sqrt{2g_c\rho(p_0 - p_b)} \tag{12-92}$$

For region I, Fig. 12-21, $0 < L/D < 3$, the liquid immediately speeds up and becomes a metastable liquid core jet where evaporation occurs from its surface, Fig. 12-23b. The flow is determined from

$$G = 0.61\sqrt{2g_c\rho(p_0 - p_c)} \tag{12-93}$$

where p_c is obtained from Fig. 12-21.

In region II, $3 < L/D < 12$, the metastable liquid core breaks up, Fig. 12-33c, resulting in high-pressure fluctuations. The flow is less than would be predicted by Eq. 12-93. Figure 12-24 shows experimental critical flows for region II.

All the above data were obtained on sharp entrance channels. In rounded-entrance channels the metastable liquid remains more in contact with the walls and flow restriction requires less vapor. For $0 < L/D < 3$ channels, such as *nozzles,* the rounded entrances result in much higher critical pressure ratios than indicated by Fig. 12-21 as well as somewhat greater flows [135]. The effect of rounded entrances is negligible for long channels ($L/D > 12$) so that the slip-equilibrium model can be used there. The effect of L/D ratio on flow diminishes between 3 and 12.

The condition of the wall surface is not believed to affect critical flow in sharp-entrance channels, since the liquid core is not in touch with the walls and evaporation occurs at the core surface or by core breakup. It will have some effect on rounded-entrance channels. The existence of gases or vapor bubbles will affect the flow also, since they will act as nucleation centers [136].

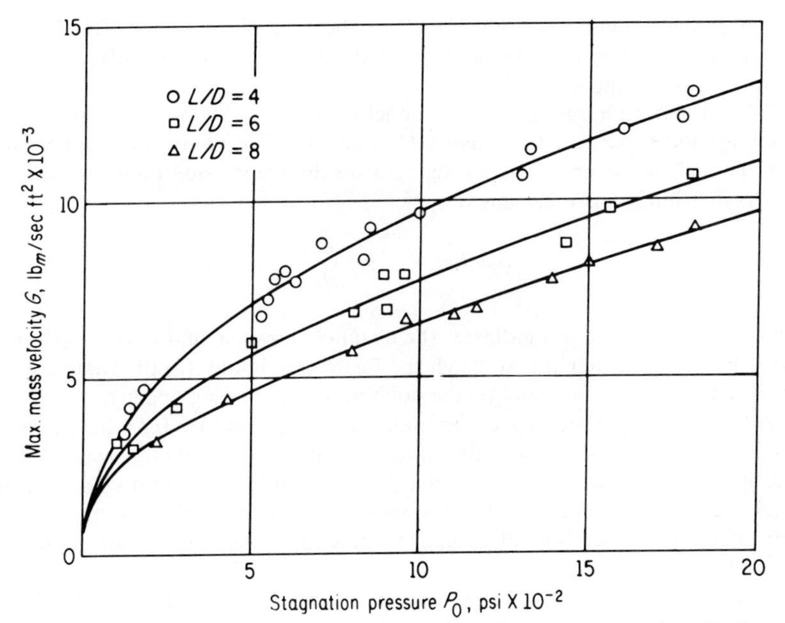

FIG. 12-24. Experimental two-phase critical flow rates for region II of Fig. 12-21.

PROBLEMS

12-1. Water enters a 6-ft-long uniformly heated 1-in. tube at 460°F and 600 psia. The entrance velocity is 2.5 fps. Neglecting pressure losses in the channel, determine the maximum heat input, in Btu/hr ft of channel length, if the exit void fraction is not to exceed 60 percent.

12-2. A boiling-water reactor operates at 600 psig. In one of its channels the inlet velocity is 1.4 fps, the heat generated is 5×10^6 Btu/hr, and the total coolant flow is 76,000 lb_m/hr. The incoming water is 11°F subcooled. Estimate the exit void fraction in that channel.

12-3. A boiling-sodium, graphite-moderated reactor core operates at an average pressure of 15.4 psia. The average slip ratio is 4. Sodium enters the core at 1140°F and leaves with an average void fraction of 80 percent. Determine the amount of heat added per pound mass of coolant.

12-4. A boiling-sodium reactor channel, 1 in.^{2} cross-sectional area, operates at an average pressure of 9.153 psia. Subcooled sodium enters the channel at 1900 °R and 20 fps. 2.37×10^6 Btu/hr are added in the channel. The exit void fraction is 96 percent. What is the slip ratio?

12-5. A 12-ft-high boiling-water reactor channel operating at 1,200 psia with sinusoidal heat generation has a nonboiling height of 4 ft. The inlet water is 27.22°F subcooled. Find the exit quality of the channel.

12-6. A 6-ft-high boiling-water reactor channel operates at an average pressure of 700 psia. Water enters the channel 23.1°F subcooled and leaves with a 6 percent steam quality. Calculate the nonboiling and boiling heights if heat is added along the channel (*a*) uniformly and (*b*) sinusoidally. Neglect the extrapolation lengths.

12-7. A 6-ft-high boiling-water channel operates at 700 psia average, 23.1°F subcooling, and 6 percent exit quality (as above). The voids in the upper part of the channel, however, cause strong neutron-flux depression there, so that the axial flux distribution is represented by

$$\varphi = Ce^{-\pi z/H} \sin \frac{\pi z}{H}$$

where C is a constant, $z = 0$ indicates the channel entrance, and H is the channel height. Find (*a*) the height z at which the flux is maximum, (*b*) the value of C in terms of the maximum flux, and (*c*) the nonboiling and boiling heights.

12-8. A 5-ft-high boiling-water channel is 4.25 in. by 0.45 in. in cross section. Heat is added sinusoidally at an average pressure of 600 psia. Water enters the channel 5 Btu/lb_m subcooled at 2 fps and leaves with a void fraction of 32.9 percent. Neglecting the extrapolation lengths, calculate the power density in kilowatts per liter of coolant volume (*a*) in the entire channel and (*b*) in the boiling height only.

12-9. A boiling-water reactor channel operates at 1,000 psia, with 19.6°F subcooling and 10 percent exit quality. The acceleration pressure drop is 0.1 psi. The slip ratio is 2. Compute the amount of heat added in the channel in Btu/hr if the channel cross-sectional area is 3 in.^{2}.

12-10. A 4-ft-high boiling-water channel is 4.5 in. by 0.5 in. in cross section. It receives heat uniformly at the rate of 3×10^5 Btu/hr ft^2 of wide sides only. The average channel pressure is 1,200 psia. Water enters the channel saturated at 2 fps. For a slip ratio of 2 and a friction factor of 0.03, calculate the friction and acceleration pressure drops in the channel.

12-11. A 5-ft-high boiling-water channel has an equivalent diameter of 0.5 in. Water enters the channel at the rate of 2×10^5 lb_m/hr at 10 fps (forced circulation), 22°F subcooled. Five Mw(t) of heat are added sinusoidally in the channel. The slip ratio is 1.8, and the average pressure is 900 psia. Neglecting the extrapolation lengths, calculate the friction and acceleration pressure drops. Consider the cladding surface to correspond to smooth-drawn tubing.

12-12. A 12-ft-high BWR channel operates at 800 psia and 18.23°F subcooling. The nonboiling height is 4 ft. The slip ratio is 2. The exit quality is 20 percent. The friction factor is 0.015. The equivalent diameter in the channel is 0.14 ft. (*a*) Find the acceleration pressure drop if the friction pressure drop is 10 lb_f/ft^2. (*b*) Find the power generated in the channel in Mw(t) per ft^2 of channel-flow area.

12-13. A boiling-water channel has a cross-sectional area of 0.025 ft^2. It operates at 1,000 psia, receives saturated inlet water and has an exit quality of 10 percent. The slip ratio is 3. The acceleration pressure drop is 0.1 psi. How much heat is generated in the channel in Btu/hr?

12-14. A 10-ft-high channel in a 1,000-psia boiling-water reactor generates 2×10^5 Btu/hr uniformly. It has an equivalent diameter of 0.15 ft. The inlet water has a velocity of 6 fps and is 24.61°F subcooled. The water mass-flow rate is 2000 lb_m/hr and the slip ratio is 3.0. Find the friction pressure drop in that channel if the friction factor is 0.015.

12-15. A vertical fuel element in a boiling-water reactor is in the form of a thin cylindrical shell 2 in. ID. Water flowing upward enters at core bottom 22°F subcooled. The inlet water speed on the inside of the cylinder is 3 fps. The average pressure within the element is 900 psia. The portion of the heat generated by the element and conducted radially inward is 10^6 Btu/hr. The slip ratio is 2.0. The element support at the top is such that there is a sudden reduction in the inside diameter to 1.7 in., followed by a sudden expansion back to 2.0 in. Calculate the net pressure change due to this obstruction. The obstruction may be considered long enough so that the pressure changes are additive.

12-16. A 2-in.-diameter, two-phase flow channel has an obstruction in it in the form of a concentric 1-in.-diam. disk. The pressure drop due to the obstruction is 0.1 psf at a total mass-flow rate of 124.3 lb_m/hr, and a pressure of 600 psia. What is the quality at the obstruction? Take $C_D = 0.6$. The fluid is water.

12-17. 100 lb_m/hr of saturated water enter a 1-in.-diam. boiling channel at 1,000 psia. At a point where 3,247 Btu/hr have been added in the channel, a restriction in the form of a 0.2-in.-diam. orifice exists. Calculate the pressure drop due to the restriction if $C_D = 0.6$.

12-18. A PWR pressure vessel is connected to a heat exchanger via a long pipe. The pressure in the vessel is 2,000 psia. A break occurred at the end of the pipe. At the break, the quality was found to be 5 percent. What is the void fraction at the same location?

12-19. A small hole, 0.02 ft^2 in area, developed in the core shroud of a natural-circulation BWR. Two-phase mixture at 800 psia and 10 percent quality spilled into the downcomer. The downcomer is at 799 psia. The coefficient of discharge through the hole can be taken as 0.6. What is the rate of spillage in lb_m/hr?

12-20. A 12-in.-diam. primary coolant pipe carries 2,000 psia, 560°F water from the pressure vessel of a PWR. A sudden clean break is presumed to have occurred 2 ft. from the vessel. Calculate the initial rate of coolant loss in lb_m/sec.

12-21. A PWR operates at 2,000 psia and 580°F average water temperature. The outlet pipe is 1 ft in diam. A sudden break occurred about 20 ft from the vessel. The break is clean and perpendicular to the pipe axis. The back pressure is atmospheric. Calculate the rate of coolant loss in lb_m/sec at the instant the break occurred.

chapter **13**

Core Thermal Design

13-1. INTRODUCTION

As was indicated previously, the amount of reactor power generation in a given reactor is limited by thermal rather than by nuclear considerations. The reactor core must be operated at such a power level, that with the best available heat-removal system, the *temperatures* of the fuel and cladding anywhere in the core must not exceed safe limits. Otherwise fuel element damage might result in release of large quantities of radioactive material into the coolant, or in core-fuel meltdown. The thermal limitations on reactor power have obvious effects on nuclear power economics.

During the first twenty years of reactor experience, a substantial increase in the ratings of various nuclear power plant types has taken place, as knowledge and experience have been gained. In many cases, in-core instrumentation have aided in the evaluation of core and fuel performance under a variety of design, manufacturing and operating (burnup) conditions.

Reactor cores are usually limited by those parameters that cause the temperatures to exceed safe limits. In liquid-cooled reactors these may be the burnout heat flux (Sec. 11-5), which affects the cladding surface. In gas-cooled reactors the relatively low heat-transfer coefficients would pose limitations on the fuel centerline or cladding temperatures.

In this chapter the general heat-transfer and thermodynamic considerations and their effects on fuel and coolant temperatures will first be discussed. The procedure for obtaining maximum fuel temperatures in the core will be presented. Finally the steps taken in core thermal design will be outlined. In a reactor in which a liquid coolant is allowed to undergo bulk boiling, such as a boiling-water reactor, other hydrodynamic considerations must be taken into account. These are presented in the next chapter.

13-2. GENERAL CONSIDERATIONS

The rate at which heat is removed in a heterogeneous reactor core may be given by the familiar equation

$$Q = UA\,\Delta t \tag{13-1}$$

where Q = heat removal rate, Btu/hr
U = overall heat-transfer coefficient, Btu/hr ft^2 °F
A = area across which heat is transferred, ft^2
Δt = mean temperature difference between heat generator and coolant, °F

A nuclear reactor core is capable of generating large quantities of heat per unit volume, approximately 10^4 times as much as in a fossil-fueled steam generator. To be able to utilize a nuclear reactor to its maximum capability, a heat-transfer system must be designed in which each of the three components of Eq. 13-1 is as large as possible.

The first factor, U, is governed by (1) the heat transfer by conduction through the fuel elements, cladding, and other structures, if any, separating these elements from the coolant and (2) the heat transfer by convection to the coolant.

1. The thermal conductivities of the fuel are dependent upon the fuel type and are more or less fixed, Table 5-1. Fuel cladding should have as little resistance as possible to heat conduction, i.e., high thermal conductivity. However, other considerations enter into the choice of cladding materials. These include low neutron absorption cross sections, workability, the ability to withstand high temperatures, and also the ability to withstand the corrosive action of the coolant fluid. These frequently conflicting considerations have necessitated that a compromise be made with the requirements of high thermal conductivity and high-temperature operation. Thus materials such as stainless steel, aluminum, zirconium, and magnesium are used (rather than, say, copper, which has a high thermal conductivity but also a relatively high neutron absorption cross section). The resistance to heat conduction in the cladding is then reduced by holding the cladding thickness to a minimum. Table 5-2 lists some of the important physical characteristics of some cladding materials.

2. The heat transfer by convection to the coolant is dependent on the coolant's physical characteristics and on operating conditions, such as coolant velocity. High values of the convective heat-transfer coefficient h (Btu/hr ft^2 °F), unusual in nonnuclear applications, have been obtained by the use of very high coolant velocities and of either ordinary heat-transfer fluids, such as water, or liquid metals.

The second factor in Eq. 13-1, A, is the heat-transfer surface, i.e., the circumferential area of all the fuel elements, that must be designed within the reactor core. Cores vary in overall size, from several thousand cubic feet for natural uranium to a few cubic feet for reactors using highly enriched fuels. Both types are capable of generating large quantities of heat. It is obvious that considerable ingenuity should be exercised in order to be able to squeeze a large heat-transfer area into a small core.

For example, the Enrico Fermi sodium-cooled fast-breeder reactor has a core volume of 11.65 ft^3 and is designed to produce heat at a rate equivalent to 268 Mw(t) corresponding to an average power density in the core of 23 Mw/ft^3. A heat-transfer surface of 1.378 ft^2 is provided within the core by dividing the fuel into a very large number of pins with a large total surface area. The total cross-sectional flow area is 1.99 ft^2 and the coolant velocity is 31.2 ft/sec. Gas-cooled reactors, on the other hand, are very large in size. The heat-transfer coefficients h of gases are relatively poor. The heat-transfer area must be made large to compensate for the poor h. This is not very difficult, because of the large number of fuel elements used in this type of reactor and also because of the possible use of claddings with extended surfaces (fins). Gas-cooled reactors are capable of power densities of the order of 10 kw/ft^3 or lower. In general, however, large-volume reactors are capable of generating larger quantities of energy because of the relative ease of providing them with larger heat-transfer areas.

The third factor, Δt, is the mean difference between the temperatures of the coolant and fuel. High rates of heat release thus require low coolant temperatures. In power reactors, however, this requirement conflicts with thermodynamic considerations since good thermal efficiency requires high working-fluid temperatures. To keep the coolant temperature as high as possible and yet retain a high Δt in the reactor, the fuel elements have to operate at the highest possible temperatures. These temperatures are limited by metallurgical and burnup considerations if the fuel is in solid form. Uranium metal, used in some early stages of nuclear power, changes phase from α to β, with subsequent increase in volume, at about 1235°F (Table 4-2), necessitating operation below this value. At about 750°F, however, the metal strength diminishes, allowing fission gases to diffuse and gather in microscopic pockets within the fuel. These grow and eventually cause the fuel to expand. This condition puts a limitation on fuel temperature and burnup—the higher the former, the shorter the latter. Burnup limits of metallic fuels are of the order of 3,000 Mw-day/ton. Alloying of metallic fuels is used to partly overcome this weakness.

Ceramic fuels (UO_2, UC) do not suffer from this low-temperature phase change. They are capable of burnups of more than 30,000 Mw-day/ton. Uranium dioxide has low thermal conductivities, causing it to

FIG. 13-1. Progressive stages of UO_2 fuel pellets under irradiation (Ref. 137).

operate at very high maximum temperatures (about 4,500°F). In power-reactor operation UO_2 pellets develop radial and circumferential cracks as well as central holes (see Fig. 13-1) [137]. This is a source of uncertainty as to the actual value of heat conductance in the fuel.

Fluid-fueled reactors [2] with fuel in molten salt form or kept in solution in water, in liquid metal, or as slurries in suspension in a liquid or a gas, do not suffer from the limitations of temperature discussed above.

Often it is the melting point of the cladding material (e.g., aluminum and magnesium; see Table 5-2) that determines the maximum permissible temperature of the fuel elements.

It can be seen that reactor design, like many other problems in engineering, is necessarily a compromise between often conflicting nuclear, metallurgical, structural, thermodynamic, and heat-transfer requirements.

13-3. AXIAL TEMPERATURE DISTRIBUTIONS OF FUEL ELEMENT AND COOLANT

In Chapters 5-8 the temperature distributions in short sections of fuel elements were obtained in the steady state, one- and two-dimensional cases, and in the unsteady state. In such sections the neutron flux φ and therefore volumetric thermal source strength q''' were, in most cases, considered constant in the fuel in the radial direction, because of the small cross-sectional dimensions of the element compared to those of the core. They were also considered constant in the axial direction because only short sections of an element were treated.

In this section, we shall deal with a whole fuel element placed in a reactor core, as in Fig. 13-2. The axial variations in φ and q''' in the fuel element follow that in the core and must be taken into account.

In the discussion that follows we shall treat the case of a single fuel element whose length equals the core height H. The element is supposed to be situated somewhere in the core where the effective neutron flux at its center is φ_c. From considerations of reactor physics (Chapter 3), the neutron flux drops to zero at $z = \pm H_e/2$, where H_e is the extrapolated height and $z = 0$ corresponds to the core (and fuel element) center plane. The coolant fluid is assumed to pass upward through the core and parallel to the fuel element (Fig. 13-3). We shall also make the following simplifying assumptions:

1. The variation of φ in the axial direction is sinusoidal (i.e., a cosine function of z). The maximum values of φ and q''' occur in a single fuel element at its center and will be designated φ_c and q'''_c. These, of course, are the maximum values for that element only. Other fuel elements

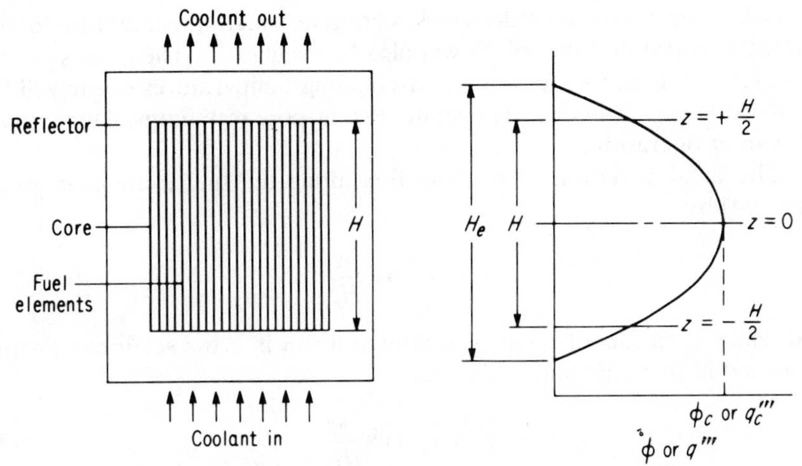

FIG. 13-2. Axial flux distribution in a reactor core. H = active core height; H_e = extrapolated height.

closer to the center of the core normally have higher values of φ_c and q_c'''. It should be indicated here that when the reactivity varies with z, because of a large axial temperature rise in a water-moderated and water-cooled core or because of a change in phase, such as in the boiling-water reactor, the axial flux may deviate appreciably from sinusoidal and some other relationship between q''' and z should be used. If the q''' variation is complicated, the problem may be solved by separately treating segments of the fuel element where the q''' variation can be approximated by a single or linear function of z.

2. An average value of the convective coefficient of heat transfer between coolant fluid and fuel element h will be used for the entire length of the fuel element. Actually a small variation in h takes place because of the dependence of h on the physical properties of the coolant, which normally vary with temperature, and on the operating characteristics, such as velocity, which vary with both temperature and coolant-channel geometry (entrance effects, etc.). In order to maintain as uniform moderating power as possible (in a water-cooled core) the coolant temperature rise is deliberately kept to a minimum (by increasing its mass rate of flow). To introduce h in the analysis as a variable function would unnecessarily complicate the analysis.

3. Similarly, the thermal conductivity of the fuel and the cladding as well as the physical properties of the coolant fluid will all be considered constant and independent of z.

We were so far able to evaluate the temperatures of fuel, cladding, and coolant at any one cross section. In the following analysis we shall evaluate the axial temperature variation of the coolant and cladding surface. The position of the cross section at which maximum cladding-surface temperature takes place will also be evaluated. The cross sections at which maximum center fuel and cladding temperatures occur will be discussed later. It is these temperatures that pose metallurgical limitations on reactor operation.

The axial variation of neutron flux along the fuel element is given (Sec. 3-4) by

$$\varphi = \varphi_c \cos \frac{\pi z}{H_e} \tag{13-2}$$

and, since each fuel element is usually uniform in cross-sectional dimensions and in fuel type and enrichment,

$$q''' = q_c''' \cos \frac{\pi z}{H_e} \tag{13-3}$$

where q''' and q_c''' are the volumetric thermal source strengths at any point z and the center of the fuel element, respectively.

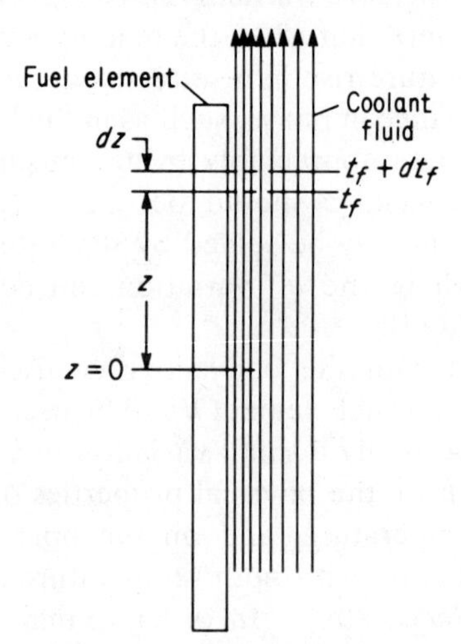

FIG. 13-3. Fuel element and associated coolant (shown on one side for simplicity).

If we consider a heat balance for a differential section of the fuel element of height dz (Fig. 13-3), at z the sensible heat gain by the coolant fluid (assuming no change in phase) is equivalent to the heat generated in the differential fuel element. Thus

$$\dot{m} c_p \, dt_f = q''' A_c \, dz \tag{13-4}$$

where $\dot{m}$ = mass-flow rate of coolant fluid per fuel element, lb_m/hr
c_p = specific heat of coolant fluid, Btu/lb_m °F
dt_f = coolant-fluid temperature rise between z and $z + dz$, °F
A_c = cross-sectional area of one fuel element, ft²

Equation 13-4 combines with 13-3 and integrates between the limits $z = -H/2$ and z as follows:

$$\dot{m}c_p \int_{t_{f_1}}^{t_f} dt_f = q_c''' A_c \int_{-H/2}^{z} \cos \frac{\pi z}{H_e} \, dz$$

where t_{f_1} and t_f are the coolant-fluid temperatures at $z = -H/2$ and z. Thus

$$t_f = t_{f_1} + \frac{q_c''' A_c H_e}{\pi c_p \dot{m}} \left(\sin \frac{\pi z}{H_e} + \sin \frac{\pi H}{2H_e} \right) \tag{13-5}$$

This equation gives the temperature of the coolant fluid t_f as a function of z. The coolant temperature at the center of the element can be obtained by putting $z = 0$. The exit temperature of the coolant fluid, t_{f_2}, can be obtained by putting $z = +H/2$ to give

$$t_{f_2} = t_{f_1} + \frac{2q_c''' A_c H_e}{\pi c_p \dot{m}} \sin \frac{\pi H}{2H_e} \tag{13-6}$$

In case the extrapolation lengths can be ignored, $H \simeq H_e$, and Eq. 13-6 reduces to

$$t_{f_2} = t_{f_1} + \frac{2q_c''' A_c H}{\pi c_p \dot{m}} \tag{13-7}$$

The axial temperature variation of the cladding surface, t_c, will now be evaluated. The heat transferred between cladding and coolant, at any point z along the fuel element, per unit area of cladding surface (circumferential) is given by $h(t_c - t_f)$. Since h has been assumed constant along the fuel element, the temperature difference $t_c - t_f$ is directly proportional to the volumetric thermal source strength q''' at that point. Since the latter is a cosine function of z, the above temperature difference also has the form

$$t_c - t_f = (t_c - t_f)_c \cos \frac{\pi z}{H_e} \tag{13-8}$$

but

$$q_c''' A_c \, dz = hC \, dz \, (t_c - t_f)_c \tag{13-9}$$

where C is the circumferential length of the clad fuel element in feet, a constant value, and $(t_c - t_f)_c$ is the temperature difference between cladding surface and coolant at $z = 0$. Combining Eqs. 13-5, 13-8, and 13-9 and rearranging give

$$t_c = t_{f_1} + q_c''' A_c \left[\frac{H_e}{\pi c_p \dot{m}} \left(\sin \frac{\pi z}{H_e} + \sin \frac{\pi H}{2H_e} \right) + \frac{1}{hC} \cos \frac{\pi z}{H_e} \right] \tag{13-10}$$

The variation of q''' (same as φ), t_f, and t_c along the coolant path (in the axial direction) is shown in Fig. 13-4. Note that $t_c - t_f$ is a cosine function of z and has a maximum value at the center.

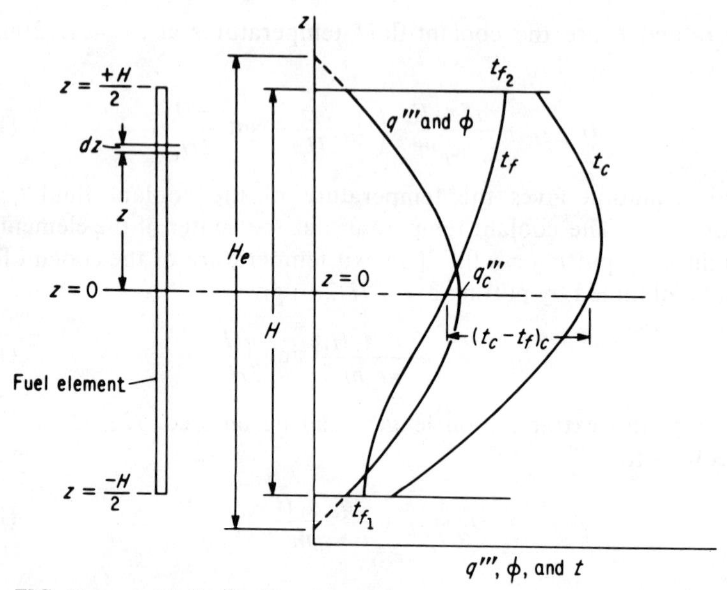

FIG. 13-4. Axial distribution of cladding-surface and coolant temperatures.

Other fuel temperatures, such as surface t_s and center t_m are evaluated by working backward into the fuel element, using the appropriate relationships given in Secs. 5-4 to 5-8. It should be remembered that this technique is not entirely correct inasmuch as the axial flux change actually causes the heat flow out of the fuel element to deviate from one-dimensional (radial). However, for a long, thin fuel element, the error involved is quite small.

13-4. MAXIMUM TEMPERATURES IN FUEL ELEMENT

The point at which the maximum cladding-surface temperature occurs, called z_c, can be obtained by differentiating Eq. 13-10 with respect to z, equating to zero, and solving for z_c as follows:

$$\frac{dt_c}{dz} = 0 + q'''A_c \left[\frac{H_e}{\pi c_p \dot{m}} \frac{\pi}{H_e}\left(\cos\frac{\pi z}{H_e} + 0\right) + \frac{1}{hC}\frac{\pi}{H_e}\left(-\sin\frac{\pi z}{H_e}\right)\right] = 0$$

Dividing by $\cos \pi z/H_e$ and rearranging give

$$\tan\frac{\pi z}{H_e} = \frac{hCH_e}{\pi c_p \dot{m}}$$

Solving for z in the above equation gives z_c as follows:

$$z_c = \frac{H_e}{\pi}\tan^{-1}\frac{hCH_e}{\pi c_p \dot{m}} \qquad (13\text{-}11a)$$

and for a cylindrical fuel element of fuel radius R and cladding thickness c, so that $C = 2\pi(R + c)$,

$$z_c = \frac{H_e}{\pi}\tan^{-1}\frac{H_e}{c_p \dot{m}\left\{\frac{1}{2}\left[\frac{1}{h(R+c)}\right]\right\}} \qquad (13\text{-}11b)$$

In this equation all components of the arc tangent are positive, indicating that it can assume values only in the first quarter, i.e., between 0 and $\pi/2$. This means that z_c is a positive value or that the point of maximum cladding-surface temperature occurs past the mid-plane ($z = 0$), as shown in Fig. 13-5. This is to be expected, since the coolant, coming in from the bottom at low temperature, helps to keep the bottom half of

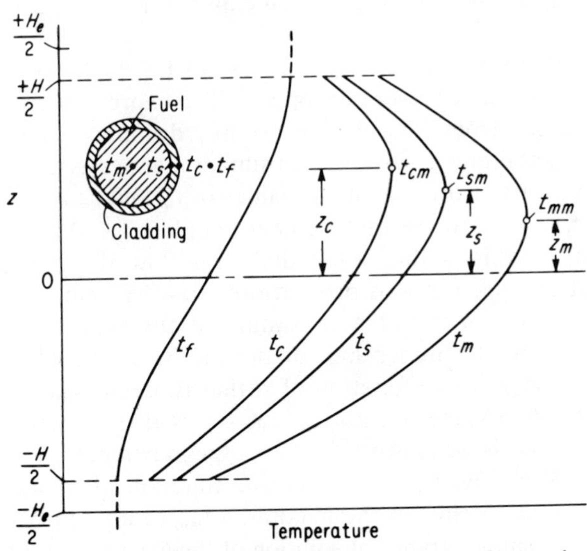

FIG. 13-5. Variation of fuel-element center, surface, and cladding temperatures with height.

the fuel element cooler than the top half. Equations 13-11 also indicate that increasing the heat-transfer coefficient or decreasing the coolant mass-flow rate shifts the point of maximum cladding-surface temperature farther away from the fuel element midplane. It also indicates that z_c is independent of q''' or φ.

The maximum cladding-surface temperature t_{cm} can now be obtained by substituting the value of z_c from Eqs. 13-11 into Eq. 13-10.

It is also of interest to evaluate the maximum values of the cladding inner temperature t_s and the fuel-center temperature t_m, since together with burnout these are the ones that pose limitations on the maximum allowable heat generation from the fuel element (Sec. 13-10). The positions of these two maxima along the fuel element can be evaluated by a technique similar to that above. Thus, for a cylindrical fuel element of fuel radius R with cladding of thickness c,

$$z_m = \frac{H_e}{\pi} \tan^{-1} \frac{H_e}{c_p \dot{m} \left\{ \frac{1}{4k_f} + \frac{1}{2} \left[\frac{1}{k_c} \ln \frac{R+c}{R} + \frac{1}{h(R+c)} \right] \right\}} \tag{13-12}$$

where k_f and k_c are the thermal conductivities of the fuel material and cladding respectively. An expression for $t_m - t_f$, such as Eq. 5-51a for a cylindrical fuel element, should be combined with Eq. 13-5 where z is replaced by z_m from Eq. 13-12 and an expression for t_{mm}, in terms of t_{f_1}, obtained.

Note that Eq. 13-11b is a special case of Eq. 13-12 in which the fuel and cladding resistances were dropped. The expression for z_s can therefore be obtained from Eq. 13-12 by simply dropping the fuel resistance term only. Obtaining t_{sm} is done in a similar manner as above.

The positions of the maximum values of t_m and t_s, t_{mm} and t_{sm}, occur at z_m and z_s, closer to the fuel-element midplane, as shown in Fig. 13-5. The reason for this is that the radial heat flow through the fuel and cladding, at any section z, is proportional to q''' at that section. This is a cosine function of z and a maximum at the fuel-element midplane ($z = 0$). Thus in the upper half of the fuel element, while t_f increases with z, $t_c - t_f$ is a cosine function of z, that is, decreases with z, giving a maximum for t_c at some position z_c. $t_s - t_c$ is also a cosine function of z. Thus when $t_c - t_f$ is added to t_c it causes the maximum t_{sm} to shift further below t_{cm}. Similarly, $t_m - t_s$ is a cosine function of z and when added to t_s causes another shift in the maximum t_{mm} to a point below t_{sm}. An interesting and simple graphical solution of the axial temperature variations has been worked out by French [138].

Example 13-1. In Example 5-1, the conditions at the cross section of maximum surface temperature in a clad plate fuel element 3.5 in. wide, 0.2 in. thick, and 4 ft long, clad in 0.005-in.-thick 304L stainless steel, were given or calculated as follows:

At z_c,

$$\begin{aligned}\varphi &= 3.0 \times 10^{13} \text{ neutrons/sec cm}^2 \\ q''' &= 2.185 \times 10^7 \text{ Btu/hr ft}^3 \\ h &= 3480 \text{ Btu/hr ft}^2 \text{ °F} \\ t_m &= 700\text{°F} \\ t_s &= 658.7\text{°F} \\ t_c &= 652.3\text{°F} \\ t_f &= 600\text{°F}\end{aligned}$$

If light water acts as coolant-moderator and infinite reflector (in the axial direction) and flows at the rate of 3 lb_m/sec per fuel element, calculate (*a*) the maximum neutron flux occurring in the fuel element, (*b*) the maximum volumetric thermal source strength, (*c*) the total heat generated by the fuel element, (*d*) the entrance and exit temperatures of the coolant, and (*e*) the minimum coolant pressure necessary to avoid coolant boiling.

Solution. It is first necessary to calculate the extrapolated height of the fuel element. This is given approximately by (Sec. 3-5)

$$H_e = H + 2\psi_e$$

where ψ_e is the extrapolated reflector savings. Since water acts as an infinite reflector in the axial direction, ψ_e is approximately equal to the thermal diffusion length L for light water. This is given in Table 3-1 as 2.88 cm or 0.0935 ft. Thus

$$H_e = 4 + 2 \times 0.0935 = 4.187 \text{ ft}$$

Next the value of z_c, the position of the cross section at which the maximum cladding-surface temperature occurs, must be found. First

$$C = 2(3.5 + 2 \times 0.005 + 0.2 + 2 \times 0.005) = 7.44 \text{ in.} = 0.62 \text{ ft}$$

c_p of light water at 600°F (assumed average overall coolant temperature) = 1.448 Btu/lb_m °F. Thus

$$\begin{aligned}z_c &= \frac{4.187}{\pi} \tan^{-1} \frac{3480 \times 0.62 \times 4.187}{\pi \times 1.448 \times (3 \times 3{,}600)} \\ &= 1.333 \tan^{-1} 0.1839 = 1.333 \times 0.1820 = 0.243 \text{ ft}\end{aligned}$$

(*a*) Since this is the value of z at which the neutron flux is 3×10^{13}, the maximum flux φ_c occurring at $z = 0$ is

$$\begin{aligned}\varphi_c &= \frac{\varphi}{\cos(\pi z/H_e)} \\ &= \frac{3 \times 10^{13}}{\cos[(\pi \times 0.243)/4.187]} = 3.050 \times 10^{13}\end{aligned}$$

(*b*) The maximum volumetric thermal source strength q_c''' can be computed as in Example 5-1, giving

$$q_c''' = 2.227 \times 10^7 \text{ Btu/hr ft}^3 \qquad \text{compared with } 2.185 \times 10^7 \text{ at } z_c$$

(*c*) Using Eq. 4-20, the total heat generated by the fuel element is

$$q_t = \frac{2}{\pi} 2.227 \times 10^7 \frac{3.5 \times 0.2}{144} 4.187 \left(\sin \frac{\pi \times 4}{2 \times 4.187} \right)$$
$$= 2.952 \times 10^5 \text{Btu/hr}$$

(*d*) The coolant inlet temperature t_f can be evaluated if the heat generated in the fuel element between $z = -H/2$ and $z = z_c$ is known. This can be computed by integrating Eq. 4-19 between these limits. However, Eq. 13-10 can be used since t_c is known at a corresponding z, namely, 652.3°F at 0.243 ft. Thus

$$t_{f_1} = 652.3 - 2.227 \times 10^7 \frac{3.5 \times 0.2}{144} \left[\frac{4.187}{\pi \times 1.448(3 \times 3{,}600)} \left(\sin \frac{0.243\pi}{4.187} + \sin \frac{4\pi}{2 \times 4.187} \right) + \frac{1}{3480 \times 0.62} \cos \frac{0.243\pi}{4.187} \right] = 652.3 - 60.1 = 592.2°\text{F}$$

The exit coolant temperature can be evaluated either from Eq. 13-6 or from the relationship

$$q_t = \dot{m} c_p (t_{f_2} - t_{f_1})$$
$$2.952 \times 10^5 = (3 \times 3{,}600) 1.448 (t_{f_2} - 592.2)$$

from which

$$t_{f_2} = 611.0°\text{F}$$

(*e*) To avoid boiling of the coolant, with consequent reactivity and burnout problems, the coolant must be pressurized beyond the saturation pressure corresponding to the highest coolant temperature along the fuel channel. This corresponds to the temperature in the water film adjacent to the cladding surface at the point where this is maximum. In other words, this equals t_c at z_c, that is, 652.3°F.

The steam tables show a saturation pressure, corresponding to 652.3°F, of approximately 2,244 psia. This is the minimum pressure that may be maintained in the above system. Most pressurized-water power reactors operate at pressures in the neighborhood of 2,200 psia. It should be mentioned here that local hot spots may develop (other than at z_c) causing local or subcooled boiling to take place. This will be discussed further in this chapter.

13-5. COOLANT-CHANNEL ORIFICING

The magnitude of the heat generated by a fuel element depends upon its position in the core. In a cylindrical unreflected large core, for example, the fuel elements situated in the center of the core generate about 10 percent more heat than fuel elements one-quarter of the radial distance away. If the mass rate of flow of the coolant and coefficient of heat transfer h are the same in all fuel channels, irrespective of their position in the core, the maximum fuel temperatures of the fuel elements, t_m, vary accordingly. The fuel elements near the center of the core may have temperatures in excess of the safe limit, while those near the core boundaries operate much below this limiting temperature. Also, the outlying fuel channels generate so little heat that the exit temperature of the coolant, t_{f_2}, is correspondingly lower than in the center channels. Thus the core average exit coolant temperature could be lower than that desired for good plant thermal efficiency.

This state of affairs can be corrected by *orificing* the core channels so that a gradual reduction in the coolant flow rate from the core center line takes place. Orificing can be done by restricting the coolant flow in the channel at some point along the axial path, as in the bottom or top grid plates, or both. This method may cause small pressure differences to occur between adjacent channels, with consequent leakage between them. Another method is to vary the coolant channel cross-sectional areas so that they become smaller with radial distance. This design could vary the fuel-moderator ratio and should be checked against the nuclear design of the core.

In orificing, coolant mass-flow rates may be adjusted as desired. However, two methods will be discussed.

1. One is to make the coolant mass-flow rate proportional to the heat generated in the channel, which would be given by the radial flux distribution in the core. In the case of a normal flux distribution, the mass-flow-rate distribution in a cylindrical core with a large number of channels would be given by

$$\dot{m} = \dot{m}_0 J_0 \left(\frac{2.4048r}{R_e} \right) \tag{13-13}$$

where $\dot{m}$ and $\dot{m}_0$ are the mass-flow rates at r and in the center channel, respectively.

This results in the exit temperature of the coolant being equal in all channels (assuming, of course, that the inlet temperature is uniform—a good assumption). However, this system results also in a reduction of the fuel temperatures with r. This can be seen when Eq. 13-10 for t_c is rewritten so that $\dot{m}$ is made to decrease in direct proportion to q_c'''. In single-phase turbulent flow, the heat-transfer coefficient h is a function of the Reynolds number to the power 0.8 (Sec. 9-5) or, other things being equal, to the mass-flow rate $\dot{m}$ to the same power. Thus if we put $h = \dot{m}^{0.8}/a$, where a is a constant, Eq. 13-10 can be rewritten in the form

$$t_c = t_{f_1} + A_c \left[\frac{q_c'''}{\dot{m}} \frac{H_e}{\pi c_p} \left(\sin \frac{\pi z}{H_e} + \sin \frac{\pi H}{2H_e} \right) + \frac{q_c'''}{\dot{m}} \dot{m}^{0.2} \frac{a}{C} \cos \frac{\pi z}{H_e} \right] \tag{13-14}$$

where t_c is the cladding temperature at height z, measured from the fuel-element center.

The net effect is that, at any height z and for given values of t_{f_1} and q_c'''/m, the first part within the brackets of Eq. 13-14 remains unchanged. The second part, however, decrease as $m^{0.2}$ (or $J_0^{0.2}$), meaning t_c decreases with r, though to a lesser degree than before orificing. This method of orificing also causes a shift in the position of maximum cladding temperature, z_c, to a point farther from the fuel-element center. This can be seen by examining Eqs. 13-11 for z_c, where the arc tangent increases when m is decreased.

2. The other method of orificing has the effect of equalizing the maximum fuel-surface temperatures, to the maximum allowable limits, throughout the core. This is desirable where the fuel or cladding temperatures are controlling. The coolant mass rate of flow in this case has to be reduced with r to a greater degree than in case 1 above. In other words, it should be reduced faster than q_c''', so that, at any point z, the total quantity within the brackets in Eq. 13-14 remains constant. In this system the exit coolant temperature t_{f_2} increases with r (for constant t_{f_1}). This can be shown by examining Eq. 13-7 in which the ratio q_c'''/m is made to increase with r. Thus the core average coolant exit temperature will be higher than from the center channels, a desirable effect from the standpoint of plant thermal efficiency.

In boiling-type reactors, orificing may be used to even out the quality of the vapor-liquid mixture emanating from the various fuel channels.

13-6. HOT-SPOT FACTORS

When a reactor is in the design stage, it may be presumed that, when constructed, it will be an exact copy of the blueprints. For example all the fuel elements will have precisely the dimensions, cladding thickness, enrichments, and general physical characteristics (such as thermal conductivity) specified in the design; the coolant passages will be perfect and therefore the coolant pressure drops and flows in the core channels are those envisioned in the design. Also the core may be presumed to be subjected to a neutron flux distribution exactly as predicted in the physics design of that core. If that is the case, the maximum temperatures (which are really the performance-limiting parameters in a power-reactor core) are those calculated strictly on the basis of such parameters.

This best of all possible worlds, of course, does not exist in reality.

Constructed fuel elements, for example, will deviate somewhat from specifications due to inevitable manufacturing tolerances. These long, slender elements will also suffer from some bowing during operation and the flow cross sections will not be exactly those planned. Together with orifice tolerances, this causes the coolant flow to have a distribution somewhat different from that required in the design. Also the neutron flux distribution, for many reasons, will not agree exactly with predictions.

In practice, although local deviations will occur on either side of predictions (see next section), it is those that cause the temperatures to *exceed* the *nominal* temperatures (those predicted on the basis of no deviations) that must be guarded against. This is done by the use of safety factors, called the *hot-spot* or *hot-channel factors*.

These factors were originally described by Tourneau and Grimble [139]. They are obtained, by estimating for each physical deviation, a factor representing the maximum change in temperature rise from the nominal value. The various factors are then combined so as to multiply the nominal temperature rise in the core, under average conditions of flux, to obtain a more realistic maximum temperature in the core. This latter must be within prescribed safe limits.

Hot-spot factors are broadly classified, according to the physical deviations that cause them, into two broad classifications: (1) *nuclear* hot-spot factors, and (2) *engineering* hot-spot factors. Each of these will have several subfactors that are largely determined by the designer. Table 13-1 contains an abbreviated list of such factors for discussion purposes. (Table 13-3, Sec. 13-9, contains a more detailed list used in an actual design.)

(1) The *Nuclear* hot-spot factors are those that occur due to the variations in neutron flux from a core *average* value. Thus these are due to overall flux distributions (cosine, Bessel) and to the many valleys and peaks superimposed on these distributions. These local deviations (aside from those due to the fuel acting as a sink and the moderator as a source of thermal neutrons) are caused by many factors, such as the presence of partially inserted control rods depressing thermal neutrons in their vicinity with consequent large peaking elsewhere (Fig. 13-6), nonhomogeneity in the moderator (e.g., when in solid form or when boiling takes place) or fuel, and the presence of structural materials or other inhomogeneities. The effect of inhomogeneity is most pronounced when the neutron mean free path is small, as in the case of a strong hydrogenous moderator (e.g., H_2O) where short-range thermalization of the fast, newly born neutrons takes place. The effect is less pronounced in fast reactors or thermal reactors using weaker moderators having long neutron-thermalization mean free paths, such as heavy water or graphite. Also, the presence of a good, infinite reflector causes peaking of the

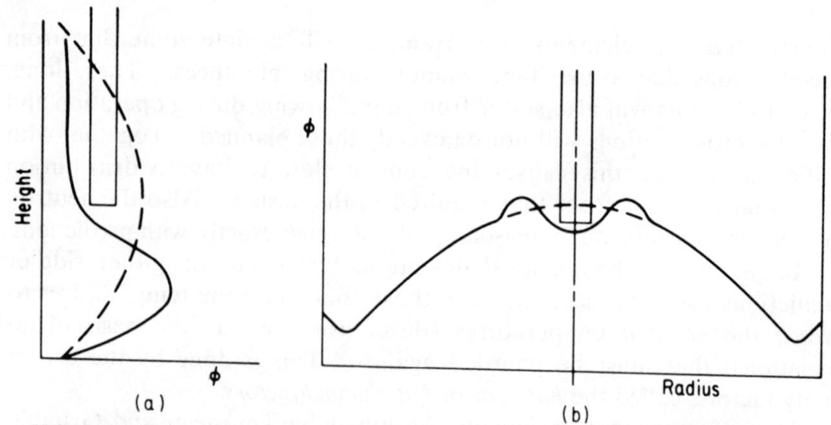

FIG. 13-6. Effect of central rod insertion on (a) axial, and (b) radial flux distributions.

thermal neutron flux at the ends of the reactor core (Fig. 13-7) and consequent high heat generation and high fuel temperatures there. The nuclear hot-spot factors therefore take into account the above deviations as well as the deviation of the overall flux distribution from the core average.

(2) *Engineering* hot-spot factors are usually subclassified into (a) *mechanical* factors, and (b) *flow distribution* factors. The *mechanical* subfactors are numerous. Some are due to deviations from nominal in fuel

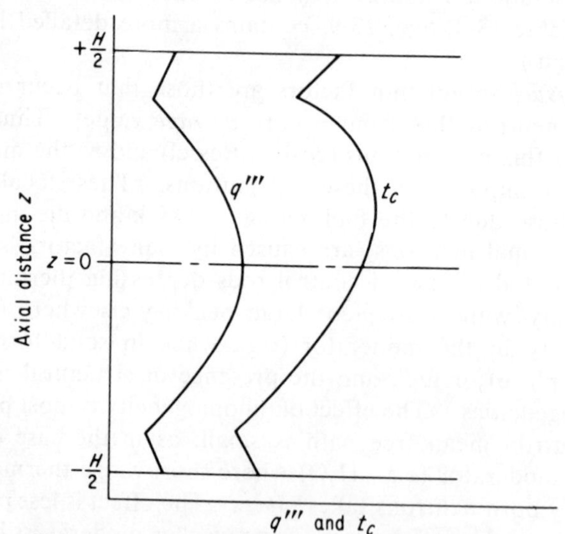

FIG. 13-7. Heat generation and cladding-surface temperature distribution in a thermal reactor with good reflector.

or clad thickness due to manufacturing tolerances. A thick spot of cladding, for example, creates higher resistance to heat conduction from the fuel element and creates a hot spot in the cladding and fuel behind it. Warping of a fuel element, as another example (which may occur due to high fuel burnup as well as mechanical causes) also results in hot spots. Other factors may be increased heat resistance in metal, deviations in fuel elements dimensions from nominal, etc. *Flow distribution* factors result from the maldistribution of coolant flow due to reactor vessel geometry or defects in orificing (if used). A decrease in coolant flow causes a decrease in the heat-transfer coefficient *h*, and in turn causes higher fuel temperatures.

TABLE 13-1
Components of Hot-Channel Factors

Factors	F_c	F_f	F_e
Nuclear:			
Neutron distribution	1.20	1.20	1.20
Fuel concentration	1.10	1.17	1.18
Engineering:			
(a) Mechanical			
Fuel-element warpage	1.04	1.03	
Fuel-element thermal conductivity	....	....	1.07
(b) Distribution			
Flow distribution	1.02	1.02	
Heat-transfer coefficient, *h*	....	1.15	
Fuel-element dimensions	1.26	1.18	1.10
Products	1.76	2.00	1.66

Hot-spot factors are also classified according to their effects on temperature or enthalpy of fuel and coolant, Table 13-1. So there are hot-spot factors for (1) the temperature rise of the coolant from core inlet conditions, F_c (this is also often called the *enthalpy rise* hot-spot factor $F_{\Delta H}$); (2) the temperature rise across the coolant film or boundary layer, F_f, and (3) the temperature rise across the fuel element, F_e. The latter may be broken down into its usual components, i.e., the temperature rise across an oxide layer, the cladding, a gas layer, and finally across the fuel material itself.

In Table 13-1 for example, if the nominal rise of the coolant from reactor inlet conditions to the point of the hot spot is 35°, the expected hot-spot temperature rise due to fuel element warpage alone would be $35 \times 1.04 = 36.4$ °F.

The individual subfactors are usually determined from statistical data.

Before this is discussed some necessary statistical principles are presented in the next section.

13-7. BASIC STATISTICAL RELATIONSHIPS

The *normal* or *Gaussian* distribution for a statistical quantity Q (Fig. 13-8) is given by

$$g(Q) = \frac{1}{S\sqrt{2\pi}} e^{-\frac{1}{2}\left(\frac{Q-Q_n}{S}\right)^2} \tag{13-15}$$

where S = *standard deviation* from Q, having the same dimensions as Q
Q_n = nominal or average value of Q

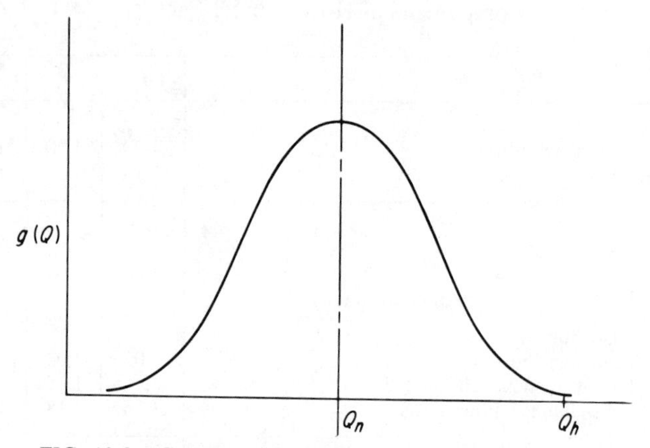

FIG. 13-8. Gaussian distribution of a statistical quantity Q.

The probability that the quantity Q would exceed a selected upper (or be less than a lower) limit Q_h is given by

$$P(Q > Q_h) = \int_{Q_h}^{\infty} \frac{1}{S\sqrt{2\pi}} e^{-\frac{1}{2}\left(\frac{Q-Q_n}{S}\right)^2} dQ \tag{13-16}$$

Using the *dimensionless standard deviation* σ, given by

$$\sigma = \frac{S}{Q_n} \tag{13-17}$$

and the dimensionless factor f, given by

$$f = \frac{Q_h}{Q_n} \tag{13-18}$$

give for constant Q_n

$$P(Q > Q_h) = \int_f^\infty \frac{1}{\sigma\sqrt{2\pi}} e^{-\frac{1}{2}\left(\frac{R-1}{\sigma}\right)^2} dR \tag{13-19a}$$

where

$$R = \frac{Q}{Q_n} \tag{13-20}$$

and since Q_n is a constant, the above indicate that

$$P(Q > Q_h) = P(R > f) \tag{13-21a}$$

and R has the same distribution as Q, Fig. 13-8. The factor f is usually equated to a linear function of the standard deviation σ, as

$$f = 1 + A\sigma \tag{13-22a}$$

where A is a constant, often, but not necessarily, an integer. The above equation which can also be written in the form

$$Q_h = Q_n + AS \tag{13-22b}$$

simply states that Q_h is chosen to exceed Q_n by a number A of standard deviations. Eq. 13-19a may be simplified by putting

$$B = \frac{R-1}{\sigma} \tag{13-23a}$$

so that

$$dR = \sigma dB \tag{13-23b}$$

and

$$P(Q > Q_h) = P(R > f) = P\left(\frac{R-1}{\sigma} > A\right) = P(B > A) \tag{13-21b}$$

and Eq. 13-19a may now be written in the form

$$P(Q > Q_h) = P(B > A) = \int_A^\infty \frac{1}{\sqrt{2\pi}} e^{-\frac{1}{2}B^2} dB \tag{13-19b}$$

The above integral results in the numerical values given in Table 13-2.

As the table suggests, a probability of .00135, for example, means that there are 1.35 chances in a thousand that Q will exceed Q_h or that R will exceed f. In our case of interest *f is the hot-spot factor,* and 1.35 elements in 1000 will be expected to have temperatures exceeding the hot-spot temperature. This corresponds to a *confidence limit* of 99.865

TABLE 13-2

Number of Standard Deviations, A	Probability $P(Q > Q_h)$	Probable Failures in 1,000	Confidence Limit percent
0.	.500000	500.000	50.0000
0.5	.308538	308.538	69.1462
1.0	.158655	158.655	84.1345
1.5	.066807	66.807	93.3193
2.0	.022750	22.750	97.7250
2.5	.006210	6.210	99.3790
3.0	.001350	1.350	99.8650
3.2	.000687	.687	99.9313
3.4	.000337	.337	99.9663
3.6	.000159	.159	99.9841
3.8	.000072	.072	99.9928
4.0	.000032	.032	99.9968
4.2	.000013	.013	99.9987
4.4	.000005	.005	99.9995
4.6	.000002	.002	99.9998
4.8	7.93×10^{-7}	.001	99.9999
5.0	2.87×10^{-7}	~ .000	~100.0000

percent. The confidence limit is obviously equal to $(1 - P)$ in percentage points.

In selecting hot-spot subfactors, the distribution of the physical deviation obtained from statistical data is fitted to a normal or Gaussian distribution. This is done by computing the sample mean, d_i, and the standard deviation S, given by

$$d_i = \frac{\Sigma d_i}{n} \tag{13-24}$$

and

$$S = \left[\frac{\Sigma \left(d_i - \frac{\Sigma d_i}{n} \right)^2}{n - 1} \right]^{0.5} \tag{13-25}$$

where d is a certain physical dimension, such as the diameter of a fuel element, the cross-sectional area of a flow channel, etc., and n is the number of samples. Figure 13-9 graphically shows a fitted normal distribution of the physical dimension. d_n is the *nominal* dimension, such as specified in the design. The sample mean d_i and the nominal dimension d_n are not necessarily equal (the probability that they are is zero), but are usually very close, provided the statistical sample is large.

A confidence limit is now decided upon, say 99.865 percent, corresponding to 3 standard deviations. The corresponding deviation in

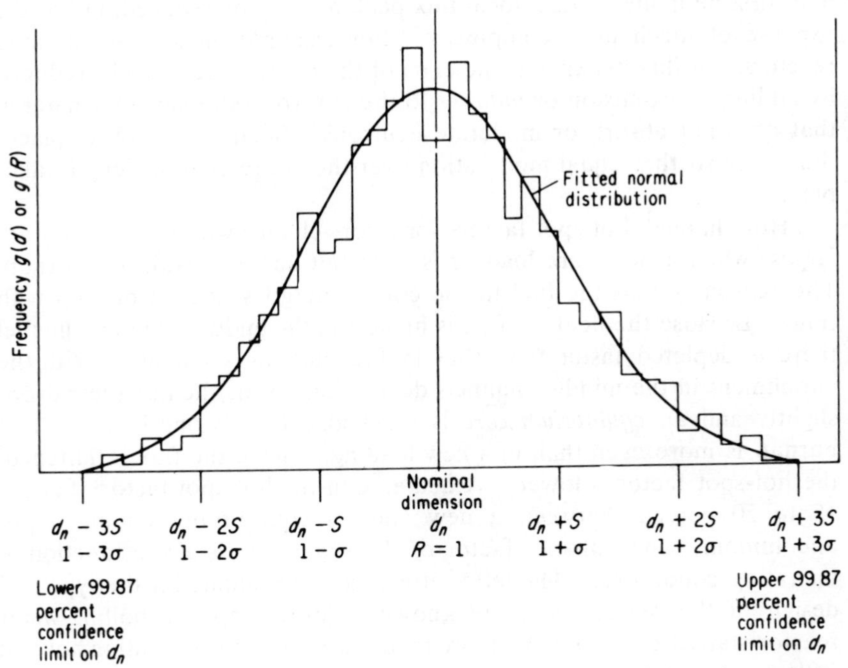

FIG. 13-9. Fitted normal distributions of a physical dimension d about the nominal dimension d_n, $R = d/d_n$.

physical dimensions is obtained and the hot-spot subfactor due to it is computed. This is illustrated in the next section.

Basic to the determination of the hot-spot subfactors, therefore, is the reliability of the input data (the deviations). These generally fall in three classes: (1) Those for which statistical distributions are well known, such as fuel-element diameters, for which data have been extensively obtained in the manufacturing process; (2) those for which the deviations are estimated from the spread of experimental data, such as the spread commonly encountered in experimental heat transfer by convection data, on which Nusselt-type correlations are based; and (3) those for which only a rough estimate can be made, such as the uncertainty due to the maldistribution of flux, or thermal conductivity of irradiated oxide fuel for which data has been obtained in out-of-pile tests. Experience will help reduce the uncertainty, but conservatism should increase with the degree of ignorance.

It sometimes may be imperative to design certain hot spot factors into a core. This occurs if only a certain quantity of fuel is available and the factors will have to be scaled down. In that case, for example, an added effort to reduce the effect of some of the manufacturing tolerances may first be made. Also, local flux peaking may be reduced, but at the expense of mechanical complexity. For example, in water-moderated reactors, the flux peaking at the ends of the control rods may be reduced by adding an extension or *follower* to the control rod, made of a material that does not absorb or moderate neutrons. Such a follower displaces the water so that equal moderation over the entire channel length takes place.

Hot-channel, hot-spot factors for a core change with time. They are largest when a new core loading is used but decrease with fuel burnup. The reason is that the fuel in the core undergoes uneven burnup with time. Because the neutron flux is highest in the middle channel, the fuel there is depleted faster than that in the outlying channels. With the enrichment in the middle channels decreasing faster, the flux there drops slightly and an *equilibrium core* is reached. This is one in which fuel burnup is more even than in a new loading. Since the flux is flattened, the hot-spot factor is lower. A decrease in the hot-spot factor of about 15 to 20 percent between a new and an equilibrium loading is not uncommon. The hot-spot factor of the same core is also a function of operating conditions. The latter effect can be evaluated if the type and design of the reactor core are known. For example, a boiling-water reactor started cold contains more moderator than one operating hot and with steam voids.

13-8. AN EXAMPLE OF DETERMINING A HOT-SPOT SUBFACTOR

An example of determining the hot-spot subfactor in the case where the actual constructed flow areas of the channels in the core deviate from the nominal (design) flow area will be presented here.

It shall be assumed that the constructed areas deviate statistically from the nominal area according to a Gaussian distribution and that a 99.865 percent confidence limit, equivalent to three times the standard deviation, was chosen to determine the hot spot subfactor (as above) and that this results in a hot-spot flow area A_h as opposed to the nominal flow area A_n. The case of the constructed area being *less* than the nominal area is the one that gives rise to hot spots. In that case, the coolant mass rate of flow and flow velocity will decrease. Since the heat generation per channel is unaffected, this will have two effects:

1. The coolant temperature rise Δt_c will increase, and therefore

$$\Delta t_{c,h}(z) > \Delta t_{c,n}(z) \tag{13-26}$$

where $\Delta t_{c,h}(z)$ and $\Delta t_{c,n}(z)$ are the hot-spot and nominal coolant temperature rises at any height z above the core inlet. Since the core inlet temperature is unaffected, $t_{c,h}(z) > t_{c,n}(z)$, and the coolant will be hotter.

2. Because of lower velocities, the heat-transfer coefficient, a function of Reynolds number, will decrease, and consequently

$$\Delta t_{f,h}(z) > \Delta t_{f,n}(z) \tag{13-27}$$

where $\Delta t_{f,h}(z)$, and $\Delta t_{f,n}(z)$ are the hot-spot and nominal temperature rises across the film (boundary layer) between coolant bulk and cladding surface at z. Thus the cladding surface temperature will increase due to two hot-spot subfactors, themselves brought about by the same physical uncertainty (deviation of flow area from nominal dimensions).

The hot-spot subfactor for the *coolant temperature rise* f_c for the deviation can be calculated by noting that the pressure drops in all core channels must be equal. Assuming turbulent flow (a good assumption in reactors), the pressure drop is obtained from the Darcy formula

$$\Delta p = f \frac{H}{D_e} \frac{\rho V^2}{2g_c} = \text{const.} \tag{13-28}$$

where f = friction factor, not to be confused with f in Eq. 13-18.
H = height of channel
D_e = equivalent diameter
V = coolant velocity
g_c = conversion factor

The parameters affected by the deviation in flow area in the above equation are V, D_e, and f. Comparing the hot and nominal cases.

$$\frac{V_h}{V_n} = \left(\frac{D_{e_h}}{D_{e_n}}\right)^{0.5} \left(\frac{f_n}{f_h}\right)^{0.5} \tag{13-29}$$

where the subscripts h, and n refer to hot and nominal conditions. For turbulent flow, it was shown (Sec. 9-2) for circular tubes that

$$f \propto (D_e V)^{-0.2} \tag{13-30}$$

Using this relationship in Eq. 13-29 gives

$$\frac{V_h}{V_n} = \left(\frac{D_{e_h}}{D_{e_n}}\right)^{0.5} \left(\frac{D_{e_h} V_h}{D_{e_n} V_n}\right)^{0.1}$$

or

$$\frac{V_h}{V_n} = \left(\frac{D_{e_h}}{D_{e_n}}\right)^{2/3} \tag{13-31}$$

But since $D_e = 4A/P$ where A is the channel-flow area and P the channel wetted perimeter, and since the wetted perimeter will not be affected (think of the cases of flow parallel to rod bundles or between flat plates), then

$$D_e \propto A \tag{13-32}$$

and therefore

$$\frac{V_h}{V_n} = \left(\frac{A_h}{A_n}\right)^{2/3} \tag{13-33}$$

The hot-spot factors are ratios of hot-to-nominal temperature rises. The coolant temperature rise in the channel is obtained from the heat generated in the channel up to z, $q(z)$, which remains unaffected, as

$$q(z) = \rho A V c_p \Delta t_c(z)$$

where ρ and c_p are the coolant density and specific heat, unaffected. Thus the desired hot-spot factor f_c is given by

$$f_c = \frac{\Delta t_{c,h}(z)}{\Delta t_{c,n}(z)} = \frac{A_n V_n}{A_h V_h} \tag{13-34}$$

Combining Eqs. 13-33 and 13-34 gives finally

$$f_c = \left(\frac{A_n}{A_h}\right)^{1.667} \tag{13-35}$$

The hot-spot subfactor for the *film temperature rise* f_f is obtained by considering the single-phase correlation (Sec. 9-5).

$$\left.\begin{aligned} \mathrm{Nu} = \frac{hD_e}{k} &= 0.023\ \mathrm{Re}^{0.8}\ \mathrm{Pr}^{0.4} \\ &= 0.023 \left(\frac{D_e V \rho}{\mu}\right)^{0.8} \mathrm{Pr}^{0.4} \end{aligned}\right\} \tag{13-36}$$

where Nu is the Nusselt number and h is the heat-transfer coefficient. Pr, k, and μ are the Prandtl number, thermal conductivity, and viscosity, all physical properties of the fluid and are largely unaffected. Thus

$$h \propto \frac{V^{0.8}}{D_e^{0.2}} \tag{13-37}$$

The fuel-element heat flux at any height z is given by

$$q''(z) = h\,\Delta t_f(z) \tag{13-38}$$

where h is assumed independent of z, and Δt_f is the temperature rise in the film. Since $q''(z)$ is unaffected, then

$$\Delta t_f(z) \propto \frac{1}{h} \tag{13-39}$$

Combining Eqs. 13-37 and 13-39 gives

$$f_f = \frac{\Delta t_{f,h}}{\Delta t_{f,n}} = \left(\frac{D_{e_h}}{D_{e_n}}\right)^{0.2} \left(\frac{V_n}{V_h}\right)^{0.8} \tag{13-40}$$

Noting that $V \propto D_e^{2/3}$, Eq. 13-31, gives

$$f_f = \left(\frac{D_{e_h}}{D_{e_n}}\right)^{0.2} \left(\frac{D_{e_n}}{D_{e_h}}\right)^{(0.8)2/3} \tag{13-41}$$

and since $D_e \propto A$, Eq. 13-32, gives finally

$$f_f = \left(\frac{A_n}{A_h}\right)^{1/3} \tag{13-42}$$

Example 13-2. Calculate the coolant and film temperature rise hot-spot subfactors due to deviations of channel flow area from nominal dimensions, based on three times the standard deviation being equal to 2 percent.

Answer:

$$\frac{A_n}{A_h} = \frac{1}{0.98} = 1.0204$$

Thus

$$f_c = (1.0204)^{1.667} = 1.034$$

and

$$f_f = (1.0204)^{1/3} = 1.007$$

13-9. THE OVERALL HOT-SPOT FACTOR

We have so far discussed hot-spot subfactors and how they are obtained. They are useful in themselves if the deviation of the temperature rise from nominal, at a given locality, due to a given physical effect is desired. In reactor core thermal design, it is often a combination of all the hot-spot subfactors, f, into a given *overall hot-spot F* that is of interest.

The hot-spot subfactors maybe combined into an overall hot-spot factor by one of two schemes (a) the *multiplicative,* and (b) the *statistical.*

In any one core, there will always be a particular channel in which a combination of circumstances cause it to be the hottest in that core. The hottest spot in the core normally would occur in that hot channel, but not necessarily so. In the *multiplicative* approach, all the factors that tend to increase the temperature are supposed to take place simultaneously and at the same point, the extreme values of the uncertainties such as the worst deviations in fuel loading and dimensions will occur in the same element which will be located in the channel with the poorest coolant

TABLE 13-3
Hot-Spot Factors. Used in the Enrico Fermi Fast-Breeder Reactor (Ref. 141).

Physical Uncertainty	Factor for Coolant Temperature Rise $\Delta t_c = 285°F$		Factor for Temperature Rise Through Film $\Delta t_f = 41.5°F$		Factor for Temperature Rise Through Oxide Layer $\Delta t_o = 15.5°F$		Factor for Temperature Rise Through Clad $\Delta t_{Zr} = 39.8°F$		Factor for Temperature Rise Through Fuel $\Delta t_U = 196°F$		Summary	
	f	3σ	f	3σ	f	3σ	f	3σ	f	3σ	$\Sigma(3\sigma)$	$[\Sigma(3\sigma)]^2$
Maldistribution of coolant (a) To subassemblies (b) Within subassemblies	1.15	42.8									42.8	1,831.8
Deviation from nominal dimensions	1.03	8.6									8.6	74.0
			1.04	1.7							1.7	2.9
					2.00	15.5					15.5	240.3
							1.02	0.8			0.8	0.6
									1.02	3.9	3.9	15.2
Maldistribution of U-235	1.02	5.7	1.02	0.8	1.02	0.3	1.02	0.8	1.02	3.9	11.5	132.3
Maldistribution of flux	1.05	14.3	1.05	2.1	1.05	0.8	1.05	2.0	1.05	9.8	29.0	841.0
Burnup in core fuel pin	1.10	28.5	1.10	4.2	1.10	1.6	1.10	4.0	1.10	19.6	57.9	3,352.4
Power measurement and control	1.08	22.8	1.08	3.3	1.08	1.2	1.08	3.2	1.08	15.7	46.2	2,134.4
Film heat-transfer coefficient			1.30	12.5							12.5	156.3
Thermal conductivity of zirconium oxide					1.20	3.1					3.1	9.6
Thermal conductivity of zirconium							1.10	4.0			4.0	16.0
Thermal conductivity of fuel alloy									1.20	39.2	39.2	1,536.6

Totals: 277.0

$\Sigma[\Sigma(3\sigma)]^2$ = 10,343.4

Square root of sum of squares = 101°F

Uranium hot spot located at fuel alloy length fraction, $z/L = 0.70$.
Maximum uranium temperature without hot-channel factors = 1134°F (50% confidence)
Maximum uranium temperature with hot-channel factors: 1168°F (1σ, 84.135% confidence), 1202°F (2σ, 97.725% confidence), 1235°F (3σ, 99.865% confidence).

flow which in turn will be located in the region of highest deviation from the core average neutron flux, and so on. In the multiplicative approach, then, the overall hot-spot factor F is equal to the gross *product* of all the hot-spot subfactors. Using the data of Table 13-1 the subfactors in each column are multiplied and the overall hot-spot factor is the gross product of the three, or

$$F = 1.76 \times 2.00 \times 1.66 = 5.84$$

Since the overall hot-spot factor is often used in core design to reduce the maximum allowable heat flux and hence increase fuel surface area (Sec. 13-11), it is needless to say that the multiplicative approach leads to ultra conservative designs.

The alternative *statistical* approach, holds that the probability of the extreme values of the deviations (many being quite independent of each other) all occurring at the same point in the core is quite remote. The multiplicative approach is then replaced by a more realistic one in which the chosen standard deviations are combined in a statistical manner. This results in a lower overall hot-spot factor, as well as a numerical value for the *probability* that the actual maximum temperature in the core will exceed the calculated hot-spot value. Lower overall hot-spot factors result in reduced fuel loading during the design stage of a reactor of given themal power output, or in an increase in the design power output of a core of given fuel loading, undoubtedly a great economic incentive.

The statistical approach was first suggested by Tourneau and Grimble [139], described in some detail by Hitchcock [140] and first used in an actual design on the Enrico-Fermi fast-breeder reactor [141]. In that work, a confidence limit of 99.865 percent, equivalent to 3 standard deviations, and 1.3 chances in 1,000 that the calculated hot-spot temperature will be exceeded (Sec. 13-9), was chosen as the basis of computations. Someone might argue that 1.3 chances in 1,000 represent too high an acceptable risk for his own type of reactor. But such a ratio does not necessarily mean failure, however, since the temperature limit itself is conservatively chosen might only mean reduction in burnup (in core residence time), or core shutdown and changing a fuel subassembly if necessary. Reactors have operated with "failed" elements in place, as long as the level of coolant contamination was within permissible limits.

In the above-cited work, the fuel centerline temperature, rather than cladding burnout, was the limiting consideration. Table 13-3 summarizes this work. For each contributing physical uncertainty (the left-hand column), hot-spot subfactors are listed, where applicable, for each of five temperature rises from coolant core-inlet temperature to maximum fuel center-line temperature at 0.70 of the element height. The quantities Δt_c, Δt_f, etc. are the nominal temperature rises. The sum of these nominal temperature rises is 577.8°F. The core inlet temperature is 556.2°F so that

$$\text{Nominal fuel center-line temperature} = 556.2 + 557.8 = 1134°\text{F}$$

The quantities f in each column are the hot-spot subfactors based on a 3σ calculation. The 3σ deviations in temperatures, equal to $[(f - 1) \times$ nominal temperature rises], are shown next to the corresponding f values. For example, $(1.15 - 1) \times 285 = 42.8°$F, and so on.

If at this stage a multiplicative approach is resorted to, the sum of all the resulting temperature deviations, such as the 42.8°F above, 277°F in this example, the total of the $\Sigma 3\sigma$ column, would be *added* to the sum of the nominal temperature rises, 577.8°F in this example, resulting in an overall hot-spot temperature rise of $577.8 + 277 = 854.8°$F, and

$$\text{Multiplicative hot-spot temperature} = 1134 + 277 = 1411°\text{F}$$

In the statistical approach, the same physical uncertainty may cause hot spots in more than one Δt. Some of the 3σ values are statistically *dependent,* such as due to the maldistribution of U^{235}, and are therefore tabulated in the same row in Table 13-3. These are added in the $\Sigma 3\sigma$ column. Other 3σ values are statistically *independent,* even though they may be due to the same physical uncertainty, as in the case of deviation from nominal dimensions, are tabulated in separate rows, and are therefore listed in the $\Sigma 3\sigma$ column separately. The numbers in the $\Sigma 3\sigma$ column are now all statistically independent and are combined statistically by taking the square root of the sum of the squares* (last column). This yields a value of 101°F, which is the statistically obtained overall hot-spot temperature deviation in 3σ or 99.865 percent confidence calculation. When added to the nominal temperature rise it results in a hot spot temperature rise of $577.8 + 101 = 678.8°$F, and

$$3\sigma \text{ statistical hot-spot temperature} = 1134 + 101 = 1235°\text{F}$$

As a matter of interest, other calculations based on 2 and 1 standard deviations, equivalent to 97.725 percent and 84.135 percent confidences respectively, would yield:

$$2\sigma \text{ statistical hot-spot temperature} = 1134 + 68 + 1202°\text{F}$$

$$1\sigma \text{ statistical hot-spot temperature} = 1134 + 34 = 1168°\text{F}$$

* In statistical theory, the standard deviation squared is called the *variance*, and the variance of two or more statistical quantities is equal to the sum of the variances of these quantities.

13-10. CORE THERMAL DESIGN

Reactor power plants generate hundreds of megawatts (electrical). Considering the thermal efficiencies of the plants, their cores are called upon to generate hundreds or thousands of megawatts (thermal)*. This must be done in such a manner that the temperatures anywhere in the core must not exceed material property limitations. This is done by adjusting the fuel quantity and surface area so that the heat fluxes do not result in temperatures exceeding these limitations. The size of a power reactor is therefore very large (many times critical) and the whole or portions of it, are prevented from becoming supercritical by proper design of the reactor-core control system. With such large sizes, the neutron leakage is small and nuclear considerations are no longer sensitive to changes in core size. Once the core basic lattice has been determined from nuclear, hydraulic, and heat-transfer considerations, the overall size of the core for a particular power output usually becomes a sole function of the thermal considerations.

The detailed design of a core, however is a compromise between nuclear, material, structural, hydraulic, heat-transfer, and economic considerations, usually involving an iterative procedure to arrive at an optimum design. The thermal (hydraulic and heat transfer) considerations, which determine overall core size, will be discussed in this section.

The most important parameter in the thermal design is the fuel temperature distribution. The specific power [kw(t)/kg_m], power density [kw(t)/liter] and heat flux (Btu/hr ft^2), all related once a basic fuel lattice or cell dimensions have been chosen, must be limited so that, together with the designed heat removal system (h) they would not result in temperatures that exceed fuel element material limitations and therefore in fuel element failure.

The limiting material property must be carefully determined. In the Enrico-Fermi example (Sec. 13-9) for example, it was the fuel center-line temperature. The same is true in gas-cooled reactors, though depending upon materials, the cladding maximum (inner) temperature may be the limiting consideration. Gas-cooled reactors usually have large temperature rises through a core and, because of relatively low heat-transfer coefficients, across the film also. In water-cooled reactors (both pressurized and boiling) the usual consideration is cladding burnout (Sec. 11-5). In this case the deviations are not linearly related to the hot-spot factors. In some cases, more than one limiting consideration may have to be taken into account, such as the fuel center-line temperature and cladding temperature, or burnout. Operational parameters

* The core of a 750 Mw(e) plant with a thermal efficiency of 30 percent must generate 750/0.3 = 2,500 Mw(t).

that cause variations during the life of the core should also be taken into account. Some of these are the variations in power distribution throughout the core due to fuel burnup and fuel management (such as out-in reloading), changes in the fuel-material structure (Sec. 13-2) and the effect of the buildup of fission gases on the gas layer between fuel and clad, the possibility of the buildup of an oxide layer or scale on the fuel element heat-transfer surface, built-in transients such as during startup and shutdown, unforeseen transients such as in a loss-of-coolant accident, thermal stresses in the fuel element due to temperature gradients, and pressure stresses due to buildup of fission gases (a function of temperature and burnup). The latter may be reduced by *venting* fission gases along with a portion of the coolant stream and trapping them later. Such a scheme is considered for high-temperature, high-specific-power operation such as in fast reactors.

In thermally designing a core one begins with several basic requirements, usually obtained from the power plant and steam cycle end. The following steps only suggest the processes involved rather than recommend a definitive design procedure.

1. The net power output in Mw(e) is determined, usually by the electric utility ordering the plant, based on its network growth requirements.
2. The type of reactor, such as pressurized-water, boiling-water, gas-cooled, fast-breeder, dual-purpose desalination-power, etc., is selected by the electric utility, usually based on competitive bids from the different manufacturers, but sometimes on a desire to gain experience in a new field or type of power plant or reactor.
3. The steam cycle is designed, including the number of feedwater heaters, reheat stages, etc., and the main and auxiliary power components selected and the main steam conditions established.
4. Based on the above, the net thermal efficiency of the power plant is determined.
5. The reactor output in Mw(t) is now computed.
6. A basic fuel lattice (cell or subassembly) is determined from nuclear, hydraulic and heat-transfer considerations. These include fuel type (UO_2, etc.) and enrichment, cladding, fuel-element diameter, moderator-to-fuel ratio, etc. Often the choice of fuel type and fuel-element diameter are predicated on the existence of manufacturing facilities for a particular type and size.
7. For a liquid-cooled core (PWR, BWR, etc.), a critical (burnout) heat flux is obtained from appropriate data or correlation (Sec.

11-7). This may be reduced by a safety margin, say 1.3, depending upon whether the correlation is for design or actual burnout values. This reduced heat flux is the maximum that can be tolerated anywhere in the core, q''_{max}. (Detail on next page.)

8. An overall hot-spot factor F is determined.
9. The core average heat flux q''_{av} may now by determined by dividing the core maximum heat flux by the overall hot-spot factor. Thus

$$q''_{av} = \frac{q''_{max}}{F} \tag{13-43}$$

10. From q''_{av}, the total heat-transfer surface (cladding surface) area A_t is determined

$$A_t = \frac{Q_t}{q''_{av}} \tag{13-44}$$

where Q_t is the total thermal energy output of the core, obtained in step 5.

11. From the known fuel-element cladding diameter d_c the total length, L_t, and therefore total mass M_t of the fuel are determined from

$$L_t = \frac{1}{\pi d_c} A_t \tag{13-45}$$

and

$$M_t = \frac{d^2 \rho}{4 d_c} A_t \tag{13-46}$$

where d and ρ are the fuel material (meat) diameter and density.

12. The overall core shape is now determined, e.g., an upright cylinder of a certain height-to-diameter ratio, commensurate with pressure drops, reactor pressure vessel diameter, etc.
13. The maximum fuel cladding and center-line temperatures are now checked using the individual hot-spot factors to ascertain that they do not exceed material limitations.
14. The number and type (blackness) of the control rods are designed into the system and checked against nuclear, control, and excess reactivity requirements for the new and equilibrium cores. For example, a certain excess reactivity should be built into the core to overcome fuel depletion effects and still maintain good control rod positions in the core.
15. The reactor vessel, core support structure, thermal shields, biological shields, etc. are now designed.

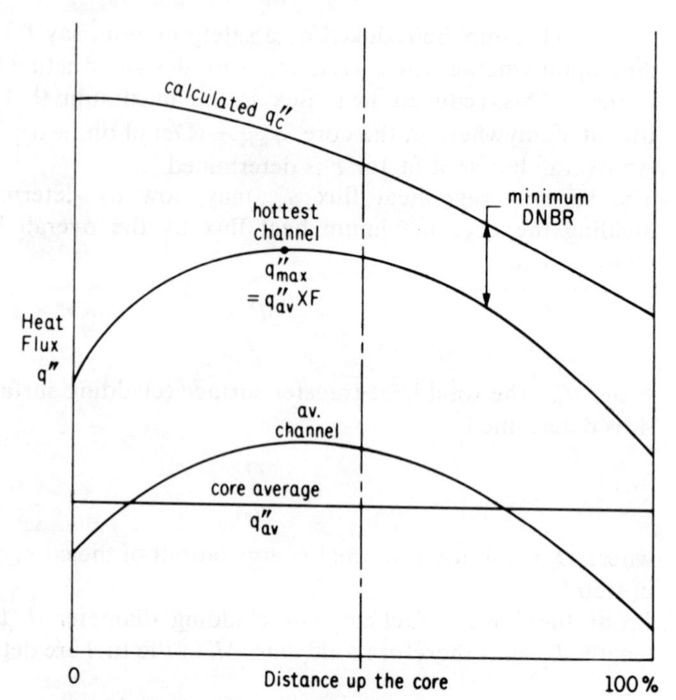

FIG. 13-10. Thermal design heat flux parameters in a burnout-limited core. DNBR = Departure from nucleate boiling ratio. Minimum DNBR = 1.3.

Figure 13-10 shows the roles the parameters discussed in 7 to 9 above play in design. For simplicity, the plot is for a single-phase liquid coolant, such as in a pressurized-water reactor. The lowest horizontal line represents the average heat flux in the core q_{av}'', obtained by dividing the total heat generated in the core by the total fuel surface area, Eq. 13-44. The curve labeled *av. channel* represents the axial distribution of heat flux in an average channel. The total heat generated in an average channel is equal to Q_t divided by the number of channels in the core. The curve is nearly sinusoidal but peaks to the left because of higher moderator density there. The average core line is also the mean of the average channel line. The hottest channel line represents the axial heat-flux distribution in the hottest channel in the core. Its point of highest heat flux is related to the average core heat flux by the overall hot-spot factor. It is the maximum permissible heat flux anywhere in the core. The top curve is the locus of computed values of the critical heat flux, and should at its closest approach to the hottest channel curve, be greater than it by a factor, such as 1.3. This is the extra factor of safety associated with the critical heat-flux correlation. It is called the minimum DNBR, or minimum departure from nucleate boiling ratio.

In the case where the fuel centerline, or cladding, temperatures, rather than cladding burnout, are limiting, the maximum permissible (hot-spot) heat flux (obtained for water-cooled reactors in step 7, above) is selected such that it would give a volumetric thermal source strength q''' that would result in the maximum permissible temperatures in question. q'' and q''' are related for a clad cylindrical fuel element of fuel radius R and gas layer (if any) and cladding thicknesses g and c by

$$q''' = 2\,\frac{R + g + c}{R^2}\,q'' \qquad (13\text{-}47)$$

In case the fuel centerline temperature is limiting, for example, it is the heat flux that results in the limiting temperature from Eqs. 5-51 for a fuel rod when t_f is obtained from Eq. 13-5 where z is replaced by z_m in Eq. 13-12 for sinusoidal heating. (A similar procedure is followed, if the cladding temperature is limiting, for t_{sm}, Sec. 13-3.) The resulting heat flux is then divided by the overall hot-spot factor to give the core average heat flux and determine the quantity of fuel, as in steps 10 and 11 above.

PROBLEMS

13-1. The temperature profiles in Fig. 13-5 are for a single-phase coolant in a channel with sinusoidal heat generation. Draw the same profiles but for a boiling channel and sinusoidal heat addition if the coolant enters the channel bottom as a subcooled liquid and begins to boil below the channel center.

13-2. A pressurized-water reactor has 12-ft-high fuel elements. They contain UO_2 pellets 0.3669 in. in diam. and clad in 0.0243-in.-thick and 0.422 in. OD. Water flow per fuel element is approximately 3080 lb_m/hr. The core inlet temperature is 551.8°F. The outlet temperature in the hottest channel is 650.7°F. The fuel enrichment is 2.35 percent in that channel. Assuming no perturbation of flux due to control rods, etc., and ignoring the hot-channel factor, find the maximum thermal flux and maximum position of cladding surface temperature. $h = 3080$ Btu/hr ft² °F.

13-3. Derive Eq. 13-12, and from it an expression for z_s.

13-4. A 7-ft-high cylindrical pressurized-water reactor core contains 25,000 cylindrical fuel elements, 0.34 in. OD, containing UO_2 cylindrical pellets of 0.3 in. diameter each. The heat generated in the fuel is 180 Mw(t); 20×10^6 lb_m/hr of coolant enters the core channels uniformly at 500°F. The heat-transfer coefficient is 5000 Btu/hr ft² °F. Neglecting the extrapolation lengths, find the minimum water pressure necessary to ensure that no boiling occurs anywhere in the core.

13-5. A cylindrical reactor core 75 in. in diameter and 102 in. high contains 20,000 cylindrical fuel rods of 0.3-in.-diameter each. The fuel material is 3 percent enriched UO_2 clad in 0.02-in.-thick stainless steel. The core is cooled and infinitely reflected in both the axial and radial directions by light water. The total heat generated in the core is 50 Mw(t). The coolant enters the core at the bottom at 485°F. The total coolant flow is 40×10^6 lb_m/hr, equally distributed over the core cross section. The heat-transfer coefficient, assumed constant over the entire core, is 6000 Btu/hr ft² °F. Determine (*a*) the minimum coolant pressure to avoid boiling and (*b*) the maximum fuel temperature in the core.

13-6. A water-cooled cylindrical reactor core is 4 ft in diameter and 5 ft high. The fuel is uranium metal 5 percent enriched in the form of flat plates 0.2 in. × 4 in. in cross section. At a point in the core 1 ft above the centerplane and 1.5 ft radial distance from the center-line, the coolant is 500°F and the fuel generates heat with a volumetric thermal source strength of 2.6×10^7 Btu/hr ft³ Find: (*a*) the maximum thermal neutron flux in the core (the extrapolation lengths may be ignored), and (*b*) the maximum temperature drop in the fuel element at the above point in °F.

13-7. In a 6-ft-high pressurized-water reactor core the coolant pressure is not to exceed 1,900 psia. The fuel is in the form of pins 0.45 in. in diameter, clad in 0.025-in.-thick Zircaloy 2. The maximum volumetric thermal source strength in the core is 4×10^7 Btu/hr ft³. The heat-transfer coefficient between coolant and cladding is 6000 Btu/hr ft² °F. The coolant mass rate of flow per fuel rod is a constant 3,370 lb_m/hr. Find the maximum temperature of the coolant entering the core if boiling is to be avoided. Neglect the extrapolation lengths.

13-8. A sodium-cooled fast-reactor core contains vertical fuel rods 0.5 in. in diameter and 5 ft high. Sodium enters the core at 27 fps and 540°F. The ratio of coolant to fuel by volume (in an infinite lattice corresponding to that of the core) is 1:1. The maximum allowable surface temperature is 1000°F. Determine (*a*) the heat-transfer coefficient and (*b*) the total heat given off by one rod having the above maximum temperature in Btu/hr. Neglect extrapolation lengths.

13-9. A 0.6-in.-diameter, 4.45-ft-long fuel rod is made of 1.3 percent enriched uranium metal. It is light-water-cooled. The coolant flow is 19,000 lb_m/hr. The heat-transfer coefficient is 7920 Btu/hr ft² °F and may be assumed constant along the entire length of the rod. The minimum coolant temperature is 350°F. The maximum fuel-surface temperature is 1100°F. Neglect the effects of cladding and extrapolated lengths and find (*a*) the position along the fuel rod at which the maximum surface temperature occurs and (*b*) the total heat given off by the rod, in Btu/hr. Assume sinusoidal heating.

13-10. A 4-ft-high fuel element is in the form of a clad cylindrical rod. The outside diameter of the cladding is 0.5 in. The maximum volumetric thermal source strength in the element is 2×10^7 Btu/hr ft³. The coolant enters the bottom of the element channel at 500°F at a rate of 2,000 lb_m/hr per element. The heat-transfer coefficient is 2000 Btu/hr ft² °F. The specific heat of the coolant is 1 Btu/lb_m °F. For sinusoidal axial heat generation find the percent of the element volume with a cladding surface temperature within 23°F of the maximum. Assume that the extrapolation lengths are negligible.

13-11. A 5-ft-long cylindrical fuel element is composed of 0.5-in.-diameter fuel and 0.003-in.-thick cladding. The element, placed horizontally in the reactor, is subjected to coolant flow perpendicular to its axis. It is assumed, for simplicity, that the heat-transfer coefficient and the temperature of the coolant are uniform around the entire surface of the element and are 500 Btu/hr ft² °F and 400°F, respectively. The maximum volumetric thermal source strength in the element is 10^7 Btu/hr ft³, and the neutron flux distribution is sinusoidal. The element is reflected at both ends by infinite heavy-water reflectors. The fuel and cladding thermal conductivities are 15 and 10 Btu/hr ft °F, respectively. Determine the maximum and minimum temperatures in the center line of the fuel element.

13-12. A thermal pebble-bed reactor has a 5-ft-diameter spherical core. It uses 1-in.-diameter spherical fuel pebbles made of 3 percent enriched UO_2. The pebbles are randomly packed in the core so that they occupy about two-thirds of the core volume. The maximum thermal-neutron flux is 10^{13}. Neglect the effects of extrapolated length and cladding, and determine (*a*) the maximum heat produced by a single fuel element and (*b*) the maximum fuel temperature in the same fuel element if its surface temperature is 1000°F. (*c*) Is this the maximum fuel temperature in the core? Why? Use 550 b for fission cross section.

13-13. A water-cooled reactor contains vertical fuel rods 0.6 in. in diameter and 136 cm long. The fuel is uranium metal 1.3 percent enriched. Water enters the core at 350°F. The maximum allowable fuel-surface temperature is 1100°F. The ratio of moderator-coolant to fuel by volume in an infinite square lattice corresponding to that of the core is 2.5:1. The inlet velocity is 20 fps. Neglecting cladding and extrapolated lengths, determine the position above the core center plane at which the above maximum fuel-surface temperature takes place.

13-14. A heterogeneous reactor core is in the form of a cube 4 ft on the side. It contains 4,900 fuel rods of 20 percent enriched uranium metal. The fuel rods are 0.50 in. in diameter, including 0.025-in.-thick zirconium cladding. The core generates 5×10^8 Btu/hr. The flux distribution is undisturbed by coolant or other materials. Determine (*a*) the hot-spot factor due to flux distribution only if the flux reaches zero at the boundaries, (*b*) same as (*a*) but with extrapolated lengths greater by 50 percent than the core sides, in all directions and (*c*) q''_{av}.

13-15. A 500-Mw(t) boiling-water reactor uses UO_2 fuel pellets 0.50 in. in diameter and 0.60 in. long. The pellets are placed end to end in a 0.032-in.-thick Zircaloy 2 can 0.57 in. OD. The gap between fuel and cladding is filled with helium gas. The maximum allowable heat flux in the core is set at 335,000 Btu/hr ft². Four percent of the thermal output of the core is produced in the moderator-coolant. Find the total number of pellets necessary if the overall hot-spot factor is 4.50.

13-16. The maximum temperature through the film (between coolant and cladding surface) in a channel with sinusoidal heat generation in the axial direction is 100°F under no hot-channel conditions. For $z_c = 4$ ft, $H = H_e = 12$ ft, and $F_f = 3$, find the maximum film temperature drop in the channel under hot-channel conditions.

13-17. A fuel element of the cross section shown is 12 ft high. It is cooled by water entering the core at 500°F with a heat transfer coefficient 2000 Btu/hr ft² °F. The core generates 1500 Mw(t) and contains 20,000 ft² of fuel surface area. The average coolant temperature rise in the core is 50°F. The hot-channel

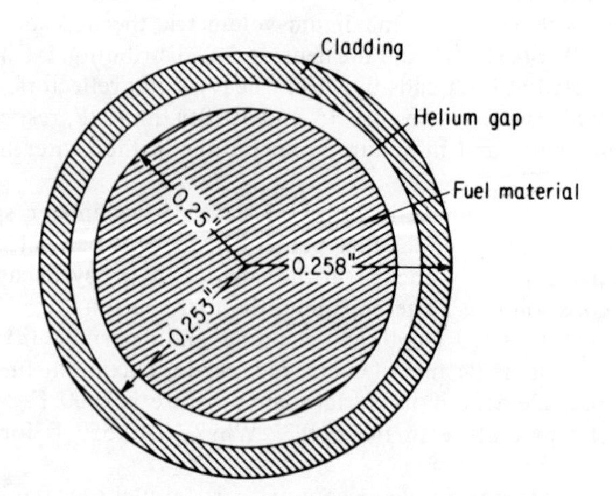

factors are: $F_c = 1.610$, $F_f = 3.510$. Calculate the highest possible cladding surface temperature in the core.

13-18. The diameter of the fuel material in a cylindrical fuel element was designed to be 0.6025 in. A check of a sample of the manufactured elements showed the following statistical distribution:

Group Number, i	Average Diameter in Group, in.	Number of Elements in Each Group
1	0.6005	4
2	0.6010	87
3	0.6015	302
4	0.6020	489
5	0.6025	1,013
6	0.6030	501
7	0.6035	296
8	0.6040	97
9	0.6045	11

Calculate the diameter corresponding to three standard deviations.

13-19. Derive an expression for the hot-spot subfactor for the temperature rise through the fuel material due to deviations in fuel material diameter. What would be the numerical value of such a subfactor, based on the information in the previous problem?

chapter **14**

The Boiling Core

14-1. INTRODUCTION

In Chapters 11 and 12, boiling and two-phase flow phenomena were discussed. Many of these phenomena apply to both boiling and pressurized liquid-cooled reactor cores, since in both the limiting heat flux anywhere in the core is determined by burnout considerations, Chapter 13.

In this chapter we shall deal with those cases where bulk boiling and two-phase flow are permitted in the core, such as in a boiling-water reactor, giving rise to interesting and unique phenomena. In particular, the heat and mass balance, hydrodynamics, and void effects on reactor operation in such cores will be treated. A design sequence for boiling cores will be outlined.

14-2. BOILING-REACTOR FLOWS

In the boiling reactor, slightly subcooled liquid enters the core at the bottom, Fig. 14-1, receives sensible heat to saturation in the nonboiling heights of each channel in the core (Sec. 12-4), then some latent heat of vaporization in the boiling heights. It leaves the top of the core as a very wet, or low-quality, two-phase mixture of liquid and vapor. The vapor is made to separate from the liquid, go to the turbine, do work, and condense in the condenser, and is then pumped back to the reactor as feedwater (after several stages of feedwater heating).

The saturated liquid that separates from the mixture at the top of the core, or in a steam separator, is recirculated back to the core entrance via *downcomers* within or outside the reactor vessel and is referred to as *recirculation* flow. It then mixes with the relatively cool feedwater forming the slightly subcooled liquid that reenters the core. We now may have either an *internal recirculation* or an *external recirculation* boiling reactor, Fig. 14-1.

Recirculation flow could be either *natural*, i.e., by a force caused by the density differential in downcomer and core and the action of the gravity field g, or *forced* by pumps (not shown) in the downcomer

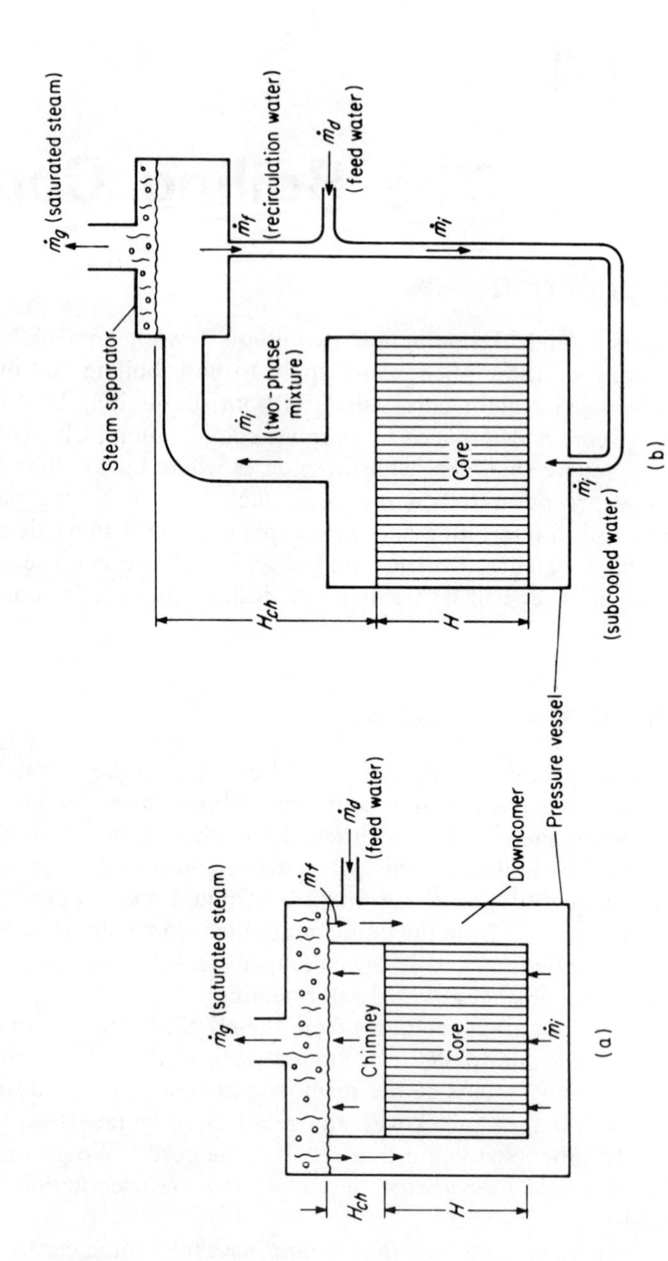

FIG. 14-1. Mass flow rates in simple boiling-water reactor with (a) internal and (b) external recirculation.

or return lines (other than the feedwater pump). Thus we may have a combination of internal, external, natural and forced recirculation flows. Natural recirculation is usually of the internal flow type because of the lower pressure losses in the system. Forced recirculation is usually of the external flow type. A combination internal-external flow with forced circulation* using jet pumps inside the reactor vessel is used in large boiling-water reactors [2].

14-3. BOILING-REACTOR MASS AND HEAT BALANCE

As shown in Fig. 14-1 for both internal and external recirculation boiling reactors, a slightly subcooled liquid enters the core bottom at a rate of $\dot{m}_i$ lb_m/hr. This liquid rises through the core and *chimney*, if any. The chimney is an unheated section above the core which helps to increase the driving pressure in natural circulation (Sec. 14-6). The resulting vapor separates and proceeds to the power plant at a rate of $\dot{m}_g$ lb_m/hr. The saturated recirculation liquid flows via the downcomer at the rate of $\dot{m}_f$ lb_m/hr. There it mixes with the relatively cold return feed liquid $\dot{m}_d$ from the power plant to form the slighty subcooled inlet liquid $\dot{m}_i$.

An *overall* mass balance in the reactor core is given by

$$\dot{m}_d = \dot{m}_g \tag{14-1}$$

$$\dot{m}_g + \dot{m}_f = \dot{m}_i \tag{14-2}$$

The *average* exit quality of the entire core $\bar{x}_e$, that is, the quality of all the vapor-liquid mixture at the core exit, is given by

$$\bar{x}_e = \frac{\dot{m}_g}{\dot{m}_g + \dot{m}_f} = \frac{\dot{m}_d}{\dot{m}_d + \dot{m}_f} = \frac{\dot{m}_d}{\dot{m}_i} \tag{14-3}$$

The *recirculation ratio* is the ratio of recirculation liquid to vapor produced. It is given by modifying Eq. 14-3 as follows:

$$\frac{\dot{m}_f}{\dot{m}_g} = \frac{1 - \bar{x}_e}{\bar{x}_e} \tag{14-4}$$

Now if the incoming feed liquid has a specific enthalpy h_i Btu/lb_m and the saturated recirculated liquid has a specific enthalpy h_f (at the system pressure), a heat balance is obtained, if we assume no heat losses to the outside (a good assumption) and neglect changes in kinetic and potential energies, as follows:

$$\dot{m}_i h_i = \dot{m}_f h_f + \dot{m}_d h_d \tag{14-5}$$

* *Circulation* and *recirculation* will be used interchangeably throughout this chapter.

where h_i is the specific enthalpy of the liquid at the reactor-core inlet. Equation 14-5 can be modified to

$$h_i = (1 - \bar{x}_e)h_f + \bar{x}_e h_d \tag{14-6}$$

Rearranging gives the following expression for $\bar{x}_e$:

$$\bar{x}_e = \frac{h_f - h_i}{h_f - h_d} \tag{14-7}$$

The condition of the liquid entering the bottom of the core is given by the *enthalpy of subcooling,*

$$\Delta h_{\text{sub}} = h_f - h_i \tag{14-8a}$$

or

$$\Delta h_{\text{sub}} = \bar{x}_e(h_f - h_d) \tag{14-8b}$$

or by the *degree of subcooling,*

$$\Delta t_{\text{sub}} = t_f - t_i \tag{14-9}$$

where t_i is the core inlet liquid temperature, corresponding to h_i.

The *total heat generated,* Q_t, can be obtained from a heat balance on the core as a system, or on the reactor as a system. Assuming negligible heat losses, the two relationships (which should yield identical results) are, respectively,

$$Q_t = \dot{m}_i[(h_f + \bar{x}_e h_{fg}) - h_i] \tag{14-10}$$

$$= \dot{m}_g(h_g - h_d) \tag{14-11}$$

Example 14-1. Find the overall heat and mass-balance parameters of a boiling-water reactor operating at 1,000 psia and producing 10^7 lb_m/hr of saturated steam. The average exit void fraction has been set by physics considerations at 40 percent. The slip ratio, assumed uniform throughout the core is 1.95. The feedwater is at 300°F.

Solution. Using the steam tables, Appendix D:

At 1,000 psia, $t_f = 544.6$°F, $h_f = 542.4$ Btu/lb_m, $h_g = 1191.8$ Btu/lb_m, $v_f = 0.0216$ ft^3/lb_m, $v_g = 0.4456$ ft^3/lb_m
At 300°F, $h_d = 269.6$ Btu/lb_m
Thus (see Eqs. 12-10 and 12-11)

$$\psi = \frac{0.0216}{0.4456} \times 1.95 = 0.0945$$

The average core exit quality $\bar{x}_e = \dfrac{1}{1 + [(1 - \bar{\alpha}_e)/\bar{\alpha}_e]1/\psi}$

$$= \frac{1}{1 + (0.6/0.4)/0.0945} = 0.0593$$

$$\text{Recirculation ratio} = \frac{1 - \bar{x}_e}{\bar{x}_e} = \frac{1 - 0.0593}{0.0593} = 15.86$$

$$\text{Recirculation rate} = \dot{m}_f = 15.86 \times 10^7 = 1.586 \times 10^8 \text{ lb}_m/\text{hr}$$

$$\text{Total core flow} = \dot{m}_i = (15.86)10^7/(1 - 0.0593) = 1.686 \times 10^8 \text{ lb}_m/\text{hr}$$

$$\Delta h_{\text{sub}} = \bar{x}_e(h_f - h_d) = 0.0593(542.4 - 269.6) = 16.2 \text{ Btu}/\text{lb}_m$$

$$h_i = h_f - \Delta h_{\text{sub}} = 542.4 - 16.2 = 526.2 \text{ Btu}/\text{lb}_m$$

Core inlet temperature $= t_i =$ saturation temperature corresponding to a saturated liquid enthalpy of 526.2 Btu/lb$_m$ = 531.7°F (by interpolation).

$$\Delta t_{\text{sub}} = t_f - t_i = 544.6 - 531.7 = 12.9°\text{F}$$

Total heat generated in the core, Eq. 14-11,

$$Q_t = 10^7(1191.8 - 269.6) = 9.222 \times 10^9 \text{ Btu/hr}$$
$$= 9.222 \times 10^9 \times 2.931 \times 10^{-7} = 2{,}703 \text{ Mw(t)}$$

The same result could be obtained from Eq. 14-10.

14-4. THE DRIVING PRESSURE IN A BOILING CHANNEL

A boiling core is usually composed of many flow channels. The individual channels do not generate the same heat, have different flows, have different pressure drops* and different driving pressures. In this section we shall evaluate the driving pressure in a single channel. The case of a multichannel core will be taken up later.

The liquid in the downcomer of a boiling reactor is either saturated or slightly subcooled, depending upon whether the feed water is added near the bottom or top of the core. In either case, the density in the downcomer, ρ_{dc}, is greater than the channel average density (in the core and chimney, if any) $\bar{\rho}_c$ where voids are found. Because of this density differential, a driving pressure, Δp_d is established. It is given, in the general case, by

$\Delta p_d =$ (hydrostatic pressure in downcomer) – (hydrostatic pressure in channel) + (pressure rise, Δp_p due to pumping, if any)

$$= [\rho_{dc}(H + H_{ch}) - \bar{\rho}_c(H + H_{ch})] \frac{g}{g_c} + \Delta p_p \tag{14-12}$$

where H = height of core, ft
H_{ch} = height of chimney, ft

* The total pressure drop across each channel must, however, be the same for all channels in a core.

g = gravitational acceleration, ft/hr²

g_c = conversion factor = 4.17×10^8 $lb_m ft/lb_f hr^2$

In the case of natural circulation, and the case of forced circulation during a power loss to the pump, $\Delta p_p = 0$. In the case of forced circulation Δp_p may be much greater than the natural circulation caused by the difference in the hydrostatic pressure terms, and the latter may be ignored. In the case of a reactor core without chimney, $H_{ch} = 0$.

The driving pressure, Δp_d, must equal the total system pressure losses, $\Sigma\Delta p$, at the desired rate of flow. The latter are given by

$$\Sigma\Delta p = \Sigma\Delta p_f + \Sigma\Delta p_a + \Sigma\Delta p_{c,e} \tag{14-13}$$

where $\Sigma\Delta p$ = total pressure losses, lb_f/ft^2

$\Sigma\Delta p_f$ = sum of frictional pressure losses in core, chimney (if any), and downcomer all computed in direction of flow (arrows in Fig. 14-2)

$\Sigma\Delta p_a$ = sum of acceleration pressure losses (these may be ignored in chimney and downcomer, since no large changes in density are encountered there)

$\Sigma\Delta p_{c,e}$ = sum of pressure losses due to area contractions and expansions such as at core entrance and exit and due to restrictions and submerged bodies such as at spacers, support plates,

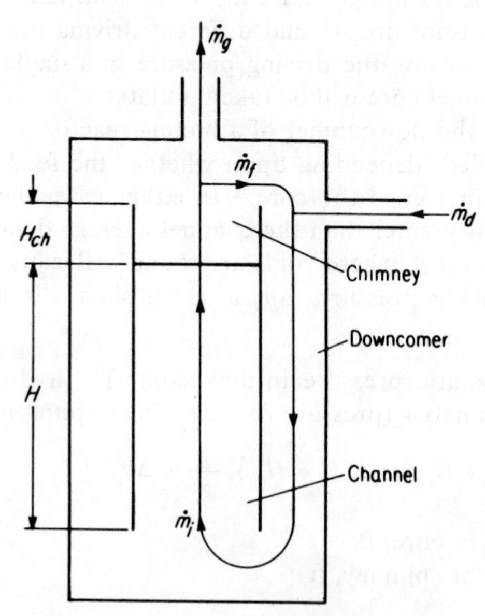

FIG. 14-2. Flow in a natural-circulation reactor composed of a single channel core, downcomer, and chimney.

feedwater distribution ring, fuel-element handles, etc., again all added up in direction of flow indicated

If the driving pressure is less than the losses at a given flow rate, a new equilibrium condition will be established at a reduced flow rate (recall that all single and two-phase flow losses are proportional to the square of the flow rate) and consequently a reduced reactor power output.

Methods of obtaining the various losses have been outlined in Chapter 12. The driving pressure will be evaluated below.

The density in the downcomer, ρ_{dc}, is equal to the density of the subcooled water at core inlet, ρ_i, if feedwater is added near the top of the core. In this case,

$$\rho_{dc} = \rho_i = \frac{1}{v_i} \tag{14-14}$$

where v_i is very nearly equal to the specific volume of the saturated liquid obtained at temperature t_i. If feedwater is added near the bottom of the core, then

$$\rho_{dc} = \rho_f = \frac{1}{v_f} \tag{14-15}$$

where v_f is the specific volume of saturated liquid at the pressure of the system. It can be seen that since ρ_i is greater than ρ_f, it is advantageous to add the feedwater near the top. Also the addition there of the relatively cool feedwater helps prevent vapor bubble carry-under to the downcomer where it would have an adverse effect on the driving pressure (because of reduced downcomer density), and to the core where it would have an adverse effect on moderation.

The average channel density in core and chimney is given by

$$\bar{\rho}_c = \frac{\bar{\rho}_0 H_0 + \bar{\rho}_B H_B + \bar{\rho}_{ch} H_{ch}}{H_0 + H_B + H_{ch}} \tag{14-16}$$

and where H_0 = core nonboiling height, ft.

H_B = core boiling height, ft.

$\bar{\rho}_0$ = average density in the nonboiling height H_0

$\bar{\rho}_B$ = average density in the boiling height H_B

$\bar{\rho}_{ch}$ = density in the chimney, of height H_{ch}

We shall now evaluated these three densities reserving the most complex, $\bar{\rho}_B$, to last.

$\bar{\rho}_0$ is that constant density that, if existed in H_0, would give the same hydrostatic pressure as the actual varying density in H_0. Thus

$$(\Delta p_h)_{H_0} = \int_0^{H_0} \rho_0(z)\, dz \, \frac{g}{g_c} = \bar{\rho}_0 H_0 \frac{g}{g_c}$$

or

$$\bar{\rho}_0 = \frac{1}{H_0} \int_0^{H_0} \rho_0(z)\, dz \tag{14-17}$$

where $\rho_0(z)$ is the local density at position z within H_0. Evaluation of the above integral depends upon the axial heat-flux distribution in H_0 and may be a complex matter. Since, however, the degree of subcooling is usually small, the density of the liquid can be considered to vary linearly with temperature, and it is therefore sufficiently accurate to use

$$\bar{\rho}_0 = \frac{1}{2}(\rho_i + \rho_f) = \frac{1}{2}\left(\frac{1}{v_i} + \frac{1}{v_f}\right) \tag{14-18}$$

The chimney contains a two-phase mixture. The density there, ρ_{ch}, therefore depends upon the void fraction. Since no heat is added in the chimney, the quality in it will be equal to the channel exit quality. If the slip ratio remains constant (it changes at the sudden entrance to a chimney but returns to the original value), then the void fraction in the chimney will be equal to the channel exit void fraction, and

$$\rho_{ch} = \rho_e \tag{14-19}$$

Note therefore that the density in the chimney is the lowest in the reactor. An addition of a chimney therefore increases the driving pressure materially, by the quantity $(\rho_{dc} - \rho_e)\, H_{ch}\, (g/g_c)$, and a chimney is therefore often resorted to in natural circulation systems. Also the chimney contains no fuel rods and contains less obstruction to flow so that the added pressure losses due to it are less, per foot, than in the core. The driving pressure is now given by

$$\Delta p_d = \left[\, \bar{\rho}_{dc}(H + H_{ch}) - (\bar{\rho}_0 H_0 + \bar{\rho}_B H_B + \rho_e H_{ch}) \right] \frac{g}{g_c} + \Delta p_p \tag{14-20}$$

There now only remains the average density in the boiling height ρ_B to be evaluated to complete the evaluation of driving pressure. This is done in the following section.

14-5. THE AVERAGE DENSITY IN A BOILING CHANNEL

The hydrostatic head in the boiling height is given by

$$(\Delta p_h)_{H_B} = \int_{H_0}^{H} \rho_B(z)\, dz\, \frac{g}{g_c} = \bar{\rho}_B H_B \frac{g}{g_c} \tag{14-21}$$

where $\rho_B(z)$ is the density of the two-phase mixture at height z within the boiling height. $\bar{\rho}_B$ may now be given by

$$\bar{\rho}_B = \frac{1}{H_B}\int_{H_0}^{H} \rho_B(z)\,dz \tag{14-22}$$

It is now necessary to obtain a relationship between $\rho_B(z)$ and z. With reference to Fig. 12-12, the derivation will involve obtaining $\rho_B(z)$ in terms of void fraction at z, α_z, which in turn will be obtained in terms of quality at z, x_z. Finally an expression involving x_z and z will be obtained and integrated. $\rho_B(z)$ is given in terms of α_z as

$$\rho_B(z) = (1 - \alpha_z)\rho_f + \alpha_z\rho_g = \rho_f - \alpha_z(\rho_f - \rho_g) \tag{14-23}$$

where ρ_f and ρ_g are constants at the system pressure. Combining the above two equations gives

$$\bar{\rho}_B = \frac{1}{H_B}\int_{H_0}^{H} [\rho_f - \alpha_z(\rho_f - \rho_g)]\,dz \tag{14-24}$$

Equation 12-9 is used to give the relationship between α_z and x_z as

$$\alpha_z = \frac{x_z}{\psi + (1 - \psi)x_z} \tag{14-25}$$

so that

$$\bar{\rho}_B = \frac{1}{H_B}\int_{H_0}^{H} \left[\rho_f - \frac{x_z}{\psi + (1 - \psi)x_z}\,(\rho_f - \rho_g)\right] dz \tag{14-26}$$

Next a relationship between x_z and z is obtained. This is done by relating z to the heat addition as follows:

$$q_z = (h_f + x_z h_{fg}) - h_i \tag{14-27}$$

$$q_t = (h_f + x_e h_{fg}) - h_i \tag{14-28}$$

where q_z is the heat generated (Btu/lb$_m$) and transferred up to the height z, and q_t is the total heat generated and transferred in the entire channel (all the fuel element), per pound mass of coolant entering the channel.

The above equations apply to all modes of heat addition. An expression between x_z and z, however, depends upon the axial heat-flux distribution. An expression will be derived for the case of *sinusoidal* axial heat flux.

In the case of sinusoidal heat addition, q_z and q_t are related by an expression similar to Eq. 12-15. Thus

$$\frac{q_z}{q_t} = \frac{1}{2}\left(1 - \cos\frac{\pi z}{H}\right) \tag{14-29}$$

Equations 14-27 to 14-29 are now combined to give

$$x_z = c_1 + c_2 \cos \frac{\pi z}{H} \tag{14-30}$$

where

$$c_1 = \frac{q_t}{2h_{fg}} - \frac{h_f - h_i}{h_{fg}} \tag{14-31a}$$

and

$$c_2 = -\frac{q_t}{2h_{fg}} \tag{14-31b}$$

Equations 14-25 and 14-30 may now be combined to give the desired relation between α_z and z. Substituting in Eq. 14-24 and rearranging give

$$\bar{\rho}_B = \frac{1}{H_B}\left[\int_{H_0}^{H} \rho_f \, dz - (\rho_f - \rho_g)\int_{H_0}^{H} \frac{c_1 + c_2 \cos(\pi z/H)}{c_3 + c_4 \cos(\pi z/H)}\right]$$

$$= \rho_f - \frac{\rho_f - \rho_g}{H_B}\int_{H_0}^{H} \frac{c_1 + c_2 \cos(\pi z/H)}{c_3 + c_4 \cos(\pi z/H)}\, dz \tag{14-32}$$

where

$$c_3 = \psi + (1 - \psi)c_1 \tag{14-31c}$$

and

$$c_4 = (1 - \psi)c_2 \tag{14-31d}$$

Equation 14-32 can be integrated by putting

$$y = \frac{\pi z}{H} \tag{14-33a}$$

and

$$dz = \frac{H}{\pi}\, dy \tag{14-33b}$$

Thus

$$\bar{\rho}_B = \rho_f - \frac{\rho_f - \rho_g}{H_B}\,\frac{H}{\pi}\int \frac{c_1 + c_2 \cos y}{c_3 + c_4 \cos y}\, dy \tag{14-34a}$$

which can be broken down into

$$\bar{\rho}_B = \rho_f - \frac{\rho_f - \rho_g}{H_B}\,\frac{H}{\pi}\left(c_1 \int \frac{dy}{c_3 + c_4 \cos y} + c_2 \int \frac{\cos y \, dy}{c_3 + c_4 \cos y}\right)$$

This reduces to

$$\bar{\rho}_B = \rho_f - \frac{\rho_f - \rho_g}{H_B}\,\frac{H}{\pi}\left(c_1 \int \frac{dy}{c_3 + c_4 \cos y} + \frac{c_2}{c_4}\, y - \frac{c_2 c_3}{c_4}\int \frac{dy}{c_3 + c_4 \cos y}\right)$$

$$= \rho_f - \frac{\rho_f - \rho_g}{H_B}\,\frac{H}{\pi}\left[\frac{c_2}{c_4}\, y + \left(c_1 - \frac{c_2 c_3}{c_4}\right)\int \frac{dy}{c_3 + c_4 \cos y}\right] \tag{14-34b}$$

The limits of the integration are

$$z = H_0 \quad \text{to} \quad z = H \tag{14-35a}$$

or

$$y = \pi \frac{H_0}{H} \quad \text{to} \quad y = \pi \tag{14-35b}$$

The integral in Eq. 14-34b has two solutions [142]. Thus:

1. For $c_3^2 > c_4^2$:

$$\bar{\rho}_B = \rho_f - \frac{\rho_f - \rho_g}{H_B} \frac{H}{\pi} \left[\frac{c_2}{c_4} y + \left(c_1 - \frac{c_2 c_3}{c_4} \right) \frac{2}{\sqrt{c_3^2 - c_4^2}} \tan^{-1} \frac{(c_3 - c_4) \tan (y/2)}{\sqrt{c_3^2 - c_4^2}} \right]$$

Introducing the limits and rearranging give

$$\bar{\rho}_B = \rho_f - (\rho_f - \rho_g) \left\{ \frac{c_2}{c_4} + \frac{c_1 c_4 - c_2 c_3}{c_4 \sqrt{c_3^2 - c_4^2}} \frac{H}{H_B} \left[1 - \frac{2}{\pi} \tan^{-1} \frac{(c_3 - c_4) \tan (\pi H_0 / 2H)}{\sqrt{c_3^2 - c_4^2}} \right] \right\} \tag{14-36}$$

2. For $c_4^2 > c_3^2$:

$$\bar{\rho}_B = \rho_f - \frac{\rho_f - \rho_g}{H_B} \frac{H}{\pi} \left[\frac{c_2}{c_4} y + \left(c_1 - \frac{c_2 c_3}{c_4} \right) \frac{1}{\sqrt{c_4^2 - c_3^2}} \ln \frac{(c_4 - c_3) \tan (y/2) + \sqrt{c_4^2 - c_3^2}}{(c_4 - c_3) \tan (y/2) - \sqrt{c_4^2 - c_3^2}} \right]$$

Again introducing the limits and rearranging give

$$\bar{\rho}_B = \rho_f - (\rho_f - \rho_g) \left[\frac{c_2}{c_4} - \frac{c_1 c_4 - c_2 c_3}{c_4 \sqrt{c_4^2 - c_3^2}} \frac{H}{H_B} \frac{1}{\pi} \ln \frac{(c_4 - c_3) \tan (\pi H_0 / 2H) + \sqrt{c_4^2 - c_3^2}}{(c_4 - c_3) \tan (\pi H_0 / 2H) - \sqrt{c_4^2 - c_3^2}} \right] \tag{14-37}$$

For the case of *uniform axial heating,* a similar and simpler procedure than the one used above for evaluating ρ_B can be employed with the help of Fig. 12-11. The result, which may be verified by the reader (Prob. 14-5), is

$$\bar{\rho}_B = \rho_f - (\rho_f - \rho_g) \frac{1}{1 - \psi} \left\{ 1 - \left[\frac{1}{\alpha_e (1 - \psi)} - 1 \right] \ln \frac{1}{1 - \alpha_e (1 - \psi)} \right\} \tag{14-38}$$

If the axial flux distribution is such that the heat addition is neither sinusoidal nor uniform, $\bar{\rho}_B$ may be evaluated by a similar analytical technique, provided that distribution is representable by a function that can be easily handled, or by a stepwise or graphical technique in which the channel is divided into several small segments in which the heat added may be considered uniform. Note, however, that Eq. 14-38 represents the average density in uniform heating of a channel portion starting with zero quality.

Example 14-2. Calculate the average density and the hydrostatic pressure for a 6-ft-high boiling-water-reactor channel (without chimney) in which heat is added sinusoidally at the rate of 50.05 Btu/lb$_m$ of incoming water. The sensible heat added is 28.1 Btu/lb$_m$. $t_i = 522°$F; S = 1.9. Channel average pressure = 1,000 psia.

Solution. The average density of the water in the nonboiling height is

$$\bar{\rho}_0 = \frac{1}{2}\left(\frac{1}{v_i} + \frac{1}{v_f}\right) = \frac{1}{2}\left(\frac{1}{0.02094} + \frac{1}{0.0216}\right)$$
$$= 47.026 \text{ lb}_m/\text{ft}^3$$

where v_i is evaluated at 522°F, the inlet temperature.

$$\phi = 1.9 \times 0.0216/0.4456 = 0.0921$$

Using Eq. 12-15,

$$\frac{H_0}{H} = 0.539$$

and

$$\frac{H_B}{H} = 0.461$$

Thus

$$c_1 = -0.00473 \quad \text{from Eq. 14-31a}$$
$$c_2 = -0.03854 \quad \text{from Eq. 14-31b}$$
$$c_3 = +0.08781 \quad \text{from Eq. 14-31c}$$
$$c_4 = -0.03499 \quad \text{from Eq. 14-31d}$$

Since $c_3^2 > c_4^2$, Eq. 14-36 applies:

$$\bar{\rho}_B = 46.296 - (46.296 - 2.244)\left[1.1015 - 2.7323\left(1 - \frac{2}{\pi}\tan^{-1} 1.7236\right)\right]$$
$$= 46.296 - 44.052[1.1015 - 2.7323(1 - 0.6652)]$$
$$= 38.07 \text{ lb}_m/\text{ft}^3$$

Thus

$$\bar{\rho}_c = \frac{1}{H}(\bar{\rho}_0 H_0 + \bar{\rho}_B H_B)$$
$$= \frac{1}{6}(47.026 \times 3.234 + 38.07 \times 2.766)$$
$$= \frac{1}{6}(257.384) = 42.897 \text{ lb}_m/\text{ft}^3$$

and

$$\Delta p_H = \bar{\rho} H \frac{g}{g_c} = 257.384 \text{ lb}_f/\text{ft}^2 = 1.787 \text{ lb}_f/\text{in}^2$$

where g/g_c is numerically equal to 1.0 in standard gravity.

14-6. THE CHIMNEY EFFECT

As indicated previously, a chimney, Fig. 14-2, is an unheated extension of the core. It usually has fewer walls, dividers, etc., and, of course, no fuel elements so that there is less friction than in the core. The quality in the chimney is substantially the same as that at the channel exit, that is, $x_{ch} = x_e$. (If the chimney is compartmentalized, the quality in each compartment is the same as the average exit quality of the fuel channels feeding into that compartment.) Since no heat is generated in or added to the chimney, x_{ch} does not vary in the axial direction. If the slip ratio in the chimney is assumed to be the same as that in the core, the chimney void fraction will be constant and equal to that at the channel exit. The density along the chimney will therefore be also equal to that at channel exit ρ_e. The addition of a chimney of height H_{ch} increases the driving pressure by the quantity

$$(\rho_{dc} - \rho_e) H_{ch} (g/g_c)$$

which may be desirable or necessary in natural circulation systems.

In the case of the computed driving pressure being less than the total system losses at any designated coolant rate of flow and core heat generation, the coolant rate of flow decreases. When this happens, the system losses, which are mostly functions of the coolant velocity to the second power, rapidly decrease. Also the driving pressure increases, since the same quantity of heat generates more voids in the reduced coolant flow. An equilibrium is reached when Δp_d and $\Sigma \Delta p$ become equal. The reverse occurs, of course, if the computed Δp_d is greater than the system losses.

In the former case, however, if it is desired to maintain a certain level of steam generation, i.e., a certain coolant rate of flow, a chimney is added in natural-circulation systems, a larger pump is used in forced-circulation systems, or a combination chimney and pump are used.

Of course the existence of a chimney increases pressure-vessel height and consequently mass and cost and may lengthen and complicate the fuel-loading mechanism and the control-rod drives (if they are actuated from the top).

In most cases the height of the chimney necessary to maintain a certain coolant rate of flow is calculated by first computing the total losses associated with the increased path length of the coolant in terms of $H + H_{ch}$, where H_{ch} is an unknown quantity. This is then equated to the necessary driving pressure which will also be a function of the unknown H_{ch}. The equation is then solved for H_{ch}. This procedure is illustrated in the following example.

Example 14-3. Calculate the chimney height necessary to maintain design flow in a 6-ft-high natural circulation, boiling-water reactor channel if the combined pressure losses $\Sigma \Delta p_{c,e}$ in the channel, chimney, and downcomer are 0.201 $lb_f/in.^2$, the two-phase friction losses in the chimney are 0.01 $lb_f/in.^2$ ft, and the friction losses in the downcomer are 0.005 $lb_f/in.^2$ ft. The feedwater is added near the top of the chimney at 522°F. Exit void fraction = 0.499. Friction drop in channel = 0.0864 $lb_f/in.^2$ Acceleration drop in channel = 0.0877 $lb_f/in.^2$. average channel density in core only $\bar{\rho} = 43.370\ lb_m/ft^3$.

Solution

$$\text{Friction drop in chimney} = 0.01H_{ch} \qquad \text{where } H_{ch} \text{ is in feet}$$

$$\text{Friction drop in downcomer} = 0.005\,(H + H_{ch}) = (0.03 + 0.005H_{ch}) \qquad lb_f/in.^2$$

$$\text{Acceleration drop in chimney and downcomer} \simeq 0$$

therefore,

$$\text{Total system losses} = (0.0864 + 0.01H_{ch} + 0.03 + 0.005H_{ch}) + 0.0877 + 0.201 = 0.4051 + 0.015H_{ch} \qquad lb_f/in.^2$$

$$\rho_e = \alpha_e \rho_g + (1 - \alpha_e)\rho_f = 0.499\,\frac{1}{0.4456} + (1 - 0.499)\,\frac{1}{0.0216} = 24.314\ lb_m/ft^3$$

$$\rho_{dc} = \frac{1}{v_i} = \frac{1}{0.02094} = 47.755\ lb_m/ft^3$$

$$\Delta p_d = [(\rho_{dc} - \bar{\rho})H + (\rho_{dc} - \rho_e)H_{ch}]\,\frac{g}{g_c}$$
$$= (47.755 - 43.370)\,6 + (47.755 - 24.314)\,H_{ch}$$
$$= 26.310 + 23.441H_{ch} \qquad lb_f/ft^2$$
$$= 0.1827 + 0.1628H_{ch} \qquad lb_f/in.^2$$

$$0.4051 + 0.015H_{ch} = 0.1827 + 0.1628H_{ch}$$

from which

$$H_{ch} = 1.50\ ft$$

In case the computed driving pressure, without chimney, is greater than the system losses, the desired coolant flow rate may be maintained by increasing flow resistance, i.e., adding obstructions to flow in the fuel channels. Otherwise, an equilibrium condition will be established in which the coolant flow rate will be higher than at design conditions. An alternative procedure is to reduce the heat generation in the core.

In a forced-circulation boiling reactor, the determination of system losses and natural driving pressure is required to determine the pumping power. The addition of a chimney above the core in this case is, however, beneficial in case of accidental loss of pumping power so that a sufficient natural driving pressure will be maintained to take care of the reactor decay heat (Sec. 4-10).

14-7. THE MULTICHANNEL BOILING CORE

Because neutron-flux distribution in the radial direction is never really uniform, the heat generated by the fuel differs from one channel to another. Also, the amount of steam generated and consequently the exit quality vary from one channel to another. In a natural-circulation reactor, or a forced circulation reactor operating under natural-circulation conditions (due to pump power failure), the driving pressure is therefore not equal in all channels, being greatest for those channels with the highest neutron flux (i.e., near the core center) and least near the core periphery. This also results in coolant flows in the different channels that vary in a similar order.

Complete calculation of the fluid-flow pattern in the usual multichannel boiling core involves a multiple iteration procedure, requiring lengthy and tedious calculations that are best performed on a digital computer. The core is first regionized, i.e., divided into a number of regions, each usually containing a number of fuel channels of approximately the same neutron-flux level. For rough calculations, the core shown in Fig. 14-3, which contains 72 fuel channels (or subassemblies), may be divided into four regions: *A*, *B*, *C*, and *D*. Note that these regions do not necessarily contain the same number of fuel channels, nor do they necessarily contain adjacent fuel channels. These regions contain the following number of fuel channels.

Region	*No. of fuel channels*
A	16
B	20
C	24
D	12

For more accuracy, and much more work, the core may be divided into 11 regions, as numbered in Fig. 14-4. Each of these 11 regions contains those fuel channels that are symmetrically situated with respect to the axis of the core. The four regions of Fig. 14-3 may allow a solution using only a desk calculator. Even this is possible only if the axial flux distribution is simple enough so that calculations of the driving pressure do not require a stepwise solution. On the other hand, a large number of regions or an axial flux distribution that requires a stepwise solution almost always requires the use of digital computers.

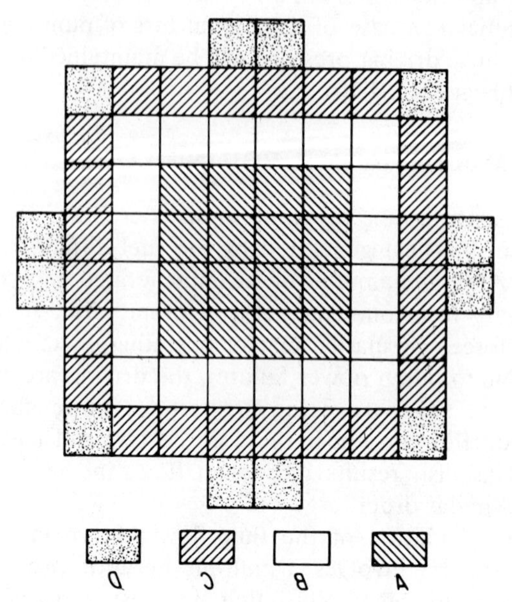

FIG. 14-3. Top view of 72-fuel subassembly core, divided into four regions.

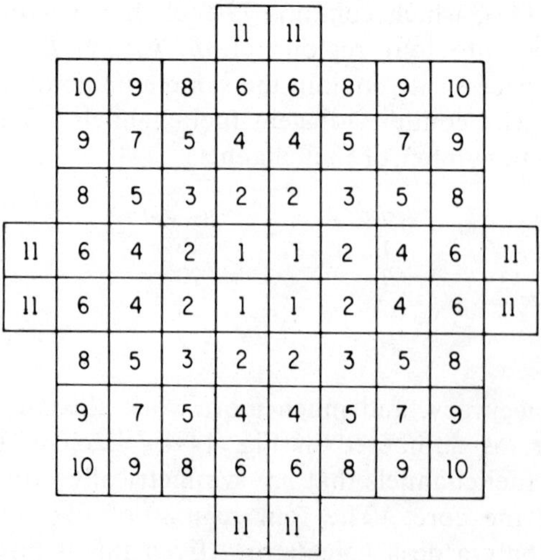

				11	11				
	10	9	8	6	6	8	9	10	
	9	7	5	4	4	5	7	9	
	8	5	3	2	2	3	5	8	
11	6	4	2	1	1	2	4	6	11
11	6	4	2	1	1	2	4	6	11
	8	5	3	2	2	3	5	8	
	9	7	5	4	4	5	7	9	
	10	9	8	6	6	8	9	10	
				11	11				

FIG. 14-4. A 72-fuel subassembly core divided into 11 regions.

We shall now outline a procedure by which a multiregion core may be analyzed as to chimney requirements, mass-flow rate, quality and void fraction distribution. Let us assume that the data given in a particular case include the power plant electric output, thermal efficiency, and therefore core thermal power, the core operating pressure, the feedwater enthalpy, (or temperature) and the general core design, i.e., fuel type and distribution, core shape (such as shown in Fig. 14-3), and general description, i.e., natural internal circulation, etc. The value of the maximum permissible void fraction in the core is also usually set by nuclear considerations, such as its effects on moderation, etc. The steps may be as follows:

1. The heat generated in each of the regions selected is calculated from the total core thermal power and known (or assumed) neutron flux distribution throughout the core.

2. A value for the core average exit quality $\bar{x}_e$ is first assumed. With this, the core recirculation ratio can be calculated and consequently the total flow in the downcomer. If the system pressure (and thus h_f and h_g) and the feedwater enthalpy h_d are known, the properties of the inlet water at the reactor bottom, such as h_i, the degree of subcooling, t_i, etc., may be found. (Alternatively, h_i may be assumed and the other quantities calculated.)

3. Starting with the region of maximum heat generation, for example A in Fig. 14-3, here called the hot region, its exit void fraction, α_{e_A}, is set equal to the maximum permissible in the core. From available experimental or theoretical correlations, an appropriate value for the slip ratio S is selected. The exit quality for the hot region, $\bar{x}_{e_A}$, is now calculated. This should naturally be greater than the assumed value for the core average exit quality $\bar{x}_e$. (The difference between the two depends upon the degree to which the radial flux in the core is flat.) If not, a lower value for $\bar{x}_e$ is chosen and steps 2 and 3 are repeated.

4. The total flow in the hot region is now calculated. From h_i and the heat generated in that region and the axial flux distribution, the nonboiling and boiling heights in that region are also calculated. (These will not be the same in all regions.) Also, all the pressure losses in that region are calculated.

5. A downcomer velocity is selected. This selection is usually predicated on two criteria. First, it is desired that no vapor carry-under into the downcomer should take place, since it reduces the downcomer density (and consequently the driving pressure) and reduces the core power by reducing the moderating power in the lower parts of the core. In this respect, a low downcomer velocity is preferred. Second, too low a downcomer velocity means a large downcomer flow area and therefore a large and consequently costly pressure vessel. Usually a

downcomer velocity of no more than 1.5 ft/sec is a good compromise in natural circulation with no artificial means of bubble separation. If larger velocities are preferred, some form of steam separator should be incorporated in the reactor design.

6. After selection of the downcomer velocity, the downcomer area, based on total downcomer flow calculated in step 2, is determined. The pressure losses in the downcomer are then calculated.

7. The driving pressure in the hot region is now computed in terms of an unknown chimney height. The chimney height is then calculated by a method similar to that given in Example 14-3.

8. The next hottest region, *B*, is now tackled. An exit quality for that region, x_{e_B}, is assumed. It should naturally be lower than x_{e_A}. The flow and the corresponding pressure losses in that region, including the already known chimney height, are also calculated. From x_{e_B}, the heat generation in that region and the axial flux distribution, the exit void, α_{e_B}, the boiling and nonboiling heights, the average density and the driving pressure for that region are calculated. This driving pressure is compared with the losses in the region plus those in the downcomer (calculated in step 6). If the two values are not equal, it is necessary to iterate by making another assumption for x_{e_B}, recalculating and repeating until agreement is attained. This fixes x_{e_B} and the corresponding α_{e_B}.

9. Step 8 is repeated for each of the other regions. This fixes values of flow, exit quality and exit void fraction for each of the regions selected.

10. With these values, the total core flow and the average core exit quality $\bar{x}_e$ are calculated. These are compared with the values assumed in step 2. Steps 2 to 10 are repeated until all the assumed and calculated values are equal.

It is important to remember here that neutron-flux distributions in the core are functions of void fractions. Note that the void fractions assumed and calculated in the sequence above were based on an assumed flux distribution (step 1). The new calculations yield new void fractions and, therefore, new flux distributions are computed from the physics of the reactor. These are used again and the entire process is repeated until a final solution which couples the hydrodynamic and nuclear parameters of the reactor is obtained.

It can be seen that the final solution yields flux and void distributions that are interdependent. Since voids abound in the upper part of the core, the moderating power is highest in the nonboiling portion of the core. This causes the peak of the neutron flux, and consequently the power-density in a boiling core to shift from the center position, as encountered in sinusoidal flux distributions, toward the bottom of the core. This is also accompanied by an increase in the peak-to-average ratio of both flux and power density. The effect is more serious in the case of large-size boiling reactors, where the dimensions are large compared with the neutron migration length. Typical flux distributions for both small and large boiling reactors are shown in Fig. 14-5.

In a multichannel core, the effect of low flux and power density in the peripheral channels causes the nonboiling heights to be predominant

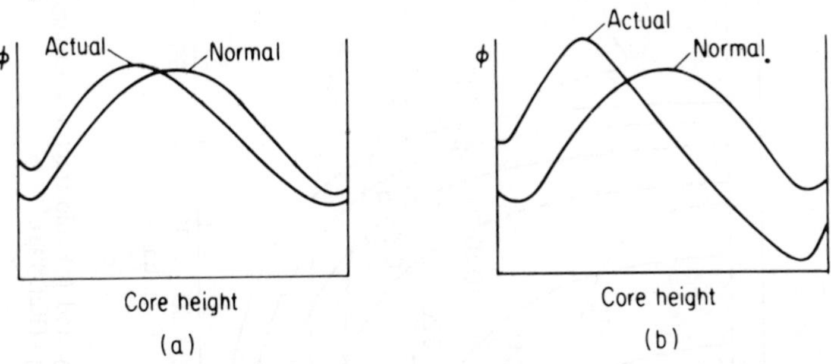

FIG. 14-5. Deviation of actual flux from normal in boiling-water reactors. (a) Small reactors; (b) large reactors.

there. Thus moderating power in such channels is greatest, a situation which tends to partially flatten the radial power and void distribution. Further radial flattening may be accomplished by orificing, control-rod and fuel enrichment programming. Figure 14-6 shows calculated relative power-density and void-fraction distributions in a large boiling core under normal and orificing conditions [143]. Note that orificing shifts the maximum power density and the nonboiling heights nearer the center of the core and has a flattening effect on the nonboiling heights and void fractions in general.

Because the power density is greatest near the core bottom in boiling reactors, the control rods are usually made to enter the core from the bottom, rather than from the top, as is usual in other reactor types. This partially corrects the skewed axial flux distribution in such reactor cores.

PROBLEMS

14-1. A boiling-sodium reactor core with a cross-sectional flow area of 100 in.^{2} operates at an average pressure of 38 psia. Sodium enters all channels 100°F-subcooled at 3 fps. It is pumped into the top of the downcomer at 1300°F. Find the amount of heat generated in the core in kw(t).

14-2. The outside diameter of the core vessel of an internal-circulation boiling-water reactor is 14 ft. The total heat generated is 600 Mw(t), the average core exit quality is 6 percent, and the average reactor pressure is 1,000 psia. Feedwater at 250°F is added to the recirculating water at the top of the downcomer. Estimate the inside diameter of the reactor pressure vessel which would result in

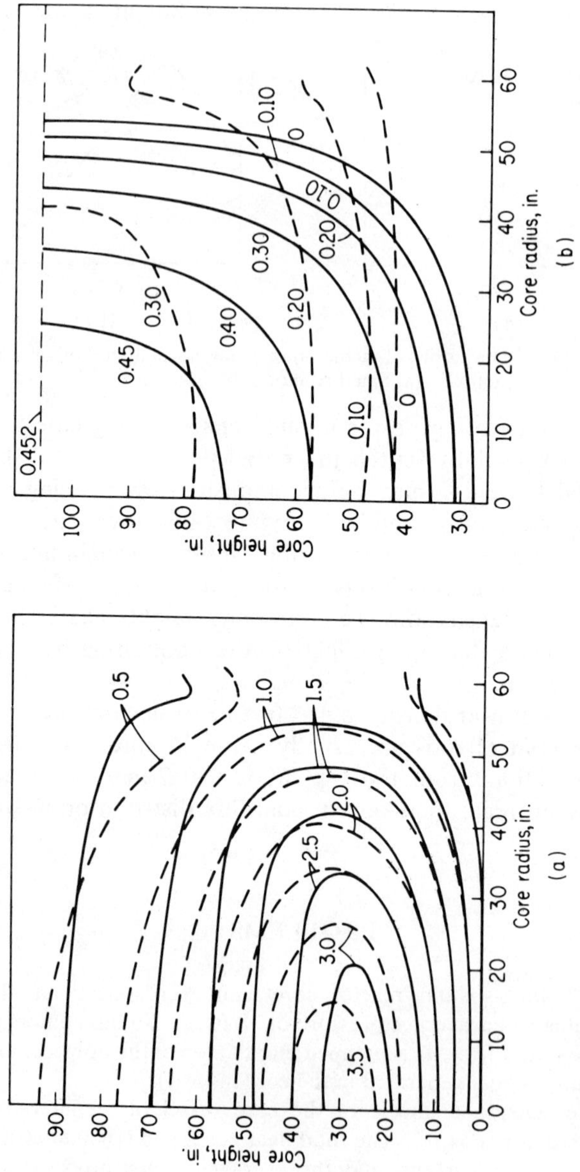

FIG. 14-6. Two-dimensional plots of (a) ratios of local to average power density and (b) void fractions for large boiling-water reactor. (Dotted lines for core with inlet orificing.) (Ref. 143).

little or no bubble carry-under into the downcomer if no bubble separators are used.

14-3. A natural-internal-circulation boiling-water-reactor core is simulated by a single 4-ft-high 4.5×0.6-in. rectangular channel. Heat is added in the channel uniformly at the rate of 10^5 Btu/hr ft^2 of the wide sides only. The channel operates at an average pressure of 1,200 psia and with an inlet velocity of 2 fps. Feedwater is at 200°F. Determine (*a*) the exit quality, (*b*) the recirculation ratio, and (*c*) the degree of subcooling in °F.

14-4. If the channel of the previous problem has a 1.5-ft-high chimney (of the same cross-sectional dimensions as the channel itself), determine the driving pressure of the system in psi. Take $S = 2.0$.

14-5. Derive the expression given by Eq. 14-38 for the average boiling density in a uniformly heated channel.

14-6. The flow into a forced-circulation boiling-water-reactor channel is 200,000 lb_m/hr of 22°F subcooled water. The inlet pressure is 900 psia. The channel generates 5 Mw(t). Other pertinent data: channel height = 5 ft, equivalent diameter = 0.5 in., inlet velocity = 10 fps, friction factor = 0.018, slip ratio = 1.8. Calculate (*a*) the pressure drop due to friction, (*b*) the acceleration pressure drop, and (*c*) the hydrostatic pressure drop in the channel.

14-7. The core of a forced-external-circulation boiling-water reactor is 6 ft high. The top of the core is 30 ft below the steam-separation drum. The reactor produces 1,000,000 lb_m/hr of steam at 1,000 psia. The recirculation ratio is 19:1, feedwater is at 230°F, the average boiling density is 41 lb_m/ft^3, and the slip ratio is 1.8. The axial flux distribution changes linearly from 0 at core bottom to maximum in 2 ft, remains at maximum for 1 more foot, then drops linearly, becoming 0 again at core top. Neglecting the variations in the radial flux, determine (*a*) the nonboiling height, (*b*) the pressure rise across the pump, and (*c*) the corresponding pump horsepower necessary to maintain flow if the contraction, expansion, submerged body, and other similar pressure losses in the loop are 3.26 psi, and the total frictional and acceleration losses are 4.53 psi.

14-8. A boiling-water reactor operates with an average exit void fraction of 0.50, an average inlet velocity of 3 ft/sec, and a slip ratio of 1.9. The average reactor pressure is 1,000 psia. One channel in the reactor, operating at these conditions, is 6 ft high and square, 4.85 in. on each side, and contains 36 fuel rods each 0.55 in. in diameter. Feedwater is at 400°F. For uniform heat flux (axial and radial) calculate the height of the chimney necessary for natural circulation in that channel. The entrance, exit, contraction, expansion, spacer, and other losses are 0.301 psi in the channel and chimney and 0.182 psi in the downcomer. The frictional losses in the downcomer are 0.008 psi/ft. The fuel-rod cladding corresponds to smooth-drawn tubing.

14-9. An internal-natural-circulation boiling-sodium reactor is composed of a single channel 4 ft high with a 2-ft chimney of the same cross-sectional dimensions. The heat, generated uniformly in the reactor, is adjusted at all conditions so that the exit quality is always 11.5 percent. Liquid sodium returns from the power plant at 1800°R and is fed to the reactor near the top of the downcomer. The reactor average pressure is 150.5 psia, and the slip ratio is 4.0. The reactor was first operated on earth and then on the lunar surface where the gravitational field is one-sixth of that on earth. Calculate (*a*) the change in driving pressure, psi, and (*b*) the percent decrease in reactor power necessary to maintain exit quality as mentioned, if the pressure losses in the core and downcomer varied as the (rate of flow)^2.

14-10. An external-natural-circulation boiling-water reactor was designed to produce 150,000 lb_m/hr of steam. The following data applies: $H_0 = 1.23$ ft, $H_B = 4.77$ ft, $H_{ch} = 12.26$ ft; $\rho_i = 46.3$, $\bar{\rho}_0 = 46.2$, $\bar{\rho}_B = 33.1$, and $\rho_e = 24.3$ (all in lb_m/ft^3). After an initial period of operation, a decision was made to reduce reactor power so that only 100,000 lb_m/hr of steam are produced. The same pressure, exit quality, and feedwater temperatures were maintained. Because the pressure losses decreased below the available driving pressure, it was decided that the simplest way to correct the situation was to add an orifice in the downcomer. Calculate the orifice diameter (coefficient of discharge = 0.85). Assume all pressure losses to be proportional to the square of the flow speed and that the downcomer diameter is much larger than the orifice diameter.

14-11. A natural internal circulation BWR operates at an average pressure of 1000 psia. The average densities in the nonboiling and boiling heights are 47 and 38 lb_m/ft^3 respectively. These heights are 2 and 3 ft respectively. The radial flux is uniform. Downcomer temperature is 520°F. The core exit quality is 8 percent. The slip ratio is 2:1. The total pressure losses are 0.527 psi (including chimney). Calculate the necessary chimney height in feet.

14-12. A boiling-internal-natural-circulation reactor has a 5-ft high core and a 2-ft chimney. The density in the downcomer is 46.5 lb_m/ft^3. The average density in the core is 40.0 lb_m/ft^3. The core exit density is 30.0 lb_m/ft^3. The pressure losses in core, chimney and downcomer may be given by $\Delta p/\text{foot} = kG^2$ where k is a constant and G is the mass velocity of coolant. Assuming densities do not change, find the percent change in mass flow rate if the chimney height is increased to 3 ft.

14-13. A forced-circulation BWR operates at 1,000 psia with a core exit quality of 7.75 percent. The inlet water to the core is at 500°F. The average core density is 43 lb_m/ft^3, $S = 2$. The core height is 8 ft *plus* chimney. The total core, chimney, and downcomer losses are 2.9 psi. A pump produces 2 psi. Calculate the height of the chimney.

14-14. A forced-circulation BWR generates 600 Mw(t) at 1000 psia. A total of 2×10^7 lb_m/hr of 44.6°F-subcooled water enter the core. The average density in the core only is 43 lb_m/ft^3. The slip ratio is 2. The core is 8 ft high. There is a chimney which is 3 ft high. The total core and downcomer losses are 2.8 psi. What should the pump pressure in psi be?

appendix **A**

Alphabetical List of the Elements

Alphabetical List of the Elements

Element	Symbol	Atomic number, Z	Element	Symbol	Atomic number, Z
Actinium	Ac	89	Mendelevium	Md	101
Aluminum	Al	13	Mercury	Hg	80
Americium	Am	95	Molybdenum	Mo	42
Antimony	Sb	51	Neodymium	Nd	60
Argon	A	18	Neon	Ne	10
Arsenic	As	33	Neptunium	Np	93
Astatine	At	85	Nickel	Ni	28
Barium	Ba	56	Niobium	Nb	41
Berkelium	Bk	97	Nitrogen	N	7
Beryllium	Be	4	Nobelium	No	102
Bismuth	Bi	83	Osmium	Os	76
Boron	B	5	Oxygen	O	8
Bromine	Br	35	Palladium	Pd	46
Cadmium	Cd	48	Phosphorus	P	15
Calcium	Ca	20	Platinum	Pt	78
Californium	Cf	98	Plutonium	Pu	94
Carbon	C	6	Polonium	Po	84
Cerium	Ce	58	Potasium	K	19
Cesium	Cs	55	Praseodymium	Pr	59
Chlorine	Cl	17	Promethium	Pm	61
Chromium	Cr	24	Protactinium	Pa	91
Cobalt	Co	27	Radium	Ra	88
Copper	Cu	29	Radon	Rn	86
Curium	Cm	96	Rhenium	Re	75
Dysprosium	Dy	66	Rhodium	Rh	45
Einsteinium	Es	99	Rubidium	Rb	37
Erbium	Er	68	Ruthenium	Ru	44
Europium	Eu	63	Samarium	Sm	62
Fermium	Fm	100	Scandium	Sc	21
Fluorine	F	9	Selenium	Se	34
Francium	Fr	87	Silicon	Si	14
Gadolinium	Gd	64	Silver	Ag	47
Gallium	Ga	31	Sodium	Na	11
Germanium	Ge	32	Strontium	Sr	38
Gold	Au	79	Sulfur	S	16
Hafnium	Hf	72	Tantalum	Ta	73
Helium	He	2	Technetium	Tc	43
Holmium	Ho	67	Tellurium	Te	52
Hydrogen	H	1	Terbium	Tb	65
Indium	In	49	Thallium	Tl	81
Iodine	I	53	Thorium	Th	90
Iridium	Ir	77	Thulium	Tm	69
Iron	Fe	26	Tin	Sn	50
Krypton	Kr	36	Titanium	Ti	22
Khurchatorium	Ku	104	Tungsten (Wolfram)	W	74
Lanthanum	La	57	Uranium	U	92
Lawrencium	Lw	103	Vanadium	V	23
Lead	Pb	82	Xenon	Xe	54
Lithium	Li	3	Ytterbium	Yb	70
Lutecium	Lu	71	Yttrium	Y	39
Magnesium	Mg	12	Zinc	Zn	30
Manganese	Mn	25	Zirconium	Zr	40

appendix **B**

Partial List of the Isotopes

Partial List of the Isotopes*

Atomic Number, Z	Element	Symbol	Element Atomic Mass, amu	Isotope Mass Number, A	Isotope Mass,† amu	Natural Abundance, Atomic %	Half-life‡	Cross Sections for 2,200 m/sec Neutrons, Barns	
								Absorp.	Scatt.
0	Neutron	N	....	.....	1.008665	....	11.7 m	....	......
1	Hydrogen	H	1.00797	.....	....	....	....	0.332	38
				1	1.007825	99.985			
				2	2.01410	0.015	....	.00050	7.0
				3	3.01605	....	12.26 y	$<6.7\times10^{-6}$	
2	Helium	He	4.0026	.....	....	....	....	<0.05	0.73
				3	3.01603	0.00013	....	5,327	2.0
				4	4.00260	99.99987	....	0	0.8
				5	5.01230	....	2×10^{-21} s		
				6	6.01888	....	0.8 s		
				7	....	....	$\sim5\times10^{-5}$ s		
				8	8.0375	....	....		
3	Lithium	Li	6.939	.....	....	....	....	70.7	1.6
				5	5.0125	....	....		
				8	6.01512	7.42		0.045,945 (n, α)	
				7	7.01600	92.58	....	0.037	
				8	8.022466	....	0.845 s		
				9	....	....	0.176 s		
4	Beryllium	Be	9.0122	.....	....	....	....	0.095	6.0
				6	6.0197	....	$>4\times10^{-21}$ s		
				7	7.0169	....	53.6 d	54,000	
				8	8.0053	....	2×10^{-4} s		
				9	9.01218	100	....	0.010	6.0
				10	10.0135	....	2.5×10^{6} y		
				11	11.0216	....	13.6 s		
5	Boron	B	10.811	.....	....	....	....	759	4.0
				8	8.0246	....	0.77 s		
				9	9.01333	....	$\geq3\times10^{-19}$ 2		
				10	10.01294	19.78	....	3,837	
				11	11.00931	80.22	....		
				12	12.0143	....	0.02 s		
				13	13.0178	....	0.019 s		

* This table, plus additional data and information, from Refs. 152 to 161.
The data is likely to change with time and the reader is advised to seek the latest information in the literature.

† Physical scale, based on mass of 0^{12} isotope = 12.000 amu

‡ μs = microsecond, s = second, m = minute, h = hour, d = day, y = year.

Partial List of the Isotopes *(continued)*

Atomic Number, Z	Element	Symbol	Element Atomic Mass, amu	Isotope Mass Number, A	Isotope Mass, amu	Natural Abundance, Atomic %	Half-life	Cross Section for 2,200 m/sec Neutrons, Barns	
								Absorp.	Scatt.
6	Carbon	C	12.01115	......	....	....	....	0.0034	4.8
				10	....	....	19.0 s		
				11	11.01141		20.3 m		
				12	12.00000	98.89	....	0.0034	
				13	13.00335	1.11	....	0.0009	
				14	14.00323	....	5,730 y	< 200	
				15	15.00939	....	2.4 s		
				16	....	....	0.74 s		
7	Nitrogen	N	14.0067	......	....	....	....	1.85	10.0
				12	12.01895		0.011 s		
				13	13.00572	....	9.96 m		
				14	14.00307	99.63	....	1.81	
				15	15.00011	0.37			
				16	16.00656	....	7.2 s		
				17	17.00862	....	4.16 s		
8	Oxygen	O	15.9994	......	....	....	....	0.0002	4.2
				14	14.00856	....	73. s		
				15	15.0030	....	124 s		
				16	15.99491	99.759	....	0.000178	
				17	16.99914	0.037	....	0.04	
				18	17.99915	0.204	....		
				19	19.00344	....	29 s		
				20	....	....	14 s		
9	Fluorine	F	18.9984	......	....	....	....	0.0098	3.9
				16	16.01171	....	$\sim 10^{-19}$ s		
				17	17.00210	....	66 s		
				18	18.00094	....	109.7 m		
				19	18.99841	100	....	0.0098	3.9
				20	19.99999	....	11.4 s		
				21	....	....	4.4 s		
10	Neon	Ne	20.183	......	....	....	....	0.032	2.38
				18	18.00546	....	1.46 s		
				19	19.00187	....	17.5 s		
				20	19.99244	90.92			
				21	20.99395	0.26	....	96 (n,α)	
				22	21.99138	8.82			
				23	22.99437	....	37.6 s		
				24	....	....	3.38 m		

Partial List of the Isotopes *(continued)*

Atomic Number, Z	Element	Symbol	Element Atomic Mass, amu	Isotope Mass Number, A	Isotope Mass, amu	Natural Abundance, Atomic %	Half-life	Cross Section for 2,200 m/sec Neutrons, Barns	
								Absorp.	Scatt.
11	Sodium	Na	22.9898	......	....	....	....	0.534	6.0
				20	20.00887	....	0.4 s		
				21	21.99760	....	23 s		
				22	21.99432	....	2.62 y	90,000	
				23	22.98977	100	....	0.534	4.0
				24	23.99102	....	15 h		
				25	24.98984	....	60 s		
				26	....	....	1.0 s		
12	Magnesium	Mg	24.312	......	....	....	....	0.063	3.60
				23	22.99380	....	12.1 s		
				24	23.98504	78.70	....	0.034	
				25	24.98584	10.13	....	0.280	
				26	25.98259	11.17	....	0.060	
				27	26.98436	....	9.5 m		
				28	27.98381	....	21.3 h		
13	Aluminum	Al	26.98153	......	....	....	....	0.2355	1.4
				23	....	....	.13 s		
				24	24.00006	....	2.1 s		
				25	24.99036	....	7.2 s		
				26	25.98793	....	7.4×10^5 y		
				27	26.98153	100		0.230	1.4
				28	27.98193	....	2.31 m		
				29	28.98053	....	6.6 m		
				30	....	....	3.3 s		
14	Silicone	Si	28.086	....	....	....	....	0.16	2.1
				26	....	....	2 s		
				27	27.98667	....	4.2 s		
				28	27.97693	92.21	....	0.080	
				29	28.97649	4.70	....	0.28	
				30	29.97376	3.09	....	0.40	
				31	30.97536	....	2.62 h		
				32	31.97396	....	~650 y		
15	Phosphorus	P		......	....	....	....	0.19	10.0
				28	27.99168	....	.28 s		
				29	28.98178	....	4.4 s		
				30	29.97863	....	2.5 m		
				31	30.97376	100	....	0.19	10.0
				32	31.97392	....	14.3 d		
				33	32.97168	....	25 d		
				34	33.97331	....	12.4 s		

Partial List of the Isotopes *(continued)*

Atomic Number, Z	Element	Symbol	Element Atomic Mass, amu	Isotope Mass Number, A	Isotope Mass, amu	Natural Abundance, Atomic %	Half-life	Cross Section for 2,200 m/sec Neutrons, Barns	
								Absorp.	Scatt.
16	Sulfur	S	32.064	...	...	...	...	0.52	1.1
				30	...	...	1.4 s		
				31	30.97901	...	2.6 s		
				32	31.97207	95.0	...	0.007	
				33	32.97146	.76	...	0.125	
				34	33.96786	4.22	...	0.270	
				35	34.96923	...	86.7 d		
				36	35.96709	0.014			
				37	36.97029	...	5.1 m		
				38	...	...	2.87 h		
17	Chlorine	Cl	35.453	...	...	...	...	33.2	16
				32	31.98601	...	0.31 s		
				33	32.97725	...	2.5 s		
				34	33.97376	...	1.56 s		
				35	34.96885	75.53	...	44	
				36	35.96852	...	3.1×10^5 y	100	
				37	36.96590	24.47	...	0.43	
				38	37.96797	...	37.3 m		
				39	38.96742	...	55.5 m		
				40	...	...	1.4 m		
18	Argon	Ar	39.948	...	...	...	...		0.62
				35	34.97459	...	1.83 s		
				36	35.96755	0.337		0.0552 (n,α)	
				37	36.96674	...	35 d		
				38	37.96272	0.063			
				39	38.96428	...	265 y		
				40	39.96238	99.60		0.61(n,α)	
				41	40.96454	...	1.83 h		
				42	...	...	>3.5 y		
19	Potassium	K	39.102	...	...	...	...	2.10	1.5
				37	36.97324	...	1.2 s		
				38	37.96905	...	7.7 m		
				39	38.96371	93.10	...	1.94	
				40	39.9740	0.0118	1.28×10^9 y	70.0	
				41	40.96184	6.88	...	1.24	
				42	41.96352	...	12.4 h		
				43	42.96066	...	22.4 h		
				44	43.96192	...	22 m		
				45	...	...	20 m		

Partial List of the Isotopes *(continued)*

Atomic Number, Z	Element	Symbol	Element Atomic Mass, amu	Isotope Mass Number, A	Isotope Mass, amu	Natural Abundance, Atomic%	Half-life	Cross Section for 2,200 m/sec Neutrons, Barns	
								Absorp.	Scatt.
20	Calcium	Ca	40.08	...	...	...	...	0.44	2.99
				38	...	...	0.7 s		
				39	38.97100	...	0.87 s		
				40	39.96259	96.97	...	0.22	
								0.0025(n,α)	
				41	40.96228	...	8×10^4 y		
				42	41.95863	0.64	...	42	
				43	42.95878	0.145	...		
				44	43.95549	2.06	...		
				45	...	...	165 d		
				46	45.9537	0.0033	...		
				47	...	...	4.7 d		
				48	47.9524	0.18	2×10^6 y		
				49	48.95559	...	...	8.7 m	
21	Scandium	Sc	44.956	...	...	...	...	25.0	24.0
				40	39.97753	...	0.18 s		
				41	40.96860	...	0.87 s		
				43	42.96106	...	66 s		
				44	43.95928	...	3.92 h		
				45	44.95592	100	...	25.0	24.0
				46	45.95487	...	83.8 d		
				47	46.95230	...	3.4 d		
				48	47.95216	...	44 h		
				49	48.94997	...	57.5 m		
				50	...	...	1.8 m		
22	Titanium	Ti	47.90	...	...	...	...	6.1	4.0
				43	...	...	.06 s		
				44	...	...	$\sim 10^3$ y		
				45	44.95797	...	3.08 h		
				46	45.95263	7.93	...	0.6	
				47	46.9518	7.28	...	1.7	
				48	47.94795	73.94	...	8.3	
				49	48.94787	5.51	...	1.9	
				50	49.9448	5.34	...	< 0.2	
				51	50.94645	...	5.80 m		

Partial List of the Isotopes *(continued)*

Atomic Number, Z	Element	Symbol	Element Atomic Mass, amu	Isotope Mass Number, A	Isotope Mass, amu	Natural Abundance, Atomic %	Half-life	Cross Section for 2,200 m/sec Neutrons, Barns	
								Absorp.	Scatt.
23	Vanadium	V	50.942	...	...	...	...	5.06	5.0
				45	...	...	~ 1 s		
				46	45.96028	...	0.4 s		
				47	46.95469	...	32 m		
				48	47.95220	...	1.61 d		
				49	48.94847	...	330 d		
				50	49.9472	0.24	6×10^{15} y	80	
				51	50.9440	99.76	...	4.9	
				52	51.94418	...	3.76 m		
				53	...	...	2.0 m		
				54	...	...	55 s		
24	Chromium	Cr	59.996	...	...	...	...	3.1	4.8
				46	...	...	1.1 s		
				47	...	...	0.4 s		
				48	...	...	23 h		
				49	48.95122	...	42 m		
				50	49.9461	4.31	...	17	
				51	50.94418	...	27.8 d		
				52	51.9405	83.76	...	0.76	
				53	52.9407	9.55	...	18.2	
				54	53.9389	2.38	...	<0.3	
				55	54.94095	...	3.5 m		
				56	...	...	5.9 m		
25	Manganese	Mn	54.9380	...	...	...	...	13.3	2.3
				50	49.95411	...	0.29 s		
				51	50.94809	...	45 m		
				52	51.94618	...	5.6 d		
				53	52.94126	...	~ 10^{6} y		
				54	53.9404	...	303 d		
				55	54.9381	100	...	13.3	2.3
				56	55.93904	...	2.58 h		
				57	...	...	1.7 m		
				58	...	...	1.1 m		
26	Iron	Fe	55.847	...	...	...	...	2.55	11
				52	51.94769	...	8 h		
				53	52.94541	...	9 m		
				54	53.9396	5.82	...	2.3	
				55	54.93856	...	2.6 y		
				56	55.9349	91.66	...	2.7	
				57	56.9354	2.19	...	2.5	
				58	57.9333	0.33	...	1.2	
				59	58.9349	...	45.1 d		
				60	...	...	~ 3×10^{5} y		
				61	...	...	6.0 m		

Partial List of the Isotopes *(continued)*

Atomic Number, Z	Element	Symbol	Element Atomic Mass, amu	Isotope Mass Number, A	Isotope Mass, amu	Natural Abundance, Atomic %	Half-life	Cross Section for 2,200 m/sec Neutrons, Barns	
								Absorp.	Scatt.
27	Cobalt	Co	58.9332	...	...	...	...	37.2	7
				54	53.94904	...	0.18 s		
				55	54.94188	...	18 h		
				56	55.93982	...	77.3 d		
				57	56.93587	...	270 d		
				58	57.93520	...	71.3 d		
				59	58.9332	100	...	37.2	7
				60	59.93344	...	5.26 y		
				61	60.93199	...	9.90 m		
				62	61.93324	...	13.9 m		
				63	...	...	52 s		
				64	...	...	7.8 m		
28	Nickel	Ni	58.71	...	...	...	...	4.6	17.5
				56	...	...	6.5 d		
				57	56.9394	...	36 h		
				58	57.9353	67.88	...	4.4	
				59	58.9342	...	8×10^4 y		
				60	59.9332	26.23	...	2.6	
				61	60.9310	1.19	...	2.0	
				62	61.9283	3.66	...		
				63	62.9286	...	92 y		
				64	63.9280	1.08	...		
				65	64.9291	...	2.56 h		
				66	...	...	55 h		
29	Copper	Cu	63.54	...	...	...	...	3.80	7.2
				58	57.9456	...	3.3 s		
				59	...	...	81 s		
				60	59.9375	...	2.4 m		
				61	60.9327	...	3.3 h		
				62	61.9316	...	9.8 m		
				63	62.9298	60.09	...	4.5	
				64	64.9288	...	12.9 h		
				65	64.9278	30.91	...	2.2	
				66	65.9288	...	5.1 m		
				67	66.9278	...	61 h		
				68	...	...	32 s		

Partial List of the Isotopes *(continued)*

Atomic Number, Z	Element	Symbol	Element Atomic Mass, amu	Isotope Mass Number, A	Isotope Mass, amu	Natural Abundance, Atomic %	Half-life	Cross Section for 2,200 m/sec Neutrons, Barns	
								Absorp.	Scatt.
30	Zinc	Zn	65.37	......	....	....	....	1.1	3.6
				60	....	....	2.1 m		
				61	....	....	89 s		
				62	61.9339	....	9.3 h		
				63	62.9330	....	38.3 m		
				64	63.9291	48.89	....	0.000015	
				65	64.9283	....	243.6 d		
				66	65.9260	27.81	....	0.000020	
				67	66.9271	4.11	....	0.000006	
				68	67.9249	18.57	....	1.1	
				69	68.9257	....	58 m		
				70	69.9253	0.62	....	0.115	
				71	70.9273	....	2.2 m		
				72	....	....	46.5 h		
31	Gallium	Ga	69.72	......	....	....	....	2.80	6.0
				64	63.9368	....	2.6 m		
				65	64.9325	....	15 m		
				66	65.9315	....	9.5 h		
				67	66.9283	....	78 h		
				68	67.9270	....	68.3 m		
				69	68.9257	60.4	....	2.1	
				70	69.9259	....	21 m		
				71	70.9249	39.6	....	5.15	
				72	71.9245	....	14.1 h		
				73	72.9248	....	4.8 h		
				74	....	....	7.8 m		
				75	....	....	2.0 m		
				76	....	....	32 s		
32	Germanium	Ge	72.59	......	....	....	....	2.45	3.0
				65	....	....	1.5 m		
				66	....	....	2.4 h		
				67	66.9330	....	19 m		
				68	....	....	280 d		
				69	68.9280	....	40 h		
				70	69.9243	20.52	....	3.68	
				71	70.9251	....	11 d		
				72	71.9217	27.43	....	0.98	
				73	72.9234	7.76	....	14.0	
				74	72.9212	36.54	....	0.45	
				75	74.9228	....	82.8 m		
				76	75.9214	7.76	....	0.2	
				77	76.9215	....	11.3 h		
				78	....	....	2.1 h		

Partial List of the Isotopes *(continued)*

Atomic Number, Z	Element	Symbol	Element Atomic Mass, amu	Isotope Mass Number, A	Isotope Mass, amu	Natural Abundance, Atomic %	Half-life	Cross Section for 2,200 m/sec Neutrons, Barns	
								Absorp.	Scatt.
35	Arsenic	As	74.9216	.....	....	....	....	4.3	5.1
				68	....	....	~7 m		
				69	....	....	15 m		
				70	....	....	50 m		
				71	70.9271	....	62h		
				72	71.9264	....	26h		
				73	72.9237	....	76d		
				74	73.9217	....	17.9d		
				75	74.9216	100	....	4.3	5.1
				76	75.9201	....	26.5h		
				77	76.9206	....	39h		
				78	77.9217	....	91m		
				79	78.9209	....	9m		
				80	....	....	15s		
				81	....	....	33s		
				85	....	....	0.43s		
34	Selenium	Se	78.96	.....	....	....	....	12.2	11
				70	....	....	44m		
				71	....	....	45m		
				72	....	....	8.4d		
				73	72.9266	....	7.1h		
				74	73.9225	0.087	....	50	
				75	74.9225	....	120.4d		
				76	75.9192	9.02	....	22	
				77	76.9199	7.58	....	42	
				78	77.9173	23.52	....	0.38	
				79	78.9185	....	6.5×10^4y		
				80	79.9165	49.82	....	0.61	
				81	80.9185	....	18.6m		
				82	81.9167	9.19	....	2.1	
				83	....	....	70s		
				84	....	....	3m		
				85	....	....	39s		
				87	....	....	16s		

Partial List of the Isotopes *(continued)*

Atomic Number, Z	Element	Symbol	Element Atomic Mass, amu	Isotope Mass Number, A	Isotope Mass, amu	Natural Abundance, Atomic %	Half-life	Cross Section for 2,200 m/sec Neutrons, Barns	
								Absorp.	Scatt.
35	Bromine	Br	79.909	......	....	....	....	6.8	6.8
				78	77.9211	....	6.4 m		
				79	78.9183	50.54	....	10.8	
				80	79.9172	....	17.6 m		
				81	80.9163	49.46	....	6.36	
				82	81.9158	....	35.5 h		
36	Krypton	Kr	83.8	......	....	....	....	24	7.5
				78	77.9204	0.35	....		
				79	78.200	....	34.9 h		
				80	79.9164	2.27	....	14	
				81	80.9165	....	2.1×10^5 y		
				82	81.9135	11.56	....	3	
				83	82.9155	11.55	....	220	
				84	83.9116	56.90	....	0.042	
				85	84.9126	....	10.76 y	< 15	
				86	85.9109	17.37	....	< 2	
				87	86.9136	....	76 m		
37	Rubidium	Rb	85.47	......	....	....	....	0.73	12.0
				84	83.9142	....	33 d		
				85	84.9117	72.15	....	0.761	
				86	85.9100	....	18.66 d		
				87	86.9019	27.85	5×10^{10} y		
				88	87.9113	....	17.8 m		
38	Strontium	Sr	87.62	......	....	....	....	1.21	10
				84	83.9134	0.56	....	1.05	
				85	84.9095	....	64 d		
				86	85.9094	9.86	....	0.8	
				87	86.9089	7.02	....		
				88	87.9056	82.56	....		
				89	88.9057	....	52 d		
				90	89.9072	....	28.1 y		
				91	90.9097	....	9.67 h		
39	Yttrium	Y	88.905	......	....	....	....	1.28	7.6
				88	87.9096	....	106.6 d		
				89	88.9054	100	....		
				90	89.9066	....	64 h		
				91	90.9069	....	58.8 d		
				92	91.9083	....	....		

Partial List of the Isotopes *(continued)*

Atomic Number, Z	Element	Symbol	Element Atomic Mass, amu	Isotope Mass Number, A	Isotope Mass, amu	Natural Abundance, Atomic %	Half-life	Cross Section for 2,200 m/sec Neutrons, Barns	
								Absorp.	Scatt.
40	Zirconium	Zr	91.22	......	....	....	....	0.185	6.4
				89	88.9086	....	78.4h		
				90	89.9043	51.46	....	0.10	
				91	90.9053	11.23	....	1.58	
				92	91.9046	17.11	....	0.25	
				93	92.9063	....	1.5×10^6 y		
				94	93.0961	17.40	....	0.075	
				95	94.9072	....	65 d		
				96	95.9082	2.80	$>3.6 \times 10^{17}$ y	0.05	
				97	96.91104	....	17 h		
41	Niobium	Nb	92.906	......	....	....	....	1.15	5
				92	91.9062	....	> 350 y		
				93	92.9060	100	....		
				94	93.9063	....	2×10^4 y	~5	
				95	94.9060	....	35.15	<7	
42	Molybdenum	Mo	95.911	......	....	....	....	2.7	7
				92	91.9063	15.84	....	<0.3	
				93	92.9057	....	> 100 y		
				94	93.9047	9.04	....		
				95	94.9046	15.72	....	14.5	
				96	95.9046	16.53	....	1.2	
				97	96.9058	9.45	....	2.2	
				98	97.9055	23.78	....	0.15	
				99	98.9069	....	66.7 h		
				100	99.9076	9.63	....	0.5	
				101	....	....	14.6 m		
				102	....	....	11 m		
43	Technetium	Te	98.97	......	....	....	....	22	
				95	94.9073	....	20 h		
				97	....	....	2.6×10^6 y		
				98	....	....	1.5×10^6 y		
				99	98.9054	....	2.12×10^5 y	22	
				100	....	....	17 s		
				101	....	....	14 m		
44	Ruthenium	Ru	101.07	......	....	....	....	2.56	6
				95	94.9095	....	1.7 h		
				96	95.9076	5.51	....		
				97	....	....	2.9 d		
				98	97.9055	1.87	....	<8	
				99	98.9061	12.72	....	10.6	
				100	99.9030	12.62	....	5.8	
				101	....	17.07	....	3.1	
				102	101.9037	31.61	....	123	
				103	102.9058	....	39.6 d		

Partial List of the Isotopes *(continued)*

Atomic Number, Z	Element	Symbol	Element Atomic Mass, amu	Isotope Mass Number, A	Isotope Mass, amu	Natural Abundance, Atomic %	Half-life	Cross Section for 2,200 m/sec Neutrons, Barns	
								Absorp.	Scatt.
				104	103.9055	18.58	....	0.47	
				105	104.9075	....	4.44h	0.2	
				106	105.94073	....	367d	0.146	
45	Rhodium	Rh	102.905	.....	....	....	....	150	5
				102	101.9064	....	206d		
				103	102.9048	100	....	150	5
				104	103.9064	....	43s		
				105	....	....	35.9h	18,000	
46	Palladium	Pd	106.4	.....	....	....	....	8	3.6
				102	101.9049	0.96	....		
				103	102.9058	....	17d		
				104	103.9046	10.97	....		
				105	104.9032	22.23	....		
				106	105.9032	27.33	....	0.292	
				107	106.9049	....	$\sim 7 \times 10^{6}$y		
				108	107.9030	26.71	....	12.26	
				109	108.9059	....	13.47h		
				110	109.9060	11.81	....	0.237	
				111	110.9076	....	22m		
47	Silver	Ag	107.87	.....	....	....	....	63.6	
				106	105.9061	....	....		
				107	106.9041	51.82	....		
				108	107.9059	....	2.2m		
				109	108.9047	45.18	....		
				110	109.9072	....	24.4s		
48	Cadmium	Cd	112.40	.....	....	....	....	2,450	
				106	105.9070	1.22			
				107	106.9064	....	6.5h		
				108	107.9040	0.88	....	2.0	
				109	108.9048	....	450d		
				110	109.9030	12.39	....	0.1	
				111	110.9042	12.75	....		
				112	111.9028	24.07	....		
				113	112.9046	12.26	....	20,000	
				114	113.9036	28.86	....	0.3	
				115	114.9070	....	53.5h		
				116	115.9050	7.58	....		
				117	116.9076	....	2.4h		

Partial List of the Isotopes *(continued)*

Atomic Number, Z	Element	Symbol	Element Atomic Mass, amu	Isotope Mass Number, A	Isotope Mass, amu	Natural Abundance, Atomic %	Half-life	Cross Section for 2,200 m/sec Neutrons, Barns	
								Absorp.	Scatt.
49	Indium	In	114.82	...	...	...	...	194	2.2
				113	112.9043	4.28	...	11.1	
				114	113.9070	...	72s		
				115	114.9041	95.72	6×10^{14}y		
				116	115.9071	...	14s		
50	Tin	Sn	118.69	...	...	...	...	0.63	4
				112	111.9040	0.95	...		
				113	...	...	115d	1.25	
				114	113.9030	0.65	...	284	
				115	114.9035	0.35	...		
				116	115.9021	14.3	...		
				117	116.9031	7.61	...		
				118	117.9018	24.03	...		
				119	118.9034	8.58	...		
				120	119.9009	32.85	...		
				121	120.9025	...	27h		
				122	121.9034	4.72	...	0.18	
				123	122.9037	...	42m		
				124	123.9052	5.94	60.3d		
				125	...	...	9.4d		
51	Antimony	Sb	121.75	...	...	...	...	5.7	4.3
				120	119.9032	...	15.9m		
				121	120.9038	57.25	...	5.9	
				122	121.9035	...	2.8d		
				123	122.9041	42.75	...	4.1	
				124	...	...	60.3d	< 20	
52	Tellurium	Te	127.60	...	...	...	...	4.7	5
				120	119.9045	0.089	...	2.34	
				121	...	...	17d		
				122	121.9030	2.46	...	2.8	
				123	122.9042	0.87	...	410	
				124	123.9028	4.61	...	4	
				125	124.9044	6.99	...	1.56	
				126	125.9032	18.71	...	1.04	
				127	126.9053	...	9.4h		
				128	127.9047	31.79	...	0.3	
				129	128.9066	...	69m		
				130	129.9067	34.48	...	0.5	
				131	130.9084	...	25m		
53	Iodine	I	126.9044	...	...	...	...	6.2	3.6
				126	125.9062	...	13d		
				127	126.9044	100	...	6.2	
				128	127.9060	...	25.08m		
				129	128.9047	...	1.7×10^{7}y	28	

Partial List of the Isotopes *(continued)*

Atomic Number, Z	Element	Symbol	Element Atomic Mass, amu	Isotope Mass Number, A	Isotope Mass, amu	Natural Abundance, Atomic %	Half-life	Cross Section for 2,200 m/sec Neutrons, Barns	
								Absorp.	Scatt.
				130	129.9065	....	12.3h	18	
				131	130.9060	....	8.07d	24.5	
54	Xenon	Xe	131.30	124	123.9061	0.096	....	100	
				125	....	....	17 h	1.5	
				126	125.9042	0.09	....		
				127	126.9055	....	36.41 d		
				128	127.9035	1.92	....	< 5	
				129	128.9048	26.44	....	21	
				130	129.9035	4.08	....	< 5	
				131	130.9051	21.18	....	110	
				132	131.9042	26.89	....	0.27	
				133	132.9054	....	5.27 d		
				134	133.9054	10.44	....	0.228	
				135	....	....	9.2 h	3.6×10^6	
				136	135.9072	8.87	....	0.281	
				137	....	....	3.9 m		
55	Cesium	Cs	132.905	......	....	....	....	31.6	20
				132	131.9060	....	6.5 d		
				133	132.9051	100	....	29	
				134	133.9064	....	2.05 y	134	
				135	....	....	3×10^6 y	8.7	
				137	138.9073	....	0.11		
56	Barium	Ba	137.34	......	....	....	....	1.2	8
				130	129.9062	0.101	....	13.5	
				131	....	....	12 d		
				132	131.9057	0.097	....	8.6	
				133	....	....	10.66 y		
				134	133.9043	2.42	....	0.158	
				135	134.9056	6.59	....	5.8	
				136	135.9044	7.81	....	0.012	
				137	136.9061	11.32	....	5.1	
				138	137.9050	71.66	....	0.35	
				139	138.9079	....	82.9 m		
				140	139.9099	....	12.8 d	~ 12	
57	Lanthanum	La	138.91	......	....	....	....	8.9	18
				137	....	....	6×10^4 y		
				138	137.9068	0.089	1.12×10^{11} y		
				139	138.9061	99.911	....		
				140	139.9085	....	40.22 h		
				141	140.9095	....	3.9 h		

Partial List of the Isotopes *(continued)*

Atomic Number, Z	Element	Symbol	Element Atomic Mass, amu	Isotope Mass Number, A	Isotope Mass, amu	Natural Abundance, Atomic %	Half-life	Cross Section for 2,200 m/sec Neutrons, Barns	
								Absorp.	Scatt.
58	Cerium	Ce	140.12	...	...	...	...	0.73	9.0
				136	...	0.193	...	0.95	
				137	...	...	9 h		
				138	137.9057	0.25	...	91.115	
				139	138.9054	...	140 d		
				140	139.9053	88.48	...	0.60	
				141	140.9069	...	33 d	29	
				142	141.9090	11.07	$> 5 \times 10^{16}$ y	0.95	
				143	142.9111	...	33 h		
				144	143.9127	...	284.9 d	1.0	
59	Praseodymium	Pr	140.907	...	...	...	...	12.0	3.5
				140	139.9079	...	3.39 m		
				141	140.9074	100	...	11.6	
				142	141.9087	...	19.2 h		
				143	142.9096	...	13.7 d	89	
60	Neodymium	Nd	144.24	...	...	...	...	48	25
				142	141.9075	27.11	...	18	
				143	142.9096	12.17	...	335	
				144	143.9099	23.85	...	5.0	
				145	144.9122	8.30	...	52	
				146	145.9127	17.62	...	10	
				147	...	...	11.1 d		
				148	147.9165	5.73	...	2.9	
				149	148.9203	...	1.73 h		
				150	149.9207	5.62	...	1.8	
				151	...	...	10 m		
61	Promethium	Pm		...	...	...	...		
				146	145.9125	...	~710 d		
				147	146.9138	...	2.5 y	200	
				148	147.9171	...	5.39 d	2.000	
				149	148.9175	...	53.1 h	1,700	
62	Samarium	Sm	150.35	...	...	...	...	5.820	5.0
				144	143.9117	3.09	...	0.7	
				145	...	...	340 d		
				146	145.9129	...	7×10^{7} y		
				147	146.9146	14.97	1.06×10^{11} y	87	
				148	147.9146	11.24	1.2×10^{13} y		
				149	148.9169	13.83	$\sim 4 \times 10^{14}$ y	41.000	
				150	149.9170	7.44	...	102	
				151	...	...	93 y	15,000	
				152	151.9195	26.72	...	210	
				153	...	...	46.8 h		
				154	153.9220	22.71			
				155	154.9242	...	22 m		

Partial List of the Isotopes *(continued)*

Atomic Number, Z	Element	Symbol	Element Atomic Mass, amu	Isotope Mass Number, A	Isotope Mass, amu	Natural Abundance, Atomic %	Half-life	Cross Section for 2,200 m/sec Neutrons, Barns	
								Absorp.	Scatt.
63	Europium	Eu	151.96					4,400	8
				151	150.9196	47.82		8,800	
				152			13 y	5,500	
				153	152.9209	52.18		390	
				154	153.9240		16 y	1,500	
				155	154.9219		1.81 y	14,000	
64	Gadolinium	Gd	157.24					49,000	
				148	147.9177		~ 130 y		
				149	148.9189		9 d		
				150	149.9185		2.1×10^6 y		
				152	151.9195	0.20	$1 \times 1 \times 10^{14}$ y		
				153			242 d		
				154	153.9207	2.15		100	
				155	154.9226	14.73		61,000	
				156	155.9221	20.47		11.5	
				157	156.9339	15.68		254,000	
				158	157.9241	24.87		3.5	
				159			18 h		
				160	159.9071	21.90		768	
				161			3.7 m		
65	Terbium	Tb	158.924					46	
				156			5.4 d		
				158			1.2×10^3 y		
				159	158.9250	100		46	
				160	159.9269		73 d		
66	Dysprosium	Dy	162.50					930	100
				156	155.9238	0.052			
				157			8.1 h		
				158	157.9240	0.090		96	
				159			144 d		
				160	159.9248	2.29		55	
				161	160.9266	18.88		600	
				162	161.9265	25.53		160	
				163	162.9284	24.97		125	
				164	163.9288	28.18		2,700	
				165	164.9300		2.3 h		
67	Holmium	Ho	164.93					67	
				164	163.9306		37 m		
				165	164.9303	100		65	
				166			26.9 h		

Partial List of the Isotopes *(continued)*

Atomic Number, Z	Element	Symbol	Element Atomic Mass, amu	Isotope Mass Number, A	Isotope Mass, amu	Natural Abundance Atomic %	Half-life	Cross Section for 2,200 m/sec Neutrons, Barns	
								Absorp.	Scatt.
68	Erbium	Er	167.26	...	...	...	...	160	12
				162	161.9288	0.136			
				163	...	...	75 m		
				164	163.9293	1.56	...	1.65	
				165	...	...	10.3 h		
				166	165.9304	33.41		45	
				167	166.9320	22.94	...	650	
				168	167.9324	27.07	...	2.03	
				169	...	...	9.4 d		
				170	169.9355	14.88	...	9	
				171	...	...	7.5 h		
69	Thulium	Tm	168.934	...	...	...	...	127	7
				168	...	...	87 d		
				169	168.9344	100	...	127	
				170	...	...	128 d		
70	Ytterbium	Yb	173.04	...	...	...	...	37.5	12
				168	167.9339	0.135	...	5,500	
				169	...	...	32 d		
				170	169.9349	3.03	...		
				171	170.9365	14.31	...	46	
				172	171.9366	21.82	...		
				173	172.9383	16.13	...	20	
				174	173.9390	31.84			
				175	...	...	101 h		
				176	175.9427	12.73			
				177	...	...	1.9 h		
71	Lutecium	Lu	174.97	...	...	...	...	108	
				173	...	...	~ 1.37 y		
				175	174.9409	97.41	...	23	
				176	175.9414	2.59	3×10^{10} y	2,100	
				177	...	...	6.7 d		
72	Hafnium	Hf	178,49	...	...	...	...	105	8
				174	173.9403	0.18	...	390	
				175	...	...	70 d		
				176	175.9435	5.20	...	15	
				177	176.9435	18.50	...	380	
				178	177.9439	27.14	...	75	
				179	178.9460	13.75	...	65	
				180	179.9468	35.24	...	14	
				181	180.9464	...	42.4 d		

Partial List of the Isotopes *(continued)*

Atomic Number, Z	Element	Symbol	Element Atomic Mass, amu	Isotope Mass Number, A	Isotope Mass, amu	Natural Abundance, Atomic %	Half-life	Cross Section for 2,200 m/sec Neutrons, Barns	
								Absorp.	Scatt.
73	Tantalum	Ta	180.948	......	....	....	....	21	5
				179	....	....	~ 600 d		
				180	179.9415	0.0123			
				181	180.9480	99.988	....	21	
				182	181.94[illegible]5	....	115 d	8,200	
74	Tungsten	W	183.85	......	....	....	....	18.5	5
				180	179.9470	0.14	....	60	
				181	....	....	140 d		
				182	181.9483	26.41	....	20.7	
				183	182.9503	14.40	....	10.2	
				184	183.9510	30.64	....	1.8	
				185	....	....	75.8 d		
				186	185.9543	28.41	....	38	
				187	186.9530	....	24 h		
				188	....	....	69 d		
75	Rhenium	Re	186.2	......	....	....	....	86	14
				185	184.9530	37.07	....	105	
				186	185.9515	....	90 h		
				187	186.9560	62.93	7×10^{10} y	73	
				188	187.9565	....	16.7 h		
76	Osmium	Os	190.20	......	....	....	....	15.3	11
				184	183.9526	0.018	....		
				185	....	....	94 d		
				186	185.9539	1.59			
				187	186.9560	1.64	....		
				188	187.9560	13.3	....		
				189	188.9586	16.1	....		
				190	189.9586	26.4	....		
				191	190.9607	....	15 d		
				192	191.9613	41.0	....		
				193	192.9650	....	31 h		
77	Iridium	Ir	192.20	......	....	....	....	440	
				191	190.9609	37.3	....	924	
				192	191.9636	....	74 d		
				193	192.9633	62.7	....	112.5	
				194	193.9647	....	17.4 h		

Partial List of the Isotopes *(continued)*

Atomic Number, Z	Element	Symbol	Element Atomic Mass, amu	Isotope Mass Number, A	Isotope Mass, amu	Natural Abundance, Atomic %	Half-life	Cross Section for 2,200 m/sec Neutrons, Barns	
								Absorp.	Scatt.
78	Platinum	Pt	195.09	…	…	…	…	9.0	10
				190	189.9600	0.0127	6×10^{11} y	150	
				191	…	…	3.0 d		
				192	191.9614	0.78	$\sim 10^{15}$ y	8	
				193	192.9640	…	< 500 y		
				194	193.9628	32.19	…	1.2	
				195	194.9648	33.8	…	27	
				196	195.9650	25.3	…	1.0	
				197	196.9666	…	18 h		
				198	197.9675	7.21	…	4.0	
				199	…	…	30 m		
79	Gold	Au	196.967	…	…	…	…	98.8	9.3
				196	195.9658	…	6.18 d		
				197	196.9666	100	…	98.8	
				198	197.9675	…	2.693 d	25,800	
				199	198.9677	…	3.15 d		
80	Mercury	Hg	200.59	…	…	…	…	375	20
				196	195.9650	0.146	…	3,100	
				197	…	…	65 h		
				198	197.9668	10.02	…	0.018	
				199	198.9683	16.84	…	2,500	
				200	199.9683	23.13	…		
				201	200.9703	13.22	…		
				202	201.9706	29.80	…	4.5	
				203	202.9719	…	46.57 d		
				204	203.9735	6.85	…		
				205	204.9751	…	5.5 m		
81	Thallium	Tl	204.37	…	…	…	…	3.4	14
				203	202.9723	29.50	…	11.4	
				204	203.9721	…	3.8 y		
				205	204.9745	70.50	…	0.80	
				206	205.9747	…	4.19 m		
82	Lead	Pb	207.18	…	…	…	…	0.17	11
				204	203.9730	1.48	…	0.655	
				205	204.9731	…	3×10^{7} y		
				206	205.9745	23.6	…	0.0305	
				207	206.9759	22.6	…	0.709	
				208	207.9766	52.3	…	< 0.03	
				209	208.9798	…	3.3 h		
				210	209.9828	…	21 y		

Partial List of the Isotopes *(continued)*

Atomic Number, Z	Element	Symbol	Element Atomic Mass, amu	Isotope Mass Number, A	Isotope Mass, amu	Natural Abundance, Atomic %	Half-life	Cross Section for 2,200 m/sec Neutrons, Barns	
								Absorp.	Scatt.
83	Bismuth	Bi	208.98	...	...	...	...	0.034	9
				208	207.9784	...	3.7×10^{5} y		
				209	208.9804	100	...	0.034	9
				210	209.9827	...	5.10 d		
				211	210.9873	...	2.15 m		
84	Polonium	Po		206	205.9805	...	8.8 d		
				207	206.9816	...	5.7 h		
				208	207.9813	...	2.93 y		
				209	208.9825	...	103 y		
				210	209.9829	...	138.4 d		
				211	210.9866	...	0.52 s		
				212	211.9889	...	0.304 s		
				213	212.9928	...	412 μs		
				214	213.9952	...	164 μs		
				215	214.9995	...	1.78×10^{-3} s		
				216	216.0019	...	0.15 s		
				218	218.0089	...	3.05 m		
85	Astantine	At	~ 210	...	...	...	...		
				209	...	...	5.5 h		
				210	209.9865	...	8.3 h		
				211	210.9875	...	7.2 h		
				212	211.9893	...	0.30 s		
				213	212.9963	...	short		
				214	213.9963	...	short		
				215	214.9987	...	~ 100 μs		
				216	216.0024	...	300 μs		
				217	217.0046	...	0.032 s		
				218	218.0086	...	~ 25		
				219	219.0114	...	0.9 m		
86	Radon	Rn		...	...	...	...	0.7	
				219	219.0095	...	4 s		
				220	220.0114	...	55 s		
				221	221.0135	...	25 m		
				222	222.0175	...	3.823 d		
87	Francium	Fr		220	220.0123	...	27.55		
				221	221.0142	...	4.8 m		
				222	222.0161	...	14.8 m		
				223	223.0198	...	22 m		
				224	224.0219	...	...		

Partial List of the Isotopes *(continued)*

Atomic Number, Z	Element	Symbol	Element Atomic Mass, amu	Isotope Mass Number, A	Isotope Mass, amu	Natural Abundance, Atomic %	Half-life	Cross Section for 2,200 m/sec Neutrons, Barns	
								Absorp.	Scatt.
88	Radium	Ra	226.05	.....	....	....	....		
				220	220.0110	....	0.0235		
				221	221.0139	....	30 s		
				222	222.0154	....	38 s	20	< 100
				223	223.0186	....	11.43 d		
				224	224.0202	....	3.64 d		
				225	225.0219	....	14.8 d		
				226	226.0245	....	1,600 y		<0.0001
				227	227.0276	....	41.2 m		
				228	228.0296	....	6.7 y		< 2
89	Actinium	Ac		225	225.0231	....	10 d		
				226	226.0246	....	29 h		
				227	227.0278	....	21.6 y	830	< 2
				228	228.0295	....	6.13 h		
				229	229.0308	....	66 m		
90	Thorium	Th	232.12	.....	....	....	....		
				228	228.0287	....	1.913 y		
				229	229.0316	....	7,340 y	$32\sigma_f$	
				230	230.0331	....	8×10^4 y	27	
				231	231.0347	....	25.5 h		
				232	232.0382	100	1.41×10^{10} y	7.56	13
				233	233.0387	....	22.2 m	1,500 (th)*	
				234	234.0421	....	24.2 d		
91	Protactinium	Pa	~ 231	.....	....	....	....		
				226	226.0278	....	1.8 m		
				227	227.0289	....	38.3 m		
				228	228.0310	....	26 h		
				229	229.0305	....	1.5 d		
				230	230.0328	....	17.7 d		
				231	231.0359	....	3.35×10^4 y	200	
				232	232.0371	....	1.31 d		
				233	233.0384	....	27 d		
				234	234.0414	....	6.75 h		
				235	235.0438	....	23.7 m		
92	Uranium	U	238.07	.....	....	....	....	7.68	4.18
				227	227.0309	....	1.3 m		
				228	228.0313	....	9.3 m		
				229	229.0332	....	58 m		
				230	230.0339	....	20.8 d		
				231	231.0347	....	4.3 d		
				232	232.0372	....	73.6 y	$78.77\sigma_f$ (th)	

* Thermal spectrum

Partial List of the Isotopes *(continued)*

Atomic Number, Z	Element	Symbol	Element Atomic Mass, amu	Isotope Mass Number, A	Isotope Mass, amu	Natural Abundance, Atomic %	Half-life	Cross Section for 2,200 m/sec Neutrons, Barns	
								Absorp.	Scatt.
				233	233.0395	....	1.65×10^5 y	573,525σ_f	
				234	234.0409	0.006	2.47×10^5 y	95	
				235	235.0439	0.720	7.1×10^8 y	678.2, 577.1σ_f	15
				236	236.0457	....	2.39×10^7 y	6	
				237	237.0469	....	6.75 d		
				238	238.0508	99.274	4.51×10^9 y	2.73	13.8
				239	239.0526	....	23.5 m	14σ_f	
				240	240.0546	....	14.1 h		
93	Neptunium	Np	237.1	231	231.0383	....	~ 50 m		
				233	233.0406	....	35 m		
				234	234.0419	....	4.4 d		
				235	235.0425	....	410 d		
				236	236.0466	....	22 h	(2,800 σ_f)	
				237	237.0480	....	2.14×10^6 y	170, 0,019σ_f	
				238	238.0494	....	2.1 d		
				239	239.0513	....	2.35 d	1,026	
				240	240.0537	....	63 m	295	
				241	241.0558	....	16 m	1,400	
94	Plutonium	Pu	~ 242	232	232.0411	....	36 m		
				234	234.0433	....	9 h		
				235	235.0441	....	26 m		
				236	236.0461	....	2.85 y		
				237	237.0483	....	45.6 d		
				238	238.0495	....	86 y	500, 16.6σ_f	
				239	239.0522	....	24,400 y	1014.5 740.6σ_f	9.6
				240	240.0540	....	6,580 y	295, <0.1σ_f	
				241	241.13154	....	13.2 y	1,375, 950σ_f	
				242	242.0587	....	3.79×10^5 y	30	
				243	243.0601	....	5 h		
				244	244.0630	....	8×10^7 y		
95	Americium	Am		240	240.0539	....	51 h		
				241	241.0567	....	458 y	585,3.1σ_f	
				242	242.0574	....	16 h	8,000,σ_f	
				243	243.0614	....	7,370 y	180, <0.075σ_f	
				244	244.0625	....	10.1 h	2,300σ_f(th)	
				245	245.0648	....	2.1 h		

Partial List of the Isotopes *(continued)*

Atomic Number, Z	Element	Symbol	Element Atomic Mass, amu	Isotope Mass Number, A	Isotope Mass, amu	Natural Abundance, Atomic %	Half-life	Cross Section for 2,200 /sec Neutrons, Barns	
								Absorp.	Scatt.
96	Curium	Cm		238	238.0530	....	2.5 h		
				240	240.0555	....	26.8 d		
				241	241.0556	....	35 d		
				242	242.0558	....	163 d	$(<5\sigma_f)$	
				243	243.0614	....	32 y	$(700\sigma_f)$	
				244	244.0629	....	17.6 y		
				245	245.0653	....	9,300 y	$(1{,}900\sigma_f)$	
				246	246.0674	....	5,500 y		
97	Berkelium	Bk		245	245.0643	....	4.98 d		
				246	246.0672	....	1.8 d		
				247	247.0702	....	~ 1,400 y		
				248	248.0718	....	16 h		
				249	249.0726	....	314 d	500	
98	Californium	Cf		244	244.0659	....	25 m		
				245	245.0666	....	44 m		
				246	246.0688	....	36 h		
				247	247.0690	....	2.5 h		
				248	248.0724	....	350 d		
				249	249.0748	....	360 y	1,735, $600\sigma_f$(th)	
				250	250.0766	....	13 y	$<350\sigma_f$	
				251	....	....	800 y	$3{,}000\sigma_f$, (th)	
99	Einsteinium	Es		251	251.0788	....	1.5 d		
				252	252.0829	....	~ 140 d		
				253	253.0847	....	20.47 d		
				254	254.0881	....	276 d	160	
				255	255.0880	....	38.3 d		
100	Fermium	Fm		250	250.0795	....	30 m		
				252	252.0827	....	23 h		
				253	253.0837	....	34 h		
				254	254.0870	....	3.2 h		
				255	255.0877	....	1 h		
				256	....	....	2.7 h		
101	Mendelevium	Mv		255	255.0906	....	0.6 h		
				256	256.0984	....	1.5 h		
102	Nobelium	No		253	253.0892	....	95 s		
				254	....	....	55 s		
103	Lawrencium	Lw		256	....	....	8 s		
				257	....	....	....		
104	Khurchatorium	Ku		260	....	....	0.3 s		

appendix C

Bessel Functions

Bessel functions are solutions of the variable-coefficient equation

$$x^2 \frac{d^2y}{dx^2} + x \frac{dy}{dx} + (x^2 - \nu^2)y = 0 \tag{C-1}$$

where ν is a constant. This equation is called Bessel's equation of order ν. It derives its name from the German mathematician and astronomer Frederich Bessel (1784-1846) who encountered it in his studies of the motion of planets. It arises in problems of diffusion of heat, electricity, and particles (such as neutrons) in bodies having cylindrical symmetry. The general solution of this equation consists of two linearly independent particular solutions, combined as follows:

$$y(x) = AJ_\nu(x) + BY_\nu(x) \tag{C-2}$$

where A, B = const
$J_\nu(x)$ = Bessel function of first kind, order ν
$Y_\nu(x)$ = Bessel function of second kind, order ν (also called Neumann function)

Thus a Bessel equation of zero order has $\nu = 0$ and a solution containing $J_0(x)$ and $Y_0(x)$. Bessel functions (also called cylindrical functions) of the first and second kind are given for integer orders in series form as follows:

$$J_\nu(x) = \sum_{k=0}^{k=\infty} \frac{(-1)^k (x/2)^{2k+\nu}}{k!(k+\nu)!} \tag{C-3}$$

and

$$Y_\nu(x) = \frac{1}{\sin \nu\pi} [J_\nu(x) \cos (\nu\pi) - J_{-\nu}(x)] \tag{C-4}$$

These two functions are real if ν is real and x is positive.

If x is replaced by ix, where $i = \sqrt{-1}$, Bessel's equation modifies to the form

$$x^2 \frac{d^2y}{dx^2} + x \frac{dy}{dx} - (x^2 + \nu^2)y = 0 \tag{C-5}$$

According to the above, the solution of this equation may be given in the form

$$y(x) = AJ_\nu(ix) + BY_\nu(ix) \tag{C-6}$$

To avoid the imaginaries in the above functions, they are modified into real functions of real variables, $I_\nu(x)$, given by

$$J_\nu(ix) = I_\nu(x) = i^{-\nu} J_\nu(x) \tag{C-7}$$

and

$$Y_v(ix) = i^v \left[iI_v(x) - \frac{2}{\pi} (-1)^v K_v(x) \right] \tag{C-8}$$

so that the general solution of Eq. C-5 is given by

$$y(x) = AI_v(x) + BK_v(x) \tag{C-9}$$

where A and B are new constants.

$I_v(x)$ and $K_v(x)$ are called the modified Bessel functions of the first and second kinds, respectively, and of order v.

The four Bessel functions of zero order are plotted in Fig. C-1,

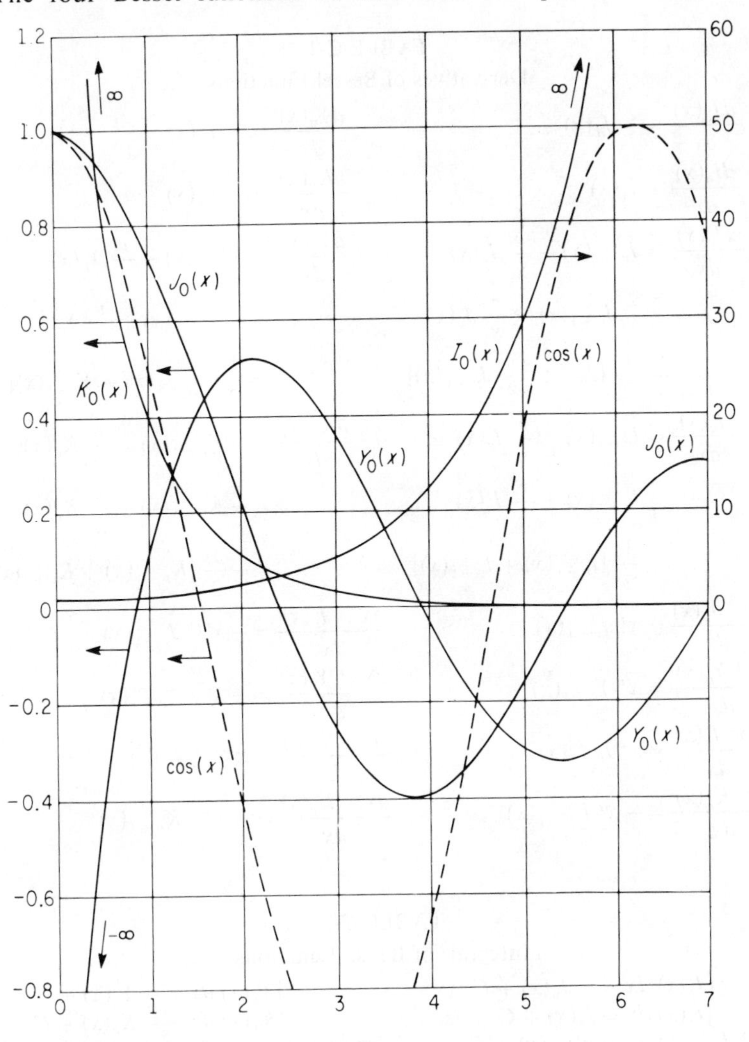

FIG. C 1

together with cos x for comparison. $J_0(x)$ and $Y_0(x)$ are both oscillatory, with the distance between the roots (the values of x at which the functions are zero) becoming larger and approaching π as x increases. The amplitudes of the two functions decrease with x, and the functions are bounded (not infinite) everywhere except for $Y_0(x)$ being $-\infty$ at $x = 0$. $I_0(x)$ and $K_0(x)$ are nonoscillatory und unbounded, the former going to ∞ at $x = \infty$, and the latter at $x = 0$.

Some useful derivatives and integrals of Bessel functions are given in Table C-1.

TABLE C-1
Derivatives of Bessel Functions

$$\frac{dJ_0(x)}{dx} = -J_1(x) \qquad \frac{dY_0(x)}{dx} = -Y_1(x)$$

$$\frac{dI_0(x)}{dx} = I_1(x) \qquad \frac{dK_0(x)}{dx} = -K_1(x)$$

$$\frac{dJ_v(x)}{dx} = J_{v-1}(x) - \frac{v}{x} J_v(x) \qquad \frac{dY_v(x)}{dx} = Y_{v-1}(x) - \frac{v}{x} Y_v(x)$$

$$= -J_{v+1}(x) + \frac{v}{x} J_v(x) \qquad = -Y_{v+1}(x) + \frac{v}{x} Y_v(x)$$

$$= \frac{1}{2}[J_{v-1}(x) - J_{v+1}(x)] \qquad = \frac{1}{2}[Y_{v-1}(x) - Y_{v+1}(x)]$$

$$\frac{dI_v(x)}{dx} = I_{v-1}(x) - \frac{v}{x} I_v(x) \qquad \frac{dK_v(x)}{dx} = -K_{v-1}(x) - \frac{v}{x} K_v(x)$$

$$= I_{v+1}(x) + \frac{v}{x} I_v(x) \qquad = -K_{v+1}(x) + \frac{v}{x} K_v(x)$$

$$= \frac{1}{2}[I_{v-1}(x) + I_{v+1}(x)] \qquad = -\frac{1}{2}[K_{v-1}(x) + K_{v+1}(x)]$$

$$\frac{dx^v J_v(x)}{dx} = x^v J_{v-1}(x) \qquad \frac{dx^{-v} J_v(x)}{dx} = -x^{-v} J_{v+1}(x)$$

$$\frac{dx^v Y_v(x)}{dx} = x^v Y_{v-1}(x) \qquad \frac{dx^{-v} Y_v(x)}{dx} = -x^{-v} Y_{v+1}(x)$$

$$\frac{dx^v I_v(x)}{dx} = x^v I_{v-1}(x) \qquad \frac{dx^{-v} I_v(x)}{dx} = x^{-v} I_{v+1}(x)$$

$$\frac{dx^v K_v(x)}{dx} = -x^v K_{v-1}(x) \qquad \frac{dx^{-v} K_v(x)}{dx} = -x^{-v} K_{v+1}(x)$$

TABLE C-2
Integrals of Bessel Functions

$$\int J_1(x)\,dx = -J_0(x) + C \qquad \int Y_1(x)\,dx = -Y_0(x) + C$$

$$\int I_1(x)\,dx = I_0(x) + C \qquad \int K_1(x)\,dx = -K_0(x) + C$$

$$\int x^v J_{v-1}(x)\,dx = x^v J_v(x) + C \qquad \int x^{-v} J_{v+1}(x)\,dx = -x^{-v} J_v(x) + C$$

TABLE C-3
Some Bessel Functions

x	$J_0(x)$	$J_1(x)$	$Y_0(x)$	$Y_1(x)$	$I_0(x)$	$I_1(x)$	$K_0(x)$	$K_1(x)$
0	1.0000	0.0000	$-\infty$	$-\infty$	1.000	0.0000	∞	∞
0.05	0.9994	0.0250	−1.979	−12.79	1.001	0.0250	3.114	19.91
0.10	0.9975	0.0499	−1.534	−6.459	1.003	0.0501	2.427	9.854
0.15	0.9944	0.0748	−1.271	−4.364	1.006	0.0752	2.030	6.477
0.20	0.9900	0.0995	−1.081	−3.324	1.010	0.1005	1.753	4.776
0.25	0.9844	0.1240	−0.9316	−2.704	1.016	0.1260	1.542	3.747
0.30	0.9776	0.1483	−0.8073	−2.293	1.023	0.1517	1.372	3.056
0.35	0.9696	0.1723	−0.7003	−2.000	1.031	0.1777	1.233	2.559
0.40	0.9604	0.1960	−0.6060	−1.781	1.040	0.2040	1.115	2.184
0.45	0.9500	0.2194	−0.5214	−1.610	1.051	0.2307	1.013	1.892
0.50	0.9385	0.2423	−0.4445	−1.471	1.063	0.2579	0.9244	1.656
0.55	0.9258	0.2647	−0.3739	−1.357	1.077	0.2855	0.8466	1.464
0.60	0.9120	0.2867	−0.3085	−1.260	1.092	0.3137	0.7775	1.303
0.65	0.8971	0.3081	−0.2476	−1.177	1.108	0.3425	0.7159	1.167
0.70	0.8812	0.3290	−0.1907	−1.103	1.126	0.3719	0.6605	1.050
0.75	0.8642	0.3492	−0.1372	−1.038	1.146	0.4020	0.6106	0.9496
0.80	0.8463	0.3688	−0.0868	−0.9781	1.167	0.4329	0.5653	0.8618
0.85	0.8274	0.3878	−0.0393	−0.9236	1.189	0.4646	0.5242	0.7847
0.90	0.8075	0.4059	−0.0056	−0.8731	1.213	0.4971	0.4867	0.7165
0.95	0.7868	0.4234	0.0481	−0.8258	1.239	0.5306	0.4524	0.6560
1.0	0.7652	0.4401	0.0883	−0.7812	1.266	0.5652	0.4210	0.6019
1.1	0.6957	0.4850	0.1622	−0.6981	1.326	0.6375	0.3656	0.5098
1.2	0.6711	0.4983	0.2281	−0.6211	1.394	0.7147	0.3185	0.4346
1.3	0.5937	0.5325	0.2865	−0.5485	1.469	0.7973	0.2782	0.3725
1.4	0.5669	0.5419	0.3379	−0.4791	1.553	0.8861	0.2437	0.3208
1.5	0.4838	0.5644	0.3824	−0.4123	1.647	0.9817	0.2138	0.2774
1.6	0.4554	0.5699	0.4204	−0.3476	1.750	1.085	0.1880	0.2406
1.7	0.3690	0.5802	0.4520	−0.2847	1.864	1.196	0.1655	0.2094
1.8	0.3400	0.5815	0.4774	−0.2237	1.990	1.317	0.1459	0.1826
1.9	0.2528	0.5794	0.4968	−0.1644	2.128	1.448	0.1288	0.1597
2.0	0.2239	0.5767	0.5104	−0.1070	2.280	1.591	0.1139	0.1399
2.1	0.1383	0.5626	0.5183	−0.0517	2.446	1.745	0.1008	0.1227
2.2	0.1104	0.5560	0.5208	−0.0015	2.629	1.914	0.0893	0.1079
2.3	0.0288	0.5305	0.5181	0.0523	2.830	2.098	0.0791	0.0950
2.4	0.0025	0.5202	0.5104	0.1005	3.049	2.298	0.0702	0.0837
2.5	0.0729	0.4843	0.4981	0.1459	3.290	2.517	0.0623	0.0739
2.6	−0.0968	0.4708	0.4813	0.1884	3.553	2.755	0.0554	0.0653
2.7	−0.1641	0.4260	0.4605	0.2276	3.842	3.016	0.0493	0.0577
2.8	−0.1850	0.4097	0.4359	0.2635	4.157	3.301	0.0438	0.0511
2.9	−0.2426	0.3575	0.4079	0.2959	4.503	3.613	0.0390	0.0453
3.0	−0.2601	0.3391	0.3769	0.3247	4.881	3.953	0.0347	0.0402
3.2	−0.3202	0.2613	0.3071	0.3707	5.747	4.734	0.0276	0.0316
3.4	−0.3643	0.1792	0.2296	0.4010	6.785	5.670	0.0220	0.0250
3.6	−0.3918	0.0955	0.1477	0.4154	8.028	6.793	0.6175	0.0198
3.8	−0.4026	0.0128	0.0645	0.4141	9.517	8.140	0.0140	0.0157
4.0	−0.3971	−0.0660	−0.0169	0.3979	11.302	9.759	0.0112	0.0125

The Bessel and modified Bessel functions of zero and first order are tabulated in Table C-3 for positive values of x up to $x = 4.0$. Some roots are given below:

Roots of $J_0(x): x = 2.4048, 5.5201, 8.6537, 11.7915, \ldots$
Roots of $J_1(x): x = 3.8317, 7.0156, 10.1735, 13.3237, \ldots$
Roots of $Y_0(x): x = 0.8936, 3.9577, 7.0861, 10.2223, \ldots$
Roots of $Y_1(x): x = 2.1971, 5.4297, 8.5960, 11.7492, \ldots$

appendix **D**

Some Thermodynamic Properties

TABLE D-1a*
Properties of Dry Saturated Steam †
Pressure

Abs. press., psia	Temp., °F	Specific volume		Enthalpy			Entropy		
		Sat. liquid	Sat. vapor	Sat. liquid	Evap.	Sat. vapor	Sat. liquid	Evap.	Sat. vapor
p	t	v_f	v_g	h_f	h_{fg}	h_g	s_f	s_{fg}	s_g
1.0	101.74	0.01614	333.6	69.70	1036.3	1106.0	0.1326	1.8456	1.9782
2.0	126.08	0.01623	173.73	93.99	1022.2	1116.2	0.1749	1.7451	1.9200
3.0	141.48	0.01630	118.71	109.37	1013.2	1122.6	0.2008	1.6855	1.8863
4.0	152.97	0.01636	90.63	120.86	1006.4	1127.3	0.2198	1.6427	1.8625
5.0	162.24	0.01640	73.52	130.13	1001.0	1131.1	0.2347	2.6094	1.8441
6.0	170.06	0.01645	61.98	137.96	996.2	1134.2	0.2472	1.5820	1.8292
7.0	176.85	0.01649	53.64	144.76	992.1	1136.9	0.2581	1.5586	1.8167
8.0	182.86	0.01653	47.34	150.79	988.5	1139.3	0.2674	1.5383	1.8057
9.0	188.28	0.01656	42.40	156.22	985.2	1141.4	0.2759	1.5203	1.7962
10	193.21	0.01659	38.42	161.17	982.1	1143.3	0.2835	1.5041	1.7876
14.696	212.00	0.01672	26.80	180.07	970.3	1150.4	0.3120	1.4446	1.7566
15	213.03	0.01672	26.29	181.11	969.7	1150.8	0.3135	1.4415	1.7549
20	227.96	0.01683	20.089	196.16	960.1	1156.3	0.3356	1.3962	1.7319
25	240.07	0.01692	16.303	208.42	952.1	1160.6	0.3533	1.3606	1.7139
30	250.33	0.01701	13.746	218.82	945.3	1164.1	0.3680	1.3313	1.6993
35	259.28	0.01708	11.898	227.91	939.2	1167.1	0.3807	1.3063	1.6870
40	267.25	0.01715	10.498	236.03	933.7	1169.7	0.3919	1.2844	1.6763
45	274.44	0.01721	9.401	243.36	928.6	1172.0	0.4019	1.2650	1.6669
50	281.01	0.01727	8.515	250.09	924.0	1174.1	0.4110	1.2474	1.6585
55	287.07	0.01732	7.787	256.30	919.6	1175.9	0.4193	1.2316	1.6509
60	292.71	0.01738	7.175	262.09	915.5	1177.6	0.4270	1.2168	1.6438
65	297.97	0.01743	6.655	267.50	911.6	1179.1	0.4342	1.2032	1.6374
70	302.92	0.01748	6.206	272.61	907.9	1180.6	0.4409	1.1906	1.6315
75	307.60	0.01753	5.816	277.43	904.5	1181.9	0.4472	1.1787	1.6259
80	312.03	0.01757	5.472	282.02	901.1	1183.1	0.4531	1.1676	1.6207
85	316.25	0.01761	5.168	286.39	897.8	1184.2	0.4587	1.1571	1.6158
90	320.27	0.01766	4.896	290.56	894.7	1185.3	0.4641	1.1471	1.6112
95	324.12	0.01770	4.652	294.56	891.7	1186.2	0.4692	1.1376	1.6068
100	327.81	0.01774	4.432	298.40	888.8	1187.2	0.4740	1.1286	1.6026
110	334.77	0.01782	4.049	305.66	883.2	1188.9	0.4832	1.1117	1.5948

* For units used in Appendix D: v, ft³/lb$_m$; h, Btu/lb$_m$; s, Btu/lb$_m$°R.

† Abridged from Joseph H. Keenan and Frederick G. Keyes, *Thermodynamic Properties of Steam*, John Wiley and Sons, Inc., New York. Copyright, 1937, by Joseph H. Keenan and Frederick G. Keyes.

TABLE D-1a
Properties of Dry Saturated Steam *(continued)*
Pressure

Abs. press., psia	Temp., °F	Specific volume		Enthalpy			Entropy		
		Sat. liquid	Sat. vapor	Sat. liquid	Evap.	Sat. vapor	Sat. liquid	Evap.	Sat. vapor
p	t	v_f	v_g	h_f	h_{fg}	h_g	s_f	s_{fg}	s_g
120	341.25	0.01789	3.728	312.44	877.9	1190.4	0.4916	1.0962	1.5878
130	347.32	0.01796	3.455	318.81	872.9	1191.7	0.4995	1.0817	1.5812
140	353.02	0.01802	3.220	324.82	868.2	1193.0	0.5069	1.0682	1.5751
150	358.42	0.01809	3.015	330.51	863.6	1194.1	0.5138	1.0556	1.5694
160	363.53	0.01815	2.834	335.93	859.2	1195.1	0.5204	1.0436	1.5640
170	368.41	0.01822	2.675	341.09	854.9	1196.0	0.5266	1.0324	1.5590
180	373.06	0.01827	2'532	346.03	850.8	1196.9	0.5325	1.0217	1.5542
190	377.51	0.01833	2.404	350.79	846.8	1197.6	0.5381	1.0116	1.5497
200	381.79	0.01839	2.288	355.36	843.0	1198.4	0.5435	1.0018	1.5453
250	400.95	0.01865	1,8438	376.00	825.1	1201.1	0.5675	0.9588	1.5263
300	417.33	0.01890	1.5433	393.84	809.0	1202.8	0.5879	0.9225	1.5104
350	431.72	0.01913	1.3260	409.69	794.2	1203.9	0.6056	0.8910	1.4966
400	444.59	0.0193	1.1613	424.0	780.5	1204.5	0.6214	0.8630	1.4844
450	456.28	0.0195	1.0320	437.2	767.4	1204.6	0.6356	0.8378	1.4734
500	467.01	0.0197	0.9278	449.4	755.0	1204.4	0.6487	0.8147	1.4634
550	476.94	0.0199	0.8424	460.8	743.1	1203.9	0.6608	0.7934	1.4542
600	486.21	0.0201	0.7698	471.6	731.6	1203.2	0.6720	0.7734	1.4454
650	494.90	0.0203	0.7083	481.8	720.5	1202.3	0.6826	0.7548	1.4374
700	503.10	0.0205	0.6554	491.5	709.7	1201.2	0.6925	0.7371	1.4296
750	510.86	0.0207	0.6092	500.8	699.2	1200.0	0.7019	0.7204	1.4223
800	518.23	0.0209	0.5687	509.7	688.9	1198.6	0.7108	0.7045	1.4153
850	525.26	0.0210	0.5327	518.3	678.8	1197.1	0.7194	0.6891	1.4085
900	531.98	0.0212	0.5006	526.6	668.8	1195.4	0.7275	0.6744	1.4020
950	538.43	0.0214	0.4717	534.6	659.1	1193.7	0.7355	0.6602	1.3957
1000	544.61	0.0216	0.4456	542.4	649.4	1191.8	0.7430	0.6467	1.3897
1100	556.31	0.0220	0.4001	557.4	630.4	1187.7	0.7575	0.6205	1.3780
1200	567.22	0.0223	0.3619	571.7	611.7	1183.4	0.7711	0.5956	1.3667
1300	577.46	0.0227	0.3293	585.4	593.2	1178.6	0.7840	0.5719	1.3559
1400	587.10	0.0231	0.3012	598.7	574.7	1173.4	0.7963	0.5491	1.3454
1500	596.23	0.0235	0.2765	611.6	556.3	1167.9	0.8082	0.5269	1.3351
2000	635.82	0.0257	0.1878	671.7	463.4	1135.1	0.8619	0.4230	1.2849
2500	668.13	0.0287	0.1307	730.6	360.5	1091.1	0.9126	0.3197	1.2322
3000	695.36	0.0346	0.0858	802.5	217.8	1020.3	0.9731	0.1885	1.1615
3206.2	705.40	0.0503	0.0503	902.7	0	902.7	1.0580	0	1.0580

TABLE D-1b
Properties of Dry Saturated Steam *(continued)*
Temperature

Temp., °F	Abs. press., psia	Specific volume		Enthalpy			Entropy		
		Sat. liquid	Sat. vapor	Sat. liquid	Evap.	Sat. vapor	Sat. liquid	Evap.	Sat. vapor
t	p	v_f	v_g	h_f	h_{fg}	h_g	s_f	s_{fg}	s_g
32	0.08854	0.01602	3306	0.00	1075.8	1075.8	0.0000	2.1877	2.1877
35	0.09995	0.01602	2947	3.02	1074.1	1077.1	0.0061	2.1709	2.1770
40	0.12170	0.01602	2444	8.05	1071.3	1079.3	0.0162	2.1435	2.1597
45	0.14752	0.01602	2036.4	13.06	1068.4	1081.5	0.0262	2.1167	2.1429
50	0.17811	0.01603	1703.2	18.07	1065.6	1083.7	0.0361	2.0903	2.1264
60	0.2563	0.01604	1206.7	28.06	1059.9	1088.0	0.0555	2.0393	2.0948
70	0.3631	0.01606	867.9	38.04	1054.3	1092.3	0.0745	1.9902	2.0647
80	0.5069	0.01608	633.1	48.02	1048.6	1096.6	0.0932	1.9428	2.0360
90	0.6982	0.01610	468.0	57.99	1042.9	1100.9	0.1115	1.8972	2.0087
100	0.9492	0.01613	350.4	67.97	1037.2	1105.2	0.1295	1.8531	1.9826
110	1.2748	0.01617	265.4	77.94	1031.6	1109.5	0.1417	1.8106	1.9577
120	1.6924	0.01620	203.27	87.92	1025.8	1113.7	0.1645	1.7694	1.9339
130	2.2225	0.01625	157.34	97.90	1020.0	1117.9	0.1816	1.7296	1.9112
140	2.8886	0.01629	123.01	107.89	1014.1	1122.0	0.1984	1.6910	1.8894
150	3.718	0.01634	97.07	117.89	1008.2	1126.1	0.2149	1.6537	1.8685
160	4.741	0.01639	77.29	127.89	1002.3	1130.2	0.2311	1.6174	1.8485
170	5.992	0.01645	62.06	137.90	996.3	1134.2	0.2472	1.5822	1.8293
180	7.510	0.01651	50.23	147.92	990.2	1138.1	0.2630	1.5480	1.8109
190	9.339	0.01657	40.96	157.95	984.1	1142.0	0.2785	1.5147	1.7932
200	11.526	0.01663	33.64	167.99	977.9	1145.9	0.2938	1.4824	1.7762
210	14.123	0.01670	27.82	178.05	971.6	1149.7	0.3090	1.4508	1.7598
212	14.696	0.01672	26.80	180.07	970.3	1150.4	0.3120	1.4446	1.7566
220	17.186	0.01677	23.15	188.13	965.2	1153.4	0.3239	1.4201	1.7440
230	20.780	0.01684	19.382	198.23	958.8	1157.0	0.3387	1.3901	1.7288
240	24.969	0.01692	16.323	208.34	952.2	1160.5	0.3531	1.3609	1.7140
250	29.825	0.01700	13.821	216.48	945.5	1164.0	0.3675	1.3323	1.6998
260	35.429	0.01709	11.763	228.64	938.7	1167.3	0.3817	1.3043	1.6860
270	41.858	0.01717	10.061	238.84	931.8	1170.6	0.3958	1.2769	1.6727
280	49.203	0.01726	8.645	249.06	924.7	1173.8	0.4096	1.2501	1.6597
290	57.556	0.01735	7.461	259.31	917.5	1176.8	0.4234	1.2238	1.6472
300	67.013	0.01745	6.466	269.59	910.1	1179.7	0.4369	1.1980	1.6350
310	77.68	0.01755	5.626	279.92	902.6	1182.5	0.4504	1.1727	1.6231
320	89.66	0.01765	4.914	290.28	894.9	1185.2	0.4637	1.1478	1.6115
330	103.06	0.01776	4.307	300.68	887.0	1187.7	0.4769	1.1233	1.6002
340	118.01	0.01787	3.788	311.13	879.0	1190.1	0.4900	1.0992	1.5891

TABLE D-1b
Properties of Dry Saturated Steam *(continued)*
Temperature

Temp., °F	Abs. press., psia	Specific volume		Enthalpy			Entropy		
		Sat. liquid	Sat. vapor	Sat. liquid	Evap.	Sat. vapor	Sat. liquid	Evap.	Sat. vapor
t	p	v_f	v_g	h_f	h_{fg}	h_g	s_f	s_{fg}	s_g
350	134.63	0.01799	3.342	321.63	870.7	1192.3	0.5029	1.0754	1.5783
360	153.04	0.01811	2.957	332.18	852.2	1194.4	0.5158	1.0519	1.5677
370	173.37	0.01823	2.625	342.79	853.5	1196.3	0.5286	1.0287	1.5573
380	195.77	0.01836	2.335	353.45	844.6	1198.1	0.5413	1.0059	1.5471
390	220.37	0.01850	2.0836	364.17	835.4	1199.6	0.5539	0.9832	1.5371
400	247.31	0.01864	1.8633	374.97	826.0	1201.0	0.5664	0.9608	1.5272
410	276.75	0.01878	1.6700	385.83	816.3	1202.1	0.5788	0.9386	1.5174
420	308.83	0.01894	1.5000	396.77	806.3	1203.1	0.5912	0.9166	1.5078
430	343.72	0.01910	1.3499	407.79	796.0	1203.8	0.6035	0.8947	1.4982
440	381.59	0.01926	1.2171	418.90	785.4	1204.3	0.6158	0.8730	1.4887
450	422.6	0.0194	1.0993	430.1	774.5	1204.6	0.6280	0.8513	1.4793
460	466.9	0.0196	0.9944	441.4	763.2	1204.6	0.6402	0.8298	1.4700
470	514.7	0.0198	0.9009	452.8	751.5	1204.3	0.6523	0.8083	1.4606
480	566.1	0.0200	0.8172	464.4	739.4	1203.7	0.6645	0.7868	1.4513
490	621.4	0.0202	0.7423	476.0	726.8	1202.8	0.6766	0.7653	1.4419
500	680.8	0.0204	0.6749	487.8	713.9	1201.7	0.6887	0.7438	1.4325
520	812.4	0.0209	0.5594	511.9	686.4	1198.2	0.7130	0.7006	1.4136
540	962.5	0.0215	0.4649	536.6	656.6	1193.2	0.7374	0.6568	1.3942
560	1133.1	0.0221	0.3868	562.2	624.2	1186.4	0.7621	0.6121	1.3742
580	1325.8	0.0228	0.3217	588.9	588.4	1177.3	0.7872	0.5659	1.3532
600	1542.9	0.0236	0.2668	610.0	548.5	1165.5	0.8131	0.5176	1.3307
620	1786.6	0.0247	0.2201	646.7	503.6	1150.3	0.8398	0.4664	1.3062
640	2059.7	0.0260	0.1798	678.6	452.0	1130.5	0.8679	0.4110	1.2789
660	2365.4	0.0278	0.1442	714.2	390.2	1104.4	0.8987	0.3485	1.2472
680	2708.1	0.0305	0.1115	757.3	309.9	1067.2	0.9351	0.2719	1.2071
700	3093.7	0.0369	0.0761	823.3	172.1	995.4	0.9905	0.1484	1.1389
705.4	3206.2	0.0503	0.0503	902.7	0	902.7	1.0580	0	1.0580

TABLE D-2
Properties of Superheated Steam*

Abs. press., psia (Sat. temp., °F.)		Temperature, °F											
		200	300	400	500	600	700	800	900	1000	1100	1200	1400
	v	392.6	452.3	512.0	571.6	631.2	690.8	750.4	809.9	869.5	929.1	988.7	1107.8
1	h	1150.4	1195.8	1241.7	1288.3	1335.7	1383.8	1432.8	1482.7	1533.5	1585.2	1637.7	1745.7
(101.74)	s	2.0512	2.1153	2.1720	2.2233	2.2702	2.3137	2.3542	2.3923	2.4283	2.4625	2.4952	2.5566
	v	78.16	90.25	102.26	114.22	126.16	138.10	150.03	161.95	173.87	185.79	197.71	221.6
5	h	1148.8	1195.0	1241.2	1288.0	1335.4	1383.6	1432.7	1482.6	1533.4	1585.1	1637.7	1745.7
(162.24)	s	1.8718	1.9370	1.9942	2.0456	2.0927	2.1361	2.1767	2.2148	2.2509	2.2851	2.3178	2.3792
	v	38.85	45.00	51.04	57.05	63.03	69.01	74.98	80.95	86.92	92.88	98.84	110.77
10	h	1146.6	1193.9	1240.6	1287.5	1335.1	1383.4	1432.5	1482.4	1533.2	1585.0	1637.6	1745.6
(193.21)	s	1.7927	1.8595	1.9172	1.9689	2.0160	2.0596	2.1002	2.1383	2.1744	2.2068	2.2413	2.3028
	v	....	30.53	34.68	38.78	42.86	46.94	51.00	55.07	59.13	63.19	67.25	75.37
14.696	h	....	1192.8	1239.9	1287.1	1334.8	1383.2	1432.3	1482.3	1533.1	1584.8	1637.5	1745.5
(212.00)	s	....	1.8160	1.8743	1.9261	1.9734	2.0170	2.0576	2.0958	2.1319	2.1662	2.1989	2.2603
	v	....	22.36	25.43	28.46	31.47	34.47	37.46	40.45	43.44	46.42	49.41	55.37
20	h	....	1191.6	1239.2	1286.6	1334.4	1382.9	1432.1	1482.1	1533.0	1584.7	1637.4	1745.4
(227.96)	s	....	1.7808	1.8396	1.8918	1.9392	1.9829	2.0235	2.0618	2.0978	2.1321	2.1648	2.2263
	v	....	11.040	12.628	14.168	15.688	17.198	18.702	20.20	21.70	23.20	24.69	27.68
40	h	....	1186.8	1236.5	1284.8	1333.1	1381.9	1431.3	1481.4	1532.4	1584.3	1637.0	1745.1
(267.25)	S	....	1.6994	1.7608	1.8140	1.8619	1.9058	1.9467	1.9850	2.0212	2.0555	2.0883	2.1498
	v	....	7.259	8.357	9.403	10.427	11.441	12.449	13.452	14.454	15.453	16.451	18.446
60	h	....	1181.6	1233.6	1283.0	1331.8	1380.9	1430.5	1480.8	1531.9	1583.8	1636.6	1744.8
(292.71)	s	....	1.6492	1.7135	1.7678	1.8162	1.8605	1.9015	1.9400	1.9762	2.0106	2.0434	2.1049
	v	....	....	6.220	7.020	7.797	8.562	9.322	10.077	10.830	11.582	12.332	13.830
80	h	....	....	1230.7	1281.1	1330.5	1379.9	1429.7	1480.1	1531.3	1583.4	1636.2	1744.5
(312.03)	s	....	....	1.6791	1.7346	1.7836	1.8281	1.8694	1.9079	1.9442	1.9787	2.0115	2.0731
	v	....	....	4.937	5.589	6.218	6.835	7.446	8.052	8.656	9.259	9.860	11.060
100	h	....	....	1227.6	1279.1	1329.1	1378.9	1428.9	1479.5	1530.8	1582.9	1635.7	1744.2
(327.81)	s	....	....	1.6518	1.7085	1.7581	1.8029	1.8443	1.8829	1.9193	1.9538	1.9867	2.0484
	v	....	....	4.081	4.636	5.165	5.683	5.195	6.702	7.207	7.710	8.212	9.214
120	h	....	....	1224.4	1277.2	1327.7	1377.8	1428.1	1478.8	1530.2	1582.4	1635.3	1743.9
(341.25)	s	....	....	1.6287	1.6869	1.7370	1.7822	1.8237	2.8625	1.8990	1.9335	1.9664	2.0281

	v	...	...	...	3.468	3.954	4.413	4.861	5.301	5.738	6.172	6.604	7.035	7.895
140	h	...	...	...	1221.1	1275.2	1326.4	1376.8	1427.3	1478.2	1529.7	1581.9	1634.9	1743.5
(353.02)	s	...	...	...	1.6087	1.6683	1.7190	1.7645	1.8063	1.8451	1.8817	1.9163	1.9493	2.0110
O	v	...	...	...	3.008	3.443	3.849	4.244	4.631	5.015	5.396	5.775	6.152	6.906
160	h	...	...	...	1217.6	1273.1	1325.0	1375.7	1426.4	1477.5	1529.1	1581.4	1634.5	1743.2
(363.53)	s	...	...	...	1.5908	1.6519	1.7033	1.7491	1.7911	1.8301	1.8667	1.9014	1.9344	1.9962
	v	...	...	...	2.649	3.044	3.411	3.764	4.110	4.452	4.792	5.129	5.466	6.136
180	h	...	...	...	1214.0	1271.0	1323.5	1374.7	1425.6	1476.8	1528.6	1581.0	1634.1	1742.9
(373.06)	s	...	...	...	1.5745	1.6373	1.6894	1.7355	1.7776	1.8167	1.8534	1.8882	1.9212	1.9831
	v	...	...	...	2.361	2.726	3.060	3.380	3.693	4.002	4.309	4.613	4.917	5.521
200	h	...	...	...	1210.3	1268.9	1322.1	1373.6	1424.8	1476.2	1528.0	1580.5	1633.7	1742.6
(381.79)	s	...	...	...	1.5594	1.6240	1.6767	1.7232	1.7655	1.8048	1.8415	1.8763	1.9094	1.9713
	v	...	...	...	2.125	2.465	2.772	3.066	3.352	3.634	3.913	4.191	4.467	5.017
220	h	...	...	...	1206.5	1266.7	1320.7	1372.6	1424.0	1475.5	1527.5	1580.0	1633.3	1742.3
(389-86)	s	...	...	...	1.5453	1.6117	1.6652	1.7120	1.7545	1.7939	1.8308	1.8656	1.8987	1.9607
	v	...	...	...	1.9276	2.247	2.533	2.804	3.068	3.327	3.584	3.839	4.093	4.597
240	h	...	...	...	1202.5	1264.5	1319.2	1371.5	1423.2	1474.8	1526.9	1579.6	1632.9	1742.0
(397.37)	s	...	...	...	1.5319	1.6003	1.6546	1.7017	1.7444	1.7839	1.8209	1.8558	1.8889	1.9510
	v	...	...	...	...	2.063	2.330	2.582	2.827	3.067	3.305	3.541	3.776	4.242
260	h	...	...	...	...	1262.3	1317.7	1370.4	1422.3	1474.2	1526.3	1579.1	1632.5	1741.7
(404.42)	s	...	...	...	...	1.5897	1.6447	1.6922	1.7352	1.7748	1.8118	1.8467	1.8799	1.9420
	v	...	...	...	...	1.9047	2.156	2.392	2.621	2.845	3.066	3.286	3.504	3.938
280	h	...	...	...	...	1260.0	1316.2	1369.4	1421.5	1473.5	1525.8	1578.6	1632.1	1741.4
(411.05)	s	...	...	...	...	1.5796	1.6354	1.6834	1.7265	1.7662	1.8033	1.8383	1.8716	1.9337
	v	...	...	...	...	1.7675	2.005	2.227	2.442	2.652	2.859	3.065	3.269	3.674
300	h	...	...	...	...	1260.0	1316.2	1368.3	1420.6	1472.8	1525.2	1578.1	1631.7	1741.0
(417.33)	s	...	...	...	...	1.5701	1.6268	1.6751	1.7184	1.7582	1.7954	1.8305	1.8638	1.9260
	v	...	...	...	...	1.4923	1.7036	1.8980	2.084	2.266	2.445	2.622	2.798	3.147
350	h	...	...	...	...	1251.5	1310.9	1365.5	1418.5	1471.1	1523.8	1577.0	1630.7	1740.3
(431.72)	s	...	...	...	...	1.5481	1.6070	1.6563	1.7002	1.7403	1.7777	1.8130	1.8463	1.9086
	v	...	...	...	...	1.2851	1.4770	1.6508	1.8161	1.9767	2.134	2.290	2.445	2.751
400	h	...	...	...	...	1245.1	1306.9	1362.7	1416.4	1469.4	1522.4	1575.8	1629.6	1739.5
(444.59)	s	...	...	...	...	1.5281	1.5894	1.6398	1.6842	1.7247	1.7623	1.7977	1.8311	1.8936

* Abridged from Joseph H. Keenan and Frederick G. Keyes, *Thermodynamic Properties of Steam,* John Wiley and Sons, Inc., New York.

TABLE D-2
Properties of Superheated Steam *(continued)*

Abs. pres., psia (Sat. temp., °F.)		Temperature, °F												
		500	550	600	620	640	660	680	700	800	900	1000	1200	1400
	v	1.1231	1.2155	1.3005	1.3332	1.3652	1.3967	1.4278	1.4584	1.6074	1.7516	1.8928	2.170	2.443
450	h	1238.4	1272.0	1302.8	1314.6	1326.2	1337.5	1348.8	1359.9	1414.3	1467.7	1521.0	1628.6	1738.7
(456.28)	s	1.5095	1.5437	1.5735	1.5845	1.5951	1.6054	1.6153	1.6250	1.6699	1.7108	1.7486	1.8177	1.8803
	v	0.9927	1.0800	1.1591	1.1893	1.2188	1.2478	1.2763	1.3044	1.4405	1.5715	1.6996	1.9504	2.197
500	h	1231.3	1266.8	1298.6	1310.7	1322.6	1334.2	1345.7	1357.0	1412.1	1466.0	1519.6	1627.6	1737.9
(467.01)	s	1.4919	1.5280	1.5588	1.5701	1.5810	1.5915	1.6016	1.6115	1.6571	1.6982	1.7363	1.8056	1.8683
	v	0.8852	0.9686	1.0431	1.0714	1.0989	1.1259	1.1523	1.1783	1.3038	1.4241	1.5414	1.7706	1.9957
550	h	1223.7	1261.2	1294.3	1306.8	1318.9	1330.8	1342.5	1354.0	1409.9	1464.3	1518.2	1626.6	1737.1
(476-94)	s	1.4751	1.5131	1.5451	1.5568	1.5680	1.5787	1.5890	1.5991	1.6452	1.6868	1.7250	1.7946	1.8575
	v	0.7947	0.8753	0.9463	0.9729	0.9988	1.0241	1.0489	1.0732	1.1899	1.3013	1.4096	1.6208	1.8279
600	h	1215.7	1255.5	1289.9	1302.7	1315.2	1327.4	1339.3	1351.1	1407.7	1462.5	1516.7	1625.5	1736.3
(486.21)	s	1.4586	1.4990	1.5323	1.5443	1.5558	1.5667	1.5773	1.5875	1.6343	1.6762	1.7147	1.7846	1.8476
	v	....	0.7277	0.7934	0.8177	0.8411	0.8639	0.8860	0.9077	1.0108	1.1082	1.2024	1.3853	1.5641
700	h	....	1243.2	1280.6	1294.3	1307.7	1320.3	1332.8	1345.0	1403.2	1459.0	1513.9	1623.5	1734.8
(503.10)	s	....	1.4722	1.5084	1.5212	1.5333	1.5449	1.5559	1.5665	1.6147	1.6573	1.6963	1.7666	1.8299
	v	....	0.6154	0.6779	0.7006	0.7223	0.7433	0.7635	0.7833	0.8763	0.9633	1.0470	1.2088	1.3662
800	h	....	1229.8	1270.7	1285.4	1299.4	1312.9	1325.9	1338.6	1398.6	1455.3	1511.0	1621.4	1733.2
(518.23)	s	....	1.4467	1.4863	1.5000	1.5129	1.5250	1.5366	1.5476	1.5972	1.6407	1.6801	1.7510	1.8146
	v	....	0.5264	0.5873	0.6089	0.6294	0.6491	0.6680	0.6863	0.7716	0.8506	0.9262	1.0714	1.2124
900	h	....	1215.0	1260.1	1275.9	1290.9	1305.1	1318.8	1332.1	1393.9	1451.8	1508.1	1619.3	1731.6
(531.98)	s	....	1.4216	1.4653	1.4800	1.4938	1.5066	1.5187	1.5303	1.5814	1.6257	1.6656	1.7371	1.8009
	v	....	0.4533	0.5140	0.5350	0.5546	0.5733	0.5912	0.6084	0.6878	0.7604	0.8294	0.9615	1.0893
1000	h	....	1198.3	1248.8	1265.9	1281.9	1297.0	1311.4	1325.3	1389.2	1448.2	1505.1	1617.3	1730.0
(544.61)	s	....	1.3961	1.4450	1.4610	1.4757	1.4893	1.5021	1.5141	1.5670	1.6121	1.6525	1.7245	1.7886
	v	....	....	0.4532	0.4738	0.4929	0.5110	0.5281	0.5445	0.6191	0.6866	0.7503	0.8716	0.9885
1100	h	....	....	1236.7	1255.3	1272.4	1288.5	1303.7	1318.3	1384.3	1444.5	1502.2	1615.2	1728.4
(556.31)	s	....	....	1.4251	1.4425	1.4583	1.4728	1.4862	1.4989	1.5535	1.5995	1.6405	1.7130	1.7775

1200	v	...	...	0.4016	0.4222	0.4410	0.4586	0.4752	0.4909	0.5617	0.6250	0.6843	0.7967	0.9046
	h	...	...	1223.5	1243.9	1262.4	1279.6	1295.7	1311.0	1379.3	1440.7	1499.2	1613.1	1726.9
(567.22)	s	...	...	1.4052	1.4243	1.4413	1.4568	1.4710	1.4843	1.5409	1.5879	1.6293	1.7025	1.7672
1400	v	...	...	0.3174	0.3390	0.3580	0.3753	0.3912	0.4062	0.4714	0.5281	0.5805	0.6789	0.7727
	h	...	...	1193.0	1218.4	1240.4	1260.3	1278.5	1295.5	1369.1	1433.1	1493.2	1608.9	1723.7
(587.10)	s	...	...	1.3639	1.3877	1.4079	1.4258	1.4419	1.4567	1.5177	1.5666	1.6093	1.6386	1.7489
1600	v	...	...	...	0.2733	0.2936	0.3112	0.3271	0.3417	0.4034	0.4553	0.5027	0.5906	0.6738
	h	...	...	...	1187.8	1215.2	1238.7	1259.6	1278.7	1358.4	1425.3	1487.0	1604.6	1720.5
(604.90)	s	...	...	...	1.3489	1.3741	1.3952	1.4137	1.4304	1.4964	1.5476	1.5914	1.6669	1.7328
1800	v	...	...	...	...	0.2407	0.2597	0.2760	0.2907	0.3502	0.3986	0.4421	0.5218	0.5968
	h	...	...	...	...	1185.1	1214.0	1238.5	1260.3	1347.2	1417.4	1480.8	1600.4	1717.3
(621.03)	s	...	...	...	...	1.3377	1.3638	1.3855	1.4044	1.4765	1.5301	1.5752	1.6520	1.7185
2000	v	...	...	...	...	0.1936	0.2161	0.2337	0.2489	0.3074	0.3532	0.3935	0.4668	0.5352
	h	...	...	...	...	1145.6	1184.9	1214.8	1240.0	1335.5	1409.2	1474.5	1596.1	1714.1
(635.82)	s	...	...	...	...	1.2945	1.3300	1.3564	1.3783	1.4576	1.5139	1.5603	1.6384	1.7055
2500	v	...	...	...	...	...	...	0.1484	0.1686	0.2294	0.2710	0.3061	0.3678	0.4244
	h	...	...	...	...	...	...	1132.3	1176.8	1303.6	1387.8	1458.4	1585.3	1706.1
(668.13)	s	...	...	...	...	...	...	1.2687	1.3073	1.4127	1.4772	1.5273	1.6088	1.6775
3000	v	...	...	...	...	...	...	...	0.0984	0.1760	0.2159	0.2476	0.3018	0.3505
	h	...	...	...	...	...	...	...	1060.7	1267.2	1365.0	1441.8	1574.3	1698.0
(695.36)	s	...	...	...	...	...	...	...	1.1966	1.3690	1.4439	1.4984	1.5837	1.6540
3206.2	v	...	...	...	...	...	...	...	...	0.1583	0.1981	0.2288	0.2806	0.3267
	h	...	...	...	...	...	...	...	...	1250.5	1355.2	1434.7	1569.8	1694.6
(705.40)	s	...	...	...	...	...	...	...	...	1.3508	1.4309	1.4874	1.5742	1.6452
3500	v	...	...	...	...	...	...	...	0.0306	0.1364	0.1762	0.2058	0.2546	0.2977
	h	...	...	...	...	...	...	...	780.5	1224.9	1340.7	1424.5	1563.3	1689.8
	s	...	...	...	...	...	...	...	0.9515	1.3241	1.4127	1.4723	1.5615	1.6336
4000	v	...	...	...	...	...	...	...	0.0287	0.1052	0.1462	0.1743	0.2192	0.2581
	h	...	...	...	...	...	...	...	763.8	1174.8	1314.4	1406.8	1552.1	1681.7
	s	...	...	...	...	...	...	...	0.9347	1.2757	1.3827	1.4482	1.5417	1.6154

TABLE D-3

Temperature, °R (Sat. press., psia)		Sat. liquid	Sat. vapor	Temperature of				
				800	900	1000	1100	1200
	v	1.7232×10^{-2}	$> 10^{10}$	...	...	...	...	...
700	h	219.7	2180.5	2203.1	2224.7	2246.4	2268.0	2289.5
(8.7472×10^{-9})	s	0.6854	3.4866	3.5169	3.5424	3.5652	3.5857	3.6043
	v	1.7548×10^{-2}	7.4375×10^{8}	...	8.3835×10^{8}	9.3168×10^{8}	1.0249×10^{9}	1.1180×10^{9}
800	h	252.3	2200.1	...	2224.4	2246.3	2267.9	2289.5
(5.0100×10^{-7})	s	0.7290	3.1637	...	3.1925	3.2155	3.2360	3.2546
	v	1.7864×10^{-2}	3.6411×10^{7}	...	...	4.0267×10^{7}	4.4718×10^{7}	4.8789×10^{7}
900	h	284.3	2217.6	...	...	2245.2	2267.7	2289.4
(1.1480×10^{-5})	s	0.7667	2.9148	...	...	2.9440	2.9653	2.9841
	v	1.8180×10^{-2}	3.323×10^{6}	...	...	...	3.6834×10^{6}	4.0254×5
1000	h	325.9	2232.7	...	...	...	2264.8	2288.6
(1.3909×10^{-4})	s	0.7999	2.7168	...	...	...	2.7474	2.7680
	v	1.8496×10^{-2}	4.7592×10^{5}	...	...	...	...	5.2512×10^{5}
1100	h	347.0	2245.1	...	...	...	...	2282.7
(1.0616×10^{-3})	s	0.8296	2.5551	...	...	...	...	2.5878
	v	1.8812×10^{-2}	9.5235×10^{4}	...	...	...	...	...
1200	h	377.7	2254.9	...	...	...	...	...
(5.7398×10^{-3})	s	0.8563	2.4207	...	...	...	...	...
	v	1.9128×10^{-2}	2.4520×10^{4}	...	...	...	...	...
1300	h	408.2	2262.8	...	...	...	...	...
(2.3916×10^{-2})	s	0.8807	2.3073	...	...	...	...	...
	v	1.9444×10^{-2}	7.6798×10^{3}	...	...	...	...	...
1400	h	438.4	2269.3	...	...	...	...	...
(8.1347×10^{-2})	s	0.9031	2.2109	...	...	...	...	...
	v	1.9760×10^{-2}	2.8334×10^{3}	...	...	...	...	...
1500	h	468.5	2274.9	...	...	...	...	...
(2.3351×10^{-1})	s	0.9239	2.1282	...	...	...	...	...
	v	2.0076×10^{-2}	1.1935×10^{3}	...	...	...	...	...
1600	h	498.5	2280.0	...	...	...	...	...
(5.8425×10^{-1})	s	0.9433	2.0567	...	...	...	...	...
	v	2.0392×10^{-2}	5.5585×10^{2}	...	...	...	...	...
1700	h	528.5	2285.3	...	...	...	...	...
(1.3170)	s	0.9615	1.9948	...	...	...	...	...
	v	2.0708×10^{-2}	2.8200×10^{2}	...	...	...	...	...
1800	h	558.6	2291.1	...	...	...	...	...
(2.7164)	s	0.9786	2.9411	...	...	...	...	...
	v	2.1024×10^{-2}	1.5512×10^{2}	...	...	...	...	...
1900	h	588.8	2297.2	...	...	...	...	...
(5.1529)	s	0.9949	1.8941	...	...	...	...	...
	v	2.1340×10^{-2}	90.914	...	...	...	...	...
2000	h	619.1	2304.1	...	...	...	...	...
(9.1533)	s	1.0105	1.8530	...	...	...	...	...
	v	2.1656×10^{-2}	56.185	...	...	...	...	...
2100	h	649.7	2312.1	...	...	...	...	...
(15.392)	s	1.0255	1.8171	...	...	...	...	...
	v	2.1972×10^{-2}	36.338	...	...	...	...	...
2200	h	680.7	2321.0	...	...	...	...	...
(24.692)	s	1.0399	1.7855	...	...	...	...	...
	v	2.2288×10^{-2}	24.454	...	...	...	...	...
2300	h	712.0	2330.7	...	...	...	...	...
(38.013)	s	1.0538	1.7576	...	...	...	...	...
	v	2.2604×10^{-2}	17.109	...	...	...	...	...
2400	h	743.8	2341.2	...	...	...	...	...
(56.212)	s	1.0673	1.7329	...	...	...	...	...
	v	2.2920×10^{-2}	12.388	...	...	...	...	...
2500	h	776.2	2352.6	...	...	...	...	...
(80.236)	s	1.0805	1.7111	...	...	...	...	...
	v	2.3236×10^{-2}	9.2328	...	...	...	...	...
2600	h	809.1	2365.1	...	...	...	...	...
(1.1116×10^{2})	s	1.0934	1.6919	...	...	...	...	...
	v	2.3552×10^{-2}	7.0380					
2700	h	842.7	2378.8					
(1.052×10^{2})	s	1.1061	1.6751					

Thermodynamic Properties of Sodium*

Superheated Vapor, °R

1400	1600	1800	2000	2200	2400	2600	2700
....	....	....	....	....	....	....	....
2332.7 3.6381	2375.9 3.6665	2419.1 3.6924	2462.3 3.7148	2505.4 3.7354	2548.6 3.7545	2591.8 3.7713	2613.4 3.7796
1.3044×10^{9} 2332.7 3.2884	1.4907×10^{9} 2375.9 3.3169	1.6771×10^{9} 2419.1 3.3428	1.8634×10^{9} 2462.3 3.3652	2.0498×10^{9} 2505.4 3.3858	2.2361×10^{9} 2548.6 3.4048	2.4224×10^{9} 2591.8 3.5217	2.5156×10^{9} 2613.4 3.4299
5.6924×10^{7} 2332.7 3.0179	6.5056×10^{7} 2375.9 3.0464	7.3188×10^{7} 2419.1 3.0723	8.1320×10^{7} 2462.3 3.0947	8.9452×10^{7} 2505.4 3.1153	9.7584×10^{7} 2548.6 3.1343	1.0572×10^{8} 2591.8 3.1511	1.0978×10^{8} 2613.4 3.1594
4.6978×10^{6} 2332.6 2.8024	5.3693×10^{6} 2375.9 2.8309	6.0406×10^{6} 2419.1 2.8568	6.7118×10^{6} 2462.3 2.8792	7.383×10^{6} 2505.4 2.8998	8.0541×10^{6} 2548.6 2.9188	8.7253×10^{6} 2591.8 2.9357	9.0609×10^{6} 2613.4 2.9439
6.1515×10^{5} 2331.7 2.6263	7.0339×10^{5} 2375.7 2.6552	7.9139×10^{5} 2419.0 2.6812	8.7935×10^{5} 2462.3 2.7036	9.6729×10^{5} 2505.4 2.7243	1.0552×10^{6} 2548.6 2.7433	1.1432×10^{6} 2591.8 2.7601	1.1871×10^{6} 2613.4 2.7684
1.1345×10^{5} 2327.6 2.4778	1.3001×10^{5} 2374.7 2.5089	1.4635×10^{5} 2418.7 2.5353	1.6263×10^{5} 2462.2 2.5578	1.7891×10^{5} 2505.4 2.5785	1.9517×10^{5} 2548.6 2.5975	2.1144×10^{5} 2591.8 2.6143	2.1957×10^{5} 2613.4 2.6226
2.6936×10^{4} 2312.2 2.3445	3.1124×10^{4} 2371.1 2.3836	3.5095×10^{4} 2417.6 2.4115	3.9019×10^{4} 2461.7 2.4343	4.2931×10^{4} 2505.1 2.4551	4.6838×10^{4} 2548.5 2.4742	5.0743×10^{4} 2591.8 2.4911	5.2695×10^{4} 2613.4 2.4993
....	9.0793×10^{3} 2359.9 2.2715	1.0292×10^{4} 2414.0 2.3040	1.1460×10^{4} 2460.3 2.3280	1.2616×10^{4} 2504.5 2.3491	1.3767×10^{4} 2548.1 2.3683	1.4916×10^{4} 2591.6 2.3853	1.5491×10^{4} 2613.2 2.3935
....	3.1025×10^{3} 2332.6 2.1651	3.5625×10^{3} 2404.7 2.2083	3.9820×10^{3} 2456.5 2.2352	4.3896×10^{3} 2502.7 2.2573	4.7929×10^{3} 2547.2 2.2769	5.1944×10^{3} 2591.0 2.2940	5.3948×10^{2} 2612.8 2.3023
....	....	1.4040×10^{3} 2384.7 2.1192	1.5823×10^{3} 2448.0 2.1523	1.7496×10^{3} 2498.7 2.1765	1.9128×10^{3} 2545.1 2.1969	2.0743×10^{3} 2589.7 2.2144	2.1548×10^{3} 2611.8 2.2228
....	....	6.0659×10^{2} 2347.7 2.0309	6.9378×10^{2} 2431.3 2.0747	7.7180×10^{2} 2490.5 2.1031	8.4601×10^{2} 2540.7 2.1252	9.1858×10^{2} 2587.1 2.1433	9.5458×10^{2} 2609.8 2.1519
....	....	....	3.2952×10^{2} 2402.3 1.9996	3.7033×10^{2} 2475.6 2.0347	4.0785×10^{2} 2532.5 2.0597	4.4385×10^{2} 2582.2 2.0792	4.6158×10^{2} 2605.9 2.0882
....	....	....	1.6838×10^{2} 2359.5 1.9259	1.9197×10^{2} 2451.8 1.9701	2.1298×10^{2} 2518.8 1.9996	2.3265×10^{2} 2573.9 2.0212	2.4224×10^{2} 2599.3 2.0209
....	....	....	....	1.0543×10^{2} 2417.4 1.9072	1.1816×10^{2} 2498.0 1.9426	1.2980×10^{2} 2560.9 1.9673	1.3539×10^{2} 2588.9 1.9780
....	....	....	....	60.665 2372.9 1.8455	68.825 2469.0 1.8876	76.167 2451.9 1.9164	79.656 2573.5 1.9284
....	....	....	....	....	41.754 2431.8 1.8340	46.622 2516.2 1.8674	48.920 2552.3 1.8811
....	....	....	....	....	26.244 2388.2 1.7820	29.585 2484.0 1.8201	31.163 2525.0 1.8356
....	....	....	....	....	....	19.460 2446.8 1.7748	20.580 2492.6 1.7922
....	....	....	....	....	....	13.219 2406.5 1.7321	14.032 2456.5 1.7501
....	....	....	....	....	....	....	9.8326 2418.1 1.7120

* From Ref. 154.

appendix E

Some Physical Properties

TABLE E-1
Physical Properties of Ordinary Liquid Water*

Temp., °F	Specific heat c_p, Btu/lb$_m$ °R			Thermal conductivity k Btu/hr ft °F			Viscosity μ, lb$_m$/hr ft			Density ρ, lb$_m$/ft^3		
	Sat. liquid	1,000 psia	2,000 psia	Sat. liquid	1,000 psia	2,000 psia	Sat. liquid	1,000 psia	2,000 psia	Sat. liquid	1,000 psia	2,000 psia
32	1.0083	1.0032	1.0004	0.3185	0.3198	0.3211	4.340	4.309	4.279	62.422	62.637	63.846
40	1.0048	1.0014	0.9986	0.3245	0.3260	0.3275	3.742	3.721	3.699	62.422	62.657	62.854
60	0.9990	0.9968	0.9939	0.3397	0.3414	0.3433	2.731	2.722	2.714	62.344	62.539	62.755
80	0.9975	0.9943	0.9912	0.3532	0.3537	0.3570	2.084	2.084	2.083	62.189	62.383	62.586
100	0.9976	0.9932	0.9897	0.3641	0.3659	0.3680	1.650	1.654	1.658	61.996	62.185	62.371
120	0.9977	0.9934	0.9895	0.3733	0.3751	0.3771	1.353	1.360	1.366	61.728	61.920	62.104
140	0.9988	0.9940	0.9897	0.3810	0.3828	0.3847	1.137	1.145	1.154	61.387	61.576	61.767
160	1.0004	0.9959	0.9913	0.3861	0.3880	0.3902	0.970	0.979	0.988	61.013	61.200	61.395
180	1.0022	0.9980	0.9931	0.3905	0.3924	0.3945	0.839	0.849	0.858	60.569	60.753	60.953
200	1.0047	1.0008	0.9958	0.3935	0.3957	0.3980	0.738	0.748	0.757	60.132	60.314	60.511
210	1.0064	1.0024	0.9974	0.3944	0.3972	0.3998	0.687	0.697	0.706	59.809	60.006	60.205
220	1.0079	1.0039	0.9988	0.3950	0.3977	0.4003	0.660	0.670	0.680	59.630	59.830	60.031
240	1.0119	1.0075	1.0023	0.3961	0.3988	0.4016	0.595	0.604	0.614	59.102	59.305	59.506
260	1.0165	1.0117	1.0061	0.3964	0.3992	0.4021	0.542	0.551	0.560	58.514	58.727	58.938
280	1.0222	1.0163	1.0102	0.3959	0.3987	0.4018	0.494	0.502	0.511	57.937	58.156	58.377

300	1.0289	1.0232	1.0166	0.3952	0.3981	0.4013	0.452	0.460	0.468	57.307	57.537	57.767
320	1.0354	1.0307	1.0235	0.3944	0.3969	0.3998	0.420	0.426	0.433	56.657	56.883	57.136
340	1.0455	1.0999	1.0322	0.3921	0.3947	0.3977	0.391	0.396	0.404	55.960	56.211	56.465
360	1.0564	1.0496	1.0411	0.3891	0.3919	0.3951	0.366	0.372	0.378	55.218	55.463	55.710
380	1.0669	1.0611	1.0510	0.3857	0.3885	0.3919	0.346	0.351	0.356	54.466	54.720	55.012
400	1.0794	1.074	1.062	0.3809	0.3840	0.3880	0.327	0.330	0.335	53.648	53.903	54.218
420	1.0941	1.087	1.075	0.3753	0.3787	0.3833	0.310	0.312	0.317	52.798	53.042	53.396
440	1.1114	1.105	1.091	0.3693	0.3728	0.3776	0.294	0.296	0.301	51.921	52.154	52.546
460	1.1319	1.124	1.109	0.3640	0.3664	0.3713	0.280	0.282	0.286	51.020	51.230	51.661
480	1.1345	1.149	1.131	0.3575	0.3595	0.3642	0.267	0.270	0.273	50.000	50.191	50.659
500	1.1861	1.176	1.154	0.3494	0.3510	0.3562	0.256	0.257	0.260	49.020	49.097	49.618
520	1.23	1.21	1.188	0.3397	0.3410	0.3475	0.246	0.246	0.249	47.847	....	48.527
540	1.28	....	1.225	0.3298	....	0.3371	0.235	0.235	0.239	46.512	....	47.181
560	1.34	....	1.278	0.3189	....	0.3256	0.225	....	0.231	45.249	....	45.905
580	1.41	....	1.341	0.3064	....	0.3118	0.217	....	0.222	43.860	....	44.492
600	1.51	....	1.448	0.2919	....	0.2962	0.210	....	0.212	43.373	....	42.913
620	1.65	....	1.62	0.2753	....	0.2778	0.200	....	0.202	40.486	....	40.950
640	1.88	....	....	0.2565	....	....	0.190	....	....	38.452		
660	2.34	....	....	0.2335	....	....	0.177	....	....	35.971		
680	3.5	....	....	0.2056	....	....	0.161	....	....	32.787		
690	5.5	....	....	0.1854	....	....	0.48	....	....	30.488		

* c_p, μ, and k data from Ref. 145. ρ data computed from Keenan and Keyes (Table D-1).

TABLE E-2
Physical Properties of Helium* (at 10 atm pressure)

T, °F	Density ρ, lb_m/ft^3	Viscosity μ, $lb_m/ft\ hr$	Specific heat † c_p $Btu/lb_m\ °F$	Thermal conductivity ‡ k, Btu/ft hr °F	Prandtl no. Pr, $c_p\mu/k$
32	0.1117	0.0457	1.248	0.083	0.687
100	0.0974	0.0495	1.248	0.090	0.687
200	0.0827	0.0555	1.248	0.100	0.687
300	0.0718	0.0605	1.248	0.110	0.686
400	0.0635	0.0653	1.248	0.119	0.684
500	0.0569	0.0700	1.248	0.128	0.682
600	0.0516	0.0743	1.248	0.136	0.679
700	0.0475	0.0780	1.248	0.145	0.675
800	0.0430	0.0821	1.248	0.153	0.671
900	0.0399	0.0859	1.248	0.160	0.667
1000	0.0373	0.0889	1.248	0.167	0.662
1100	0.0351	0.0918	1.248	0.175	0.656

* From Ref. 145.
† Extrapolated.
‡ Atmospheric pressure.

TABLE E-3
Physical Properties of Carbon Dioxide*

T, °F	Density† ρ, lb_m/ft^3	Viscosity‡ μ, $lb_m/ft\ hr$	Specific heat† c_p, $Btu/lb_m\ °F$	Thermal conductivity‡ k, Btu/ft hr °F	Prandtl no‡ Pr, $c_p\mu/k$
32	1.3190	0.03318	0.2187	0.008415	0.782
100	1.1277	0.03739	0.2202	0.009962	0.768
200	0.9373	0.04332	0.2262	0.01261	0.749
300	0.8051	0.04892	0.2342	0.01533	0.729
400	0.7071	0.05419	0.2423	0.01818	0.710
500	0.6310	0.05920	0.2503	0.02117	0.691
600	0.5701	0.06397	0.2476	0.02425	0.672
700	0.5202	0.06851	0.2643		
800	0.4784	0.07288	0.2704		
900	0.4428	0.07709	0.2760		
1000	0.4125	0.0811	0.2812		
1100	0.3850	0.08511	0.2858		
1200	0.3626	0.8891	0.2901		

* From Ref. 155.
† At 10 atm pressure.
‡ At atmospheric pressure.

appendix **F**

Moody Friction Factor Chart

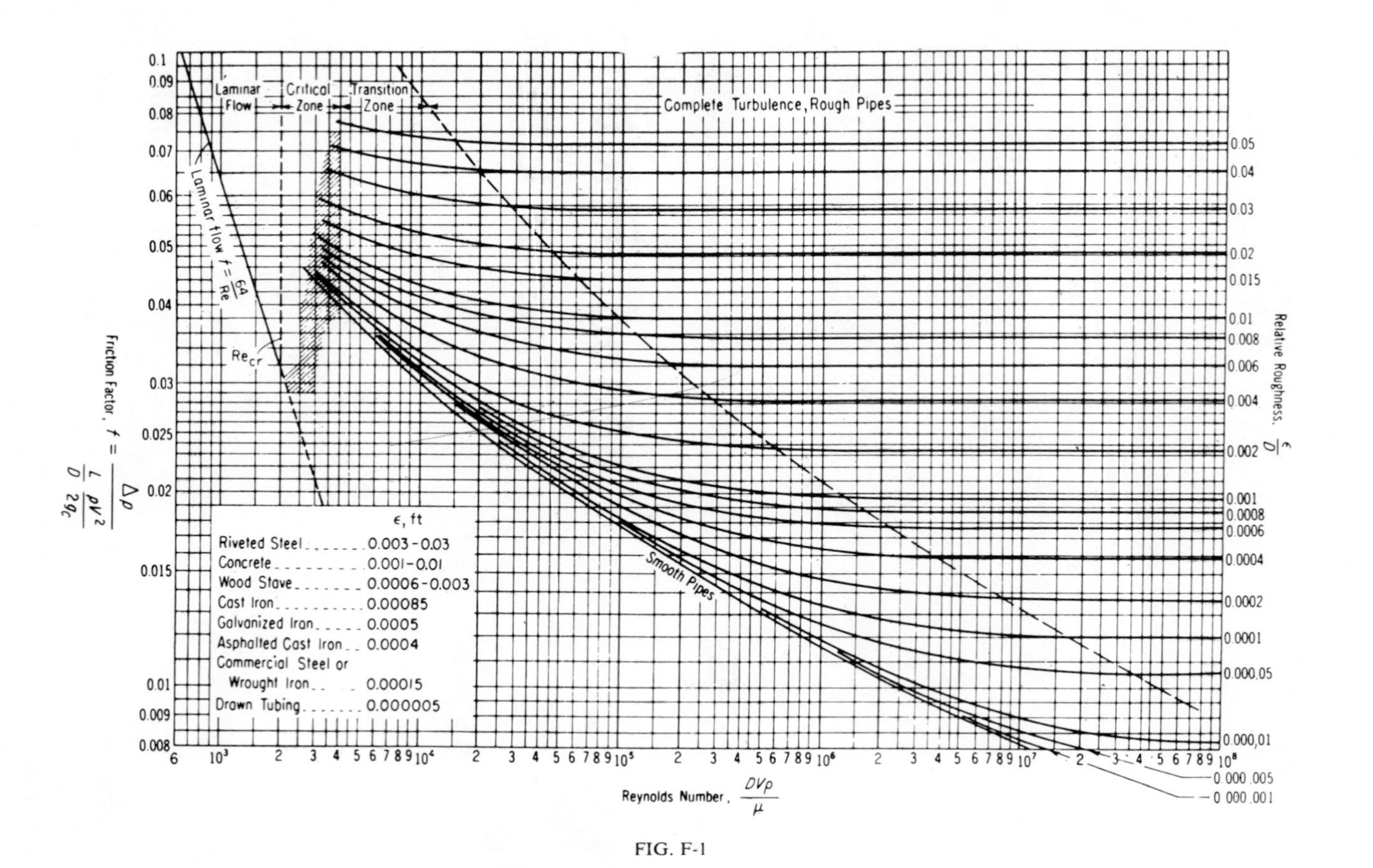
Laminar Flow
Critical Zone
Transition Zone
Complete Turbulence, Rough Pipes
Laminar flow f = 64/Re
Re cr
Smooth Pipes
Relative Roughness, ε/D
ε, ft
Riveted Steel 0.003-0.03
Concrete 0.001-0.01
Wood Stave 0.0006-0.003
Cast Iron 0.00085
Galvanized Iron 0.0005
Asphalted Cast Iron 0.0004
Commercial Steel or Wrought Iron 0.00015
Drawn Tubing 0.000005
Reynolds Number, DVρ/μ
Friction Factor, f = Δp / (L/D · ρV²/2g c)

FIG. F-1

appendix G

Some Useful Constants

Avogadro's number, Av	0.602252×10^{24} molecules/g_m mole 2.731769×10^{26} molecules/lb_m mole
Barn	10^{-24} cm^2, 1.0765×10^{-27} ft^2
Boltzmann's constant, $k = \bar{R}/Av$ ..	1.38054×10^{-16} erg/°K, 8.61747×10^{-5} ev/°K
Curie	3.70×10^{10} dis/sec
Electron charge	4.80298×10^{-10} esu, 1.60210×10^{-20} emu
g_c conversion factor	1.0 g_m cm^2/erg sec^2, 32.17 lb_m ft/lb_fsec^2 4.17×10^8 lb_m ft/lb_fhr^2, 0.9648×10^{18} amu cm^2/Mev sec^2
Gravitational acceleration .... (standard)	32.1739 ft/sec^2, 980.665 cm/sec^2
Joule's equivalent	778.16 ft-lb_f/Btu
Mass-energy conversion	1 amu = 931.478 Mev = 1.41492×10^{-13} Btu = 4.1471×10^{-17} kwhr lg_m = 5.60984×10^{26} Mev = 2.49760×10^{7}kwhr = 1.04067 Mwd 1 lb_m = 2.54458×10^{32} Mev = 3.86524×10^{16} Btu
Molecular volume	2,2413.6 cm^3/g_m mole, 359.0371 ft^3/lb_m mole, at 1 atm and 0°C
Neutron energy	0.0252977 ev at 2,200 m/sec, $\frac{1}{40}$ ev at 2,187.017 m/sec
Planck's constant, h	6.6256×10^{-27} erg sec, 4.13576×10^{-15} ev sec
Rest masses:	
Electron, m_e	5.48597×10^{-4} amu, 9.10909×10^{-28} g_m, 2.00819×10^{-30} lb_m
Neutron, m_n	1.0086654 amu, $1.6748228 \times 10^{-24}$ g_m 3.692314×10^{-27} lb_m
Proton, m_p	1.0072766 amu, 1.672499×10^{-24} g_m 3.687192×10^{-27} lb_m
Universal gas constant, $\bar{R}$	1545.08 ft-lb_f/lb_m mole °R, 1.98545 cal/g_m mole °K, 1.98545 Btu/lb_m mole °R, 8.31434×10^7 erg/g_m mole °K
Velocity of light, c	2.997925×10^{10} cm/sec, 9.83619×10^8 ft/sec

appendix **H**

Some Conversion Factors

TABLE H-1
Length

Centimeters, cm	Meters, m	Inches in.	Feet, ft	Miles	Microns, μ	Angstroms, A	Light-year
1	0.01	0.3937	0.03281	6.214×10^{-6}	10^{4}	10^{8}	1.0564×10^{-18}
100	1	39.37'	3.281	6.214×10^{-4}	10^{6}	10^{10}	1.0564×10^{-16}
2.540	0.0254	1	0.08333	1.578×10^{-5}	2.54×10^{4}	2.54×10^{8}	2.683×10^{-18}
30.48	0.3048	12	1	1.894×10^{-4}	0.3048×10^{6}	0.3048×10^{10}	3.2199×10^{-17}
1.6003×10^{5}	1609.3	6.336×10^{4}	5280	1	1.6093×10^{9}	1.6093×10^{13}	1.7001×10^{-13}
10^{-4}	10^{-6}	3.937×10^{-5}	3.281×10^{-6}	6.2139×10^{-10}	1	10^{4}	1.0564×10^{-22}
10^{-8}	10^{-10}	3.937×10^{-9}	3.281×10^{-10}	6.2139×10^{-14}	10^{-4}	1	1.0564×10^{-26}
9.4663×10^{17}	9.4663×10^{15}	3.727×10^{17}	3.1068×10^{16}	5.8822×10^{12}	9.4663×10^{21}	9.4663×10^{25}	1

TABLE H-2
Area

cm²	m²	in²	ft²	mile²	acre	barn	millibarn
1	10^{-4}	0.155	1.0764×10^{-3}	3.861×10^{-11}	2.4711×10^{-8}	10^{24}	10^{27}
10^{4}	1	1550	10.764	3.861×10^{-7}	2.4711×10^{-4}	10^{28}	10^{31}
6.4516	6.4516×10^{-4}	1	6.944×10^{-3}	2.491×10^{-10}	1.5944×10^{-7}	64517×10^{24}	6.4517×10^{27}
929	0.0929	144	1	3.587×10^{-8}	2.2957×10^{-5}	9.29×10^{26}	9.2903×10^{29}
2.59×10^{10}	2.59×10^{6}	4.0144×10^{11}	2.7878×10^{7}	1	640	2.59×10^{34}	2.59×10^{37}
4.0469×10^{7}	4.0469×10^{3}	6.2726×10^{6}	4.356×10^{4}	1.5625×10^{-3}	1	4.0469×10^{31}	4.0469×10^{34}
10^{-24}	10^{-28}	1.55×10^{-25}	1.0764×10^{-27}	3.861×10^{-35}	2.4711×10^{-32}	1	10^{3}
10^{-27}	10^{-31}	1.55×10^{-28}	1.0764×10^{-30}	3.861×10^{-38}	2.4711×10^{-35}	10^{-3}	1

TABLE H-3
Volume

cm³	liters	m³	in³	ft³	cubic yards	U.S. (liq.) gallons	Imperial gallons	acre-feet
1	10^{-3}	10^{-6}	0.06102	3.532×10^{-5}	1.308×10^{-6}	2.642×10^{-4}	2.20×10^{-4}	8.105×10^{-10}
10^{3}	1	10^{-3}	61.02	0.03532	1.308×10^{-3}	0.2642	0.220	8.105×10^{-7}
10^{6}	10^{3}	1	6.102×10^{4}	35.31	1.308	264.2	220.0	8.105×10^{-4}
16.39	0.01639	1.639×10^{-5}	1	5.787×10^{-4}	2.143×10^{-5}	4.329×10^{-3}	3.605×10^{-3}	1.328×10^{-8}
2.832×10^{4}	28.32	0.02832	1728	1	0.03704	7.481	6.229	2.295×10^{-5}
7.646×10^{5}	764.6	0.7646	4.666×10^{4}	27.0	1	202.0	168.2	6.196×10^{-4}
3785	3.785	3.785×10^{-3}	231.0	0.1337	4.951×10^{-3}	1	0.8327	3.068×10^{-6}
4546	4.546	4.546×10^{-3}	277.4	0.1605	5.946×10^{-3}	1.201	1	3.684×10^{-6}
1.234×10^{9}	1.234×10^{6}	1.234×10^{3}	7.529×10^{7}	4.357×10^{4}	1.614×10^{3}	3.259×10^{5}	2.714×10^{5}	1

TABLE H-4
Mass

grams, g_m	kilograms, kg_m	pounds, lb_m	tons (short)	tons (long)	tons (metric)	atomic mass units, amu
1	0.001	2.2046×10^{-3}	11.102×10^{-6}	9.842×10^{-7}	10^{-6}	0.60225×10^{24}
1,000	1	2.2046	0.001102	9.842×10^{-4}	10^{-3}	60.0225×10^{26}
453.6	0.4536	1	5.0×10^{-4}	4.464×10^{-4}	4.536×10^{-4}	2.7318×10^{26}
9.072×10^{5}	907.2	2,000	1	0.8929	0.9072	5.4636×10^{29}
1.016×10^{6}	1,016	2,240	1.12	1	1.016	6.1192×10^{29}
10^{6}	1,000	2,204.7	1.1023	0.9843	1	6.0225×10^{29}
1.6604×10^{-24}	1.6604×10^{-27}	3.6606×10^{-27}	1.8303×10^{-30}	1.6343×10^{-30}	1.6604×10^{-30}	1

TABLE H-5
Density

g_m/cm^3	kg_m/m^3	$lb_m/in.^3$	lb_m/ft^3	lb_m/U.S. gal	lb_m/Imp. gal
1	10^3	0.03613	62.43	8.345	10.02
10^{-3}	1	3.613×10^{-5}	0.06243	8.345×10^{-3}	0.01002
27.68	2.768×10^4	1	1728	231	277.4
0.01602	16.02	5.787×10^{-4}	1	0.1337	0.1605
0.1198	119.8	4.329×10^{-3}	7.481	1	1.201
0.09978	99.78	4.605×10^{-3}	6.229	0.8327	1

TABLE H-6
Time

microseconds μsec	seconds, sec	minutes, min	hours, hr	days	years, yr
1	10^{-6}	1.667×10^{-8}	2.778×10^{-10}	1.157×10^{-11}	3.169×10^{-14}
10^6	1	1.667×10^{-2}	2.778×10^{-4}	1.157×10^{-5}	3.169×10^{-8}
6×10^7	60	1	1.667×10^{-2}	6.944×10^{-4}	1.901×10^{-6}
3.6×10^9	3,600	60	1	0.04167	1.141×10^{-24}
8.64×10^{10}	8.64×10^4	1,440	24	1	2.737×10^{-3}
3.1557×10^{13}	3.1557×10^7	5.259×10^5	8,766	365.24	1

TABLE H-7
Flow

cm^3/sec	ft^3/min	U.S.gal/min	Imperial gal/min
1	0.002119	0.01585	0.01320
472.0	1	7.481	6.229
63.09	0.1337	1	0.8327
75.77	00.1605	1.201	1

TABLE H-8
Pressure

kg_f/cm^2	$lb_f/in.^2$	lb_f/ft^2	cm Hg (0°C)	in. Hg (32°F)	in. H_2O (60°F)	atm
1	14.22	2,048	73.56	28.96	394.1	0.9678
0.07031	1	144	5.171	2.036	27.71	0.06805
4.882×10^{-4}	0.006944	1	0.03591	0.01414	0.1924	4.725×10^{-6}
0.01360	0.1934	27.85	1	0.3937	5.358	0.01316
0.03453	0.4912	70.73	2.540	1	13.61	0.03342
0.002538	0.03609	7.197	0.1866	0.07348	1	0.002456
1.033	14.70	2,116	76.0	29.92	407.2	1

TABLE H-9
Energy

Ergs $g_m cm^2/sec^2$	Joules, watt sec	kwhr	g_m-cal	ft-lb_f	hp-hr	Btu	ev	Mev
1	10^{-7}	2.778×10^{-14}	2.388×10^{-8}	7.376×10^{-8}	3.725×10^{-14}	9.478×10^{-11}	6.2421×10^{11}	6.242×10^{5}
10^{7}	1	2.778×10^{-7}	0.2388	0.7376	3.725×10^{-7}	9.478×10^{-4}	6.2421×10^{18}	6.242×10^{12}
3.6×10^{13}	3.6×10^{6}	1	8.598×10^{5}	2.655×10^{6}	1.341	3412	2.25×10^{25}	2.25×10^{19}
4.187×10^{7}	4.187	1.163×10^{-6}	1	3.088	1.56×10^{-6}	3.968×10^{-3}	2.616×10^{19}	2.613×10^{13}
1.356×10^{7}	1.356	3.766×10^{-7}	0.3238	1	5.051×10^{-7}	1.285×10^{-3}	8.462×10^{18}	8.462×10^{12}
2.685×10^{13}	2.685×10^{6}	0.7457	6.412×10^{5}	1.98×10^{6}	1	2545	1.677×10^{25}	1.677×10^{19}
1.055×10^{10}	1.055	2.931×10^{-4}	252	778.2	3.93×10^{-4}	1	6.584×10^{21}	5.584×10^{15}
1.0021×10^{-12}	1.6021×10^{-19}	4.44×10^{-26}	3.826×10^{-20}	1.178×10^{-19}	5.95×10^{-26}	1.519×10^{-22}	1	10^{-6}
1.6021×10^{-6}	1.6021×10^{-13}	4.44×10^{-20}	3.826×10^{-14}	1.178×10^{-13}	5.95×10^{-20}	1.519×10^{-16}	10^{6}	1

TABLE H-10
Power

Ergs/sec	Joule/sec watt	kw	Btu/hr	hp	ev/sec
1	10^{-7}	10^{-10}	3.412×10^{-7}	1.341×10^{-10}	6.241×10^{11}
10^{7}	1	10^{-3}	3.412	0.001341	6.2421×10^{18}
10^{10}	10^{3}	1	3412	1.341	6.2421×10^{21}
2.931×10^{6}	0.2931	2.931×10^{-4}	1	3.93×10^{-4}	1.8294×10^{18}
7.457×10^{9}	745.7	0.7457	2545	1	4.6548×10^{21}
$1.602! \times 10^{-12}$	1.6021×10^{-19}	1.6021×10^{-22}	5.4664×10^{-19}	2.1483×10^{-22}	1

TABLE H-11
Power Density, Volumetric Thermal Source Strength

watt/cm³, kw/lit	cal/sec cm³	Btu/hr in³	Btu/hr ft³	Mev/sec cm³
1	0.2388	55.91	9.662×10^{4}	6.2420×10^{12}
4.187	1	234.1	4.045×10^{5}	2.613×10^{13}
0.01788	4.272×10^{-3}	1	1728	1.1164×10^{11}
1.035×10^{-5}	2.472×10^{-6}	5.787×10^{-4}	1	6.4610×10^{7}
1.602×10^{-13}	3.826×10^{-14}	8.9568×10^{-12}	1.5477×10^{-8}	1

TABLE H-12
Heat Flux

watt/cm²	cal/sec cm²	Btu/hr ft²	Mev/sec cm²
1	0.2388	3170.2	6.2420×10^{12}
4.187	1	1.3272×10^{4}	2.6134×10^{13}
3.155×10^{-4}	7.535×10^{-5}	1	1.9691×10^{9}
1.602×10^{-13}	3.826×10^{-14}	5.0785×10^{-10}	1

TABLE H-13
Thermal Conductivity

watt/cm °C	cal/sec cm °C	Btu/hr ft °F	Btu in./hr ft² °F	Mev/sec cm °C
1	0.2388	57.78	693.3	6.2420×10^{12}
4.187	1	241.9	2903	2.6134×10^{13}
0.01731	4.134×10^{-3}	1	12	1.0805×10^{11}
1.441×10^{-3}	3.445×10^{-4}	0.08333	1	9.004×10^{9}
1.602×10^{-13}	3.8264×10^{-14}	9.2551×10^{-12}	1.111×10^{-10}	1

TABLE H-14
Viscosity

Centipoise	Poise	kg_m/sec m	lb_m/sec ft	lb_m/hr ft	lb_fsec/ft²
1	0.01	0.001	6.720×10^{-4}	2.419	2.089×10^{-5}
100	1	0.1	0.06720	241.9	2.089×10^{-3}
1,000	10	1	0.6720	2,419	0.02089
1,488	14.88	1.488	1	3,600	0.03108
0.4134	4.134×10^{-3}	4.134×10^{-4}	2.778×10^{-4}	1	8.634×10^{-6}
4.788×10^{4}	478.8	47.88	32.17	1.158×10^{5}	1

REFERENCES

1. El-Wakil, M. M. *Nuclear Power Engineering,* New York: McGraw-Hill, 1962.
2. El-Wakil, M. M. *Nuclear Energy Conversion,* Scranton, Pa.; International Textbook, 1971.
3. Kaplan, I. *Nuclear Physics,* 2d. ed., Reading, Mass.: Addison-Wesley, 1962.
4. Meyerhof, W. E. *Elements of Nuclear Physics,* New York: McGraw-Hill, 1967.
5. Beckurtz, K. H., and K. Wirtz *Neutron Physics,* Berlin: Springer-Verlag, 1964.
6. Post, R. F. "Controlled Fusion Research; An Application of the Physics of High Temperature Plasmas," *Revs. Modern Phys.,* Vol. 28, no. 3, July, 1956; pp. 338-362. See also R. F. Post, "Fusion Power," *Sci. American,* Vol. 197, No. 6, December, 1957, pp. 73-88.
7. Katcoff, S. "Fission-product Yields from U, Th, and Pu," *Nucleonics,* Vol. 16, No. 4, April, 1958, p. 78.
8. Ferziger, J. H., and P. F. Zweifel *The Theory of Neutron Slowing Down in Nuclear Reactors,* Cambridge, Mass: The M.I.T. Press, 1966.
9. Williams, M.M.R. *The Slowing Down and Thermalization of Neutrons,* North Holland Publishing Co., 1966.
10. Lamarsh, J. R. *Introduction to Nuclear Reactor Theory,* Reading, Mass: Addison-Wesley, 1966.
11. Galanin, A. D. *Thermal Reactor Theory,* Translated from the Russian by J. B. Sykes, Elmsford, N.Y.: Pergamon Press, 1960.
12. Murray, R. L. *Nuclear Reactor Physics,* Englewood Cliffs, N.J.: Prentice-Hall, 1957.
13. Glasstone, S., and M. C. Edlund. *The Elements of Nuclear Reactor Theory,* Princeton, N.J.: Van Nostrand 1952.
14. Weinberg, A. M. and E. P. Wigner. *The Physical Theory of Neutron Chain Reactors,* Chicago, Ill.: University of Chicago Press, 1958.
15. Glasstone, S., and Sesonske. *Principles of Nuclear Reactor Engineering,* Princeton, N.J.: Van Nostrand, 1955.
16. Watt, B. E. "Energy Spectrum of Neutrons from Thermal Fission of U^{235}," *Phys. Rev.,* Vol. 87 (1952), p. 1037.
17. Blatt, J. M., and V. F. Weisskopf. "Theoretical Nuclear Physics," New York; Wiley, 1952.
18. Hughes, D. J., and R. B. Schwartz. "Neutron Cross Sections," *U.S. Atomic Energy Comm. Rept. ANL-325,* 2d Ed., 1958.
19. Arnold, W. R., J. A. Phillips, G. A. Sawyer, E. J. Stovall, Jr., and J. L. Tuck. "Cross Sections for the Reactions D(*d,p*)T, D(*d,n*)He^3, D(*d,n*)He^4 and He^3(*d,p*)He^4 below 120 kev," *Phys. Rev.,* Vol. 93, No. 3 (Feb. 1, 1954), pp. 483-497.
20. Breit, G., and E. Wigner. "Capture of Slow Neutrons," *Phys. Rev.,* Vol. 49, No. 7.
21. Westcott, C. H. "Effective Cross Sections for Thermal Spectra," *Nucleonics,* Vol. 16, No. 10 (1958), p. 108.
22. Weinberg, A. M., and L. C. Norderer. "Second order Diffusion Theory," *U.W. Atomic Energy Comm. Rept.* AECD–3410 (1952).
23. Amster, H. J., *U.S. AEC Report* WAPD–185 (1958).

24. Golden, G. H. Elementary Neutronics Considerations in LMFBR Design, *ANL Report No. 7532* (March 1969).
25. Patterson, D. R., and R. J. Schlitz. The Determination of Heat Generation in Irradiated Uranium, Expt., No. 11; School of Nuclear Science and Engineering, Argonne National Laboratory, August 1955.
26. Reactor Physics Constants, *Report No. ANL–5800,* 2nd Ed. Washington, D. C.: U.S. Atomic Energy Commission (July 1963).
27. Von Ubisch, H., S. Hall, and R. Srivastav. "Thermal Conductivities of Fission Product Gases with Helium and with Argon," Second United Nations International Conference on the Peaceful Uses of Atomic Energy, Paper A/Conf. 15/P/143. (1958).
28. Tipton, C. R., Jr. (ed.) *Reactor Handbook,* Vol. 1, *Materials* 2d ed. Prepared under contract with U. S. Atomic Energy Commission Inc., New York; Interscience, 1960.
29. Rockwell, T. (ed.) *Reactor Shielding Design Manual.* Princeton, N.J.: Van Nostrand, 1956.
30. Price, B. T., C. C. Horton, and K. T. Spinney. *Radiation Shielding.* London: Pergamon, 1957.
31. Goldstein, H. *Fundamental Aspects of Reactor Shielding.* Reading, Mass.: Addison-Wesley, 1959.
32. Irey, R. K. "Errors in the One-Dimensional Fin Solution," *Journal of Heat Transfer, Trans. ASME* (February 1968), p. 175.
33. Minkler, W. S., and W. T. Rouleau. "The Effects of Internal Heat Generation on Heat Transfer in Fins," *Nuclear Sci and Eng.,* Vol. 7, No. 5 (May 1960), pp. 400-406.
34. Jakob, M. *Heat Transfer,* Vol. I. Inc., New York: Wiley, 1949.
35. Eckert, E.R.G., and R. M. Drake, Jr. *Heat and Mass Transfer* 2d. ed. New York: McGraw-Hill, 1959.
36. Kreith, F. *Principles of Heat Transfer.* Scranton, Pa; International Textbook, 1958.
37. Choudhury, W. U., and M. M. El-Wakil. "On the Use of Porous Fuel Elements in Nuclear Reactors," ASME Paper No. 68-WA/NE-7 (1968).
38. Choudhury, W. U.: "Heat Transfer and Flow Characteristics in Conductive Porous Media with Energy Generation," Ph.D. thesis, University of Wisconsin, 1968.
39. Marr, W. W. "Hybrid Computer Solution of the Energy Equations in Conductive Porous Media," Ph.D. thesis, The University of Wisconsin, 1969.
40. Arpaci, V. *Conduction Heat Transfer.* Reading, Mass.: Addison-Wesley, 1966.
41. Moore, A. D. Mapping Techniques Applied to Fluid Mapper Patterns, *Trans. AIEE,* Vol. 71, Pt. I (1952), pp. 1-6.
42. Prandtl, L. Zur Torsiou von Prismatischen Staben, *Z. Phys.,* Vol. 4 (1902-03), pp. 758-759.
43. Schneider, P. J. The Prandtl Membrane Analogy for Temperature Fields with Permanent Heat Sources or Sinks, *J. Aero. Sci.,* Vol. 19 (1952), pp. 644-645.

44. Korn, G. A., and T. M. Korn. *Electronic Hybrid and Analog Computers.* New York: McGraw-Hill, 1964.
45. Boelter, L. M. K., V. H. Cherry, and H. A. Johnson. *Heat Transfer,* 3d. ed. New York: McGraw-Hill, 1942.
46. Gringnel, U.: *Die Grundgestze der Warmeubertragung.* 3d. ed. Berlin Springer Verlag, 1955.
47. Myers. G. E. and D. J. Kotecki. "Effect of Container Capacitance on Thermal Transients in Plane Wall, Cylinders and Spheres," *ASME Paper* No. 68-HT-8 (1968).
48. Gaumer, G. R. "The Stability of Three Finite Difference Methods of Solving for Transient Temperatures, "*Proc. Fifth U.S. Navy Symposium on Aeroballistics,* Vol. I, Paper No. 32. White Oak, Md.; U.S. Naval Ordinance Laboratory, October 1961.
49. Waggener, J. P. Friction Factors for Pressure-drop Calculations, *Nucleonics,* Vol. 19, No. 11 (November 1961), pp. 145-147.
50. McAdams, W. H. *Heat Transmission.* 3d. ed. New York: McGraw-Hill, 1954.
51. Martinelli, R. C. "Heat Transfer to Molten Metals," *Trans. ASME,* Vol. 69 (1947), pp. 947-959.
52. Dittus, F. W., and L. M. K. Boelter. *University California Publs. Eng.,* Vol. 2 (1930), p. 443.
53. Brown, A. I., and S. M. Marco. *Introduction to Heat Transfer,* 3d. ed. New York: McGraw-Hill, 1958.
54. Silberberg, M., and D. A. Huber. "Forced Convection Heat Characteristics of Polyphenyl Reactor Coolants," Unclassified *AEC Rept. NAA-SR-2796,* Jan. 15, 1959.
55. McAdams, W. H., W. E. Kennel, and J. N. Emmons. "Heat Transfer to Superheated Steam at High Pressures Addendum," *Trans. ASME,* (May 1950), pp. 421-428.
56. Deissler, R. G., and M. F. Taylor. "Reactor Heat Transfer Conference," New York, TID-7529, Pt. 1, Book 1 (November 1957), pp. 416-461.
57. Kays, W. M. *Convective Heat and Mass Transfer.* New York: McGraw-Hill, 1966.
58. Weisman, J. "Heat Transfer to Water Flowing Parallel to Tube Bundles," Letter to the Editor, *Nuclear Sci. and Eng.,* Vol. 6, No. 1 (July 1959), pp. 78-79.
59. Bialokoz, I. G., and O. A. Saunders. "Heat Transfer in Pipe Flow at High Speeds," *Proc. Instit. Mech. Engrs. (London),* Vol. 170 (1956), pp. 389-406.
60. R. N. Lyon (ed.). *Liquid Metals Handbook,* sponsored by Committee on the Basic Properties of Liquid Metals, Office of Naval Research, Department of the Navy, in collaboration with U.S. Atomic Energy Commission and Bureau of Ships, Department of the Navy, June, 1952. See also "Liquid Metals Handbook, Na-Nak Supplement," U.S. Atomic Energy Commission, July 1955.
61. Civilian Power Reactor Program Part III, Status Report on Sodium Graphite Reactors as of 1959, *U.S. Atomic Energy Comm. Rept.* TID-8518 (6), 1960.

62. Davis, M., and A. Draycott. Compatibility of Reactor Materials in Flowing Sodium, *Proc. Second United Nations Intern. Conf. on Peaceful Used of Atomic Energy,* Geneva, Vol. 7 (1958), pp. 94-110.
63. Bernal, J. D. "The Structure of Liquids," *Proc. Royal Inst. G. Brit.,* Vol. 37, No. 168 (1959), pp. 355-393. See also J. D. Bernal, "The Structure of Liquids," *Sci. American,* Vol. 203, No. 2 (August 1960), pp. 124-134.
64. Seban, R. A., and T. Shimazaki. "Heat Transfer to a Fluid Flowing Turbulently in a Smooth Pipe with Walls at Constant Temperature," *ASME Paper* 50-A-128 (1950).
65. Dwyer, O.E. "Analytical Study of Heat Transfer to Liquid Metals Flowing In-Line Through Closely Packed Rod Bundles," *Nuclear Science and Engineering,* Vol. 25, No. 4 (August 1966), pp. 343-358.
66. Borishansky, V. M., and E. V. Firsova. "Heat Exchange in the Longitudinal Flow of Metallic Sodium Past a Tube Bank," *At. Energ.,* Vol. 14 (1963), p. 584.
67. Hoe, R. J., D. Dropkin, and O. E. Dwyer. "Heat Transfer Rates to Crossflowing Mercury in a Staggered Tube Bank-I," *Trans. ASME,* Vol. 79 (1957), pp. 899-908.
68. Richards, C. L., O. E. Dwyer, and D. Dropkin. "Heat Transfer Rates to Crossflowing Mercury in a Staggered Tube Bank—II", Paper No. 57-HT-11, *ASME-AICHE* Heat Conference, 1957.
69. Seban, R. A. "Heat Transfer to a Fluid Flowing Turbulently between Parallel Walls with Asymmetric Wall Temperatures," *Trans. ASME,* Vol. 72 (1950), p. 789.
70. Bailey, R. V. "Heat Transfer to Liquid Metals in Concentric Annuli," *U.S. Atomic Energy Comm. Rept.* ORNL-531, Oak Ridge National Laboratory, 1950.
71. Werner, R. C., E. C. King, and R. A. Tidball. "Forced Convection Heat Transfer with Liquid Metals," paper presented at 42d Annual Meeting, American Institute of Chemical Engineers, December 1949.
72. Dwyer, O. E. "Heat Transfer to Liquid Metals Flowing Turbulently Between Parallel Plates," *Nuclear Sci. Eng.,* Vol. 21, No. 1 (January 1965), pp. 79-89.
73. Yu, W. S., and O. E. Dwyer. "Heat Transfer to Liquid Metals Flowing Turbulently in Electric Annuli–I," *Nuclear Sci. Eng.* Vol. 24, No. 2 (February 1966), pp. 105-117.
74. Sleicher, C. A., and M. Tribus. *Heat Transfer and Fluid Mechanics Institute,* Stanford University Press (1956), p. 59.
75. Burchill, W. E., B. G. Jones, and R. P. Stein. "Influence of Axial Heat Diffusion in Liquid Metal-Cooled Ducts with Specified Heat Flux," *Trans. ASME* Vol. 90, Series C, No. 3, *J. Heat Trans.* (August 1968), pp. 283-290.
76. Muller, G. L. "Experimental Forced Convection Heat Transfer with Adiabatic Walls and Internal Heat Generation in a Liquid Metal," presented at the ASME-AICE Joint Heat Transfer Conference, Chicago, Ill., (August 1968), *ASME Paper* 58-HT-17.
77. For example, D. C. Hamilton and F. E. Lynch. "Free Convection Theory

and Experiment in Fluids Having a Volume Heat Source," *U.S. Atomic Energy Comm. Rept.* ORNL 1888, August 1955.

78. Poppendiek, H. F. "Forced Convection Heat Transfer in Pipes with Volume-Heat Sources within the Fluid, *Chem. Eng. Progr. Symposium Ser.*, Vol. 50, No. 11, (1954), pp. 93-104.
79. Treible, J. "Unipolar Generators," *Marquette Engr.* (April 1955).
80. Jaross, R. A., and A. H. Barnes. "Design and Operation of a 10,000-gpm dc Electromagnetic Sodium Pump and 250,000-ampere Homopolar Generator," *Proc. Second United Nations Intern. Conf. on Peaceful Uses of Atomic Energy,* Geneva, *Paper* P/2157, Vol. 7 (1958), pp. 82-87.
81. Hsu, Y. Y. "On the Size Range of Active Nucleation Cavities on a Heating Surface," *J. Heat Trans., Trans. ASME,* Series C, Vol. 84, No. 3 (August 1962), pp. 207-216.
82. Chen, J. C. "Incipient Boiling Superheats in Liquid Metals," *J. of Heat Trans., Trans. ASME,* Series C, Vol. 96, No. 3 (August 1968), pp. 303-312.
83. Nukiyama, S. "Maximum and Minimum Values of Heat Transmitted from Metal to Boiling Water under Atmospheric Pressure," J. Soc. Mech. Engrs., Japan, Vol. 37 (1934), p. 367.
84. McAdams, W. H., J. N. Addams, P. M. Rinaldo, and R. S. Day. "Vaporization inside Horizontal Tubes," *Trans. ASME,* Vol. 63 pp. 545-552 (1941).
85. Farber, E. A., and R. L. Scorah, "Heat Transfer to Water Boiling under Pressure," *Trans. ASME,* Vol. 70 (1948), pp. 369-384.
86. Rohsenow, W. M. A Method of Correlating Heat Transfer Data for Surface Boiling Liquids, *Trans. ASME,* Vol. 74 (1952), pp. 969-975.
87. Kreith, F., and A. S. Faust. "Remarks on Mechanism and Stability at Surface Boiling Heat Transfer," ASME paper No. 54-A-146 (1954).
88. Pasint, D., and R. H. Pai. "Empirical Correlation of Factors Influencing Departure from Nucleate Boiling in Steam-Water Mixtures Flowing in Vertical Round Tubes," *J. Heat Trans., Trans. ASME,* Series C, Vol. 88, No. 4 (November 1966), pp. 367-375.
89. Zuber, N., M. Tribus, and J. W. Westwater. "The Hydrodynamic Crisis in Pool Boiling of Saturated and Subcooled Liquids," *Proc. International Conference on Developments in Heat Transfer,* ASME, New York, 1962.
90. Cichelli, M. T., and C. F. Bonilla. "Heat Transfer to Liquids Boiling under Pressure," *Trans, Am. Inst. Chem. Eng.,* Vol. 41 (1945), pp. 755-787.
91. Kazakova, E. A. "Maximum Heat Transfer to Boiling Water at High Pressures," *Izvest. Adak. Nauk* S.S.S.R., No. 9 (September 1950), pp. 1377-1387. Review of paper in *Eng. Dig.,* Vol. 12 (1951), pp. 81-85.
92. Lukomskii, S. M. *Doklady Akad. Nauk* S.S.S.R., Vol. 80 (1951), p. 53.
93. Tong, L. S. *Boiling Heat Transfer and Two-Phase Flow.* New York: Wiley, 1965.
94. McAdams, W. M., W. E. Kennel, C. S. Minden, P. Carl, P. M. Picarnell, and J. E. Drew. "Heat Transfer at High Rates to Water with Surface Boiling," *Ind. Eng. Chem.,* Vol. 41 (1945).
95. Macbeth, R. V. "Burnout Analysis, Part 2, The Basic Burnout Curve," U.K. Report AEEW-R-167, Winfrith, 1963.
96. Rohsenow, W. M., and P. Griffith. "Correlation of Maximum Heat Flux

Data for Boiling of Saturated Liquids," ASME-AICE Heat Transfer Symposium, Paper 9, Louisville, Ky., 1955.

97. Usiskin, C. M., and R. Siegel. "An Experimental Study of Boiling in Reduced and Zero Gravity Fields," *Trans. ASME,* Sec. C, Vol. 83 (1961), pp. 243-250.
98. Kutateladze, S. S. "Heat Transfer in Condensation and Boiling," USAEC Report AEC-tr-3770, 1952.
99. Kutateladze, S. S., and L. L. Schneiderman. "Experimental Study of Influence of Temperature of Liquid on Change in Rate, USAEC Report AEC-tr-3405 (1953), pp. 95-100.
100. Ivey, H. J., and D. J. Morris. "On the Relevance of the Vapor-Liquid Exchange Mechanism for Subcooled Boiling Heat Transfer at High Pressure" U.K. Report AEEW-R-137, Winfrith, 1962.
101. Jens, W. H., and D. A. Lottes. Analysis of Heat Transfer, Burnout, Pressure Drop and Density Data for High Pressure Water, *Argonne Natl. Lab. Rept.* ANL-4627, May 1951.
102. Tong L. S. "Prediction of Departure from Nucleate Boiling for an Acially Non-uniform Heat Flux Distribution," *J. Nuclear Energy,* Parts A and B, 1967.
103. Bernath, L.: "A Theory of Local-Boiling Burnout and Its Application to Existing Data," *Chem. Eng. Prog. Report,* Symp. Ser., Vol. 56, No. 30 (1960), pp. 95-116.
104. Zenkevich, B. A., V. I. Subbatin, and M. F. Troianov. "Critical Heat Load for Longitudinal Wetting of a Tube Bundle for Water Heated to Saturation Temperature, *Soviet J. of Atomic Energy,* Vol. 4 (1958), pp. 485-487.
105. Janssen, E., and S. Levy. "Burnout Limit Curves for Boiling Water Reactors," *General Electric Company Report* APED-3892, 1962.
106. Noyes R. C. "An Experimental Study of Sodium Pool Boiling Heat Transfer," *Trans.* ASME, Sec. C, *J. Heat Trans.,* Vol. 85 (1963), pp. 125-129.
107. Balzhiser, R. E., et al. "Investigation of Liquid Metal Boiling Heat Transfer," *Wright Patterson Air Force Base Report* RTD-TDR-63-4130, 1963.
108. Kutateladze, S. S., et al. "Liquid Metal Heat Transfer Media," *Consultants Bureau,* New York, 1959, Translated from *Suppl. of Soviet J. of At. Energy,* No. 2, 1958.
109. Hoffman, H. W. "Recent Experimental Results in ORNL Studies with Boiling Potassium," Third Annual Conference on High-Temperature Liquid Metal Heat-Transfer Technology, Oak Ridge National Laboratory, September 1963.
110. Lowdermilk, W. H., C. D. Lanzo, and B. L. Siegel. "Investigation of Boiling Burnout and Flow Stability for Water Flowing in Tubes." *Natl. Advisory Comm. Aeronaut., Tech. Note* TN-4382 (September, 1958).
111. Core, T. C., and K. Sato. Determination of Burnout Limits of Polyphenyl Coolants, *U.S. Atomic Energy Comm. Rept.* IDO-28007, February 1958.
112. Jordan, D. P., and G. Leppert. "Nucleate Boiling Characteristics of Organic Reactor Coolants." *Nuclear Sci, and Eng.,* Vol. 5, No. 6 (June, 1959), pp. 349-359.
113. Hoegerton, J. F., and R. C. Grass (ed.). Reactor Handbook, Vol. 3, Engi-

neering, Selected Reference Material, U.S. Atomic Energy Commission, August 1955.

114. Nusselt, W.: "Die Oberflächen Kondensation des Wasserdampfes," *Z. Ver. Deutsch. Ing.*, Vol. 60 (1916), pp. 541-569.

115. Rohsenow, W. M., J. H. Webber, and A. T. Ling. "Effect of Vapor Velocity on Laminar and Turbulent-Film Condensation," *Trans. ASME,* Vol. 78 (1956), p. 1637.

116. Rohsenow, W. M. "Heat Transfer and Temperature Distribution in Laminar-Film Condensation," *Trans. ASME,* Vol. 78 (1956), pp. 1645-1648.

117. Kirkbride, C. G. "Heat Transfer by Cooling Vapors on Vertical Tubes," *Trans. AICh.E,* Vol. 30 (1934), p. 170.

118. Carpenter, E. F., and A. P. Colburn. "The Effect of Vapor Velocity on Condensation Inside Tubes," *ASME, Proceedings of General Discussion on Heat Transfer* (1951), pp. 20-26.

119. Subbotin, V. I., M. N. Invanovskei, V. P. Sorokin, and B. A. Chulkov. "Heat Transfer During the Condensation of Potassium Vapor," *Trans. Teplofizika Vysokikh Temperatur,* Vol. 2 (1964), p. 616.

120. Sukhatme, S. P., and W. M. Rohsenow, "Heat Transfer During Film Condensation of Liquid Metal Vapor," *Report No. 9167-27, M.I.T.,* 1964.

121. Shea, F. J., Jr., and N. W. Krase, *Trans. AICh.E,* Vol. 36, (1940), p. 463.

122. Marchaterre, J. F., and M. Petrick. "The Prediction of Steam Volume Fractions in Boiling Systems," *Nuclear Sci. Eng.,* Vol. 7, No. 6 (June, 1960), p. 525.

123. Kutateladze, S. S. "Heat Transfer in Condensation and Boiling (translated from Russian Publication of the State Scientific and Technical Publishers of Literature on Machinery, Moscow-Leningrad, 1952), *U.S. Atomic Energy Comm. Rept.* AEC-tr-3770, August 1959.

125. von Glahn, U. H. "An Empirical Relation for Predicting Void Fraction with Two-Phase Steam-Water Flow," *Natl. Aeronaut and Space Admin., Tech. Note* D-1189 (Jan. 1962).

124. "EBWR, The Experimental Boiling Water Reactor," *U.S. Atomic Energy Comm. Nuclear Technol. Ser. Rept.* ANL-5607, prepared by Argonne National Laboratory, May 1957.

126. Martinelli, R. C., and D. B. Nelson. "Prediction of Pressure Drop During Forced-circulation Boiling of Water," *Trans. ASME,* Vol. 70, (1948), p. 695.

127. Huang, M., and M. M. El-Wakil. A Visual and Frictional Pressure-Drop Study of Natural-Circulation Single-Component Two-Phase Flow at Low Pressures, Nuclear Science and Engineering, Vol. 28 (1967), pp. 12-19.

128. Lottes, P. A., and W. S. Flinn. "A Method of Analysis of Natural Circulation Boiling Systems," *Nuclear Sci. Eng.,* Vol. 1, (1956), pp. 461-476.

129. Lottes, P. A., J. F. Marchaterre, R. Viskanta, J. A. Thie, M. Petrick, R. J. Weatherhead, B. M. Hoglund, and W. S. Flinn. "Experimental Studies of Natural Circulation Boiling."

130. Lottes, P. A. Nuclear Reactor Heat Transfer, *Argonne Natl. Lab. Rept.* ANL-6469 (December 1961).

131. ASME Test Code: Supplement of Instruments and Apparatus, Flow Measurement, Part 5, Chapter 4 (1959).
132. Murdock, J. W., "Two-Phase Flow Measurement with Orifices," Trans. *ASME, J. Basic Eng.,* Vol. 84 (1962), pp. 419-433.
133. Fauske, H. K., "Contribution to the Theory of Two-Phase One Component Critical Flow," *USAEC Report* ANL-6633 (1962).
134. Fauske, H. K., "Two-Phase Critical Flow," Paper presented at the M.I.T. Two-Phase Gas-Liquid Flow Special Summer Program, 1964.
135. Friedrich, H., and G. Vetter. "Einfluss der Dusenform auf das Dürchflussverhalten von Düsen für Wasser bei verschiedenen Thermodynamischen Zuständen," *Energie,* 14, No. 1 (1962).
136. Zalondek, F. R., "The Critical Flow of Hot Water Through Short Tubes," HW-77594, 1963.
137. Imhoff, D. H., W. H. Cook, R. T. Pennington, and C. N. Spalaris. "Experimental Programs in the Vallecitos Boiling Water Reactor," paper presented at the Annual Meeting of the American Nuclear Society, Gatlinburg, Tennessee, June 1959.
138. French, P. R. J. A Vector Diagram for Reactor Channel Temperatures, *Nuclear Eng.,* Vol. 5, No. 46 (March 1960), pp. 120-122.
139. Tourneau, B. W., and R. E. Grimble. Engineering Hot-Channel Factors for Nuclear Reactor Design, *Nuclear Sci. Eng.* Vol. I (1956), p. 359.
140. Hitchcock, A. Statistical Methods for the Estimation of Maximum Full Element Temperatures, 1 GR-TN/R-760, U. K. Atomic Energy Authority, January 1958.
141. APDA Introduces Statistical Hot Spot Factors, *Nucleonics* (August 1959), pp. 92-96.
142. Dwight, H. B. *Tables of Integrals and Other Mathematical Data.* Macmillan, New York: 1947.
143. Weil, John W. "Void-induced Power Distortion in Boiling Water Reactors," *Nucleonics* (June 1958), pp. 90-94.
144. Etherington, H. (ed.). *Nuclear Engineering Handbook.* New York: McGraw-Hill, 1958.
145. Hogerton, J. F. and R. C. Grass (eds.). *The Reactor Handbook.* Vol. 2. *Physics.* Washington, D. C.: U.S. Atomic Energy Commission, August 1955.
146. Stehn, J. F. Table of Radioactive Nuclides, *Nucleonics,* Vol. 18, No. 11, (November 1960), pp. 186-195.
147. Baucom, H. H. Nuclear Data for Reactor Studies, *Nucleonics,* Vol. 18, No. 11 (November 1960), pp. 198-200.
148. Davis, M. V., and D. T. Hauser. Thermal-Neutron Data for the Elements, *Nucleonics,* Vol. 16, No. 3 (March 1958), pp. 87-89.
149. Wyatt, E. I., S. H. Reynolds, T. H. Handley, W. S. Lyon, and H. A. Parker. Half-lives of Nuclides—II, *Nuclear Sci. Eng.,* Vol. 11, No. 1 (September 1961), pp. 74-75.
150. Hintenberger, H. (ed.). "Nuclear Masses and Their Determination," Proceedings of Conference held in the Max-Planck-Institut fur Chemie, Mainz, July, 1956. London: Pergamon, 1957.

151. Goldberg, M. D., S. F. Mughabghab, B. J. Magurno, and V. M. May. "Neutron Cross Sections," BNL 325, 2d ed., Suppl. No. 2. Washington, D. C.: U.S. Atomic Energy Commission, February 1966.
152. Reynolds, S. A., J. F. Emery, and E. I. Wyatt. Half-lives of Radionuclides—III, *Nuclear Sci. Eng.* Vol. 32, No. 1 (April 1968), pp. 46-48.
153. Lederer, C. M., J. M. Hollander, and I. Perlman. *Table of Isotopes,* 6th ed. New York: Wiley, 1967.
154. Meisl, C. J., and A. Shapiro. "Thermodynamic Properties of Alkali Metal Vapors and Mercury." 2d ed. *Gen. Elec. Flight Propulsion Lab. Rept. R60FPD358-A* (November, 1960).
155. Tables of Thermal Properties of Gases, *Natl. Bur. Standards* (U.S.), Circ. 564 (November 1955).
156. Moody, L. F. Friction Factors For Pipe Flows, *Trans. ASME,* Vol. 66 (1944), pp. 671-694.

Index